Undergraduate Texts in Mathematics

Editors:
S. Axler
K.A. Ribet

Undergraduate Texts in Mathematics

For other titles published in this series, go to
http://www.springer.com/series/666

Steven Givant • Paul Halmos

Introduction to Boolean Algebras

 Springer

Steven Givant
Mills College
Department of Mathematics
and Computer Science
5000 MacArthur Blvd
Oakland CA 94613-1301
USA
givant@mills.edu

Paul Halmos
(Deceased)

ISSN: 0172-6056
ISBN: 978-1-4419-2324-0   e-ISBN: 978-0-387-68436-9
DOI: 10.1007/978-0-387-68436-9

Mathematics Subject Classification (2000): 06Exx

Contents

Preface

The theory of Boolean algebras was created in 1847 by the English mathematician George Boole. He conceived it as a calculus (or arithmetic) suitable for a mathematical analysis of logic. The form of his calculus was rather different from the modern version, which came into being during the period 1864–1895 through the contributions of William Stanley Jevons, Augustus De Morgan, Charles Sanders Peirce, and Ernst Schröder. A foundation of the calculus as an abstract algebraic discipline, axiomatized by a set of equations, and admitting many different interpretations, was carried out by Edward Huntington in 1904.

Only with the work of Marshall Stone and Alfred Tarski in the 1930s, however, did Boolean algebra free itself completely from the bonds of logic and become a modern mathematical discipline, with deep theorems and important connections to several other branches of mathematics, including algebra, analysis, logic, measure theory, probability and statistics, set theory, and topology. For instance, in logic, beyond its close connection to propositional logic, Boolean algebra has found applications in such diverse areas as the proof of the completeness theorem for first-order logic, the proof of the Łoś conjecture for countable first-order theories categorical in power, and proofs of the independence of the axiom of choice and the continuum hypothesis in set theory. In analysis, Stone's discoveries of the Stone–Čech compactification and the Stone–Weierstrass approximation theorem were intimately connected to his study of Boolean algebras. Countably complete Boolean algebras (also called σ-algebras) and countably complete fields of sets (also called σ-fields) play a key role in the foundations of measure theory. Outside the realm of mathematics, Boolean algebra has found applications in such diverse areas as anthropology, biology, chemistry, ecology, economics, sociology, and especially computer science and philosophy. For example, in computer science, Boolean algebra is used in electronic circuit design (gating networks), programming languages, databases, and complexity theory.

Most books on Boolean algebra fall into one of two categories. There are elementary texts that emphasize the arithmetic aspects of the subject (in particular, the laws that can be expressed and proved in the theory), and that often explore applications to propositional logic, philosophy, and electronic circuit design. There are also advanced treatises that present the deeper mathematical aspects of the theory at a level appropriate for graduate students and professional mathematicians (in terms of the mathematical background and level of sophistication required for understanding the presentation).

This book, a substantially revised version of the second author's *Lectures on Boolean Algebras*, tries to steer a middle course. It is aimed at undergraduates who have studied, say, two years of college-level mathematics, and have gained enough mathematical maturity to be able to read and write proofs. It does not assume the usual background in algebra, set theory, and topology that is required by more advanced texts. It does attempt to guide readers to some of the deeper aspects of the subject, and in particular to some of the important interconnections with topology. Those parts of algebra and topology that are needed to understand the presentation are developed within the text itself. There is a separate appendix that covers the basic notions, notations, and theorems from set theory that are occasionally needed.

The first part of the book, through Chapter 28, emphasizes the arithmetical and algebraic aspects of Boolean algebra. It requires no topology, and little set theory beyond what is learned in the first two years of college-level mathematics, with two important exceptions. First, two of the proofs use a form of mathematical induction that extends beyond the natural numbers to what are sometimes called "transfinite ordinal numbers". Transfinite ordinals and transfinite induction are discussed in Appendix A, but the key ideas of the two proofs can already be grasped in the context of the natural numbers and standard mathematical induction. Second, Chapter 10 presents an important example of a Boolean algebra that is based on topological notions. These notions are discussed in Chapter 9. The example itself, and the requisite topology, are not needed to understand the remaining chapters of the first part of the book. (Some of the more advanced exercises in the chapters do require an understanding of this material, but these exercises may be ignored by readers who wish to skip Chapters 9 and 10.) The second part of the book, in particular Chapters 29, 34–41, and 43, emphasizes the interconnections between Boolean algebra and topology, and consequently does make extensive use of topological ideas and results. The necessary topological background is provided in Chapters 9, 29, 32, and 33.

Some of the important results discussed in the first part of the book are the normal form theorem (which gives a description of the Boolean subalgebra generated by a set of elements, Chapter 11), and its analogue for Boolean ideals (Chapter 18); the homomorphism extension theorem (Chapter 13) and its application to the proofs of the isomorphism theorem for countable, atomless Boolean algebras (Chapter 16) and the existence theorem for free algebras (Chapter 28); the representation theorem for atomic Boolean algebras (every atomic Boolean algebra can be mapped isomorphically to a field of sets in a way that preserves all existing suprema as unions, Chapter 14); the maximal ideal theorem (every proper ideal can be extended to a maximal ideal, Chapter 20), and its application to the celebrated representation theorem (every Boolean algebra is isomorphic to a field of sets, Chapter 22); the existence and uniqueness theorems for completions (every Boolean algebra has a minimal complete extension that is unique up to isomorphisms, Chapter 25); the isomorphism of factors theorem (two countably complete Boolean algebras that are factors of one another must be isomorphic) and the counterexamples demonstrating that the theorem cannot be extended to all Boolean algebras, or even to all countable Boolean algebras (Chapters 27 and 45).

Many of the highlights of the second part of the book center on the fundamental duality theorems for Boolean algebras and Boolean spaces: to every Boolean algebra there corresponds a Boolean space that is uniquely determined up to homeomorphism, and, conversely, to every Boolean space there corresponds a Boolean algebra that is uniquely determined up to isomorphism (Chapter 34). These theorems imply that every notion or theorem concerning Boolean algebras has a "dual" topological counterpart concerning Boolean spaces, and conversely. For instance, ideals correspond to open sets (Chapter 35), homomorphisms to continuous functions (Chapter 36), quotient algebras to closed subspaces and subalgebras to Boolean quotient spaces (Chapter 37), direct products of Boolean algebras to Stone–Čech compactifications of unions of Boolean spaces (Chapter 43), and complete Boolean algebras to extremally disconnected spaces (Chapter 38). A related result, discussed in Chapter 40, is the representation theorem for σ-algebras (every σ-algebra is isomorphic to a σ-field of sets modulo a σ-ideal).

It is not necessary to read all the chapters in the order in which they appear, since there is a fair degree of independence among them. The diagram at the end of the preface shows the main chapter dependencies. Three examples may serve to demonstrate how the diagram is to be understood. First, Chapter 28 depends on Chapters 1–8 and 11–13. Second, Chapter 24 depends on Chapters 1–8, 11–12, and 17–19. Finally, Chapter 31 depends on Chapters 1–12, 17–18, and 29–30. These remarks do not apply to the exercises, some of which depend on earlier chapters for which no dependency is indicated in the diagram. Also, minor references to earlier chapters are not indicated in the diagram. For instance, an application in Chapter 36 of the principal result of that chapter depends on the definition of a free algebra (given in Chapter 28), but not on any of the results about free algebras. Similarly, a corollary at the end of Chapter 21 depends on the notion of a maximal ideal and the easily comprehended statement of the maximal ideal theorem (given in Chapter 20).

A large number of exercises of varying levels of difficulty have been included in the text. There are routine problems that help readers understand the basic definitions and theorems; intermediate problems that extend or enrich material developed in the text; and difficult problems that often present important results not covered in the text. The harder exercises are labeled as such, and hints for their solutions are given in Appendix B. Some of the exercises are formulated, not as assertions, but as questions that readers are invited to ponder.

There is an instructor's manual that contains complete solutions to the exercises. It may serve as a guide to instructors, and in particular it may help them select problems at an appropriate level of difficulty for their students. Instructors may also wish to assign the solutions of some of the more difficult problems to individual students or groups of students for independent study or as class projects.

Historical remarks are sprinkled throughout the text. We are indebted to Don Monk for his help in tracking down the authorship of some of the main results. Regrettably, it has not been feasible to determine the origin of every theorem.

The book can serve as a basis for a variety of courses. A one-semester course that focuses on the algebraic material might cover some subset of Chapters 1–28, for instance Chapters 1–8, 11–14, and 17–27. A one-semester course that includes some of the interconnections with topology might cover Chapters 1–8, 11–12, 14, 17–22, parts of 9 and 29, and 32–36. Most of the text could be covered in a one-year course.

A quick word about terminology. In this book, the phrase "just in case" is used as a variant of the phrase "in this case, and only in this case". In other words, it is a synonym for "if and only if".

This revision of Halmos's book was planned and initially executed by both authors. Due to declining health, however, Halmos was not able to review the later versions of the manuscript. He died on October 2, 2006. Whatever imperfections remain in the text are my sole responsibility.

Steven Givant
San Francisco, California
August, 2007

Chapter dependence diagram

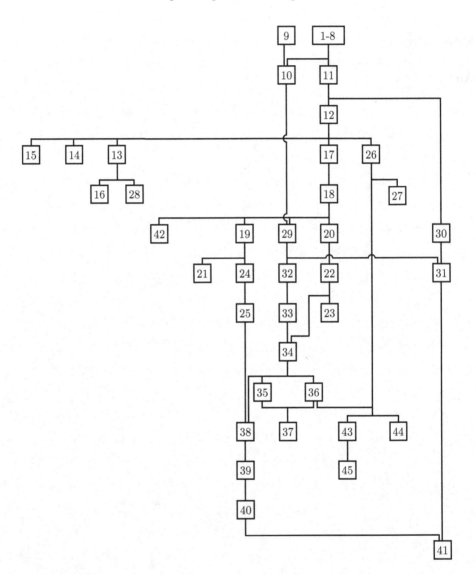

Chapter 1

Boolean Rings

A ring is an abstract version of arithmetic, the kind of thing you studied in school. The prototype is the ring of integers. It consists of a universe — the set of integers — and three operations on the universe: the binary operations of addition and multiplication, and the unary operation of negation (forming negatives). There are also two distinguished integers, zero and one. The ring of integers satisfies a number of basic laws that are familiar from school mathematics: the associative laws for addition and multiplication,

(1).
$$p + (q + r) = (p + q) + r,$$
(2)
$$p \cdot (q \cdot r) = (p \cdot q) \cdot r,$$

the commutative laws for addition and multiplication,

(3)
$$p + q = q + p,$$
(4)
$$p \cdot q = q \cdot p,$$

the identity laws for addition and multiplication,

(5)
$$p + 0 = p,$$
(6)
$$p \cdot 1 = p,$$

the inverse law for addition,

(7)
$$p + (-p) = 0,$$

and the distributive laws for multiplication over addition,

(8)
$$p \cdot (q + r) = p \cdot q + p \cdot r,$$

S. Givant, P. Halmos, *Introduction to Boolean Algebras*,
Undergraduate Texts in Mathematics, DOI: 10.1007/978-0-387-68436-9_1,
© Springer Science+Business Media, LLC 2009

(9) $$(q + r) \cdot p = q \cdot p + r \cdot p.$$

The difference between the ring of integers and an arbitrary ring is that, in the latter, the universe may be an arbitrary non-empty set of elements, not just a set of numbers, and the operations take their arguments and values from this set. The associative, commutative, identity, and inverse laws for addition, the associative law for multiplication, and the distributive laws are required to hold: they are the ring axioms. The commutative law for multiplication is not required to hold in an arbitrary ring; if it does, the ring is said to be *commutative*. Also, a ring is not always required to have a unit, an element 1 satisfying (6); if it does, it is called a ring *with unit*.

There are other natural examples of rings besides the integers. The most trivial is the ring with just one element in its universe: zero. It is called the *degenerate* ring. The simplest non-degenerate ring with unit has just two elements, zero and one. The operations of addition and multiplication are described by the arithmetic tables

+	0	1
0	0	1
1	1	0

and

·	0	1
0	0	0
1	0	1

.

An examination of the tables shows that the two-element ring has several special properties. First of all, every element is its own additive inverse: ·

(10) $$p + p = 0.$$

Therefore, the operation of negation is superfluous: every element is its own negative. Rings satisfying condition (10) are said to have *characteristic* 2. Second, every element is its own square:

(11) $$p \cdot p = p.$$

Elements with this property are called *idempotent*. When every element is idempotent, the ring itself is said to be idempotent.

A *Boolean ring* is an idempotent ring with unit. (Warning: some authors define a Boolean ring to be just an idempotent ring, which may or may not have a unit. They call the concept we have defined a "Boolean ring with unit".) The two-element ring is the simplest non-degenerate example of a Boolean ring. It will be denoted throughout by the same symbol as the ordinary integer 2. The notation is not commonly used, but it is very convenient. It is in accordance with von Neumann's definition of the ordinal numbers (under which the ordinal number 2 coincides with the set $\{0, 1\}$), with sound general principles of notational economy, and (in logical expressions such as "two-valued") with idiomatic linguistic usage.

The condition of idempotence in the definition of a Boolean ring has quite a strong influence on the structure of such rings. Two of its most surprising consequences are that (a) a Boolean ring always has characteristic 2 and (b) a Boolean ring is always commutative. For the proof, compute $(p+q)^2$, and use idempotence to conclude that

$$(12) \qquad 0 = q \cdot p + p \cdot q.$$

In more detail,

$$p + q = (p+q)^2 = p^2 + q \cdot p + p \cdot q + q^2 = p + q \cdot p + p \cdot q + q,$$

by the distributive and idempotent laws. Add the inverse of p to the left sides of the first and last terms, add the inverse of q to the right sides, and use the laws governing addition, in particular the inverse and identity laws, to arrive at (12).

This result implies the two assertions, one after another, as follows. Put $p = q$ in (12) and use idempotence to get (a):

$$0 = p^2 + p^2 = p + p.$$

Assertion (a) implies that every element is equal to its own negative, so

$$(13) \qquad p \cdot q = -(p \cdot q).$$

Add the left and right sides of (13) to the left and right sides of (12) respectively, and apply the inverse and identity laws for addition to obtain (b):

$$p \cdot q = q \cdot p + p \cdot q + -(p \cdot q) = q \cdot p + 0 = q \cdot p.$$

Since, as we now know, negation in Boolean rings is the identity operation, it is never necessary to use the minus sign for additive inverses, and we shall never again do so. (A little later we shall meet another natural use for it.) Only a slight modification in the set of axioms is needed: the identity (7) should be replaced by (10). From now on, the official axioms for a Boolean ring are (1)–(3), (5), (6), and (8)–(11).

Boolean rings are the only rings that will be considered in this book, so it is worth looking at another example. The universe of this example consists of ordered pairs (p, q) of elements from 2. In other words, it consists of the four ordered pairs

$$(0,0), \quad (0,1), \quad (1,0), \quad (1,1).$$

This set will be denoted by 2^2, in agreement with the notation $\mathbb{R}^2$ that is used to denote the set of ordered pairs of real numbers. To add or multiply two pairs in 2^2, just add or multiply the corresponding coordinates in 2:

$$(p_0, p_1) + (q_0, q_1) = (p_0 + q_0, p_1 + q_1)$$

and

$$(p_0, p_1) \cdot (q_0, q_1) = (p_0 \cdot q_0, p_1 \cdot q_1).$$

These equations make sense: their right sides refer to the elements and operations of 2. The zero and unit of the ring are the pairs $(0,0)$ and $(1,1)$.

It is a simple matter to check that the axioms for Boolean rings are true in 2^2. In each case, the verification of an axiom reduces to its validity in 2. For example, here is the verification of the commutative law for addition:

$$(p_0, p_1) + (q_0, q_1) = (p_0 + q_0, p_1 + q_1)$$
$$= (q_0 + p_0, q_1 + p_1) = (q_0, q_1) + (p_0, p_1).$$

The first and last equalities use the definition of addition of ordered pairs, and the middle equality uses the commutative law for addition in 2.

The preceding example can easily be generalized to each positive integer n. The universe of the ring is the set 2^n of n-termed sequences

$$(p_0, \ldots, p_{n-1})$$

of elements from 2. The sum and product of two such n-tuples are defined coordinatewise, just as in the case of ordered pairs:

$$(p_0, \ldots, p_{n-1}) + (q_0, \ldots, q_{n-1}) = (p_0 + q_0, \ldots, p_{n-1} + q_{n-1})$$

and

$$(p_0, \ldots, p_{n-1}) \cdot (q_0, \ldots, q_{n-1}) = (p_0 \cdot q_0, \ldots, p_{n-1} \cdot q_{n-1}).$$

The zero and unit are the n-tuples $(0, \ldots, 0)$ and $(1, \ldots, 1)$. Verifying the axioms for Boolean rings is no more difficult in this example than it is in the example 2^2.

To generalize the example still further, it is helpful to look at the set 2^n another way, namely, as the set of functions with domain $\{0, \ldots, n-1\}$ and with values in 2, that is, with possible values 0 and 1. Let X be an arbitrary set, and 2^X the set of all functions from X into 2. The elements of 2^X will be called 2-*valued functions* on X. The distinguished elements and the operations of 2^X are defined pointwise. This means that 0 and 1 in 2^X are the constant functions defined, for each x in X, by

$$0(x) = 0 \quad \text{and} \quad 1(x) = 1,$$

and if p and q are 2-valued functions on X, then the functions $p + q$ and $p \cdot q$ are defined by

$$(p + q)(x) = p(x) + q(x) \quad \text{and} \quad (p \cdot q)(x) = p(x) \cdot q(x).$$

Again, these equations make sense; their right sides refer to elements and operations of 2.

Verifying that 2^X is a Boolean ring is conceptually the same as verifying that 2^2 is a Boolean ring, but notationally it looks a bit different. Consider, as an example, the verification of the distributive law (8). In the context of 2^X, the left and right sides of (8) denote functions from X into 2. It must be shown that these two functions are equal. They obviously have the same domain X, so it suffices to check that the values of the two functions at each element x in the domain agree, that is,

(14) $$(p \cdot (q + r))(x) = (p \cdot q + p \cdot r)(x).$$

The left and right sides of (14) evaluate to

(15) $$p(x) \cdot (q(x) + r(x)) \quad \text{and} \quad p(x) \cdot q(x) + p(x) \cdot r(x)$$

respectively, by the definitions of addition and multiplication in 2^X. Each of these terms denotes an element of 2. Since the distributive law holds in 2, the two terms in (15) are equal. Therefore, equation (14) is true. The other Boolean ring axioms are verified for 2^X in a similar fashion.

For another example of a Boolean ring let A be the set of all idempotent elements in a commutative (!) ring R with unit, with addition redefined so that the new sum of p and q in A is $p + q - 2pq$. The distinguished elements of A are the same as those of R, and multiplication in A is just the restriction of multiplication in R. The verification that A becomes a Boolean ring in this way is an amusing exercise in ring axiomatics. Commutativity is used repeatedly; it is needed, for instance, to prove that A is closed under multiplication.

Exercises

1. Verify that 2 satisfies ring axioms (1)–(9).

2. Verify that 2^3 satisfies ring axioms (1)–(9).

3. Verify that 2^X satisfies ring axioms (1)–(9) for any set X. What ring do you get when X is the empty set?

4. Essentially, what ring is 2^X when X is a set consisting of just one element? Can you make this statement precise?

5. A *group* is a non-empty set, together with a binary operation $+$ (on the set), a unary operation $-$, and a distinguished element 0, such that the associative law (1), the identity laws

$$p + 0 = p \qquad \text{and} \qquad 0 + p = p,$$

and the inverse laws

$$p + -p = 0 \qquad \text{and} \qquad -p + p = 0$$

are all valid. Show that in a group the cancellation laws hold: if

$$p + q = p + r \qquad \text{or} \qquad q + p = r + p,$$

then $q = r$. Conclude that in a group, the inverse element is unique: if $p + q = 0$, then $q = -p$.

6. Prove that in an arbitrary ring,

$$p \cdot 0 = 0 \cdot p = 0 \qquad \text{and} \qquad p \cdot (-q) = (-p) \cdot q = -(p \cdot q)$$

for all elements p and q.

7. Let A be the set of all idempotent elements in a commutative ring R with unit. Define the sum $p \oplus q$ of two elements p and q in A by

$$p \oplus q = p + q - 2pq,$$

where the right-hand term is computed in R (and pq means $p \cdot q$). The distinguished elements of A are the same as those of R, and multiplication in A is the restriction of multiplication in R. Show that A is a Boolean ring.

8. A *Boolean group* is a group in which every element has order two (in other words, the law (10) is valid). Show that every Boolean group is commutative (that is, the commutative law (3) is valid).

9. A *zero-divisor* in a ring is a non-zero element p such that $p \cdot q = 0$ for some non-zero element q. Prove that a Boolean ring (with or without a unit) with more than two elements has zero-divisors. (This observation is due to Stone [66].)

10. (Harder.) Prove that every Boolean ring without a unit can be extended to a Boolean ring with a unit. To what extent is this extension procedure unique? (This result is due to Stone [66].)

11. (Harder.) Does every finite Boolean ring have a unit? (The answer to this question is due to Stone [66].)

12. Give an example of a Boolean ring that has no unit. Exercise 10 implies that your example can be extended to a Boolean ring with unit; describe the elements of that extension.

13. (Harder.) Can every non-degenerate Boolean ring with unit be obtained by adjoining a unit to a Boolean ring without a unit?

14. (Harder.) Is every Boolean group the additive group of some Boolean ring?

Chapter 2

Boolean Algebras

Let X be an arbitrary set and let $\mathcal{P}(X)$ be the class of all subsets of X (the *power set* of X). Three natural set-theoretic operations on $\mathcal{P}(X)$ are the binary operations of union and intersection, and the unary operation of complementation. The union $P \cup Q$ of two subsets P and Q is, by definition, the set of elements that are either in P or in Q, the intersection $P \cap Q$ is the set of elements that are in both P and Q, and the complement P' is the set of elements (of X) that are not in P. There are also two distinguished subsets: the empty set $\varnothing$, which has no elements, and the universal set X. The class $\mathcal{P}(X)$, together with the operations of union, intersection, and complementation, and the distinguished subsets $\varnothing$ and X, is called the *Boolean algebra* (or *field*) *of all subsets* of X, or the *power set algebra* on X.

The arithmetic of this algebra bears a striking resemblance to the arithmetic of Boolean rings. Some of the most familiar and useful identities include the laws for forming the complements of the empty and the universal sets,

$$(1) \qquad\qquad \varnothing' = X, \qquad\qquad X' = \varnothing,$$

the laws for forming an intersection with the empty set and a union with the universal set,

$$(2) \qquad\qquad P \cap \varnothing = \varnothing, \qquad\qquad P \cup X = X,$$

the identity laws,

$$(3) \qquad\qquad P \cap X = P, \qquad\qquad P \cup \varnothing = P,$$

the complement laws,

$$(4) \qquad\qquad P \cap P' = \varnothing, \qquad\qquad P \cup P' = X,$$

S. Givant, P. Halmos, *Introduction to Boolean Algebras*,
Undergraduate Texts in Mathematics, DOI: 10.1007/978-0-387-68436-9_2,
© Springer Science+Business Media, LLC 2009

the double complement law,

(5) $$(P')' = P,$$

the idempotent law,

(6) $$P \cap P = P, \qquad P \cup P = P,$$

the De Morgan laws,

(7) $$(P \cap Q)' = P' \cup Q', \qquad (P \cup Q)' = P' \cap Q',$$

the commutative laws,

(8) $$P \cap Q = Q \cap P, \qquad P \cup Q = Q \cup P,$$

the associative laws,

(9) $$P \cap (Q \cap R) = (P \cap Q) \cap R, \qquad P \cup (Q \cup R) = (P \cup Q) \cup R,$$

and the distributive laws,

(10) $$P \cap (Q \cup R) = (P \cap Q) \cup (P \cap R),$$
$$P \cup (Q \cap R) = (P \cup Q) \cap (P \cup R).$$

Each of these identities can be verified by an easy set-theoretic argument based on the definitions of the operations involved. Consider, for example, the verification of the first De Morgan law. It must be shown that each element x of X belongs to $(P \cap Q)'$ just in case it belongs to $P' \cup Q'$. The argument goes as follows:

$$
\begin{aligned}
x \in (P \cap Q)' \quad &\text{if and only if} \quad x \notin P \cap Q, \\
&\text{if and only if} \quad x \notin P \text{ or } x \notin Q, \\
&\text{if and only if} \quad x \in P' \text{ or } x \in Q', \\
&\text{if and only if} \quad x \in P' \cup Q'.
\end{aligned}
$$

The first and third equivalences use the definition of complementation, the second uses the definition of intersection, and the last uses the definition of union.

While (1)–(10) bear a close resemblance to laws that are true in Boolean rings, there are important differences. Negation in Boolean rings is the identity operation, whereas complementation is not. Addition in Boolean rings is not an idempotent operation, whereas union is. The distributive law for addition over multiplication fails in Boolean rings, whereas the distributive law for union over intersections holds.

Boolean rings are an abstraction of the ring 2. The corresponding abstraction of $\mathcal{P}(X)$ is called a Boolean algebra. Specifically, a *Boolean algebra* is a non-empty set A, together with two binary operations $\wedge$ and $\vee$ (on A), a unary operation $'$, and two distinguished elements 0 and 1, satisfying the following axioms, the analogues of identities (1)–(10):

(11) $\qquad\qquad\qquad 0' = 1, \qquad\qquad\qquad\qquad 1' = 0,$

(12) $\qquad\qquad\qquad p \wedge 0 = 0, \qquad\qquad\qquad p \vee 1 = 1,$

(13) $\qquad\qquad\qquad p \wedge 1 = p, \qquad\qquad\qquad p \vee 0 = p,$

(14) $\qquad\qquad\qquad p \wedge p' = 0, \qquad\qquad\qquad p \vee p' = 1,$

(15) $\qquad\qquad\qquad\qquad\qquad (p')' = p,$

(16) $\qquad\qquad\qquad p \wedge p = p, \qquad\qquad\qquad p \vee p = p,$

(17) $\qquad (p \wedge q)' = p' \vee q', \qquad\qquad\qquad (p \vee q)' = p' \wedge q',$

(18) $\qquad\quad p \wedge q = q \wedge p, \qquad\qquad\qquad\quad p \vee q = q \vee p,$

(19) $\quad p \wedge (q \wedge r) = (p \wedge q) \wedge r, \qquad p \vee (q \vee r) = (p \vee q) \vee r,$

(20) $\quad p \wedge (q \vee r) = (p \wedge q) \vee (p \wedge r), \qquad p \vee (q \wedge r) = (p \vee q) \wedge (p \vee r).$

This set of axioms is wastefully large, more than strong enough for the purpose. The problem of selecting small subsets of this set of conditions that are strong enough to imply them all is one of dull axiomatics. For the sake of the record: one solution of the problem, essentially due to Huntington [28], is given by the *identity laws* (13), the *complement laws* (14), the *commutative laws* (18), and the *distributive laws* (20). To prove that these four pairs imply all the other conditions, and, in particular, to prove that they imply the *De Morgan laws* (17) and the *associative laws* (19), involves some non-trivial trickery.

There are several possible widely adopted names for the operations $\wedge$, $\vee$, and $'$. We shall call them *meet, join,* and *complement* (or *complementation*), respectively. The distinguished elements 0 and 1 are called *zero* and *one*. One is also known as the *unit*.

Equations (1)–(10) imply that the class of all subsets of an arbitrary set X is an example of a Boolean algebra. When the underlying set X is empty, the resulting algebra is *degenerate* in the sense that it has just one element. In this case, operations of join, meet, and complementation are all constant, and $0 = 1$. The simplest non-degenerate Boolean algebra is the class of all subsets of a one-element set. It has just two elements, 0 (the empty set) and 1 (the one-element set). The operations of join and meet are described by the arithmetic tables

$\vee$	0	1
0	0	1
1	1	1

and

$\wedge$	0	1
0	0	0
1	0	1

,

and complementation is the unary operation that maps 0 to 1, and conversely. We shall see in a moment that this algebra and the two-element Boolean ring are interdefinable. For that reason, the same symbol 2 is used to denote both structures.

Here is a comment on notation, inspired by the associative laws (19). It is an elementary consequence of those laws that if $p_1, \ldots, p_n$ are elements of a Boolean algebra, then $p_1 \vee \cdots \vee p_n$, makes sense. The point is, of course, that since such joins are independent of how they are bracketed, it is not necessary to indicate any bracketing at all. The element $p_1 \vee \cdots \vee p_n$ may alternatively be denoted by $\bigvee_{i=1}^{n} p_i$, or, in case no confusion is possible, simply by $\bigvee_i p_i$.

If we make simultaneous use of both the commutative and the associative laws, we can derive a slight but useful generalization of the preceding comment. If E is a non-empty finite subset of a Boolean algebra, then the set E has a uniquely determined join, independent of any order or bracketing that may be used in writing it down. (In case E is a singleton, it is natural to identify that join with the unique element in E.) We shall denote the join of E by $\bigvee E$.

Both the preceding comments apply to meets as well as to joins. The corresponding symbols are, of course,

$$\bigwedge_{i=1}^{n} p_i, \quad \text{or} \quad \bigwedge_i p_i, \quad \text{and} \quad \bigwedge E.$$

The conventions regarding the order of performing different operations in the absence of any brackets are the following: complements take priority over meets and joins, while meets take priority over joins. Example: the expression $p' \vee q \wedge p$ should be read as $(p') \vee (q \wedge p)$. It is convenient to write successive applications of complement without any bracketing, for instance p'' instead of $(p')'$.

Exercises

1. Verify that the identities (1)–(10) are true in every Boolean algebra of all subsets of a set.

2. (Harder.) Show that the identities in (13), (14), (18), and (20) together form a set of axioms for the theory of Boolean algebras. In other words, show that they imply the identities in (11), (12), (15), (16), (17), and (19). (This result is essentially due to Huntington [30].)

3. Prove directly that the two-element structure 2 defined in the chapter is a Boolean algebra, by showing that axioms (13), (14), (18), and (20) are all valid in 2.

4. In analogy with the construction, for each set X, of the Boolean ring 2^X in Chapter 1, define operations of join, meet, and complementation on 2^X, and distinguished constants zero and one, and prove that the resulting structure is a Boolean algebra.

5. (Harder.) A member of a set of axioms is said to be *independent* of the remaining axioms if it is not derivable from them. One technique for demonstrating the independence of a given axiom is to construct a model in which that axiom fails while the remaining axioms hold. The given axiom cannot then be derivable from the remaining ones, since if it were, it would have to hold in the model as well. The four pairs of identities (13), (14), (18), and (20) constitute a set of eight axioms for Boolean algebras.

    (a) Show that the distributive law for join over meet in (20) is independent of the remaining seven axioms.

    (b) Show that the distributive law for meet over join in (20) is independent of the remaining seven axioms.

    (c) Show that each of the complement laws in (14) is independent of the remaining seven axioms.

   (These proofs of independence are due to Huntington [28].)

6. (Harder.) A set of axioms is said to be *independent* if no one of the axioms can be derived from the remaining ones. Do the four pairs of identities (13), (14), (18), and (20) constitute an independent set of axioms for Boolean algebras?

7. (Harder.) The operation of meet and the distinguished elements zero and one can be defined in terms of join and complement by the equations

$$p \wedge q = (p' \vee q')', \qquad 0 = (p \vee p')', \qquad 1 = p \vee p'.$$

A Boolean algebra may therefore be thought of as a non-empty set together with two operations: join and complement. Prove that the following identities constitute a set of axioms for this conception of Boolean algebras: the commutative and associative laws for join, and

(H) $$(p' \vee q')' \vee (p' \vee q)' = p.$$

(This axiomatization, and the proof of its equivalence to the set of axioms (13), (14), (18), and (20), is due to Huntington [30]. In fact, (H) is often called *Huntington's axiom.*)

8. (Harder) Prove that the three axioms in Exercise 7 are independent. (The proof of independence is due to Huntington [30].)

9. (Harder.) Prove that the following identities constitute a set of axioms for Boolean algebras:

$$p'' = p, \quad p \vee (q \vee q')' = p, \quad p \vee (q \vee r)' = ((q' \vee p)' \vee (r' \vee p)')'.$$

(This axiomatization, and the proof of its equivalence with the axiom set in Exercise 7, is due to Huntington [30].)

10. (Harder.) Prove that the three axioms in Exercise 9 are independent. (The proof of independence is due to Huntington [30].)

11. (Harder.) Prove that the commutative and associative laws for join, and the equivalence

$$p \vee q' = r \vee r' \qquad \text{if and only if} \qquad p \vee q = p,$$

together constitute a set of axioms for Boolean algebras. (This axiomatization, and the proof of its equivalence with the axiom set in Exercise 7, is due to Byrne [11].)

Chapter 3

Boolean Algebras Versus Rings

The theories of Boolean algebras and Boolean rings are very closely related; in fact, they are just different ways of looking at the same subject. More precisely, every Boolean algebra can be turned into a Boolean ring by defining appropriate operations of addition and multiplication, and, conversely, every Boolean ring can be turned into a Boolean algebra by defining appropriate operations of join, meet, and complement. The precise way of accomplishing this can be elucidated by comparing the Boolean algebra $\mathcal{P}(X)$ of all subsets of X and the Boolean ring 2^X of all 2-valued functions on X. Each subset P of X is naturally associated with a function p from X into 2, namely the *characteristic function* of P, defined for each x in X by

$$p(x) = \begin{cases} 1 & \text{if } x \in P, \\ 0 & \text{if } x \notin P. \end{cases}$$

The correspondence that maps each subset to its characteristic function is a *bijection* (a one-to-one, onto function) from $\mathcal{P}(X)$ to 2^X. The inverse correspondence maps each function q in 2^X to its *support*, the set of elements x in X for which $q(x) = 1$.

How should the operations of addition and multiplication, and the distinguished elements zero and the unit, be defined in $\mathcal{P}(X)$ so that it becomes a Boolean ring? To answer this question, it is helpful to analyze more closely the definitions of the ring operations in 2^X, and to translate these definitions (via the bijective correspondence) into the language of $\mathcal{P}(X)$. Suppose P and Q are subsets of X, and let p and q be their characteristic functions.

S. Givant, P. Halmos, *Introduction to Boolean Algebras*,
Undergraduate Texts in Mathematics, DOI: 10.1007/978-0-387-68436-9_3,
© Springer Science+Business Media, LLC 2009

The sum $p + q$ and the product $p \cdot q$ are defined pointwise: for any x in X,

$$(p + q)(x) = p(x) + q(x) = \begin{cases} 1 & \text{if} \quad p(x) \neq q(x), \\ 0 & \text{if} \quad p(x) = q(x), \end{cases}$$

and

$$(p \cdot q)(x) = p(x) \cdot q(x) = \begin{cases} 1 & \text{if} \quad p(x) = q(x) = 1, \\ 0 & \text{otherwise}, \end{cases}$$

as is clear from the arithmetic tables for the ring 2. The values $p(x)$ and $q(x)$ are different just in case one of them is 1 and the other is 0, that is to say, just in case x is in P but not in Q, or vice versa. The values $p(x)$ and $q(x)$ are both 1 just in case x is in both P and Q. These observations suggest the following definitions of ring addition and multiplication in $\mathcal{P}(X)$:

(1) $P + Q = (P \cap Q') \cup (P' \cap Q)$ and $P \cdot Q = P \cap Q$.

(The Boolean sum $P + Q$ is usually called the *symmetric difference* of P and Q.) A similar analysis suggests the definitions

(2) $0 = \varnothing$ and $1 = X$

for the distinguished ring elements zero and one in $\mathcal{P}(X)$.

   With these operations and distinguished elements, the set $\mathcal{P}(X)$ becomes a Boolean ring: it satisfies axioms (1.1)–(1.3), (1.5), (1.6), and (1.8)–(1.11). In fact, the correspondence h that takes each function in 2^X to its support is what is usually called an *isomorphism* between the two rings: it maps 2^X one-to-one onto $\mathcal{P}(X)$, and it *preserves* the ring operations and distinguished elements in the sense that

$$h(p + q) = h(p) + h(q), \quad h(p \cdot q) = h(p) \cdot h(q), \quad h(0) = 0, \quad h(1) = 1.$$

The operations and distinguished elements on the left sides of the equations are those of the ring 2^X, while the ones on the right are those of the ring $\mathcal{P}(X)$. These equations just express, in a slightly different form, the definitions in (1) and (2) of the ring operations and distinguished elements for $\mathcal{P}(X)$. The whole state of affairs can be summarized by saying that the Boolean rings 2^X and $\mathcal{P}(X)$ are *isomorphic* via the correspondence that takes each function in 2^X to its support. The two rings are structurally the same (which is what really matters); they differ only in the "shape" of their elements.

   It is also possible to turn the ring 2^X into a Boolean algebra. To understand how the Boolean operations and distinguished elements should be defined in 2^X, it is helpful to analyze the definitions of these operations in $\mathcal{P}(X)$ and to translate these definitions into the language of 2^X. Suppose once more that P and Q are subsets of X, and that p and q are their characteristic functions. Then

$$
\begin{aligned}
x \in P \cup Q \quad &\text{if and only if} \quad x \in P \text{ or } x \in Q \\
&\text{if and only if} \quad p(x) = 1 \text{ or } q(x) = 1 \\
&\text{if and only if} \quad p(x) \neq q(x) \text{ or } p(x) = q(x) = 1 \\
&\text{if and only if} \quad p(x) + q(x) + p(x) \cdot q(x) = 1 \\
&\text{if and only if} \quad (p + q + p \cdot q)(x) = 1.
\end{aligned}
$$

The first equivalence uses the definition of union, the second uses the definitions of the characteristic functions, the third uses the fact that 2 has just two elements, the fourth uses the arithmetic of 2, and the last uses the definitions of the ring operations of 2^X. Similarly,

$$
\begin{aligned}
x \in P' \quad &\text{if and only if} \quad x \notin P \\
&\text{if and only if} \quad p(x) \neq 1 \\
&\text{if and only if} \quad p(x) \neq 1(x) \\
&\text{if and only if} \quad p(x) + 1(x) = 1 \\
&\text{if and only if} \quad (p + 1)(x) = 1.
\end{aligned}
$$

The occurrence of the symbol "1" in the second equivalence, and its rightmost occurrence in the fourth and fifth equivalences, denote the unit of 2; its occurrence in the third equivalence, and its leftmost occurrences in the fourth and fifth equivalences, denote the unit function of 2^X. To justify the equivalences, use the definition of complementation, the definition of the characteristic function, the definition of the unit function, the arithmetic of 2, and the definition of addition in the ring 2^X.

The preceding observations suggest the following definitions for the operations of join and complement in 2^X:

$$
p \vee q = p + q + p \cdot q \qquad \text{and} \qquad p' = p + 1.
$$

A similar but simpler analysis implies that the distinguished Boolean elements zero and one should coincide with the distinguished ring elements zero and one, and that meet should coincide with ring multiplication. With these operations and distinguished elements, the set 2^X becomes a Boolean algebra: it satisfies axioms (2.11)–(2.20). In fact, the Boolean algebras $\mathcal{P}(X)$ and 2^X are isomorphic via the correspondence g that takes each subset of X to its characteristic function. In more detail, the correspondence g maps $\mathcal{P}(X)$ one-to-one onto 2^X, and it preserves the Boolean operations and distinguished elements in the sense that

$$g(P \cup Q) = g(P) \vee g(Q), \qquad g(P \cap Q) = g(P) \wedge g(Q),$$
$$g(P') = g(P)', \qquad g(\varnothing) = 0, \qquad g(X) = 1.$$

Motivated by this set-theoretic example, we can introduce into every Boolean algebra A operations of addition and multiplication very much like symmetric difference and intersection; just define

(3) $$p + q = (p \wedge q') \vee (p' \wedge q) \qquad \text{and} \qquad p \cdot q = p \wedge q.$$

Under these operations, together with 0 and 1 (the zero and unit of the Boolean algebra), A becomes a Boolean ring. Conversely, every Boolean ring can be turned into a Boolean algebra with the same zero and unit; just define operations of join, meet, and complement by

(4) $$p \vee q = p + q + p \cdot q, \qquad p \wedge q = p \cdot q, \qquad p' = p + 1.$$

Start with a Boolean algebra, turn it into a Boolean ring (with the same zero and unit) using the definitions in (3), and then convert the ring into a Boolean algebra using the definitions in (4); the result is the original Boolean algebra. Conversely, start with a Boolean ring, convert it into a Boolean algebra using the definitions in (4), and then convert the Boolean algebra into a Boolean ring using the definitions in (3); the result is the original ring.

The customary succinct way of summarizing the preceding discussion is to say that the theories of Boolean algebras and Boolean rings are *definitionally equivalent*. The precise way of proving this statement is to derive the Boolean algebra axioms (2.11)–(2.20) and the definitions in (3) from the Boolean ring axioms (1.1)–(1.3), (1.5), (1.6), (1.8)–(1.11), and the definitions in (4), and, conversely, to derive the Boolean ring axioms and the definitions in (4) from the Boolean algebra axioms and the definitions in (3). In this book we shall use the two terms "Boolean ring" and "Boolean algebra" almost as if they were synonymous, selecting on each occasion the one that seems intuitively more appropriate. Since our motivation comes from set theory, we shall speak of Boolean algebras much more often than of Boolean rings.

The point of view of Boolean algebras makes it possible to give a simple and natural description of an example (due to Sheffer [54]) that would be quite awkward to treat from the point of view of Boolean rings. Let m be a positive integer, and let A be the set of all positive integral divisors of m. Define the Boolean structure of A by the equations

$$0 = 1,$$
$$1 = m,$$
$$p \wedge q = \gcd\{p, q\},$$
$$p \vee q = \operatorname{lcm}\{p, q\},$$
$$p' = m/p.$$

It turns out that, with the distinguished elements and operations so defined, A forms a Boolean algebra if and only if m is square-free (that is, m is not divisible by the square of any prime). Query: what are the number-theoretic expressions of the ring operations in this Boolean algebra? And, while we are on the subject, what are the expressions for the Boolean operations in the Boolean ring A consisting of the idempotent elements of an arbitrary commutative ring R with unit? (See Chapter 1.) The answer to this question is slightly different from (4); those equations give the answer in terms of the ring operations in A, and what is wanted is an answer in terms of the ring operations in R.

The theory of Boolean algebras was created by Boole in the 1840s, and subsequently refined by De Morgan, Jevons, Schröder, Whitehead, and others. The name "Boolean algebra" was suggested by Sheffer [53] in 1913. It was Stone who realized, in the mid-1930s, that Boolean algebras could be treated as rings in which the operation of multiplication is idempotent. He introduced in [66] the notion of a Boolean ring, and developed the basic algebraic theory of such rings. In particular, he proved that the class of Boolean rings is definitionally equivalent to the class of Boolean algebras.

Exercises

1. Prove that the Boolean algebras $\mathcal{P}(X)$ and 2^X are isomorphic via the mapping that takes each subset of X to its characteristic function.

2. The purpose of this exercise is to demonstrate the definitional equivalence of the theories of Boolean algebras and Boolean rings.

   (a) Show that every Boolean algebra becomes a Boolean ring with the same zero and unit under the operations defined by the equations in (3).

(b) Show that, conversely, every Boolean ring becomes a Boolean algebra with the same zero and unit under the operations defined by the equations in (4).

(c) If a Boolean algebra is converted into a Boolean ring using the definitions in (3), and if that Boolean ring is then converted into a Boolean algebra using the definitions in (4), prove that the result is just the original Boolean algebra.

(d) If a Boolean ring is converted into a Boolean algebra using the definitions in (4), and if that Boolean algebra is then converted into a Boolean ring using the definitions in (3), prove that the result is just the original Boolean ring.

3. Prove that the set of positive integral divisors of a positive integer m is a Boolean algebra (under the operations defined at the end of the chapter) just in case m is square-free. What are the number-theoretic expressions of the ring operations in this algebra?

4. What are the expressions for the Boolean operations in the Boolean ring consisting of the idempotent elements of an arbitrary commutative ring (Exercise 1.7)?

5. (Harder.) Prove that every law of Boolean algebra in which the unit and complement do not occur is valid for Boolean rings without unit when meet and join in the ring are defined by

$$p \wedge q = p \cdot q \qquad \text{and} \qquad p \vee q = p + q + p \cdot q.$$

Chapter 4

The Principle of Duality

Every Boolean polynomial has a *dual*: it is defined to be the polynomial that results from interchanging 0 and 1, and at the same time interchanging $\wedge$ and $\vee$. For example, the polynomials

$$(p \vee q') \wedge (p' \vee 1) \wedge 0 \qquad \text{and} \qquad (p \wedge q') \vee (p' \wedge 0) \vee 1$$

are duals of one another. (The definition of Boolean polynomials is the same as that of ordinary polynomials, except that the admissible operations are not addition and multiplication but meet, join, and complement.) Every Boolean equation also has a dual, obtained by forming the duals of the polynomials on each side of the equation. The identities

$$p \wedge p' = 0 \qquad \text{and} \qquad p \vee p' = 1$$

are duals of one another, as are the identities

$$p \wedge (q \vee r) = (p \wedge q) \vee (p \wedge r) \qquad \text{and} \qquad p \vee (q \wedge r) = (p \vee q) \wedge (p \vee r).$$

A technical reason for preferring the language of Boolean algebras to that of Boolean rings is the so-called *principle of duality*.

The principle consists in observing that the axioms (2.11)–(2.20) for Boolean algebras come in dual pairs. It follows that the same is true for all the consequences of those axioms: the general theorems about Boolean algebras, and, for that matter, their proofs also, come in dual pairs.

The *absorption laws*

(1) $$p \wedge (p \vee q) = p \qquad \text{and} \qquad p \vee (p \wedge q) = p,$$

and their derivations, may serve as an example. Here is a proof of the first law, using (in order) the second identity in (2.13), the second identity in (2.20),

S. Givant, P. Halmos, *Introduction to Boolean Algebras*,
Undergraduate Texts in Mathematics, DOI: 10.1007/978-0-387-68436-9_4,
© Springer Science+Business Media, LLC 2009

the first identity in (2.18), the first identity in (2.12), and the second identity in (2.13):

(2) $p \wedge (p \vee q) = (p \vee 0) \wedge (p \vee q) = p \vee (0 \wedge q) = p \vee (q \wedge 0) = p \vee 0 = p.$

The dual derivation is

(3) $p \vee (p \wedge q) = (p \wedge 1) \vee (p \wedge q) = p \wedge (1 \vee q) = p \wedge (q \vee 1) = p \wedge 1 = p.$

The axioms used in this derivation are, step by step, the duals of the axioms used in the derivation of (2). The identity that is proved by (3) is just the second law in (1), which is the dual of the first law.

A practical consequence of the principle of duality, often exploited in what follows, is that in the theory of Boolean algebras it is sufficient to state and to prove only half the theorems; the other half come gratis from the principle.

A slight misunderstanding can arise about the meaning of duality, and often does. It is well worthwhile to clear it up once and for all, especially since the clarification is quite amusing in its own right. If an experienced Boolean algebraist is asked for the dual of a Boolean polynomial, such as say $p \vee q$, his answer might be $p \wedge q$ one day and $p' \wedge q'$ another day; the answer $p' \vee q'$ is less likely but not impossible. What is needed here is some careful terminological distinction. Let us restrict attention to the completely typical case of a polynomial $f(p, q)$ in two variables. The *complement* of $f(p, q)$ is by definition $(f(p, q))'$, abbreviated $f'(p, q)$; the *dual* of $f(p, q)$ is $f'(p', q')$; the *contradual* of $f(p, q)$ is $f(p', q')$.

The polynomial

(4) $p \wedge (q \vee (p' \wedge 0))$

may serve as an example. Its complement is formed by applying the operation $'$ to the entire expression, and then simplifying the result with the help of the De Morgan laws and the double complement law:

$$p' \vee (q' \wedge (p \vee 1)).$$

The contradual is formed by replacing p and q in (4) with their complements, and then simplifying:

$$p' \wedge (q' \vee (p \wedge 0)).$$

The dual is the complement of the contradual, appropriately simplified:

$$p \vee (q \wedge (p' \vee 1)).$$

What goes on here is that there is a group (see Exercise 1.5) acting on the set of polynomials, and it is not the two-element group, but a group with four elements. In more detail, there are four functions mapping the set of Boolean polynomials one-to-one onto itself: the identity function (taking each polynomial to itself), the complement function (taking each polynomial to its complement), the contradual function, and the dual function. The set of these functions is closed under composition. For instance, the composition of the complement and contradual functions is the dual function, while the composition of the contradual and dual functions is the complement function. There is an identity element — the identity function — and each function is its own inverse with respect to composition. Consequently, the four functions form a group under composition, and in fact a Boolean group that is often called the *Klein four-group* (see Exercise 1.8).

This comment was made by Gottschalk [18], who describes the situation by speaking of the *principle of quaternality*.

A word of warning: the word "duality" is frequently used in contexts startlingly different from one other and from the one we met above. This is true even within the theory of Boolean algebras, where, for instance, a topological duality theory turns out to play a much more important role than the elementary algebraic one just described. If the context alone is not sufficient to indicate the intended meaning, great care must be exercised to avoid confusion.

Exercises

1. Form the duals of the following polynomials.

   (a) $(p \wedge q)' \vee (0 \wedge 1) \vee (q' \vee p')$.

   (b) $(p \wedge (q \vee r)) \vee ((p \wedge p') \vee r)$.

2. Form the duals of the following equations.

   (a) $p \wedge (p' \vee q) = p \wedge q$.

   (b) $(p \vee q) \wedge (p' \wedge q') = (p \wedge (p' \wedge q')) \vee (p' \wedge (q' \wedge q))$.

3. Form the complements, the contraduals, and the complements of the contraduals of the following polynomials, and simplify the results using the De Morgan laws and the double complement law.

   (a) $(p \wedge (q \vee r)) \vee ((p \wedge p') \vee r)$.

   (b) $(p \wedge (p' \wedge q')) \vee (p' \wedge (q' \wedge q))$.

4. Derive the identity

$$p \wedge (p' \vee q) = p \wedge q$$

from axioms (2.13), (2.14), (2.18), and (2.20). Write out the dual form of the derivation. What identity does the dual derivation prove?

5. Let c, k, and d denote the functions on the set of Boolean polynomials of forming the complement, the contradual, and the dual. Denote the identity function on the set by i. Complete the following table for composing these operations:

$\circ$	i	c	k	d
i				
c			d	
k				c
d				

Chapter 5

Fields of Sets

To form $\mathcal{P}(X)$ is not the only natural way to make a Boolean algebra out of a non-empty set X. A more general way is to consider an arbitrary non-empty subclass A of $\mathcal{P}(X)$ that is closed under intersection, union, and complement; in other words, if P and Q are in A, then so are $P \cap Q$, $P \cup Q$, and P'. Since A contains at least one element, it follows that A contains $\varnothing$ and X (cf. (2.4)), and hence that A is a Boolean algebra. Every Boolean algebra obtained in this way is called a *field* (of sets). There is usually no danger in denoting a field of sets by the same symbol as the class of sets that go to make it up. This does not, however, justify the conclusion (it is false) that set-theoretic intersection, union, and complement are the only possible operations that convert a class of sets into a Boolean algebra.

To show that a class A of subsets of a set X is a field, it suffices to show that A is non-empty and closed under union and complement. In fact, if A is closed under these two operations, then it must also be closed under intersection, since

$$P \cap Q = (P' \cup Q')'$$

for any two subsets P and Q of X. Dually, A is a field whenever it is non-empty and closed under intersection and complement.

A subset P of a set X is *cofinite* (in X) if its complement P' is finite; in other words, P is cofinite if it can be obtained from X by removing finitely many elements. For instance, the set of integers greater than 1000 or less than -10 is cofinite (in the set of all integers), and so is the set of all integers with the numbers 2, -75, and 1037 removed; the set of even integers is not cofinite. The class A of all those subsets of a non-empty set X that are either finite or cofinite is a field of subsets of X. The proof is a simple cardinality

S. Givant, P. Halmos, *Introduction to Boolean Algebras*,
Undergraduate Texts in Mathematics, DOI: 10.1007/978-0-387-68436-9_5,
© Springer Science+Business Media, LLC 2009

argument. The union of two finite subsets is finite, while the union of a cofinite subset with any subset is cofinite; also, the complement of a finite subset is cofinite, and conversely. If the set X itself is finite, then A is simply $\mathcal{P}(X)$; if X is infinite, then A is a new example of a Boolean algebra and is called the *finite–cofinite algebra* (or *field*) of X.

The preceding construction can be generalized. Call a subset P of X *cocountable* (in X) if its complement P' is countable. For example, the set of irrational numbers is cocountable in the set of real numbers, since the set of rational numbers is countable. Also, any cofinite subset of reals is cocountable, since any finite subset is countable. The class of all those subsets of X that are either countable or cocountable is a field of subsets of X, the so-called *countable–cocountable algebra* (or *field*) of X. Different description of the same field: the class of all those subsets P of X for which the cardinal number of either P or P' is less than or equal to $\aleph_0$ (the first infinite cardinal). A further generalization is obtained by using an arbitrary infinite cardinal number in place of $\aleph_0$.

Let X be the set of all integers (positive, negative, or zero), and let m be an arbitrary integer. A subset P of X is *periodic* of *period* m if it coincides with the set obtained by adding m to each of its elements. The class A of all periodic sets of period m is a field of subsets of X. If $m = 0$, then A is simply $\mathcal{P}(X)$. If $m = 1$, then A consists of just the two sets $\varnothing$ and X. In all other cases A is a new example of a Boolean algebra. For example, if $m = 3$, then A consists of eight sets, namely the eight possible unions that can be formed using one, two, three, or none of the sets

$$\{3n : n \in X\}, \qquad \{3n + 1 : n \in X\}, \qquad \text{and} \qquad \{3n + 2 : n \in X\}.$$

It is obvious in this case that A is closed under the operation of union, and almost obvious that A is closed under complementation. Consequently, A is a field of sets. Warning: a period of a periodic set is not unique; for example, the sets of integers of period 3 also have period 6 and period 12; in fact they have infinitely many periods.

Let X be the set of all real numbers. A *left half-closed interval* (or, for brevity, since this is the only kind we shall consider for the moment, a *half-closed interval*) is a set of one of the forms

$$[a, b) = \{x \in X : a \leq x < b\},$$
$$[a, \infty) = \{x \in X : a \leq x\},$$
$$(-\infty, b) = \{x \in X : x < b\},$$
$$(-\infty, \infty) = X,$$

where, of course, a and b themselves are real numbers and $a < b$. The class A of all finite unions of half-closed intervals is a field of subsets of X. Here is the proof. The empty set is the union of the empty family of half-closed intervals, so it belongs to A. The closure of A under union is obvious: the union of two finite unions of half-closed intervals is itself a finite union of half-closed intervals.

To establish the closure of A under complement, it is helpful to make two observations. First, the intersection of two half-closed intervals is either a half-closed interval or empty. For instance, if $a_1 < b_2$ and $a_2 < b_1$, then

$$[a_1, b_1) \cap [a_2, b_2) = [c, d) \quad \text{and} \quad [a_1, b_1) \cap (-\infty, b_2) = [a_1, d),$$

where

$$c = \max\{a_1, a_2\} \quad \text{and} \quad d = \min\{b_1, b_2\}.$$

(The diagram illustrates the three possibilities in this case.) The intersection

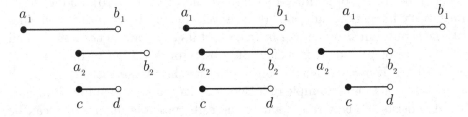

of the intervals is empty if the relevant inequalities fail. The second observation is that the complement of a half-closed interval is the union of at most two half-closed intervals. For instance, the complement of the interval $[a, b)$ is the union

$$(-\infty, a) \cup [b, +\infty),$$

while the complement of the interval $[a, +\infty)$ is $(-\infty, a)$.

If P is a finite union of half-closed intervals, then P' is a finite intersection of complements of half-closed intervals, by the De Morgan laws (2.7). Each of these complements is the union of at most two half-closed intervals, by the second observation, so P' is the intersection of a finite family of sets, each of which is the union of at most two half-closed intervals. The distributive laws (2.10) therefore imply that P' may be written as a finite union of finite intersections of half-closed intervals. Each of the latter intersections is again a half-closed interval or else empty, by the first observation, so P' may be written as a finite union of half-closed intervals. The class A is thus closed under complementation.

Any (linearly) ordered set can be used instead of the set of real numbers, though some details of the construction may require slight modification. For instance, a useful variant uses the closed unit interval $[0, 1]$ in the role of X. In this case it is convenient to stretch the terminology so as to include the closed intervals $[a, 1]$ and the degenerate interval $\{1\}$ among half-closed intervals. (The elements of this field are just the intersections with $[0, 1]$ of the finite unions of half-closed intervals discussed in the preceding paragraph. Notice, in this connection, that

$$[a, 1] = [a, b) \cap [0, 1] \qquad \text{and} \qquad \{1\} = [1, b) \cap [0, 1]$$

whenever $0 \leq a \leq 1 < b$.) The field of sets constructed from an ordered set in the manner illustrated by these two examples is usually called the *interval algebra* of the ordered set. (Interval algebras were first introduced by Mostowski and Tarski in [46].)

Valuable examples of fields of sets can be defined in the unit square (the set of ordered pairs (x, y) with $0 \leq x, y \leq 1$), as follows. Call a subset P of the square *vertical* if along with each point in P, every point of the vertical segment through that point also belongs to P. In other words, P is vertical if the presence of (x_0, y_0) in P implies the presence in P of (x_0, y) for every y in $[0, 1]$. If A is any field of subsets of the square, then the class of all vertical sets in A is another, and in particular the class of all vertical sets is a field of sets. Indeed, A contains the empty set, and is closed under union and complement, so it suffices to check that the empty set is a vertical set (this is vacuously true), and that the union of two vertical sets and the complement of a vertical set are again vertical sets. Let P and Q be vertical sets, and suppose that a point (x_0, y_0) belongs to their union. If the point belongs to P, then every point on the vertical segment through (x_0, y_0) also belongs to P (since P is vertical); consequently, every such point belongs to the union $P \cup Q$. An analogous argument applies if (x_0, y_0) belongs to Q. It follows that $P \cup Q$ is a vertical set. Suppose, next, that a point (x_0, y_0) belongs to the complement of P. No point on the vertical segment through (x_0, y_0) can then belong to P, for the presence of one such point in P implies the presence of every such point in P (since P is vertical), and in particular it implies that (x_0, y_0) is in P. It follows that every point on the vertical segment through (x_0, y_0) belongs to P', so that P' is a vertical set.

Here are two comments that are trivial but sometimes useful: (1) the *horizontal* sets (whose definition may safely be left to the reader) constitute just as good a field as the vertical sets, and (2) the Cartesian product of any two non-empty sets is, for these purposes, just as good as the unit square.

Exercises

1. Prove that the class of all sets of integers that are either finite sets of even integers, or else cofinite sets that contain all odd integers, is a field of sets.

2. Prove that the class of all sets of real numbers that are either countable or cocountable is a field of sets.

3. Prove that the class of all periodic sets of integers of period 2 is a field of sets. How many sets are in this class? Describe them all. Do the same for the class of all period sets of integers of period 3.

4. Prove that the class of all periodic sets of integers of period m is a field of sets. How many sets are in this class? Describe them all.

5. (Harder.) If m is a positive integer, and if A is the class of all those sets of integers that are periodic of some period greater than m, is A a field of sets?

6. The precise formulation and proof of the assertion that the intersection of two half-closed intervals is either half-closed or empty can be simplified by introducing some unusual but useful notation. Extend the natural linear ordering of the set X of real numbers to the set $X \cup \{-\infty, \infty\}$ of *extended real numbers* by requiring

$$-\infty < a < \infty$$

for every real number a, and $-\infty < \infty$. Thus, for example,

$$\max\{-\infty, a\} = a \quad \text{and} \quad \min\{-\infty, a\} = -\infty.$$

Extend the notation for half-closed intervals by defining

$$[-\infty, a) = (-\infty, a) = \{b \in X : b < a\}$$

for any a in $X \cup \{\infty\}$. In particular,

$$[-\infty, \infty) = (-\infty, \infty) = X.$$

Also, extend the notation $[a, b)$ to the case $a \geq b$ by writing

$$[a, b) = \{x \in R : a \leq x < b\} = \varnothing.$$

The notation $[a, b)$ thus makes sense for any two extended real numbers a and b. The assertion that the intersection of two half-closed intervals is either half-closed or empty can be expressed in this notation by the following simple statement that avoids case distinctions:

$$[a_1, b_1) \cap [a_2, b_2) = [c, d),$$

where

$$c = \max\{a_1, a_2\} \quad \text{and} \quad d = \min\{b_1, b_2\};$$

moreover, this intersection is non-empty if and only if $a_1, a_2 < b_1, b_2$. Prove the simple statement.

7. This exercise continues with the notation for extended real numbers introduced in Exercise 6. Prove that every set P in the interval algebra A of the real numbers can be written in exactly one way in the form

$$P = [a_1, b_1) \cup [a_2, b_2) \cup \cdots \cup [a_n, b_n),$$

where n is a non-negative integer and

$$a_1 < b_1 < a_2 < b_2 < \cdots < a_n < b_n.$$

(The case $n = 0$ is to be interpreted as the empty family of half-closed intervals.) Show that if P has this form, then

$$P' = [-\infty, a_1) \cup [b_1, a_2) \cup \cdots \cup [b_{n-1}, a_n) \cup [b_n, +\infty);$$

the first interval is empty (and hence should be omitted) if $a_1 = -\infty$; the second interval is empty (and hence should be omitted) if $b_n = +\infty$.

8. Prove that the class of all finite unions of left half-closed intervals of rational numbers is a countable field of sets.

9. Prove that the class of all finite unions of left half-closed subintervals of the interval $[0, 1]$ is a field of sets (where the intervals $[a, 1]$, for $0 \leq a \leq 1$, are considered left half-closed).

10. Let X be the set of real numbers. A *right half-closed interval* of real numbers is a set of one of the forms

$$(-\infty, b] = \{x \in X : x \leq b\},$$
$$(a, b] = \{x \in X : a < x \leq b\},$$

$$(a, +\infty) = \{x \in X : a < x\},$$
$$(-\infty, +\infty) = X,$$

where a and b are real numbers and $a < b$. Show that the class of all finite unions of right half-closed intervals is a field of subsets of the set of real numbers.

11. An *interval* of real numbers is a set P of real numbers with the property that whenever c and d are P, then so is every number between c and d. In other words, an interval is a set of one of the forms

$$(-\infty, a), \qquad (-\infty, a], \qquad (a, +\infty), \qquad [a, +\infty), \qquad (-\infty, +\infty),$$
$$(a, b), \qquad (a, b], \qquad [a, b), \qquad [a, b], \qquad \varnothing,$$

where a and b are real numbers and $a < b$. Prove that the class of all finite unions of intervals is a field of sets.

12. Define the notion of a *horizontal set* for the unit square, and prove that the class of all such sets is a field of sets.

13. Give an example of a Boolean algebra whose elements are subsets of a set, but whose operations are not the usual set-theoretic ones.

Chapter 6

Elementary Relations

The least profound among the properties of an algebraic system are usually the relations among its elements (as opposed to the relations among subsets of it and functions on it). In this chapter we shall discuss some of the elementary relations that hold in Boolean algebras. Since we shall later meet a powerful tool (namely, the representation theorem for Boolean algebras) the use of which reduces all elementary relations to set-theoretic trivialities, the purpose of the present discussion is more to illustrate than to exhaust the subject. An incidental purpose is to establish some notation that will be used freely throughout the sequel.

Throughout this chapter p, q, r, ... are elements of an arbitrary but fixed Boolean algebra A.

Lemma 1. *If $p \vee q = p$ for all p, then $q = 0$; if $p \wedge q = p$ for all p, then $q = 1$.*

Proof. To prove the first assertion, put $p = 0$, and use (2.18) and (2.13); the second assertion is the dual of the first.

Lemma 2. *If p and q are such that $p \wedge q = 0$ and $p \vee q = 1$, then $q = p'$.*

Proof. The assertion follows from (2.13), (2.14), (2.20), and the assumptions of the lemma, together with some implicit applications of (2.18):

$$q = 1 \wedge q = (p \vee p') \wedge q = (p \wedge q) \vee (p' \wedge q)$$
$$= 0 \vee (p' \wedge q) = (p' \wedge p) \vee (p' \wedge q)$$
$$= p' \wedge (p \vee q) = p' \wedge 1 = p'.$$

S. Givant, P. Halmos, *Introduction to Boolean Algebras*,
Undergraduate Texts in Mathematics, DOI: 10.1007/978-0-387-68436-9_6,
© Springer Science+Business Media, LLC 2009

These two lemmas can be expressed by saying that (2.13) uniquely determines 0 and 1, and (2.14) uniquely determines p'. In a less precise but more natural phrasing we may simply say that 0 and 1 are unique, and so are complements.

Lemma 3. *For all p and q,*

$$p \vee (p \wedge q) = p \qquad and \qquad p \wedge (p \vee q) = p.$$

Proof. Use (2.13), (2.20), and (2.12) (together with an implicit application of (2.18)):

$$p \vee (p \wedge q) = (p \wedge 1) \vee (p \wedge q) = p \wedge (1 \vee q) = p \wedge 1 = p.$$

The second equation is the dual of the first.

The identities of Lemma 3 are called the laws of *absorption*.

Often the most concise and intuitive way to state an elementary property of Boolean algebras is to introduce a new operation. The *difference* of two sets P and Q is the set of elements in P that are not in Q; these set-theoretic considerations suggest an operation of *subtraction* in arbitrary Boolean algebras. We write

(1) $p - q = p \wedge q'.$

The "symmetrized" version of the *difference $p - q$* is the Boolean sum:

(2) $p + q = (p - q) \vee (q - p);$

it is the analogue, for Boolean algebras, of the symmetric difference of two sets (see (3.1) and (3.3)). As a sample of the sort of easily proved relations that the notation suggests, consider the distributive laws

$$p \wedge (q - r) = (p \wedge q) - (p \wedge r)$$

and

$$(q \vee r) - p = (q - p) \vee (r - p).$$

One reason why Boolean algebras have something to do with logic is that the familiar sentential connectives *and*, *or*, and *not* have properties similar to the Boolean connectives $\wedge$, $\vee$, and $'$. Instead of meet, join, and complement, the logical terminology uses *conjunction*, *disjunction*, and *negation*. Motivated by the analogy, we now introduce into the study of Boolean algebra the operations suggested by logical *implication*,

(3) $$p \Rightarrow q = p' \vee q,$$

and *biconditional*,

(4) $$p \Leftrightarrow q = (p \Rightarrow q) \wedge (q \Rightarrow p).$$

The source of these operations suggests an unintelligent error that it is important to avoid. The result of the operation $\Rightarrow$ on the elements p and q of the Boolean algebra A is another element of A; it is not an assertion about or a relation between the given elements p and q. (The same is true of $\Leftrightarrow$.) It is for this reason that logicians sometimes warn against reading $p \Rightarrow q$ as "p implies q" and suggest instead the reading "if p, then q". Observe incidentally that if $\vee$ is read as "or", the disjunction $p \vee q$ must be interpreted in the non-exclusive sense (either p, or q, or both). The exclusive "or" (either p, or q, but not both) corresponds to the Boolean sum $p + q$.

The operations $\Rightarrow$ and $\Leftrightarrow$ would arise in any systematic study of Boolean algebra even without any motivation from logic. The reason is duality: the dual of $p - q$ is $q \Rightarrow p$, and the dual of $p + q$ is $p \Leftrightarrow q$. The next well-known Boolean operation that deserves mention here could not have been discovered through considerations of duality alone. It is called *stroke*, or *Sheffer stroke* (because it was introduced by Sheffer in [53]), and it is defined by

(5) $$p \mid q = p' \wedge q'.$$

In logical contexts this operation is known as *binary rejection* (neither p nor q), and among computer scientists it is often referred to as *nor*.

The chief theoretical application of the Sheffer stroke is the remark (due to Sheffer [53], but anticipated 33 years earlier by Peirce in the unpublished paper [49]) that a single operation, namely the stroke, is enough to define Boolean algebras. To establish this remark, it is sufficient to show that complement, meet, and join can be expressed in terms of the stroke, and indeed,

(6) $$p' = p \mid p,$$
(7) $$p \wedge q = (p \mid p) \mid (q \mid q),$$
(8) $$p \vee q = (p \mid q) \mid (p \mid q).$$

Exercises

1. Prove that the following distributive laws hold in all Boolean algebras.

(a) $(q \wedge r) - p = (q - p) \wedge (r - p)$.

(b) $(q \vee r) - p = (q - p) \vee (r - p)$.

(c) $p \wedge (q - r) = (p \wedge q) - (p \wedge r)$.

2. For each of the following identities, either prove that it is true in all Boolean algebras (by deriving it from the Boolean laws that have been established so far) or show that it fails to hold in the two-element Boolean algebra.

(a) $p \wedge (q - r) = (p \wedge q) - (q \wedge r)$.

(b) $p \vee (q - r) = (p \vee q) - (q \vee r)$.

(c) $p - (q - r) = (p - q) - r$.

(d) $p \wedge q = p - (p - q)$.

(e) $p \vee q = p \vee (q - p)$.

(f) $p - (q - r) = (p - q) \vee (p \wedge q \wedge r)$.

(g) $p + q = p' + q'$.

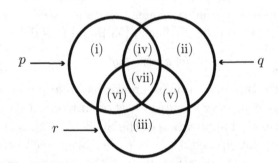

3. Prove that the seven elements

$$\text{(i)} \quad p - (q \vee r), \quad \text{(ii)} \quad q - (p \vee r), \quad \text{(iii)} \quad r - (p \vee q),$$
$$\text{(iv)} \quad (p \wedge q) - r, \quad \text{(v)} \quad (q \wedge r) - p, \quad \text{(vi)} \quad (p \wedge r) - q,$$
$$\text{(vii)} \quad p \wedge q \wedge r$$

are mutually disjoint (the meet of any two of them is 0) and the join of all of them is $p \vee q \vee r$ (see the diagram). Write each of

$$p - q, \qquad q - p, \qquad p - r, \qquad r - p, \qquad q - r, \qquad r - q$$

as a join of elements (i)–(vi). Then write each of

$$p + q, \qquad p + r, \qquad q + r$$

as a join of elements (i)–(vi).

4. Prove that $q \Rightarrow p$ and $p \Leftrightarrow q$ are the duals of $p - q$ and $p + q$, respectively.

5. Derive the following identities.

   (a) $p \Rightarrow p = 1$.
   (b) $p \Rightarrow q = q' \Rightarrow p'$.
   (c) $p \Rightarrow (p \Leftrightarrow q) = p \Rightarrow q$.
   (d) $(p \Leftrightarrow q) \Rightarrow p = p \vee q$.

6. Derive the following identities.

   (a) $p \Leftrightarrow q = (p \wedge q) \vee (p' \wedge q')$.
   (b) $p \Leftrightarrow q = (p + q)'$.

   Give geometric interpretations of these identities for fields of sets.

7. Derive the following identities concerning Boolean addition and biconditional.

   (a) $p' + q' = p + q$.
   (b) $p' \Leftrightarrow q' = p \Leftrightarrow q$.
   (c) $(p + q)' = p' \Leftrightarrow q'$.
   (d) $(p \Leftrightarrow q)' = p' + q'$.

   Notice that the last two identities express an analogue, for Boolean addition and the biconditional, of the De Morgan laws.

8. Prove that the set of elements in a Boolean algebra, under the operation $\Leftrightarrow$, is a Boolean group with the identity element 1. In other words, derive the following laws.

   (a) $p \Leftrightarrow (q \Leftrightarrow r) = (p \Leftrightarrow q) \Leftrightarrow r$.
   (b) $p \Leftrightarrow 1 = p$.
   (c) $p \Leftrightarrow p = 1$.

   (The function that maps each element of the Boolean algebra to its complement is actually an isomorphism from the Boolean group with the operation $+$ to the Boolean group with the operation $\Leftrightarrow$.)

9. The distributive law

$$p \wedge (q + r) = (p \wedge q) + (p \wedge r)$$

holds automatically in all Boolean algebras; it is just the distributive law (1.8) for Boolean rings. Prove the following related law:

$$p' \wedge (q + r) = (p \vee q) + (p \vee r).$$

Conclude that the elements

$$(p \vee q) + (p \vee r) \qquad \text{and} \qquad (p \wedge q) + (p \wedge r)$$

are disjoint, and join to $q + r$.

10. Prove that the theory of Boolean algebras is definitionally equivalent to the theory of the binary operation stroke, axiomatized by the following three identities:

$$(p \mid p) \mid (p \mid p) = p,$$
$$(p \mid (q \mid (q \mid q))) = p \mid p,$$
$$(p \mid (q \mid r)) \mid (p \mid (q \mid r)) = ((q \mid q) \mid p) \mid ((r \mid r) \mid p).$$

(This theorem is due to Sheffer [53].)

11. Enumerate all possible binary operations on $2 = \{0, 1\}$ (that is, enumerate all mappings from 2×2 into 2). Identify each of these 16 operations in terms of operations introduced in the chapter.

12. Show that all binary operations on 2 are definable in terms of the operations $\wedge$, $\vee$, and $'$. Conclude that they are definable in terms of the operations $\wedge$ and $'$ alone.

13. (Harder.) Show that not all binary operations on 2 are definable in terms of the operation $\Leftrightarrow$.

14. Show that all binary operations on 2 are definable in terms of stroke. Is there another binary operation on 2 besides stroke in terms of which the other binary operations are all definable?

15. A ternary operation on 2 is a mapping from $2 \times 2 \times 2$ into 2. Such an operation is conveniently represented by a table in which the arguments are listed on the left, and the corresponding values on the right. For example, the ternary operation $f(p, q, r)$ described by the table below maps the triples $(0, 1, 0)$, $(1, 0, 1)$, and $(1, 1, 1)$ to 1, and maps all other

triples to 0. It is not difficult to verify that f is definable in terms of meet, join, and complement. In fact,

$$f(p,q,r) = (p' \wedge q \wedge r') \vee (p \wedge q' \wedge r) \vee (p \wedge q \wedge r).$$

Prove that every ternary operation on 2 is definable in terms of meet, join, and complement.

p	q	r	f
0	0	0	0
0	0	1	0
0	1	0	1
0	1	1	0
1	0	0	0
1	0	1	1
1	1	0	0
1	1	1	1

16. Prove that every operation on 2 with n arguments is definable in terms of meet, join, and complement, for any positive integer n. (This result is due to Post [50].)

17. The form used in Exercise 15 to write the operation f in terms of meet, join, and complement is called, in the context of logic, *disjunctive normal form* because the expression on the right side is a join (disjunction) of meets of arguments and their complements. Show that f can also be written in *conjunctive normal form*, that is, as a meet of joins of arguments and their complements.

18. Prove that every ternary operation on 2 can be written in conjunctive normal form.

19. (Harder.) Show that if a ternary Boolean operation g is defined by

$$g(p,q,r) = (p \wedge q) \vee (q \wedge r) \vee (r \wedge p),$$

then that operation, together with complement and one, are enough to define Boolean algebras. Exhibit a set of axioms stated in terms of g and complement. (This approach to Boolean algebra is due to Grau [19].)

Chapter 7

Order

We continue to work with an arbitrary but fixed Boolean algebra A.

Lemma 1. $p \wedge q = p$ *if and only if* $p \vee q = q$.

Proof. If $p \wedge q = p$, then $p \vee q = (p \wedge q) \vee q$, and the conclusion follows from the appropriate law of absorption. The converse implication is obtained from this one by interchanging the roles of p and q and forming duals.

For sets, either one of the equations

$$P \cap Q = P \qquad \text{and} \qquad P \cup Q = Q$$

is equivalent to the inclusion $P \subseteq Q$. This observation motivates the introduction of a binary relation $\leq$ in every Boolean algebra; we write

$$p \leq q \qquad \text{or} \qquad q \geq p$$

in case $p \wedge q = p$, or, equivalently, $p \vee q = q$. A convenient way of expressing the relation $p \leq q$ in words is to say that p is *below* q, and that q is *above* p, or that q *dominates* p.

The relation of inclusion between sets is a *partial order*. In other words, the relation is *reflexive* in the sense that the inclusion $P \subseteq P$ always holds; it is *antisymmetric* in the sense that the inclusions $P \subseteq Q$ and $Q \subseteq P$ imply $P = Q$; and it is *transitive* in the sense that the inclusions $P \subseteq Q$ and $Q \subseteq R$ imply $P \subseteq R$. The next lemma says that the same thing is true of the relation $\leq$ in an arbitrary Boolean algebra.

Lemma 2. *The relation* $\leq$ *is a partial order.*

S. Givant, P. Halmos, *Introduction to Boolean Algebras*,
Undergraduate Texts in Mathematics, DOI: 10.1007/978-0-387-68436-9_7,
© Springer Science+Business Media, LLC 2009

Proof. Reflexivity follows from the idempotent law (2.16): $p \wedge p = p$, and therefore $p \leq p$. Antisymmetry follows from the commutative law (2.18): if $p \leq q$ and $q \leq p$, then

$$p = p \wedge q = q \wedge p = q.$$

Transitivity follows from the associative law (2.19): if $p \leq q$ and $q \leq r$, then

$$p \wedge q = p \qquad \text{and} \qquad q \wedge r = q,$$

and consequently

$$p \wedge r = (p \wedge q) \wedge r = p \wedge (q \wedge r) = p \wedge q = p.$$

It is sound mathematical practice to re-examine every part of a structure in the light of each new feature soon after the novelty is introduced. Here is the result of an examination of the structure of a Boolean algebra in the light of the properties of order.

Lemma 3.   (1) $0 \leq p$ and $p \leq 1$.

(2) *If $p \leq q$ and $r \leq s$, then $p \wedge r \leq q \wedge s$ and $p \vee r \leq q \vee s$.*

(3) *If $p \leq q$, then $q' \leq p'$.*

(4) $p \leq q$ *if and only if $p - q = 0$, or, equivalently, $p \Rightarrow q = 1$.*

The proofs of all these assertions are automatic. It is equally automatic to discover the dual of $\leq$; according to any reasonable interpretation of the phrase it is $\geq$. The inequalities in (2) are called the *monotony laws..*

If E is any subset of a partially ordered set such as our Boolean algebra A, we can consider the set F of all *upper bounds* of E and ask whether F has a smallest element. In other words: an element q belongs to F in case $p \leq q$ for every p in E; to say that F has a smallest element means that there exists an element q_1 in F such that $q_1 \leq q$ for every q in F. A smallest element in F, if it exists, is obviously unique: if q_1 and q_2 both have this property, then $q_1 \leq q_2$ and $q_2 \leq q_1$, so that $q_1 = q_2$, by antisymmetry. We shall call the smallest upper bound of the set E (if it has one) the *least upper bound,* or the *supremum,* of E. All these considerations have their obvious duals. The *greatest lower bound* of E is also called the *infimum* of E.

If the set E is empty, then every element of A is vacuously an upper bound of E (p in E implies $p \leq q$ for each q in A), and, consequently, E

has a supremum, namely 0. Similarly (dually), if E is empty, then E has an infimum, namely 1.

Consider next the case of a singleton, say $\{p\}$. Since p itself is an upper bound of this set, it follows that the set has a supremum, namely p, and, similarly, that it has an infimum, namely p again.

The situation becomes less trivial when we pass to sets of two elements.

Lemma 4. *For each p and q, the set $\{p, q\}$ has the supremum $p \vee q$ and the infimum $p \wedge q$.*

Proof. The element $p \vee q$ dominates both p and q, by the absorption laws, so it is one of the upper bounds of $\{p, q\}$. It remains to prove that $p \vee q$ is the least upper bound, or, in other words, that if both p and q are dominated by some element r, then $p \vee q \leq r$. This is easy; by (2) and idempotence,

$$p \vee q \leq r \vee r = r.$$

The assertion about the infimum follows by duality.

Lemma 4 generalizes immediately to arbitrary non-empty finite sets (instead of sets with only two elements). We may therefore conclude that if E is a non-empty finite subset of A, then E has both a supremum and an infimum, namely $\bigvee E$ and $\bigwedge E$ respectively. Motivated by these facts we hereby extend the interpretation of the symbols used for joins and meets to sets that may be empty or infinite. If a subset E of A has a supremum, we shall denote that supremum by $\bigvee E$ regardless of the size of E, and, similarly, we shall use $\bigwedge E$ for all infima. In this notation what we know about very small sets can be expressed as follows:

$$\bigvee \varnothing = 0, \qquad \bigwedge \varnothing = 1, \qquad \bigvee \{p\} = \bigwedge \{p\} = p.$$

The notation used earlier for the join or meet of a finite sequence of elements is also extendable to the infinite case. Thus if $\{p_i\}$ is an infinite sequence with a supremum (properly speaking, if the range of the sequence has a supremum), then that supremum is denoted by $\bigvee_{i=1}^{\infty} p_i$. If, more generally, $\{p_i\}$ is an arbitrary family with a supremum, indexed by the elements i of a set I, the supremum is denoted by $\bigvee_{i \in I} p_i$, or, in case no confusion is possible, simply by $\bigvee_i p_i$.

The perspective of Boolean algebras as partially ordered sets goes back to Jevons [31] and suggests a natural generalization. A *lattice* is a partially ordered set in which, for any elements p and q, the set $\{p, q\}$ has both a supremum and an infimum. Two binary operations called *join* and *meet* may be introduced into a lattice as follows: the join of elements p and q, written $p \vee q$, is defined to be the supremum of the set $\{p, q\}$, and the *meet* of p and q, written $p \wedge q$, is the infimum of $\{p, q\}$. The two operations satisfy the idempotent laws (2.16), the commutative laws (2.18), the associative laws (2.19), and the absorption laws (4.1). Consequently, the analogue of Lemma 1 also holds for lattices, and one can easily show that $p \leq q$ if and only if $p \wedge q = p$, or, equivalently, if and only if $p \vee q = p$.

There is an alternative approach to lattices: they may be defined as algebraic structures consisting of a non-empty set and two binary operations — denoted by $\vee$ and $\wedge$ — that satisfy the idempotent, commutative, associative, and absorption laws. A binary relation $\leq$ may then be defined in exactly the same way as it is defined for Boolean algebras, and the analogues of Lemmas 2 and 4 may be proved. The situation may be summarized by saying that the two conceptions of a lattice are definitionally equivalent. Notice that the axioms for lattices in this alternative approach come in dual pairs. Consequently, there is a principle of duality for lattices that is the exact analogue of the principle of duality for Boolean algebras (Chapter 4).

A lattice may or may not have a smallest or a greatest element. The smallest element, if it exists, is called the *zero* of the lattice and is usually denoted by 0; the largest element, if it exists, is called the *one*, or the *unit*, of the lattice and is denoted by 1. It is easy to check that the zero and unit of a lattice (if they exist) satisfy the identities in (2.12) and (2.13). The distributive laws in (2.20) may fail in a lattice; however, if one of them does hold, then so does the other, and the lattice is said to be *distributive*.

A lattice in which every subset has an infimum and a supremum is said to be *complete*. The infimum and supremum of a set E, if they exist, are denoted by $\bigwedge E$ and $\bigvee E$ respectively. A complete lattice automatically has a least and a greatest element, namely, the infimum and the supremum of the set of all elements in the lattice.

Every Boolean algebra, under the operations of join and meet, is a lattice, and in fact a distributive lattice. The converse is not necessarily true. For instance, the set of natural numbers under the standard ordering relation is a distributive lattice, but it is not a Boolean algebra.

The reason for mentioning lattices in this book is that they naturally arise in the study of Boolean algebras. We shall encounter some important examples later.

Exercises

1. Prove that a Boolean algebra is degenerate if and only if $1 = 0$.

2. Prove Lemma 3.

3. Prove the converse to part (3) of Lemma 3: if $q' \le p'$, then $p \le q$.

4. Prove that $p \ge q$ is the dual of $p \le q$.

5. If $p \le q$ and $p \ne 0$, prove that $q - p \ne q$.

6. The concept of divisibility makes sense in every ring: p is *divisible* by q in case $p = q \cdot r$ for some r. Show that in a Boolean ring an element p is divisible by q if and only if $p \le q$.

7. True or false: if $p \le q$ and $r \le s$, then
$$p + r \le q + s \quad \text{and} \quad p \Leftrightarrow r \le q \Leftrightarrow s.$$

8. Prove that $p - r \le (p - q) \vee (q - r)$, and equality holds if $r \le q \le p$.

9. Derive the following inequalities.

   (a) $p + r \le (p + q) \vee (q + r)$.
   (b) $(p \vee q) + (r \vee s) \le (p + r) \vee (q + s)$.

10. Derive the following inequalities.

    (a) $(p \Rightarrow q) \wedge (q \Rightarrow r) \le p \Rightarrow r$.
    (b) $(p \Rightarrow r) \wedge (q \Rightarrow s) \le (p \wedge q) \Rightarrow (r \wedge s)$.

11. Derive the following inequalities.

    (a) $(p \Leftrightarrow q) \wedge (q \Leftrightarrow r) \le p \Leftrightarrow r$.
    (b) $(p \Leftrightarrow r) \wedge (q \Leftrightarrow s) \le (p \wedge q) \Leftrightarrow (r \wedge s)$.

12. Give a precise definition of the notion of the infimum of a set E in a Boolean algebra.

13. Prove that in every lattice, conceived as a partially ordered set, the defined operations of join and meet satisfy the idempotent, commutative, associative, and absorption laws. Use these laws to show that Lemma 1 and part (2) of Lemma 3 hold for lattices and that $p \le q$ if and only if $p \wedge q = p$.

14. Suppose lattices are conceived as algebraic structures with two binary operations satisfying the idempotent, commutative, associative, and absorption laws. Prove that the binary relation $\leq$ defined by

$$p \leq q \qquad \text{if and only if} \qquad p \wedge q = p$$

is a partial order and the monotony laws (part (2) of Lemma 3) hold. Show further that Lemma 4 remains true. (Exercises 13 and 14 say, together, that the conceptions of a lattice as a partially ordered set, and as a structure with two binary operations satisfying the idempotent, commutative, associative, and absorption laws, are definitionally equivalent.)

15. Show that in a lattice with zero and one, the laws (2.12) and (2.13) are valid.

16. Show, conversely, that if a lattice has elements 0 and 1 satisfying the laws in (2.12), or else the laws in (2.13), then 0 is the least element, and 1 is the greatest element, in the lattice. (Exercises 15 and 16 say, together, that the two conceptions of a lattice with zero and one — the first as a partially ordered set with a greatest and a least element, the second as a structure with two binary operations satisfying the idempotent, commutative, associative, absorption, and either (2.12) or the identity laws — are definitionally equivalent.)

17. Prove that a part of each of the distributive laws, namely the inequalities

$$(p \wedge q) \vee (p \wedge r) \leq p \wedge (q \vee r) \quad \text{and} \quad (p \vee q) \wedge (p \vee r) \geq p \vee (q \wedge r),$$

holds in every lattice.

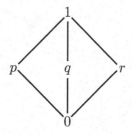

18. The preceding diagram determines a five-element lattice in which

$$p \vee q = p \vee r = q \vee r = 1 \quad \text{and} \quad p \wedge q = p \wedge r = q \wedge r = 0.$$

Show that both distributive laws fail in this lattice.

19. Prove that in every lattice the validity of one of the two distributive laws

$$p \wedge (q \vee r) = (p \wedge q) \vee (p \wedge r) \quad \text{and} \quad p \vee (q \wedge r) = (p \vee q) \wedge (p \vee r)$$

implies the validity of the other.

20. In a lattice with zero and one, a *complement* of an element p is an element q such that

$$p \wedge q = 0 \quad \text{and} \quad p \vee q = 1.$$

Prove that in a distributive lattice with zero and one, a complement of an element, if it exists, is unique.

21. Show that any linearly ordered set with a least and a greatest element, and with more than two elements, is an example of a distributive lattice with zero and one that is not a Boolean algebra.

22. A lattice with zero and one is said to be *complemented* if every element has a complement (see Exercise 20). Interpret and prove the assertion that complemented distributive lattices are the same thing as Boolean algebras.

23. Prove that a lattice in which every set of elements has a least upper bound is complete.

Chapter 8

Infinite Operations

An infinite subset of a Boolean algebra may fail to have a supremum. For example, let A be the finite–cofinite algebra of integers, and consider the set E of singletons of even integers. If P is an upper bound for E in A, then P is infinite, and therefore cofinite. The set obtained from P by removing a single odd integer is a proper subset that is still an upper bound for E, so P cannot be the least upper bound of E.

A Boolean algebra with the property that every subset of it has both a supremum and an infimum is called a *complete* (Boolean) *algebra*. Similarly, a field of sets with the property that both the union and the intersection of every class of sets in the field are again in the field is called a *complete field* of sets. The simplest example of a complete field of sets (and hence of a complete algebra) is the field of all subsets of a set.

Many laws about union, intersection, and complement have infinite versions. For example,

$$(1) \qquad \left(\bigcap_i P_i \right)' = \bigcup_i P_i', \qquad\qquad \left(\bigcup_i P_i \right)' = \bigcap_i P_i',$$

and

$$(2) \qquad P \cap \left(\bigcup_i Q_i \right) = \bigcup_i (P \cap Q_i), \qquad P \cup \left(\bigcap_i Q_i \right) = \bigcap_i (P \cup Q_i),$$

are infinite versions of the De Morgan laws (2.7) and distributive laws (2.10) for sets. The formulations of analogous laws for arbitrary Boolean algebras must always come with some sort of proviso, since the suprema and infima of infinite sets of elements may not exist. Here are the infinite versions of the De Morgan laws.

S. Givant, P. Halmos, *Introduction to Boolean Algebras*,
Undergraduate Texts in Mathematics, DOI: 10.1007/978-0-387-68436-9_8,
© Springer Science+Business Media, LLC 2009

Lemma 1. *If $\{p_i\}$ is a family of elements in a Boolean algebra, then*

$$\left(\bigwedge_i p_i\right)' = \bigvee_i p_i' \qquad and \qquad \left(\bigvee_i p_i\right)' = \bigwedge_i p_i'.$$

The equations are to be interpreted in the sense that if either term in either equation exists, then so does the other term of that equation, and the two terms are equal.

Proof. To prove the second equation, suppose $p = \bigvee_i p_i$. Since $p_i \leq p$ for every i, it follows that $p' \leq p_i'$ for every i, by Lemma 7.3(3). In other words, p' is a lower bound for the family $\{p_i'\}$. To prove that it is the greatest lower bound, assume $q \leq p_i'$ for every i. The assumption implies that $p_i \leq q'$ for every i, and hence, from the definition of supremum, that $p \leq q'$. Consequently, $q \leq p'$. A dual argument justifies the passage from the left side of the first equation to the right. To justify the reverse passages in both equations, apply the results already proved to the families of complements. $\blacksquare$

Corollary 1. *If every subset of a Boolean algebra has a supremum (or else if every subset has an infimum), then that algebra is complete.*

Proof. Suppose every subset of a Boolean algebra A has a supremum. To show that every subset also has an infimum, consider an arbitrary family of elements $\{p_i\}$ in A. The supremum q of the family $\{p_i'\}$ exists in A, by assumption. Write $p = q'$. Then

$$p = q' = \left(\bigvee_i p_i'\right)' = \bigwedge_i p_i'' = \bigwedge_i p_i.$$

The third equality uses Lemma 1, and the fourth equality uses the double complement law. Thus, p is the infimum of the family $\{p_i\}$. $\blacksquare$

It will usually not be sufficient to know merely that certain infinite suprema exist; the algebraic properties of those suprema (such as commutativity, associativity, and distributivity) are also needed.

It is almost meaningless to speak of infinite commutative laws. An infinite supremum is something associated with a set of elements, and, by definition, it is independent of any possible ordering of that set.

A reasonable verbal formulation of an infinite associative law might go like this. Form each of several suprema and then form their supremum; the result should be equal to the supremum of all the elements that originally

contributed to each separate supremum. It is worthwhile to state and prove
this in a more easily quotable form.

Lemma 2. *If $\{I_j\}$ is a family of sets with union I, and if p_i, for each i in I,
is an element of a Boolean algebra, then*

$$\bigvee_j \left(\bigvee_{i \in I_j} p_i \right) = \bigvee_{i \in I} p_i.$$

*The equation is to be interpreted in the sense that if the suprema on the left
side exist, then so does the supremum on the right, and the two are equal.*

Proof. Write $q_j = \bigvee_{i \in I_j} p_i$ and $q = \bigvee_j q_j$. We are to prove that q is an upper
bound of the family $\{p_i : i \in I\}$, and that, in fact, it is the least upper bound.
Since each i in I belongs to at least one I_j, it follows that for each i there is
a j with $p_i \leq q_j$; since, moreover, $q_j \leq q$, it follows that q is indeed an upper
bound. Suppose now that $p_i \leq r$ for every i. Since, in particular, $p_i \leq r$ for
every i in I_j, it follows from the definition of supremum that $q_j \leq r$. Since
this is true for every j, we may conclude, similarly, that $q \leq r$, and this
completes the proof.

The preceding comments on infinite commutativity and associativity were
made for suprema; it should go without saying that the corresponding (dual)
comments for infima are just as true. The most interesting infinite laws are
the ones in which suprema and infima occur simultaneously. These are the
distributive laws, to which we now turn. They too come in dual pairs; we
shall take advantage of the principle of duality and restrict our attention
to only one member of each such pair. We begin with the simplest infinite
distributive law.

Lemma 3. *If p is an element and $\{q_i\}$ a family of elements in a Boolean
algebra, then*

$$p \wedge \bigvee_i q_i = \bigvee_i (p \wedge q_i).$$

*The equation is to be interpreted in the sense that if the supremum on the
left side exists, then so does the one on the right, and the two are equal.*

Proof. Write $q = \bigvee_i q_i$. The meet $p \wedge q$ is clearly an upper bound for the
family $\{p \wedge q_i\}$, since

$$p \wedge q_i \leq p \wedge q$$

for every i, by monotony. To show that $p \wedge q$ is the least upper bound, it must be proved that if $p \wedge q_i \leq r$ for every i, then $p \wedge q \leq r$. Observe that

$$q_i = 1 \wedge q_i = (p \vee p') \wedge q_i = (p \wedge q_i) \vee (p' \wedge q_i) \leq r \vee p';$$

the last step uses monotony. Hence, by the definition of a supremum,

$$q \leq r \vee p'.$$

Form the meet of both sides of this inequality with p, and apply the distributive, complement, identity, and monotony laws, to get

$$p \wedge q \leq p \wedge (r \vee p') = (p \wedge r) \vee (p \wedge p') = (p \wedge r) \vee 0 = p \wedge r \leq r.$$

Corollary 2. *If $\{p_i\}$ and $\{q_j\}$ are families of elements in a Boolean algebra, then*

$$\left(\bigvee_i p_i\right) \wedge \left(\bigvee_j q_j\right) = \bigvee_{i,j}(p_i \wedge q_j).$$

The equation is to be interpreted in the sense that if the suprema on the left side exist, then so does the supremum on the right, and the two are equal.

To motivate the most restrictive distributive law, consider a long infimum of long suprema, such as

$$(p_{11} \vee p_{12} \vee p_{13} \vee \cdots) \wedge (p_{21} \vee p_{22} \vee p_{23} \vee \cdots) \wedge$$
$$(p_{31} \vee p_{32} \vee p_{33} \vee \cdots) \wedge \cdots .$$

Algebraic experience suggests that this ought to be equal to a very long supremum, each of whose terms is a long infimum like $p_{12} \wedge p_{23} \wedge p_{31} \wedge \cdots$. The way to get all possible infima of this kind is to pick one term from each original supremum in all possible ways. The picking is done by a function that associates with each value of the first index some value of the second index; the "very long" supremum has one term corresponding to each such function.

We are now ready for a formal definition. Suppose that I and J are two index sets and that $p(i, j)$ is an element of a Boolean algebra A whenever i is in I and j in J. Let J^I be the set of all functions from I to J. We say that the family $\{p(i, j)\}$ satisfies the *complete distributive law* in case

$$(3) \qquad \bigwedge_{i \in I} \bigvee_{j \in J} p(i, j) = \bigvee_{a \in J^I} \bigwedge_{i \in I} p(i, a(i)).$$

The assertion of the equation is intended here to imply, in particular, the existence of the suprema and infima that occur in it. The algebra A is called *completely distributive* if it has the following property: whenever all of the suprema $\bigvee_{j \in J} p(i,j)$ and infima $\bigwedge_{i \in I} p(i, a(i))$ exist for any given family $\{p(i,j)\}$, then the existence of the left side of (3) implies that the right side also exists and that the two are equal.

There is a special case of (3) that is often quite useful. When the index sets I and J are finite, the suprema and infima that occur in the equation always exist; in this case, (3) holds without any additional hypotheses, by the distributive laws (2.20).

The field of all subsets of a set is always completely distributive, and so is every complete field of sets. However, a complete Boolean algebra need not be completely distributive. We shall encounter an example in a moment.

Exercises

1. Prove (1) and (2).

2. Suppose that $\{I_j\}$ is a family of sets with union I, and that P_i, for each i in I, is a subset of a fixed set X. Show that

$$\bigcup_j \left(\bigcup_{i \in I_j} P_i \right) = \bigcup_{i \in I} P_i.$$

3. Suppose that I and J are two index sets and that $P(i,j)$ is a subset of a fixed set X for each i in I and j in J. Let J^I be the set of all functions from I to J. Prove that

$$\bigcap_{i \in I} \bigcup_{j \in J} P(i,j) = \bigcup_{a \in J^I} \bigcap_{i \in I} P(i, a(i))$$

and

$$\bigcup_{i \in I} \bigcap_{j \in J} P(i,j) = \bigcap_{a \in J^I} \bigcup_{i \in I} P(i, a(i)).$$

4. Write a complete proof of Lemma 1, supplying all missing details.

5. Prove that if every subset of a Boolean algebra has an infimum, then the algebra is complete.

6. There is another possible interpretation of the equation in Lemma 2: if the suprema $\bigvee_{i \in I} p_i$ and $\bigvee_{i \in I_j} p_i$, for each j, exist, then the supremum $\bigvee_j \left(\bigvee_{i \in I_j} p_i \right)$ exists and

$$\bigvee_j \left(\bigvee_{i \in I_j} p_i \right) = \bigvee_{i \in I} p_i.$$

Prove that under this interpretation of the equation the lemma remains true of all Boolean algebras.

7. Discuss another possible interpretation of the equation in Lemma 3 besides the one stated there.

8. Formulate and prove the dual of Lemma 2.

9. Formulate and prove the dual of Exercise 6.

10. Formulate and prove the dual of Lemma 3.

11. Prove Corollary 2.

12. Formulate and prove the dual of Corollary 2.

13. Formulate and prove a general version of the finite distributive law for meet over join.

14. Is a complete field of subsets of a set X the same as the field of all subsets of X?

15. (Harder.) Give an example of a field of sets that happens to be a complete Boolean algebra but not a complete field of sets.

16. If a Boolean algebra is such that every subset of it has either a supremum or an infimum, is it necessarily complete?

17. Interpret and prove the equation

$$p \vee \bigvee_i p_i = \bigvee_i (p \vee p_i).$$

18. Interpret and prove the equation

$$\left(\bigvee_i p_i \right) \vee \left(\bigvee_i q_i \right) = \bigvee_i (p_i \vee q_i).$$

19. Interpret and prove the following assertion: if for every i there is a j such that $p_i \leq q_j$, then

$$\bigvee_i p_i \leq \bigvee_j q_j.$$

20. Interpret and prove the following assertion: if $I \subseteq J$ then

$$\bigvee_{i \in I} p_i \leq \bigvee_{i \in J} p_i.$$

21. Let $\{p_i\}$ be a (finite or infinite) sequence of elements in a Boolean algebra (indexed by positive integers), and write $q_j = p_1 \vee \cdots \vee p_j$. Prove that

$$\bigvee_j q_j = \bigvee_i p_i.$$

The equation is to be interpreted in the sense that if either of the two suprema exists, then both suprema exist, and the two are equal.

22. Formulate and prove the dual of Exercise 21.

23. Suppose a sequence $\{p_n\}$ of elements in a Boolean algebra has a supremum. Prove that the sequence $\{q_n\}$ defined by

$$q_n = p_n - (p_1 \vee p_2 \vee \cdots \vee p_{n-1}),$$

for $n \geq 1$, consists of mutually disjoint elements and has the same supremum as $\{p_n\}$. (Notice that $q_1 = p_1$, since the supremum of the empty sequence is 0.)

24. Suppose a sequence $\{p_n\}$ of elements in a Boolean algebra is increasing (in the sense that $p_n \leq p_{n+1}$ for all n) and has a supremum. Prove that the sequence $\{q_n\}$ defined by

$$q_1 = p_1, \qquad \text{and} \qquad q_n = p_n - p_{n-1}$$

for $n \geq 2$, consists of mutually disjoint elements and has the same supremum as $\{p_n\}$.

25. Interpret and prove the following laws:

    (a) $q - \bigwedge_i p_i = \bigvee_i (q - p_i)$,

    (b) $q - \bigvee_i p_i = \bigwedge_i (q - p_i)$,

    (c) $(\bigvee_i p_i) - q = \bigvee_i (p_i - q)$,

(d) $\left(\bigwedge_i p_i\right) - q = \bigwedge_i (p_i - q)$.

26. Prove that $\left(\bigvee_i p_i\right) + \left(\bigvee_i q_i\right) \leq \bigvee_i (p_i + q_i)$. The inequality is to be interpreted in the sense that if all three suprema exist, then the inequality holds.

27. (Harder.) Let p_1, p_2, p_3, ..., be a sequence of elements in a Boolean algebra with the property that $p_1 \geq p_2 \geq p_3 \geq \cdots$, and write

$$q_n = p_n - p_{n+1} = p_n \wedge p'_{n+1}$$

for $n = 1, 2, 3, \ldots$. Prove that if the infimum of the family $\{p_n\}$ exists — call it q_0 — then the elements $q_0, q_1, q_2, \ldots$ are mutually disjoint and have p_1 as their supremum. (See the diagram.)

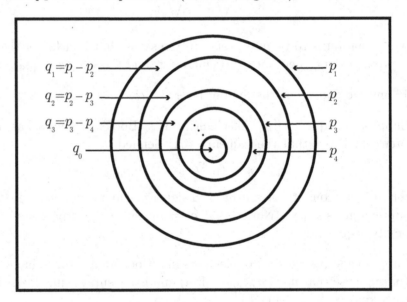

28. Prove that the interval algebra of the real numbers is not a complete field of sets. In fact, show that there is a family of elements in the algebra that possesses an infimum, but that infimum is not the intersection of the family.

29. (Harder.) Prove that the interval algebra of the real numbers is not a complete Boolean algebra.

Chapter 9

Topology

Valuable examples of Boolean algebras, and in particular of complete Boolean algebras, can be constructed using topological spaces. The purpose of this chapter is to go over some of the basic topological notions that will be needed in the construction of these algebras.

A topology is a very general kind of geometry that is suitable for studying continuous functions. The basic notion is that of an open set, an abstraction and generalization of the notion of an open interval of real numbers. A *topological space* is a set X, together with a class $\mathcal{T}$ of subsets of X that satisfies three conditions: first, $\varnothing$ and X are in $\mathcal{T}$; second, $\mathcal{T}$ is closed under finite intersections in the sense that the intersection of any finite family of sets from $\mathcal{T}$ is again in $\mathcal{T}$; third, $\mathcal{T}$ is closed under arbitrary unions in the sense that the union of an arbitrary family of sets from $\mathcal{T}$ is again in $\mathcal{T}$. The elements of X are called *points*, and the members of $\mathcal{T}$ are called *open sets*. An open set containing a point x is called a *neighborhood* of x. The set X is often called the *space*, and the class $\mathcal{T}$ the *topology* of the space. The three conditions on $\mathcal{T}$ say that $\varnothing$ and X are open sets, that the intersection of finitely many open sets is open, and that the union of an arbitrary family of open sets is open.

The classical example of a topological space is the n-dimensional *Euclidean space* $\mathbb{R}^n$. Its points are the n-termed sequences of real numbers. The topology of the space is defined in terms of the notion of distance. The *distance* between two points

$$x = (x_1, \ldots, x_n) \qquad \text{and} \qquad y = (y_1, \ldots, y_n)$$

is the length of the line segment between them:

S. Givant, P. Halmos, *Introduction to Boolean Algebras*,
Undergraduate Texts in Mathematics, DOI: 10.1007/978-0-387-68436-9_9,
© Springer Science+Business Media, LLC 2009

$$d(x, y) = \sqrt{(x_1 - y_1)^2 + (x_2 - y_2)^2 + \cdots + (x_n - y_n)^2}.$$

For a given positive real number ϵ and a point x, the (*open*) *ball* with *radius* ϵ and *center* x is defined to be the set of all points whose distance from x is less than ϵ:

$$\{y \in \mathbb{R}^n : d(x, y) < \epsilon\}.$$

In two-dimensional Euclidean space, this is just the interior of the circle of radius ϵ centered at x, and in three-dimensional Euclidean space it is the interior of the sphere of radius ϵ centered at x. A subset P of $\mathbb{R}^n$ is defined to be *open* if for every point x in P, some open ball centered at x is included in P.

The empty set vacuously satisfies the condition for being open. The whole space $\mathbb{R}^n$ satisfies the condition for being open because it includes every open ball. It is easy to check that the union of an arbitrary family of open sets in $\mathbb{R}^n$ is open, as is the intersection of a finite family of open sets. For the proof, consider such a family $\{P_i\}$ of open sets. If a point x is in the union of the family, then x is in one of the sets P_i, by the definition of union. Consequently, some open ball centered at x is included in P_i (because P_i is open) and that same open ball must be included in the union of the family $\{P_i\}$. It follows that this union is an open set. If the family is finite, and if a point x is in the intersection of the family, then x belongs to each set P_i. Consequently, for each index i there is a positive real number ϵ_i such that the open ball of radius ϵ_i centered at x is included in P_i. Let ϵ be the minimum of the (finitely many) radii ϵ_i. The open ball of radius ϵ centered at x is included in each set P_i, so it is included in the intersection of the family $\{P_i\}$. It follows that this intersection is an open set. Conclusion: the class of open subsets of $\mathbb{R}^n$ is a topology on $\mathbb{R}^n$.

In the special case of dimension one, the points of the Euclidean space are identified with real numbers. The open balls are the *open intervals*

$$(-\infty, a) = \{x \in \mathbb{R} : x < a\}, \qquad (a, \infty) = \{x \in \mathbb{R} : x > a\},$$
$$(a, b) = \{x \in \mathbb{R} : a < x < b\}, \quad (-\infty, \infty) = \mathbb{R},$$

where a and b are real numbers with $a < b$. It is not difficult to see that a set is open in this space if and only if it can be written as a countable union of mutually disjoint open intervals. Examples of open sets in $\mathbb{R}^2$ are also easy to manufacture: the entire space with finitely many points removed is an example of an open set, and so is the set obtained from $\mathbb{R}^2$ by removing any straight line (the set is the union of two open half-planes).

To generalize the notion of a Euclidean space, we must introduce the notion of a metric. A *metric* on a set X is a real-valued function d of two arguments such that for all x, y, and z in X,

$d(x, y) \geq 0$, and $d(x, y) = 0$ if and only if $x = y$ (*strict positivity*),

$d(x, y) = d(y, x)$ (*symmetry*),

$d(x, z) \leq d(x, y) + d(y, z)$ (*triangle inequality*).

A *metric space* is a set X together with a metric d on X. (The prototypical examples are the Euclidean spaces $\mathbb{R}^n$ with their distance functions.) For each positive real number ϵ, and each point x in X, the *open ball* of radius ϵ centered at x is defined to be the set of points

$$\{y \in X : d(x, y) < \epsilon\}.$$

A subset P of X is defined to be open if for every point x in P, some open ball centered at x is included in P. The resulting class of open sets constitutes a topology called the *metric topology* (induced by d) on X. The proof that the conditions for being a topology really are satisfied by this class of sets is virtually the same as in the case of Euclidean spaces.

Some types of topologies can be defined on arbitrary sets X. One example is the *discrete topology*: every subset of X is declared to be open. Under this topology, X is called a *discrete space*. Another example is the *indiscrete* or *trivial topology*: the only sets declared to be open are, by definition, $\varnothing$ and X. A third example is the *cofinite topology*: a subset of X is defined to be open if it is empty or the complement of a finite set. (It is a simple matter to check that finite intersections and arbitrary unions of cofinite sets are cofinite.) The discrete and cofinite topologies on X coincide when X is finite, but they are obviously different when X is infinite.

A subset Y of a topological space X may be endowed with the *inherited topology* by declaring a subset P of Y to be open ife it can be written in the form

$$P = Y \cap Q$$

for some open subset Q of X. It is easy to check that under this definition the empty set and Y are both open, and that finite intersections and arbitrary unions of open sets are open. (The proof that arbitrary unions of open sets are open uses the infinite distributive law (8.2).) The resulting space Y is said to be a *subspace* of X.

For a concrete example, let X be the space of real numbers and Y the closed interval $[0, 1]$. The open intervals of Y, under the inherited topology, are the subintervals

$$[0, a), \qquad (a, b), \qquad (a, 1], \qquad \text{and} \qquad [0, 1]$$

with $0 \leq a < b \leq 1$. The open subsets of Y turn out to be the countable unions of pairwise disjoint open intervals. Notice that some of these sets are not open in the topology of X.

The dual of the notion of an open set is that of a closed set. A set of points in a topological space is said to be *closed* if it is the complement of an open set. The intersection of an arbitrary family of closed sets is closed. Indeed, if $\{Q_i\}$ is a family of closed sets, then the complements Q_i' are open, by definition, and therefore the union of the family of complements is open. It follows that the complement of this union is a closed set; since that complement is just the intersection of the family $\{Q_i\}$,

$$\bigcap_i Q_i = \left(\bigcup_i Q_i' \right)',$$

the intersection of the family is closed. An analogous argument shows that a finite union of closed sets is closed. It is an elementary theorem of analysis that a subset P of a metric space (but not a subset of an arbitrary topological space) is closed just in case the limit of a convergent sequence of points from P always belongs to P. More precisely, if $\{x_n\}$ is an infinite sequence of points in P that converges to a limit x in the metric space, then x belongs to P.

As in the case of open sets, it is helpful to gain a sense of what closed sets may look like. In $\mathbb{R}$, the intervals

$$(-\infty, a] = \{x \in \mathbb{R} : x \leq a\}, \qquad\qquad [a, \infty) = \{x \in \mathbb{R} : x \geq a\},$$
$$[a, b] = \{x \in \mathbb{R} : a \leq x \leq b\}, \qquad (-\infty, \infty) = \mathbb{R}$$

are closed, as is any finite union of them (where a and b are real numbers with $a \leq b$). In particular, the sets $[a, a]$ — which are just the singletons $\{a\}$ — are closed. Consequently, any finite set of real numbers is closed (every finite set is the union of a finite class of singletons). The set of positive integers is closed, but the set of the reciprocals of positive integers is not closed; it becomes closed when the integer 0 is adjoined. In $\mathbb{R}^2$, the line segment $\{(x, 0) : 0 < x < 1\}$ is neither open nor closed, but the segment $\{(x, 0) : 0 \leq x \leq 1\}$ is closed. Every subset of a discrete space is of course closed. A subset of a space with the cofinite topology is closed if and only if it is finite or the whole space.

Sets that are both open and closed are called *clopen*. The whole space and the empty set are always clopen. Every subset of a discrete space is clopen.

In what follows, let P be an arbitrary subset of a topological space. The *interior* of P is defined to be the union of the open sets that are included in P. This interior is clearly an open set, and in fact it is the largest open set that is included in P. We shall denote it by P°. Examples: in $\mathbb{R}$, the interior of the closed interval $[0, 1]$ is the open interval $(0, 1)$ and the interior of the set of rational numbers is the empty set; in $\mathbb{R}^2$, the interior of the unit square is the unit square with its perimeter removed, while the interior of the line segment $\{(x, 0) : 0 \leq x \leq 1\}$ is the empty set. In a discrete space every subset coincides with its own interior. In a space with the cofinite topology, the interior of a cofinite subset is itself, and the interior of every other subset is empty.

The dual of the notion of an interior is that of a closure: the *closure* of P is defined to be the intersection of all closed sets that include P. This closure is of course a closed set, and in fact it is the smallest closed set that includes P. It is denoted by P^- (not $\overline{P}$) for typographic reasons. The closure of P can be characterized as the set of points x such that every neighborhood of x has a non-empty intersection with P. For the proof in one direction, suppose some neighborhood of x, say Q, is disjoint from P. The complement Q' is a closed set that includes P, so it belongs to the family of closed sets whose intersection is P^-. Since the point x is not in Q', it cannot belong to this intersection, and therefore it cannot belong to P^-. To establish the reverse implication, suppose every neighborhood of x has a non-empty intersection P. If Q is a closed set that includes P, then Q' is an open set that is disjoint from P, so Q' cannot contain x. It follows that x is in Q. In other words, x belongs to every closed set that includes P; therefore x must belong to the intersection of all such closed sets, and this intersection is just P^-. The closure of set P in a metric space (but not in a arbitrary topological space) can be characterized as the set of all limit points of convergent infinite sequences of points from P.

Here are some examples of closures. In $\mathbb{R}$, the closure of the set of rational points is the entire space, and the closure of the set

$$P = \{1/n : n \text{ is a positive integer}\}$$

is the set $P \cup \{0\}$. In $\mathbb{R}^2$, the closure of an open disk is the open disk together with its perimeter, and the closure of the set of points

$$Q = \{(x, y) : y = \sin(1/x) \text{ and } 0 < x \leq 1\}$$

is the union of Q with the set

$$\{(0,y) : -1 \leq y \leq 1\}.$$

In a discrete space, every subset is its own closure. In a space with the cofinite topology, the closure of a finite subset is itself, and the closure of an infinite subset is the whole space.

It is not difficult to see that the closure of a set P is the complement of the interior of the complement of P. In other words,

$$(1) \qquad\qquad P^- = P'^{o'}.$$

Indeed, P'^o is an open set that is included in P', by the definition of the interior of a set. Form the complements of both sets to conclude that P'' — which is just P — is included in the closed set $P'^{o'}$. Since P^- is the smallest closed set that includes P, it follows that P^- is a subset of $P'^{o'}$. To establish the reverse inclusion, consider any closed set Q that includes P. The complement Q' is an open set that is included in P'. Since P'^o is the largest open set that is included in P', it follows that Q' is included in P'^o. Form the complements of both sets to conclude that Q'' — which is just Q — includes $P'^{o'}$. This argument shows that every closed set that includes P also includes $P'^{o'}$. In particular, the closure P^- includes $P'^{o'}$.

Replace P by P' in (1), and form the complement of both sides to conclude that the interior of P is equal to the complement of the closure of the complement of P, that is,

$$(2) \qquad\qquad P^o = P'^{-'}.$$

We shall need two properties about the closure operator. First, it preserves inclusion:

$$(3) \qquad\qquad P \subseteq Q \quad \text{implies} \quad P^- \subseteq Q^-.$$

This is immediately evident from the definition of closure. Second, it preserves union:

$$(4) \qquad\qquad (P \cup Q)^- = P^- \cup Q^-.$$

The inclusion from right to left follows from (3): the sets P and Q are both included in $P \cup Q$, so their closures are included in $(P \cup Q)^-$. The reverse inclusion follows from the simple observation that $P^- \cup Q^-$ is a closed set and it includes $P \cup Q$, so it must include the smallest closed set that includes $P \cup Q$, namely $(P \cup Q)^-$.

A set P is said to be *dense* if its closure is the entire space. This means that every non-empty open set contains points of P. More generally, a set P is said to be *dense in an open set* Q if the closure of P includes Q. For instance, the set of points in $\mathbb{R}^2$ with rational coordinates is dense, while the set of points with positive rational coordinates is dense in the first quadrant of $\mathbb{R}^2$, but not in all of $\mathbb{R}^2$. In a discrete space, no proper subset of the space is dense. In an infinite space with cofinite topology, every infinite subset of the space is dense.

The opposite of being dense is being nowhere dense. A set P is defined to be *nowhere dense* if it is not dense in any non-empty open set. This means that the interior of the closure of P is empty, or what amounts to the same thing, no non-empty open set is included in the closure of P. To say that no non-empty open set is included in P^- is equivalent to saying that every non-empty open set has a non-empty intersection with $P^{-\prime}$. Thus, P is nowhere dense if and only if $P^{-\prime}$ is dense.

Examples of nowhere dense sets are not hard to manufacture. Every finite set of points is nowhere dense in $\mathbb{R}^n$. The integers are nowhere dense in $\mathbb{R}$, and any straight line is nowhere dense in $\mathbb{R}^2$. In a discrete space only the empty set is nowhere dense. In an infinite space with the cofinite topology, a set is nowhere dense just in case it is finite.

A finite union of nowhere dense sets is again nowhere dense, but a countable union of such sets may in fact be dense. For example, the set of points with rational coordinates is dense in $\mathbb{R}^n$, and yet it is a countable union of nowhere dense sets, namely the singletons of points with rational coordinates. A set is said to be *meager* if it is the countable union of nowhere dense sets. (In classically clumsy nomenclature, meager sets are also called *sets of the first category*.) An important result in analysis known as the Baire category theorem says that the interior of a meager set of points in $\mathbb{R}^n$ is always empty. (See Theorem 28 for one version of this theorem.) In a discrete space only the empty set is meager. In an infinite space with the cofinite topology, a set is meager just in case it is countable. In particular, if the space is countably infinite, then every subset is meager.

A point x is a *boundary point* of a set P if every neighborhood of x contains points of P and points of P'. In other words, x is a boundary point of P just in case it belongs to the closure of both P and P'. The *boundary* of P is the set of its boundary points, that is to say, it is the set $P^- \cap P'^-$.

The boundary of a set is always closed, because it is the intersection of two closed sets. In a metric space (but not in an arbitrary topological space), a point x is in the boundary of a set P just in case there is a sequence of points in P that converges to x and a sequence of points in P' that converges to x. If a set P is open, then its complement P' is closed, and therefore $P'^- = P'$; in this case, the boundary of P is just the set-theoretic difference $P^- - P$.

Here are some examples. The boundary of an open ball in $\mathbb{R}^3$ of radius ϵ and center x is the set of points whose distance from x is exactly ϵ. The boundary of the set of points

$$P = \{(1/n, 1/n) : n \text{ is a positive integer}\}$$

in $\mathbb{R}^2$ is the set $P \cup \{(0,0)\}$. More generally, the boundary of any nowhere dense set P in a topological space is just the closure P^-. Indeed, in this case the set $P^{-\prime}$ is dense; since $P^{-\prime} \subseteq P'$, the set P' must also be dense, so that P'^- is the whole space, and therefore $P^- \cap P'^- = P^-$. In a discrete space every set of points has an empty boundary. In an infinite space with the cofinite topology, the boundary of a finite set is itself, the boundary of a cofinite set is its complement, and the boundary of an infinite set with an infinite complement is the whole space.

The closure of a set P ought to be, and is, the union of P with its boundary. To prove this assertion, write $Q = P^- \cap P'^-$ for the boundary. Obviously, P and Q are both included in P^-, and therefore so is $P \cup Q$. To establish the reverse inclusion, consider a point x in P^-. If x is in P, then certainly x is in the union $P \cup Q$. If x is not in P, then it is in P'. In this case, every neighborhood of x contains a point in P', namely x, and also a point in P, since x is in the closure of P. It follows that x belongs to the boundary Q, by definition, and therefore it belongs to $P \cup Q$.

An open set is said to be *regular* if it coincides with the interior of its own closure. In other words, P is regular if and only if

$$P = P^{-\prime-\prime},$$

by (2). It is convenient, in this connection, to write $P^\perp = P^{-\prime}$; in these terms, P is regular if and only if

$$P = P^{\perp\perp}$$

(where $P^{\perp\perp}$ denotes $(P^\perp)^\perp$). Every open ball in $\mathbb{R}^n$ is regular. So is the open unit square in $\mathbb{R}^2$.

To construct an example of an open set that is not regular, start with a non-empty, nowhere dense set P in a topological space X, and form $P^\perp$. Certainly $P^\perp$ is open, for it is the complement of a closed set. Since P^- is not empty, its complement $P^\perp$ cannot be the whole space X. To prove that $P^\perp$ is not regular, it therefore suffices to show that

(5) $$(P^\perp)^{\perp\perp} = X.$$

The assumption that P is nowhere dense implies that the complement of its closure, which is just the set $P^\perp$, is dense. Consequently, the closure of $P^\perp$ is X, and this directly implies (5):

$$(P^\perp)^{\perp\perp} = (P^\perp)^{-/-/} = (P^{\perp-})^{/-/} = X^{/-/} = \varnothing^{-/} = \varnothing' = X.$$

This example reveals the intuition behind regular open sets: they are the open sets without "cracks". Concrete examples of open sets that are not regular can be obtained by taking for P in the preceding construction any finite set of points or any straight line in $\mathbb{R}^n$ ($n \geq 2$); the closure P^- then coincides with P, so that $P^\perp$ is the space $\mathbb{R}^2$ with finitely many points, or with a straight line, removed. In a discrete space, every set of points is a regular open set. In an infinite space with the cofinite topology, there are only two regular open sets: the empty set and the whole space.

Note incidentally that a set P is open (nothing is said about regularity here) if and only if it has the form $Q^\perp$ for some set Q. Indeed, if $P = Q^\perp$, then P is the complement of the closed set Q^-, and so it must be open. Conversely, if P is open, and if Q is the complement of P, then Q is closed and therefore

$$Q^\perp = Q^{-/} = Q' = P'' = P.$$

Exercises

1. Prove that a subset P of a topological space is open just in case every point in P belongs to an open set that is included in P.

2. Prove, using the definition of an open set in the space $\mathbb{R}^n$, that if finitely many points are removed from $\mathbb{R}^n$, the resulting set is open.

3. Prove, using the definition of an open set in the space $\mathbb{R}^n$, that if all the points on some straight line are removed from $\mathbb{R}^n$, the resulting set is open.

4. Show that in a topological space, a finite union of closed sets is always closed.

5. Prove that the class of clopen sets in a topological space is a field of sets.

6. Show that the inherited topology on a subset Y of a topological space X is in fact a topology. In other words, show that the class of sets of the form $Y \cap U$, where U ranges over the open subsets of X, satisfies the three defining conditions for a topology. Show further that a subset P of Y is closed in the inherited topology if and only if there is a closed subset Q of X such that $P = Y \cap Q$.

7. Suppose Y is an open subset of a topological space X, and P an arbitrary subset of Y. Prove that P is open in the inherited topology on Y (Exercise 6) just in case it is open in X.

8. Formulate and prove a version of Exercise 7 for closed sets.

9. Suppose Y is a subspace of a topological space X. Show that if P is a subset of Y, then the closure of P in Y is equal to the intersection with Y of the closure of P in X.

10. (Harder.) Show that the distance function defined on $\mathbb{R}^n$ satisfies the three conditions for being a metric.

11. Show that the class of open sets in a metric space satisfies the conditions for being a topology.

12. Show that a subset of a metric space is open if and only if it is a union of open balls.

13. Let P and Q be subsets of a topological space. Prove the following assertions.

    (a) If $P \subseteq Q$, then $P^\circ \subseteq Q^\circ$.
    (b) $(P \cap Q)^\circ = P^\circ \cap Q^\circ$.
    (c) $(P \cup Q)^\perp = P^\perp \cap Q^\perp$.
    (d) $P \cap Q^- \subseteq (P \cap Q)^-$ whenever P is open.
    (e) $P \cap Q^- = (P \cap Q)^-$ whenever P is clopen.

14. Give a direct proof of equation (2), without using (1).

15. If P and Q are open sets, is the equation

$$(P \cap Q)^- = P^- \cap Q^-$$

true?

16. (Harder.) If P and Q are open sets, is the equation
$$(P \cap Q)^{-\circ} = (P^- \cap Q^-)^\circ$$
true?

17. Prove that every subset of a discrete space has an empty boundary.

18. In an infinite space with the cofinite topology, prove that the boundary of a finite set is the set itself, the boundary of a cofinite set is the complement of the set, and the boundary of an infinite set with an infinite complement is the whole space.

19. Prove that a set of points and its complement always have the same boundary.

20. Prove that the boundary of a nowhere dense set is just the closure of the set.

21. Prove that the boundary of the union of two sets of points is included in the union of the boundaries of the two sets.

22. Prove that the complement of the boundary of an open set P is equal to $P \cup P^\perp$.

23. Prove that in a topological space, the class of sets with countable boundaries is a field of sets.

24. Prove that a finite union of nowhere dense sets is nowhere dense.

25. Prove that a subset of a nowhere dense set is nowhere dense.

26. Prove that in a topological space, the class of sets with nowhere dense boundaries is a field of sets. (This example of a field of sets is due to Stone [67].)

27. Prove that a subset of a meager set is meager.

28. Prove that the union of a countable sequence of meager sets is meager.

29. Prove that in a topological space, the sets with meager boundaries form a field.

30. Prove that every clopen set in a topological space is a regular open set. Conclude that every subset of a discrete space is a regular open set.

31. Prove that in an infinite space with the cofinite topology only the empty set and the whole space are regular open sets.

32. Let X be a set of uncountable cardinality. Define a subset of X to be open if it is empty or the complement of a countable set.

    (a) Prove that the class of open sets so defined satisfies the three conditions for being a topology on X. (It is called the *cocountable topology*.)
    (b) Describe the closed sets.
    (c) Describe the interior of each set.
    (d) Describe the closure of each set.
    (e) Describe the nowhere dense sets.
    (f) Describe the meager sets.
    (g) Describe the boundary of each set.
    (h) Describe the regular open sets.

33. A *linear order* (also called a *total order*) on a set X is a partial order $\leq$ on X (Chapter 7) such that any two elements x and y in X are *comparable*: either $x \leq y$ or $y \leq x$. (The set X itself is said to be *linearly ordered* or *totally ordered*.) Write $x < y$ to mean that $x \leq y$ and $x \neq y$.

    Given a linear order $\leq$ on a set X, define the open intervals of X to be the subsets

    $$(-\infty, a) = \{x \in X : x < a\}, \qquad (a, \infty) = \{x \in X : x > a\},$$
    $$(a, b) = \{x \in X : a < x < b\}, \quad (-\infty, \infty) = X,$$

    for a, b in X, and define the open sets of X to be the unions of arbitrary families of open intervals. Prove that the class of sets so defined is a topology for X. (It is called the *order topology* on X.)

34. Prove that a subset of the (Euclidean) space of real numbers is open if and only if it can be written as a countable union of mutually disjoint open intervals.

35. (Harder.) Prove that there are $2^{\aleph_0}$ open sets in the space of real numbers.

36. Let Q be an open ball in $\mathbb{R}^n$ and P a non-empty, nowhere dense subset of Q. Prove that the set $Q \cap P^{\perp} = Q - P^{-}$ is open, but not regular.

37. Can every open set in $\mathbb{R}^n$ be written as the union of a family of regular open sets?

38. Can every open set in an arbitrary topological space be written as the union of a family of regular open sets?

39. For every subset P of a topological space, prove that $P^\perp$ is the largest open set that is disjoint from P. Conclude that P is a regular open set if and only if it is the largest open set that is disjoint from the largest open set that is disjoint from P.

40. (Harder.) What is the largest number of distinct sets obtainable from a subset of $\mathbb{R}^n$ by repeated applications of closure and complementation? Construct an example for which this largest number is attained. (The question and its answer are both due to Kuratowski [36].)

Chapter 10

Regular Open Sets

The purpose of this chapter is to discuss one more example of a Boolean algebra. This example, the most intricate of all the ones so far, is one in which the elements of the Boolean algebra are subsets of a set. However, the operations are not the usual set-theoretic ones, so the Boolean algebra is not a field of sets. Artificial examples of this kind are not hard to manufacture; the example that follows arises rather naturally and plays an important role in the general theory of Boolean algebras.

Recall (Chapter 9) that an open set in a topological space X is regular if it coincides with the interior of its own closure. The next theorem (due to MacNeille [43] and Tarski [75]) asserts that the regular open sets constitute a complete Boolean algebra of sets, the *regular open algebra of X*.

Theorem 1. *The class of all regular open sets of a topological space X is a complete Boolean algebra with respect to the distinguished Boolean elements and operations defined by*

(1) $$0 = \varnothing,$$
(2) $$1 = X,$$
(3) $$P \wedge Q = P \cap Q,$$
(4) $$P \vee Q = (P \cup Q)^{\perp\perp},$$
(5) $$P' = P^{\perp}.$$

The infimum and the supremum of a family $\{P_i\}$ of regular open sets are, respectively,

$$\left(\bigcap_i P_i \right)^{\perp\perp} \quad and \quad \left(\bigcup_i P_i \right)^{\perp\perp}.$$

S. Givant, P. Halmos, *Introduction to Boolean Algebras*,
Undergraduate Texts in Mathematics, DOI: 10.1007/978-0-387-68436-9_10,
© Springer Science+Business Media, LLC 2009

The proof of the theorem depends on several small lemmas of some independent interest. The first thing to prove is that the right sides of (1)–(5) are regular open sets. For (1) and (2) this is obvious, but for (3), for instance, it is not. To say that the intersection of two regular open sets is regular may sound plausible (this is what is involved in (3)), and it is true. It is, however, just as plausible to say that the union of two regular open sets is regular, but that is false. Example: let P and Q be disjoint open half-planes in $\mathbb{R}^2$ separated by a line (a nowhere dense set), say P consists of the points to the right of the y-axis, and Q consists of the points to the left; then $P \cup Q$ is open, but not regular, since

$$(P \cup Q)^{\perp\perp} = \mathbb{R}^2.$$

In intuitive terms, an open set is regular if there are no cracks in it; the trouble with the union of two regular open sets is that there might be a crack between them. This example helps to explain the necessity for the possibly surprising definition (4). It is obvious that something unusual, such as (5) for instance, is needed in the definition of complementation; the set-theoretic complement of an open set (regular or not) is quite unlikely to be open.

Lemma 1. *If $P \subseteq Q$, then $Q^{\perp} \subseteq P^{\perp}$.*

Proof. Closure preserves inclusions and complementation reverses them.

Lemma 2. *If P is open, then $P \subseteq P^{\perp\perp}$.*

Proof. Since $P \subseteq P^-$, it follows, by complementation, that $P^{\perp} \subseteq P'$. Now apply closure: since P' is closed, it follows that $P^{\perp-} \subseteq P'$, and this is the complemented version of what is wanted.

Lemma 3. *If P is open, then $P^{\perp} = P^{\perp\perp\perp}$.*

Proof. Apply Lemma 1 to the conclusion of Lemma 2 to get $P^{\perp\perp\perp} \subseteq P^{\perp}$, and apply Lemma 2 to the open set $P^{\perp}$ (in place of P) to get the reverse inclusion.

It is an immediate consequence of Lemma 3 that if P is open, and all the more if P is regular, then $P^{\perp}$ is regular; this proves that the right side of (5) belongs to the class of regular open sets. Since $(P \cup Q)^{\perp}$ is always open, the same thing is true for (4). To settle (3), one more argument is needed.

Lemma 4. *If P and Q are open, then $(P \cap Q)^{\perp\perp} = P^{\perp\perp} \cap Q^{\perp\perp}$.*

Proof. The set $P \cap Q$ is included in P and in Q, so the set $(P \cap Q)^{\perp\perp}$ is included in $P^{\perp\perp}$ and in $Q^{\perp\perp}$, by two applications of Lemma 1. Consequently,

$$(P \cap Q)^{\perp\perp} \subseteq P^{\perp\perp} \cap Q^{\perp\perp}.$$

The reverse inclusion depends on the general topological fact that if P is open, then

$$P \cap Q^- \subseteq (P \cap Q)^-.$$

(It must be checked that every neighborhood U of a point x in $P \cap Q^-$ has a non-empty intersection with $P \cap Q$. The point x is in Q^-, by assumption, and $U \cap P$ is a neighborhood of x, so $U \cap P$ must intersect Q in some point. Of course, U meets $P \cap Q$ in the same point.) Complementing this relation, we get

$$(P \cap Q)^\perp \subseteq P' \cup Q^\perp.$$

Apply the operations of closure and complement to arrive at

$$(P' \cup Q^\perp)^{-\prime} \subseteq (P \cap Q)^{\perp\perp}.$$

Closure distributes over unions, and P' is closed (whence $P'^{-\prime} = P'' = P$), so the preceding inclusion may be written in the form

(6) $$P \cap Q^{\perp\perp} \subseteq (P \cap Q)^{\perp\perp}.$$

An application of (6) with $P^{\perp\perp}$ in place of P, followed by an application of (6) with the roles of P and Q interchanged, yields

$$P^{\perp\perp} \cap Q^{\perp\perp} \subseteq (P^{\perp\perp} \cap Q)^{\perp\perp} \subseteq (P \cap Q)^{\perp\perp\perp\perp}.$$

The desired conclusion follows from Lemma 3.

Lemma 4 implies immediately that the intersection of two regular open sets is regular, and hence that the right side of (3) belongs to the class of regular open sets.

So far it has been shown that the class of regular open sets of a topological space X is closed under the operations defined by (1)–(5). To complete the proof of the first assertion of Theorem 1, it must now be shown that these operations satisfy some system of axioms for Boolean algebras. It is less trouble to verify every one of the conditions (2.11)–(2.20) than to prove that some small subset of them is sufficient to imply the rest. In the verifications of (2.11), (2.12), (2.13), (2.15), (2.16), (2.17), (2.18), and (2.19), nothing is needed beyond the definitions and some straightforward computations involving Lemma 3 and the equation

(7) $$(P \cup Q)^{\perp} = P^{\perp} \cap Q^{\perp}$$

(valid for any two sets P and Q). The proof of (7) is quite easy. Closure distributes over union, by (9.3), so

$$(P \cup Q)^{-} = P^{-} \cup Q^{-}.$$

Form the complement of both sides of this equation to arrive at (7).

The validity of the distributive axioms (2.20) in the algebra of regular open sets depends on Lemma 4. Here is the verification of the first of these axioms:

$$P \wedge (Q \vee R) = P \cap (Q \cup R)^{\perp\perp} = P^{\perp\perp} \cap (Q \cup R)^{\perp\perp}$$
$$= (P \cap (Q \cup R))^{\perp\perp} = ((P \cap Q) \cup (P \cap R))^{\perp\perp}$$
$$= (P \wedge Q) \vee (P \wedge R).$$

The first and last equalities follow from definitions (3) and (4), the second from the assumed regularity of P, the third from Lemma 4, and the fourth from the distributive law (2.10) for intersection over union.

It remains to verify the complement laws (2.14); this amounts to showing that

$$P \cap P^{\perp} = \varnothing \quad \text{and} \quad (P \cup P^{\perp})^{\perp\perp} = X.$$

The first identity is obvious, since $P^{\perp} \subseteq P'$. The second one is not; one way to proceed is by means of a little topological lemma that has other applications also.

Lemma 5. *The boundary of an open set is a nowhere dense closed set.*

Proof. The boundary of an open set P is the set $P^{-} \cap P'$ (see p. 60). If the boundary of P included a non-empty open set, then that open set would have a non-empty intersection (namely itself) with P^{-}, and, at the same time, it would be disjoint from P (because it is included in P'). This contradicts the fundamental property of closure (often used as the definition — see p. 57).

Lemma 5 implies that if P is open, and all the more if it is regular, then the complement of the boundary of P, that is, $P \cup P^{\perp}$, is a dense open set. It follows that $(P \cup P^{\perp})^{-} = \varnothing$ and hence that $(P \cup P^{\perp})^{\perp\perp} = X$. This completes the proof of the first assertion of Theorem 1. The second assertion of the theorem follows from the next lemma and its dual.

Lemma 6. *The supremum of a family $\{P_i\}$ of regular open sets is $(\bigcup_i P_i)^{\perp\perp}$.*

Proof. Write $P = (\bigcup_i P_i)^{\perp\perp}$. Each of the sets P_i is included in their union, so Lemma 2 implies that $P_i \subseteq P$ for every i. (Since the meet of two regular open sets is the same as their intersection, the Boolean order relation for regular open sets is the same as ordinary set-theoretic inclusion.) To prove that the upper bound P is the least possible one, suppose Q is a regular open set such that $P_i \subseteq Q$ for every i. The proof that then $P \subseteq Q$ is quite easy: just observe that $\bigcup_i P_i \subseteq Q$ and apply Lemma 1 twice to obtain $P \subseteq Q^{\perp\perp} = Q$.

It is worth pointing out again that the Boolean algebra of regular open sets is not a field of sets, much less a complete field of sets. The example preceding Lemma 1 shows that the join of two regular open sets may be different from their union.

It is also worth mentioning that, in general, this algebra fails to be completely distributive. Consider, for instance, the regular open algebra of the open unit interval $(0, 1)$. (Warning to the would-be expert. Compactness, or its absence, has nothing to do with this example; the endpoints were omitted for notational convenience only.) Let I be the set of non-negative integers and let J be the set consisting of the two numbers $+1$ and -1. To define $P(i, j)$, split the interval into 2^i open intervals of length 2^{-i}; let $P(i, -1)$ be the union of the open left halves of these intervals and let $P(i, +1)$ be the union of their open right halves.

For example, when $i = 1$, the interval $(0, 1)$ is split into the two open intervals $(0, 1/2)$ and $(1/2, 1)$ of length $1/2$. The open left halves of these intervals are $(0, 1/4)$ and $(1/2, 3/4)$, and the open right halves are $(1/4, 1/2)$ and $(3/4, 1)$. The regular open sets $P(1, -1)$ and $P(1, +1)$ are defined by

$$P(1, -1) = (0, 1/4) \cup (1/2, 3/4) \quad \text{and} \quad P(1, +1) = (1/4, 1/2) \cup (3/4, 1).$$

The union of these two sets is the entire space $(0, 1)$ with the points $1/4$, $1/2$, and $3/4$ removed. Consequently, the join of the two sets is the entire space.

In general, the union of the two sets $P(i, -1)$ and $P(i, +1)$ is the entire space with the points of the form $k/2^{i+1}$ removed for $0 < k < 2^{i+1}$. These latter points form a nowhere dense subset of the space, so the join

$$P(i, -1) \vee P(i, +1)$$

is equal to the entire space $(0, 1)$ for each i. It follows that the left side of equation (8.3) is the unit element of the algebra under consideration.

For each function a in J^I, the intersection $\bigcap_i P(i, a(i))$ coincides with the intersection of a nested sequence of open intervals whose lengths go to zero; consequently, the intersection contains at most one point, whatever the function a may be. In fact, the only point that can be in the intersection is the real number whose binary representation has 0 or 1 in the $(i+1)$th place according as $a(i)$ is -1 or $+1$. It follows that the infimum $\bigwedge_i P(i, a(i))$ is the zero element of our algebra for every function a; hence, so is the right side of equation (8.3).

This last argument can be clarified with an example. Suppose a is a function from I to J whose first four values are

$$a(0) = -1, \quad a(1) = +1, \quad a(2) = +1, \quad \text{and} \quad a(3) = -1.$$

Then,

$$P(0, a(0)) = (0, 1/2),$$
$$P(1, a(1)) = (1/4, 2/4) \cup (3/4, 4/4),$$
$$P(2, a(2)) = (1/8, 2/8) \cup (3/8, 4/8) \cup (5/8, 6/8) \cup (7/8, 8/8),$$
$$P(3, a(3)) = (0/16, 1/16) \cup (2/16, 3/16) \cup \cdots \cup (14/16, 15/16),$$

$$\vdots$$

The intersection of these sets coincides with the intersection of the open intervals

$$(0, 1/2), \quad (1/4, 2/4), \quad (3/8, 4/8), \quad (6/16, 7/16), \quad \dots .$$

If there is a point in this intersection, it can only be the real number whose binary representation begins with .0110 . . . . Consequently, the infimum of the family $\{P(i, a(i))\}$ is the empty set.

Exercises

1. Prove that a subset Q of a topological space is regular and open if and only if $Q = P^{\perp\perp}$ for some set P.

2. Prove, for an arbitrary subset P of a topological space, that

$$P^{\perp-} = P^{\perp\perp\perp-}.$$

3. Show that Boolean axioms (2.11), (2.12), (2.13), (2.15), (2.16), and (2.18) are valid in the algebra of regular open sets.

4. Show that the De Morgan laws (2.17) are valid in the algebra of regular open sets.

5. Show that the associative laws (2.19) are valid in the algebra of regular open sets.

6. Show that the distributive law for join over meet in (2.20) is valid in the algebra of regular open sets.

7. Describe the Boolean algebra of regular open subsets of a discrete space.

8. Describe the Boolean algebra of regular open subsets of an infinite space with the cofinite topology.

9. (Harder.) A closed subset P of a topological space X is called *regular* if it is equal to the closure of its interior: $P = P^{\circ -}$. Define operations of join, meet, and complement on the class of regular closed subsets of X, and prove that the resulting algebra is a complete Boolean algebra. (This dual formulation of the regular open algebra is due to Tarski [75].)

10. (Harder.) Prove, using the last assertion of Theorem 1 and the infinite version of the De Morgan laws (Lemma 8.1), that if $\{P_i\}$ is a family of regular open sets, then

$$\left(\bigcap_i P_i \right)^{-\prime-\prime} = \left(\bigcap_i P_i^- \right)^{\prime-\prime}.$$

Show that this is not necessarily true for arbitrary open sets, and give a direct topological proof for regular open sets.

11. This exercise refers to the notation introduced in the final example of the chapter. Consider the function a from the set of non-negative integers into the set $\{-1, +1\}$ defined by

$$a(i) = \begin{cases} +1 & \text{if } i \text{ is even,} \\ -1 & \text{if } i \text{ is odd.} \end{cases}$$

(a) Write out explicitly the sets $P(0, a(0))$, $P(1, a(1))$, $P(2, a(2))$, and $P(3, a(3))$.

(b) The intersection of the family $\{P(i, a(i))\}$ coincides with the intersection of which family of open intervals?

(c) What is the binary representation of the only real number that can be in this intersection?

(d) Is that real number in the intersection?

12. Repeat the preceding exercise for the function a defined by

$$a(i) = \begin{cases} +1 & \text{if } i > 0, \\ -1 & \text{if } i = 0. \end{cases}$$

Chapter 11

Subalgebras

A (*Boolean*) *subalgebra* of a Boolean algebra A is a subset B of A such that B, together with the distinguished elements and operations of A (restricted to the set B), is a Boolean algebra. The algebra A is called a (*Boolean*) *extension* of B.

Warning: the distinguished elements 0 and 1 are essential parts of the structure of a Boolean algebra. A subring of a ring with unit may or may not have a unit, and if it has one, its unit may or may not be the same as the unit of the whole ring. For Boolean algebras this indeterminacy is defined away: a subalgebra must contain the element 1. The insistence on the role of 1 is not an arbitrary convention, but a theorem. Since complementation is indubitably an essential part of the structure of a Boolean algebra, the presence of 1 in every subalgebra can be proved. Proof: a subalgebra contains, along with each element p, the complement p' and the join $p \vee p'$. The latter element is just 1. This proof made implicit use of the fact that a subalgebra is not empty. If 0 and 1 are not built into the definition of a Boolean subalgebra, then non-emptiness must be explicitly assumed.

To illustrate the situation, let Y be a non-empty subset of a set X. Both $\mathcal{P}(X)$ and $\mathcal{P}(Y)$ are Boolean algebras in a natural way (Chapter 2), and clearly every element of $\mathcal{P}(Y)$ is an element of $\mathcal{P}(X)$. Since, however, the unit of $\mathcal{P}(X)$ is X, whereas the unit of $\mathcal{P}(Y)$ is Y, it is not true that $\mathcal{P}(Y)$ is a Boolean subalgebra of $\mathcal{P}(X)$. Another reason why it is not true is, of course, that complementation in $\mathcal{P}(Y)$ is not the restriction of complementation in $\mathcal{P}(X)$.

There is another possible source of misunderstanding, but one that is less likely to lead to error. (Reason: it is not special to Boolean algebras,

S. Givant, P. Halmos, *Introduction to Boolean Algebras*,
Undergraduate Texts in Mathematics, DOI: 10.1007/978-0-387-68436-9_11,
© Springer Science+Business Media, LLC 2009

but has its analogue in almost every algebraic system.) To be a Boolean subalgebra it is not enough to be a subset that is a Boolean algebra in its own right, however natural the Boolean operations may appear. The Boolean operations of a subalgebra, by definition, must be the restrictions of the Boolean operations of the whole algebra. The situation is illuminated by the regular open algebra A of a topological space X (Chapter 10). Clearly A is a subclass of the field $\mathcal{P}(X)$, but, equally clearly, A is not a subalgebra of $\mathcal{P}(X)$: the join and complement operations of A are not union and set-theoretic complementation.

Every non-degenerate Boolean algebra A includes a *trivial* subalgebra, namely 2; all other subalgebras of A will be called *non-trivial*. Every Boolean algebra A includes an *improper* subalgebra, namely A; all other subalgebras will be called *proper*.

The definition of a field of subsets of a set X may be formulated by saying that it is a Boolean subalgebra of the special field $\mathcal{P}(X)$. In general a Boolean subalgebra of a field of sets is called a *subfield*. Here are three examples of subalgebras (and in fact of subfields): the finite–cofinite algebra of a set X is a subalgebra of the countable–cocountable algebra of X; the algebra of (periodic) sets of integers of period 3 is a subalgebra of the algebra of sets of integers of period 6; and the interval algebra of finite unions of left half-closed intervals (of real numbers) with endpoints that are rational (or $\pm\infty$) is a subalgebra of the interval algebra of finite unions of arbitrary left half-closed intervals of real numbers.

If a non-empty subset B of a Boolean algebra A is closed under some Boolean operations, and if there are enough of those operations that all other Boolean operations can be defined by them, then B is a subalgebra of A. Example: if B is closed under joins and complements, then B is a subalgebra, since meet is definable in terms of join and complement; alternatively, if B is closed under the Sheffer stroke, then B is a subalgebra.

A moment's thought shows that the intersection of every family $\{B_i\}$ of subalgebras of a Boolean algebra A is again a subalgebra of A. (The intersection of the empty family is, by convention, the improper subalgebra A.) For the proof, suppose p and q are elements of the intersection $\bigcap_i B_i$. These elements then belong to every subalgebra B_i. Since subalgebras are closed under the operations of join, meet, and complement, the elements

$$p \wedge q, \qquad p \vee q, \qquad \text{and} \qquad p'$$

must belong to every subalgebra B_i, and therefore they must belong to the intersection of these subalgebras. The intersection is not empty, because it contains 0.

It follows that if E is an arbitrary subset of A, then the intersection of all those subalgebras that happen to include E is a subalgebra. (There is always at least one subalgebra that includes E, namely the improper subalgebra A.) That intersection, say B, is the smallest subalgebra of A that includes E; in other words, B is included in every subalgebra that includes E. The subalgebra B is said to be *generated* by E, and E is called a set of *generators* of B.

Here is a trivial example: if E is empty and A is not degenerate, then the subalgebra generated by E is the smallest possible subalgebra of A, namely 2. As a less trivial example, consider the field A of finite and cofinite subsets of a set X. We shall prove that it is generated in $\mathcal{P}(X)$ by the set E of one-element (that is, singleton) subsets of X. Let B be the subalgebra of $\mathcal{P}(X)$ generated by E. It is to be shown that B coincides with A. A singleton is a finite set, and therefore is an element of A; the set E is thus included in A. It follows that B (the smallest subalgebra that includes E) must also be included in A. To establish the reverse inclusion, notice that a finite subset of X is a finite union of singletons, and therefore must belong to every subalgebra that includes E. It follows that every finite subset of X belongs to B, and hence so does the complement of every finite subset. This proves that A is included in B.

A simple but useful remark for subsets E and F of a Boolean algebra A is that if F is included in E, then the subalgebra generated by F is included (as a subalgebra) in the subalgebra generated by E. This follows directly from the definition of generation, since every subalgebra that includes E must also include F.

The relation of one Boolean algebra being a subalgebra of another is a partial order on the class of all Boolean algebras. In other words, it is a reflexive, antisymmetric, transitive (binary) relation between Boolean algebras. Reflexivity means, in this case, that every Boolean algebra is a subalgebra of itself. Antisymmetry means that if two Boolean algebras are subalgebras of one another, then they are equal. Transitivity means that, for any three Boolean algebras A, B, and C, if C is a subalgebra of B, and B a subalgebra of A, then C is a subalgebra of A. These properties all follow easily from the definition of a subalgebra.

It is an interesting and occasionally useful observation that the class of all subalgebras of a Boolean algebra A is a complete lattice under the relation of being a subalgebra. The infimum of any family $\{B_i\}$ of subalgebras of A is the intersection of the family; it is the largest subalgebra of A that is a subalgebra of B_i for every i. The supremum of the family is the subalgebra generated by the union of the family, that is to say, it is the intersection of those subalgebras of A that include every B_i; it is the smallest subalgebra of A that includes B_i as a subalgebra for every i. (There is always one subalgebra that includes every B_i as a subalgebra, namely the improper subalgebra A.)

The supremum of a family $\{B_i\}$ of subalgebras is in general not the union of the family, because that union is usually not a subalgebra. There is an exception, however. The family is said to be *directed* if any two members B_i and B_j of the family are always subalgebras of some third member B_k.

Lemma 1. *The union of a non-empty, directed family of subalgebras is again a subalgebra.*

Proof. Let $\{B_i\}$ be a non-empty, directed family of subalgebras of a Boolean algebra A. It is to be demonstrated that the union B of this family is also a subalgebra of A. Certainly, B is not empty, since the family is non-empty. To prove that B is closed under the operations of A, consider any two elements, say p and q, in B. Each of these elements belongs to some subalgebra of the family, by the definition of a union; say p is in B_i and q in B_j. The two subalgebras are included in some third subalgebra B_k of the family, by the assumption that the family is directed. The elements p and q are then both in B_k, so their join, meet, and complements are also in B_k. It follows that their join, meet, and complements are also in B, as desired.

The lemma applies, in particular, to non-empty families of subalgebras that are linearly ordered by inclusion. Such families are called chains. More precisely, a family $\{B_i\}$ of subalgebras is a *chain* if for any two members B_i and B_j of the family, either B_i is a subalgebra of B_j, or vice versa.

A subalgebra of A is said to be *finitely generated* if it is generated by a finite subset of A. One consequence of the lemma is that every Boolean algebra is the directed union of its finitely generated subalgebras. In fact, a stronger statement is true.

Corollary 1. *Let A be a Boolean algebra generated by a set E, and for each finite subset F of E, let B_F be the subalgebra of A generated by F. The family*

$$\{B_F : F \subseteq E \text{ and } F \text{ is finite}\}$$

is directed, and its union is A.

Proof. It is easy to see that the given family of subalgebras is directed: if F_1 and F_2 are finite subsets of E, then so is $F_3 = F_1 \cup F_2$; since F_1 and F_2 are included in F_3, the subalgebras generated by F_1 and F_2 are included in the subalgebra generated by F_3. It now follows from Lemma 1 that the union of this directed family — call it B — is a subalgebra of A.

Since A is generated by the set E, the subalgebras B_F generated by the finite subsets F of E must be included in A; consequently, the union B of these subalgebras is included in A. On the other hand, each element p in E is contained in the subalgebra generated by the finite subset $F = \{p\}$, so p must belong to the union B. It follows that E is included in B, and therefore so is the subalgebra generated by E, namely A. Conclusion: $A = B$.

For any Boolean algebra A, apply the preceding corollary to the set $E = A$ to conclude that every Boolean algebra is the (directed) union of its finitely generated subalgebras.

The definition of the subalgebra generated by a set is "top-down" and non-constructive. One advantage of this approach is that it generalizes, practically without change, to arbitrary algebraic structures. A disadvantage is that it gives no hint which elements belong to the subalgebra. There is a "bottom-up" approach that gives precise information about the elements in the subalgebra.

Let's begin with a description of the subalgebra generated by a finite subset E of a Boolean algebra A. It is convenient to introduce some notation that will prove useful in other situations as well. For each element i in A and each j in $2 = \{0,1\}$, write

$$p(i, j) = \begin{cases} i & \text{if } j = 1, \\ i' & \text{if } j = 0. \end{cases}$$

Finally, write 2^E for the set of 2-valued functions on E, that is to say, the set of functions from E to 2. Given such a function a, the value of $p(i, a(i))$, for each i in E, is either i or i'; denote the meet of these values by p_a, so that

$$p_a = \bigwedge_{i \in E} p(i, a(i)).$$

For example, suppose $E = \{q, r, s\}$. If $a(q) = a(s) = 1$ and $a(r) = 0$, then

$$p_a = p(q, a(q)) \wedge p(r, a(r)) \wedge p(s, a(s))$$
$$= p(q, 1) \wedge p(r, 0) \wedge p(s, 1) = q \wedge r' \wedge s,$$

and if $a(q) = a(r) = a(s) = 0$, then

$$p_a = p(q, 0) \wedge p(r, 0) \wedge p(s, 0) = q' \wedge r' \wedge s'.$$

We first show that the elements p_a are mutually *disjoint* in the sense that the meet of p_a and p_b, for $a \neq b$, is always 0, and that the join of all these elements is 1. If a and b are distinct, then they differ on some index i: one of $a(i)$ and $b(i)$ is 1 and the other is 0. Consequently,

$$(1) \qquad p_a \wedge p_b \leq i \wedge i' = 0.$$

On the other hand,

$$(2) \qquad 1 = \bigwedge_{i \in E} (i \vee i') = \bigvee_{a \in 2^E} \bigwedge_{i \in E} p(i, a(i)) = \bigvee_{a \in 2^E} p_a,$$

by the distributive law (8.3). (That law is applicable because E is finite.)

Take K to be the set of those functions a in 2^E such that $p_a \neq 0$, and for each subset X of K, write

$$(3) \qquad p_X = \bigvee_{a \in X} p_a.$$

These joins are elements of A, and they obey the following laws:

$$(4) \qquad p_\emptyset = 0,$$
$$(5) \qquad p_K = 1,$$
$$(6) \qquad p_X \wedge p_Y = p_{X \cap Y},$$
$$(7) \qquad p_X \vee p_Y = p_{X \cup Y},$$
$$(8) \qquad p'_X = p_{X'},$$

where X' denotes the (set-theoretic) complement of X in K. Equation (4) holds because the supremum of the empty set is 0, equation (5) is a direct consequence of (2) and the definition of K, and equation (7) follows immediately from definition (3). The proof of equation (6) involves an easy computation:

$$p_X \wedge p_Y = \left(\bigvee_{a \in X} p_a \right) \wedge \left(\bigvee_{b \in Y} p_b \right) = \bigvee_{\substack{a \in X \\ b \in Y}} (p_a \wedge p_b)$$

$$= \bigvee_{a \in X \cap Y} (p_a \wedge p_a) = \bigvee_{a \in X \cap Y} p_a = p_{X \cap Y}.$$

The first and last equalities use definition (3), while the second equality uses the distributive law in the form of Corollary 8.2, the third equality uses (1), and the fourth equality uses the idempotent law for meet in (2.16). To prove equation (8), observe that

$$p_X \wedge p_{X'} = p_{X \cap X'} = p_\emptyset = 0 \qquad \text{and} \qquad p_X \vee p_{X'} = p_{X \cup X'} = p_K = 1,$$

by (4)–(7). These equations imply that $p_{X'}$ is the complement of p_X, by Lemma 6.2.

Take C to be the set of all elements of A of the form p_X, for some subset X of K. Equations (4)–(8) imply that C is a subalgebra of A: it is non-empty, by (4), and it is closed under the operations of A, by (6)–(8). Definition (3) and equations (4) and (6) imply that the elements p_a (for a in K) are minimal non-zero elements in C. Such elements are usually called *atoms*. Every element of C can be written in exactly one way as a join of these atoms. Indeed, suppose $p_X = p_Y$. A simple computation, using (4)–(8) and the definition of Boolean addition in (3.3), shows that

$$0 = p_X + p_X = p_X + p_Y = p_{X+Y},$$

where $X+Y$ is the symmetric difference of the sets X and Y. This symmetric difference must be empty, for otherwise p_{X+Y} would be a non-empty join of non-zero elements, and hence would be different from 0. It follows that $X = Y$.

Consider an arbitrary subset F of E, and an arbitrary 2-valued function b on F. In analogy with the notation introduced above, write

$$p_b = \bigwedge_{i \in F} p(i, b(i)).$$

We shall show that p_b is in C by proving that

(9) $$p_b = \bigvee \{p_a : a \in K \text{ and } a \text{ extends } b\}.$$

(A function a in K *extends* b if $a(i) = b(i)$ for each i in F.) Let L be the set of functions in K that extend b. If a is in L, then $a(i) = b(i)$, and therefore

$$p(i, b(i)) = p(i, a(i)),$$

for every i in F; consequently,

$$\{p(i, b(i)) : i \in F\} \subseteq \{p(i, a(i)) : i \in E\}.$$

It follows that

$$p_b = \bigwedge_{i \in F} p(i, b(i)) \geq \bigwedge_{i \in E} p(i, a(i)) = p_a.$$

In other words,

(10) $$p_b \wedge p_a = p_a \qquad \text{when} \qquad a \in L.$$

On the other hand, if a is not in L, then a and b must disagree on some argument i in F. One of $p(i, a(i))$ and $p(i, b(i))$ is therefore i, and the other is i'. Consequently,

$$p_b \wedge p_a \leq p(i, b(i)) \wedge p(i, a(i)) = i \wedge i' = 0.$$

Thus,

(11) $p_b \wedge p_a = 0$ when $a \in K - L$.

The demonstration of (9) is now straightforward:

$$p_b = p_b \wedge 1 = p_b \wedge p_K = p_b \wedge \bigvee_{a \in K} p_a$$

$$= \bigvee_{a \in K} (p_b \wedge p_a) = \bigvee_{a \in L} (p_b \wedge p_a) = \bigvee_{a \in L} p_a.$$

The first equality uses the identity law for meet in (2.13), the second equality uses (5), the third uses (3) and the definition of K, the fourth uses the distributive law in Lemma 8.3, the fifth uses (11), and the sixth uses (10).

Let B be the subalgebra of A generated by E. We shall show that $B = C$. To prove that B is included in C, consider an arbitrary element i in E; if b is the function from $\{i\}$ to 2 defined by $b(i) = 1$, then

$$p_b = p(i, b(i)) = i,$$

so that i is in C, by (9). This argument proves that E is included in C, so the subalgebra generated by E, namely B, is included in C. To prove that C is included in B, recall that each atom p_a in C is the (finite) meet of elements and complements of elements in E; consequently, p_a belongs to the subalgebra B generated by E. Every element p_X in C is the (finite) join of such atoms, and is therefore also in B.

The following *normal form theorem* summarizes what has been proved so far.

Theorem 2. *Let B be the subalgebra generated by a finite subset E of a Boolean algebra. The atoms of B are the non-zero elements of the form*

$$p_a = \bigwedge_{i \in E} p(i, a(i)),$$

and the elements of B are the joins of these atoms. Every element of B can be written in one and only one way as a join of atoms. The distinguished elements and operations of B are determined by equations (4)–(8).

How big can a finitely generated Boolean algebra A be? If a generating set E has n elements, then there are 2^n functions from E to the set $\{0, 1\}$. As a result, there can be at most 2^n atoms, by the preceding theorem. Suppose A has m atoms, where $m \leq 2^n$. The join of every set of atoms is an element of A, and every element of A can be written in exactly one way as the join of a set of atoms, by the preceding theorem. There are 2^m subsets of the set of atoms, so there must be 2^m elements in A.

Corollary 2. *Every finitely generated Boolean algebra A is finite, and the number of its elements is 2^m, where m is the number of atoms in A. If a generating set of A has n elements, then A has at most 2^n atoms, and hence it has at most 2^{2^n} elements.*

The description of the subalgebra generated by an arbitrary subset of a Boolean algebra A is obtained rather easily from the normal form theorem and Corollary 1. Suppose E is a (possibly infinite) subset of A. For each finite subset F of E, let B_F be the subalgebra of A generated by F, and let K_F be the set of 2-valued functions b on F such that the meet

$$p_b = \bigwedge_{i \in F} p(i, b(i))$$

is not zero. The elements of B_F are precisely the joins

$$p_X = \bigvee_{b \in X} p_b,$$

where X ranges over the subsets of K_F, by Theorem 2. The subalgebra B generated by the set E is the directed union of the finitely generated subalgebras B_F, by Corollary 1. Conclusion: the elements of B are just the elements p_X, for various finite subsets F of E, and various subsets X of K_F. These remarks establish the following theorem.

Theorem 3. *An element of a Boolean algebra is in the subalgebra generated by a set E if and only if it can be written as a finite join of finite meets of elements and complements of elements from E.*

Equations (4)–(8) may still be used to describe the distinguished elements and operations of the subalgebra B generated by a set E, but some caution is needed. Equations (6) and (7) do not say that

$$p_{X_1} \wedge p_{X_2} = p_{X_1 \cap X_2} \qquad \text{and} \qquad p_{X_1} \vee p_{X_2} = p_{X_1 \cup X_2},$$

whenever X_1 is a subset of K_{F_1} and X_2 a subset of K_{F_2}. In fact, the domains of the functions in X_1 and in X_2 (namely, the finite subsets F_1 and F_2 of E) are in general not equal.

To overcome this obstacle, write $F = F_1 \cup F_2$, and take Y_1, respectively Y_2, to be the set of functions in K_F that extend some function in X_1, respectively X_2. Then

$$p_{X_1} = p_{Y_1} \quad \text{and} \quad p_{X_2} = p_{Y_2},$$

by (9), so that

$$p_{X_1} \wedge p_{X_2} = p_{Y_1} \wedge p_{Y_2} = p_{Y_1 \cap Y_2} \quad \text{and} \quad p_{X_1} \vee p_{X_2} = p_{Y_1} \vee p_{Y_2} = p_{Y_1 \cup Y_2},$$

by (6) and (7).

It should be pointed out that an element of the form p_b is not, in general, an atom of B, even though it is an atom of the subalgebra B_F when F is the domain of b. Indeed, if G is a finite subset of E that properly includes F, and if a is a 2-valued function on G that extends b, then it may well happen that

$$0 < p_a < p_b.$$

An important special case of the preceding theorem, the *subalgebra extension lemma*, describes how to extend a subalgebra by adjoining a single element.

Lemma 2. *Let B be a Boolean subalgebra of A, and r an element in A. The subalgebra generated by $B \cup \{r\}$ consists of the elements in A that can be written in the form*

$$(p \wedge r) \vee (q \wedge r')$$

for some p and q in B.

Proof. The elements of the subalgebra C generated by the set $B \cup \{r\}$ are the finite joins of finite meets of elements and complements of elements from $B \cup \{r\}$, by the preceding theorem. Meets of elements and complements of elements from B are again elements in B, because B is a subalgebra. The elements of C are therefore the finite joins of elements of B, elements of the form $p \wedge r$, with p in B, and elements of the form $q \wedge r'$, with q in B. The form

$$(12) \qquad\qquad (p \wedge r) \vee (q \wedge r'),$$

with p and q in B, comprehends each of the other forms as special cases:

$$p \wedge r = (p \wedge r) \vee (0 \wedge r'),$$
$$q \wedge r' = (0 \wedge r) \vee (q \wedge r'),$$
$$p = (p \wedge r) \vee (p \wedge r'),$$

where p and q are elements of B. Consequently, the elements of C are finite joins of elements of the form (12). The join of two elements of the form (12) (and hence the join of any finite number of such elements) is again an element of the form (12), as a simple computation shows:

(13) $[(p_1 \wedge r) \vee (q_1 \wedge r')] \vee [(p_2 \wedge r) \vee (q_2 \wedge r')]$
$$= [(p_1 \wedge r) \vee (p_2 \wedge r)] \vee [(q_1 \wedge r') \vee (q_2 \wedge r')]$$
$$= [(p_1 \vee p_2) \wedge r] \vee [(q_1 \vee q_2) \wedge r'],$$

where $p_1 \vee p_2$ and $q_1 \vee q_2$ are elements of B. Conclusion: the elements of C are just the elements of the form (12). The proof of the lemma is complete.

The lemma can also be proved directly, without recourse to Theorem 3. The set D of all elements of the form (12) is certainly included in the subalgebra of A generated by B and r, so it suffices to show that D is itself a subalgebra of A. The closure of D under join follows from (13), while closure under meet and complement follows from the identities

(14) $[(p_1 \wedge r) \vee (q_1 \wedge r')] \wedge [(p_2 \wedge r) \vee (q_2 \wedge r')]$
$$= [(p_1 \wedge p_2) \wedge r] \vee [(q_1 \wedge q_2) \wedge r']$$

and

(15) $[(p \wedge r) \vee (q \wedge r')]' = (p' \wedge r) \vee (q' \wedge r').$

The definition of a Boolean subalgebra B says nothing about the infinite suprema and infima that may be formable in the whole algebra A. Anything can happen: suprema or infima can be gained or lost or change value as we pass back and forth between A and B. Everything that can happen can be illustrated in the theory of complete Boolean algebras. If B is a subalgebra of a complete algebra A, and if the supremum (in A) of every subset of B belongs to B, we say that B is a *complete subalgebra* of A. (Warning: this is stronger than requiring merely that B be a complete Boolean algebra in its own right.) Note that a complete subalgebra B of A contains the infima (in A) of all its subsets as well as their suprema. Indeed, if E is a subset of B, then so is $\{p' : p \in E\}$; the supremum of the latter subset is in B, by assumption, so the complement of this supremum, namely the infimum

$$\bigwedge E = \left(\bigvee \{p' : p \in E\}\right)',$$

is also in B. In the case of fields we speak of *complete subfields*. For complete algebras the concept of a generated complete subalgebra is defined the same way as when completeness was not yet mentioned; all that is necessary is to replace "subalgebra" by "complete subalgebra" throughout the discussion.

There is an intermediate notion, stronger than "subalgebra" but weaker than "complete subalgebra", that is sometimes useful. It does not require the algebra A to be complete. A *regular subalgebra* of A is a subalgebra B with the additional property that whenever a subset E of B has a supremum p in B, then p is the supremum of E in A also. (Warning: a subset of B may have a supremum in A without having a supremum in B; the definition says nothing about such subsets.)

A necessary and sufficient condition for a subalgebra B of A to be regular is that whenever E is a subset of B with $\bigvee E = 1$ in B, then $\bigvee E = 1$ in A. The necessity of the condition is obvious; it is part of the definition of regularity. To prove sufficiency, suppose the condition is satisfied, and let E_0 be an arbitrary subset of B that has a supremum p in B. It is to be proved that p is the supremum of E_0 in A. The supremum of the set $E = \{p'\} \cup E_0$ in B is 1, as the computation

$$1 = p' \vee p = p' \vee \bigvee E_0 = \bigvee(\{p'\} \cup E_0) = \bigvee E$$

demonstrates. It follows from the assumed condition that 1 is also the supremum of E in A. Form the meet (in A) of p with the first and last terms of the preceding string of equalities to arrive at the desired conclusion:

$$p = p \wedge 1 = p \wedge \bigvee E = p \wedge \bigvee(\{p'\} \cup E_0)$$
$$= (p \wedge p') \vee \bigvee_{q \in E_0} (p \wedge q) = 0 \vee \bigvee_{q \in E_0} q = \bigvee E_0.$$

The fourth equality uses the distributive law from Lemma 8.3. The fifth equality uses the fact that, for each q in E_0, the inequality $q \leq p$ holds in B, and therefore also in A.

Exercises

1. Show that the field of periodic sets of integers of period 3 (see Chapter 5) is a subfield of the field of periodic sets of integers of period 6.

2. Suppose that B and A are fields of periodic sets of integers of periods m and n respectively. Formulate and prove a theorem characterizing when B is a subfield of A.

3. Show that the relation of being a subalgebra is a partial ordering on the class of all Boolean algebras.

4. Show that a subring of a Boolean ring need not be a Boolean subalgebra. What if the subring contains 1?

5. Every subset of a partially ordered set inherits a partial order from the whole set. If a non-empty subset of a Boolean algebra is construed as a partially ordered set in this way, and if it turns out that with respect to this partial order it is a complemented distributive lattice, does it follow that it is a Boolean subalgebra of the original algebra? (See Exercise 7.22.)

6. If a subset B of a Boolean algebra A contains 0 and 1 and is closed under the formation of meets and joins, does it follow that B is a subalgebra of A?

7. Suppose $E = \{q, r, s\}$ is a subset of a Boolean algebra. For each i in E, write

$$p(i, j) = \begin{cases} i & \text{if } j = 1, \\ i' & \text{if } j = 0. \end{cases}$$

There are eight 2-valued functions a on E. Write out the eight corresponding meets

$$p_a = \bigwedge_{i \in E} p(i, a(i)).$$

8. Formulate the dual version of Theorem 2. Prove this dual directly, without using Theorem 2.

9. Formulate and prove the dual version of Theorem 3.

10. Prove identities (14) and (15), and then use them to give a direct proof of Lemma 2.

11. Let A be the interval algebra of finite unions of left half-closed intervals of real numbers, and let E be the subset of A consisting of the intervals $[n, n+1)$, where n ranges over the integers. Describe the elements of the subalgebra of A generated by E.

12. Prove directly, without using Lemma 1, that the union of a non-empty chain of Boolean subalgebras is again a Boolean subalgebra.

13. A family of Boolean algebras $\{B_i\}$ (not necessarily subalgebras of some fixed Boolean algebra) is called a (*subalgebra*) *chain* if for any two algebras B_i and B_j in the family, one of them is a subalgebra of the other. Let B be the union of such a chain. For any two elements p and q in B, there must be an algebra B_i in the chain that contains both of them. (Why?) Define the join and meet of p and q in B to be their join and meet in B_i, and define the complement of p in B to be its complement in B_i. Prove that these operations are well defined in the sense that they do not depend on the particular choice of the algebra B_i to which p and q both belong. Prove also that under these operations the union B is a Boolean algebra.

14. A family $\{B_i\}$ of Boolean algebras (not necessarily subalgebras of some fixed Boolean algebra) is said to be *directed* if for any two algebras B_i and B_j in the family, there is always a third algebra B_k in the family such that B_i and B_j are both subalgebras of B_k. Define operations of join, meet, and complement in the union of the directed family, and show that these operations are well defined. Prove that the union of the family, under these operations, is a Boolean algebra.

15. Give an example of a subalgebra B of a complete Boolean algebra A and of a subset E of B such that the supremum of E in A does not belong to B, and in fact E has no supremum in B.

16. (Harder.) Give an example of a subalgebra B of a Boolean algebra A and of a subset E of B such that E has a supremum in B but not in A.

17. (Harder.) Give an example of a complete Boolean algebra A and a subalgebra B of A such that some subset of B has a supremum in B and a supremum in A, but these two suprema are different.

18. (Harder.) Give an example of complete Boolean algebras A and B such that B is a subalgebra, but not a complete subalgebra, of A.

19. Prove that an infinite Boolean algebra with m generators has m elements.

20. Let B be a subalgebra of A. If a family $\{p_i\}$ in B has a supremum p in A, and if p belongs to B, prove that p is also the supremum of $\{p_i\}$ in B.

21. Prove that a subalgebra B of a Boolean algebra A is a regular subalgebra if and only if every subset of B that has an infimum in B has the same infimum in A.

22. Prove that a necessary and sufficient condition for a subalgebra B of a Boolean algebra A to be regular is that whenever E is a subset of B with
$$\bigwedge E = 0 \quad \text{in } B, \quad \text{then also} \quad \bigwedge E = 0 \quad \text{in } A.$$

23. Is every finite subalgebra a regular subalgebra?

24. Is the field of finite and cofinite sets of integers a regular subalgebra of the field of all sets of integers?

25. Suppose A is a complete Boolean algebra. Is a complete subalgebra of A necessarily a regular subalgebra? Is a regular subalgebra of A necessarily a complete subalgebra?

26. Suppose A is a complete Boolean algebra. If a regular subalgebra of A happens to be complete (considered as an algebra in its own right), is it necessarily a complete subalgebra of A?

27. Show that the relation of being a regular subalgebra is a partial order on the class of Boolean algebras. In other words, show that it is reflexive, antisymmetric, and transitive.

28. Suppose C is a (Boolean) subalgebra of B, and B a subalgebra of A. If C is a regular subalgebra of A, must C also be a regular subalgebra of B?

29. Prove that a regular subalgebra of a completely distributive Boolean algebra is completely distributive.

Chapter 12

Homomorphisms

A *Boolean homomorphism* is a mapping f from a Boolean algebra B, say, to a Boolean algebra A such that

(1) $$f(p \wedge q) = f(p) \wedge f(q),$$
(2) $$f(p \vee q) = f(p) \vee f(q),$$
(3) $$f(p') = (f(p))',$$

whenever p and q are in B. In a somewhat loose but brief and suggestive phrase, a homomorphism is a structure-preserving mapping between Boolean algebras. A convenient synonym for "homomorphism from B to A" is "A-valued homomorphism on B". Such expressions will be used most frequently in case $A = 2$. A word about notation: we shall usually write $f(p)'$ instead of $(f(p))'$.

When a homomorphism f maps B *onto* A (in the sense that every element in A is equal to $f(p)$ for some p in B), the stipulation that A be a Boolean algebra is unnecessary; it is a consequence of identities (1)–(3) and the assumption that B is a Boolean algebra. To prove this assertion, let A be an arbitrary algebraic structure with two binary operations $\wedge$ and $\vee$, and a unary operation $'$. Suppose f is a mapping from a Boolean algebra B onto A that satisfies identities (1)–(3). It must be checked that the Boolean algebra axioms (2.11)–(2.20) are true of A when zero and one are interpreted as $f(0)$ and $f(1)$ respectively. As an example, here is the verification of the distributive law for meet over join in (2.20). Let u, v, and w be elements of A. The assumption that the range of f is A implies that there are elements p, q, and r in B such that

$$f(p) = u, \qquad f(q) = v, \qquad \text{and} \qquad f(r) = w.$$

S. Givant, P. Halmos, *Introduction to Boolean Algebras*,
Undergraduate Texts in Mathematics, DOI: 10.1007/978-0-387-68436-9_12,
© Springer Science+Business Media, LLC 2009

A straightforward computation using identities (1) and (2), and the validity of the distributive law in B, now yields the desired result:

$$
\begin{aligned}
u \wedge (v \vee w) &= f(p) \wedge (f(q) \vee f(r)) \\
&= f(p) \wedge f(q \vee r) \\
&= f(p \wedge (q \vee r)) \\
&= f((p \wedge q) \vee (p \wedge r)) \\
&= f(p \wedge q) \vee f(p \wedge r) \\
&= (f(p) \wedge f(q)) \vee (f(p) \wedge f(r)) \\
&= (u \wedge v) \vee (u \wedge w).
\end{aligned}
$$

In most situations, the algebra A that constitutes the range of a Boolean homomorphism is known a priori to be a Boolean algebra. In those cases in which this is not known, the previous argument can be used to prove it. We shall see an example in a moment.

The distinguished elements 0 and 1 play a special role for homomorphisms, just as they do for subalgebras. Indeed, if f is a Boolean homomorphism (between Boolean algebras) and p is an element in its domain ($p = 0$ will do), then

$$
f(p \wedge p') = f(p) \wedge f(p)',
$$

and therefore

(4) $$f(0) = 0.$$

This much would be expected by a student of ring theory. What is important is that the dual argument proves the dual fact,

(5) $$f(1) = 1.$$

The mapping that sends every element of one non-degenerate Boolean algebra onto the zero element of another is simply not a Boolean homomorphism; between non-degenerate Boolean algebras there is no such thing as a "trivial" homomorphism.

Equations (4) and (5) imply that 0 and 1 belong to the range of every homomorphism; a glance at equations (1)–(3) is sufficient to convince oneself that the range of every homomorphism, from B into A say, is closed under the meet, join, and complement operations of A, and is therefore a Boolean subalgebra of A. The range of a homomorphism with domain B is called a *homomorphic image* of B.

Since every Boolean operation (e.g., $+$ and $\Rightarrow$) can be defined in terms of $\wedge$, $\vee$, and $'$, it follows that a Boolean homomorphism preserves all such operations. For instance, if f is a Boolean homomorphism and if p and q are elements of its domain, then

$$\begin{aligned}
f(p+q) &= f((p \wedge q') \vee (p' \wedge q)) \\
&= f(p \wedge q') \vee f(p' \wedge q) \\
&= (f(p) \wedge f(q')) \vee (f(p') \wedge f(q)) \\
&= (f(p) \wedge f(q)') \vee (f(p)' \wedge f(q)) \\
&= f(p) + f(q)
\end{aligned}$$

and

$$f(p \Rightarrow q) = f(p' \vee q) = f(p') \vee f(q) = f(p)' \vee f(q) = f(p) \Rightarrow f(q).$$

It follows, in particular, that every Boolean homomorphism is a ring homomorphism, and also that every Boolean homomorphism is order-preserving. The last assertion means that if $p \leq q$, then $f(p) \leq f(q)$.

The crucial fact in the preceding paragraph was the definability of Boolean operations and relations in terms of meet, join, and complement. Thus, more generally, if a mapping f from a Boolean algebra B to a Boolean algebra A preserves enough Boolean operations so that all others are definable in terms of them, then f is a homomorphism. Example: if f preserves $\vee$ and $'$ (that is, if f satisfies the identities (2) and (3)), then f is a homomorphism; alternatively, if f preserves the Sheffer stroke, then f is a homomorphism.

We proceed to consider some examples of Boolean homomorphisms. For the first example let B be an arbitrary Boolean algebra, and let p_0 be an arbitrary element of B. Take A to be the set of all *subelements* of p_0, that is, the set of elements p with $p \leq p_0$, or, equivalently, the set of all elements of the form $p \wedge p_0$. Define elements and operations for A as follows: 0, meet, and join in A are the same as in B, but 1 and p' in A are defined to be the elements p_0 and $p_0 - p$ of B. The mapping f defined by

$$f(p) = p \wedge p_0$$

is a homomorphism from B onto A. The proof consists of a series of computations verifying conditions (1)–(3). Let p and q be elements of B. Then

$$f(p \wedge q) = p \wedge q \wedge p_0 = p \wedge p_0 \wedge q \wedge p_0 = f(p) \wedge f(q),$$

by the definition of f and the idempotent and commutative laws;

$$f(p \vee q) = (p \vee q) \wedge p_0 = (p \wedge p_0) \vee (q \wedge p_0) = f(p) \vee f(q),$$

by the definition of f and the distributive laws; and

$$f(p') = p' \wedge p_0 = (p_0 \wedge p') \vee 0 = (p_0 \wedge p') \vee (p_0 \wedge p_0') = p_0 \wedge (p' \vee p_0')$$
$$= p_0 \wedge (p \wedge p_0)' = p_0 - (p \wedge p_0) = f(p)',$$

by the definition of f, the identity and commutative laws, the complement laws, the distributive laws, the De Morgan laws, the definition of subtraction, and the definition of complementation in A. The algebra A is the image of the Boolean algebra B under the homomorphism f, and is therefore itself a Boolean algebra with zero and unit

$$f(0) = 0 \wedge p_0 = 0 \qquad \text{and} \qquad f(1) = 1 \wedge p_0 = p_0.$$

It is called the *relativization* of B to p_0, and will be denoted by $B(p_0)$; the function f is called the *relativizing homomorphism* (*induced by* p_0).

For a concrete example of a relativization, consider an arbitrary set X. If Y is a subset of X, then Y is an element of the field $\mathcal{P}(X)$, and the relativization of $\mathcal{P}(X)$ to Y is just the field $\mathcal{P}(Y)$. The relativizing homomorphism is the correspondence that takes each set P in $\mathcal{P}(X)$ to the set $P \cap Y$.

For the next example, consider a field B of subsets of a set X, and let x_0 be an arbitrary point of X. For each set P in B, let $f(P)$ be 1 or 0 according as x_0 is, or is not, in P. To prove that the mapping f is a 2-valued homomorphism on B, it suffices to verify identities (1) and (3). The definition of f, and the definitions of the Boolean operations in a field of sets and in the Boolean algebra 2, justify the following equivalences:

$$
\begin{aligned}
f(P \cap Q) = 1 \qquad &\text{if and only if} \qquad x_0 \in P \cap Q, \\
&\text{if and only if} \qquad x_0 \in P \text{ and } x_0 \in Q, \\
&\text{if and only if} \qquad f(P) = 1 \text{ and } f(Q) = 1, \\
&\text{if and only if} \qquad f(P) \wedge f(Q) = 1;
\end{aligned}
$$

so

$$f(P \cap Q) = f(P) \wedge f(Q).$$

Similarly,

$$
\begin{aligned}
f(P') = 1 \qquad &\text{if and only if} \qquad x_0 \in P', \\
&\text{if and only if} \qquad x_0 \notin P, \\
&\text{if and only if} \qquad f(P) = 0, \\
&\text{if and only if} \qquad f(P)' = 1;
\end{aligned}
$$

so

$$f(P') = f(P)'.$$

Observe that $f(P)$ is equal to the value of the characteristic function of P at x_0.

For one more example, let ϕ be an arbitrary mapping from a non-empty set X into a set Y, and let A and B be fields of subsets of X and Y respectively. Write $f = \phi^{-1}$, or, more explicitly, for each P in B, let $f(P)$ be the *inverse image* of P under ϕ:

$$f(P) = \{x \in X : \phi(x) \in P\}.$$

In general, the set $f(P)$ will not belong to the field A. If $f(P)$ is in A whenever P is in B, then f is an A-valued homomorphism on B. The proof is very similar to the one just given; it depends on the fact that intersections and complements are preserved under the formation of inverse images. Here are the details:

$$\begin{aligned}
x \in f(P \cap Q) \quad &\text{if and only if} \quad \phi(x) \in P \cap Q, \\
&\text{if and only if} \quad \phi(x) \in P \text{ and } \phi(x) \in Q, \\
&\text{if and only if} \quad x \in f(P) \text{ and } x \in f(Q), \\
&\text{if and only if} \quad x \in f(P) \cap f(Q);
\end{aligned}$$

so

$$f(P \cap Q) = f(P) \cap f(Q).$$

Similarly,

$$\begin{aligned}
x \in f(P') \quad &\text{if and only if} \quad \phi(x) \in P', \\
&\text{if and only if} \quad \phi(x) \notin P, \\
&\text{if and only if} \quad x \notin f(P), \\
&\text{if and only if} \quad x \in f(P)';
\end{aligned}$$

so

$$f(P') = f(P)'.$$

For purposes of reference we shall call the homomorphisms described in these three examples the homomorphisms *induced* by p_0, x_0, and ϕ, respectively.

Special kinds of Boolean homomorphisms may be described in the same words as are used elsewhere in algebra. A *monomorphism*, also called an *embedding*, is a homomorphism that is one-to-one: if $f(p) = f(q)$, then $p = q$. An *epimorphism* is a homomorphism that is onto: every element of A is equal to $f(p)$ for some p in B. A homomorphism that is a *bijection*, that is to say, it is both one-to-one and onto, is called an *isomorphism*. If there is an isomorphism from one Boolean algebra onto another, the two algebras are said to be *isomorphic*. An isomorphism from a Boolean algebra onto itself is called an *automorphism*.

The existence of an isomorphism between Boolean algebras implies that the algebras are structurally identical; they differ only in the "shape" of their elements. For example, consider two sets X and Y with the same number of elements; no assumptions are made about what the elements in X and Y look like. The hypothesis implies the existence of a one-to-one mapping ϕ from X onto Y. The homomorphism induced by ϕ is an isomorphism from the field $\mathcal{P}(Y)$ onto the field $\mathcal{P}(X)$; it maps each subset Q of Y to the subset

$$\phi^{-1}(Q) = \{x \in X : \phi(x) \in Q\}$$

of X. Occasionally, it is more natural to use the inverse function ϕ^{-1} instead of ϕ. In this case, since ϕ^{-1} maps Y one-to-one onto X, the induced isomorphism maps $\mathcal{P}(X)$ onto $\mathcal{P}(Y)$, and takes each subset P of X to the subset

$$\phi(P) = \{\phi(x) : x \in P\}$$

of Y. Conclusion: if two sets X and Y have the same number of elements, the corresponding fields $\mathcal{P}(X)$ and $\mathcal{P}(Y)$ are isomorphic. The fact that the fields are structurally identical is intuitively obvious; the intuition is substantiated by the construction of a concrete isomorphism.

Here is another example of an isomorphism between Boolean algebras, one that is at first glance perhaps less obvious. Consider the field of periodic sets of integers of period 2. It has four elements, all but one of which are infinite sets: the set E of even integers, the set O of odd integers, the set X of all integers, and the empty set $\varnothing$. The arithmetic of this algebra is given by the following tables for union, intersection, and complement:

$\cup$	$\varnothing$	E	O	X
$\varnothing$	$\varnothing$	E	O	X
E	E	E	X	X
O	O	X	O	X
X	X	X	X	X

$\cap$	$\varnothing$	E	O	X
$\varnothing$	$\varnothing$	$\varnothing$	$\varnothing$	$\varnothing$
E	$\varnothing$	E	$\varnothing$	E
O	$\varnothing$	$\varnothing$	O	O
X	$\varnothing$	E	O	X

$'$	
$\varnothing$	X
E	O
O	E
X	$\varnothing$

.

The field of all subsets of the two-element set $\{0, 1\}$ also has four elements, all of them finite sets: the singletons $\{0\}$ and $\{1\}$, the pair $\{0, 1\}$, and the empty set $\varnothing$. Its arithmetic is given by the tables

∪	∅	{0}	{1}	{0,1}
∅	∅	{0}	{1}	{0,1}
{0}	{0}	{0}	{0,1}	{0,1}
{1}	{1}	{0,1}	{1}	{0,1}
{0,1}	{0,1}	{0,1}	{0,1}	{0,1}

,

∩	∅	{0}	{1}	{0,1}
∅	∅	∅	∅	∅
{0}	∅	{0}	∅	{0}
{1}	∅	∅	{1}	{1}
{0,1}	∅	{0}	{1}	{0,1}

,

′	
∅	{0,1}
{0}	{1}
{1}	{0}
{0,1}	∅

.

The forms of the elements of the two Boolean algebras are certainly quite different. A comparison of the arithmetic tables, however, reveals that from a structural point of view the two algebras are identical. More precisely, the correspondence

$$\varnothing \to \varnothing, \qquad E \to \{0\}, \qquad O \to \{1\}, \qquad X \to \{0,1\}$$

transforms the tables of the first algebra into the tables of the second algebra; it is therefore an isomorphism between the two algebras.

Isomorphisms preserve all structural properties of algebras. To show that two Boolean algebras are not isomorphic, it suffices to find a structural property of one of the algebras that is not shared by the other. For instance, the field of finite and cofinite sets of rational numbers is not isomorphic to the field of all subsets of the rational numbers because the first algebra is countable, while the second has the power of the continuum. For a less trivial example, consider the field of finite and cofinite subsets of the rational numbers, and the interval algebra of the rational numbers. Both algebras are countably infinite, but they are not isomorphic: the first algebra has an infinite number of atoms, while the second algebra has none.

If B is a subalgebra of an algebra A, then the identity mapping — that is, the mapping f defined for every p in B by $f(p) = p$ — is a monomorphism from B into A, and in particular the identity mapping on A is an automorphism of A. There is a natural way to define the product of (some) pairs of homomorphisms, and it turns out that the identity mappings just mentioned indeed act as multiplicative identities. The *product* (or *composition*) $f \circ g$

of two homomorphisms f and g is defined in case A, B, and C are Boolean algebras, f maps B into A, and g maps C into B; the value of $f \circ g$ at each element p of C is given by

$$(f \circ g)(p) = f(g(p)).$$

If, moreover, h is a homomorphism from D, say, to C, then

$$f \circ (g \circ h) = (f \circ g) \circ h,$$

that is, the operation of composition is associative.

If f is a Boolean isomorphism f from B to A, then its inverse, the function f^{-1} from A to B defined by

$$f^{-1}(r) = p \qquad \text{if and only if} \qquad f(p) = r$$

for every r in A, is a Boolean isomorphism from A to B. For instance, to show that f^{-1} preserves meet, let r and s be two elements in A. There are unique elements p and q in B such that $f(p) = r$ and $f(q) = s$. Consequently,

$$f^{-1}(r \wedge s) = f^{-1}(f(p) \wedge f(q)) = f^{-1}(f(p \wedge q)) = p \wedge q = f^{-1}(r) \wedge f^{-1}(s).$$

The second equality holds because f preserves meet; the third and fourth hold by the definition of the inverse of f. The arguments that f^{-1} preserves join and complement are similar.

Occasionally, one would like to show that a given Boolean algebra B_0 can be extended to a Boolean algebra B with certain desirable properties. The actual construction, however, may not yield an extension of B_0, but rather something weaker: a Boolean algebra A with the desired properties and an isomorphism f_0 from B_0 onto a subalgebra of A. The isomorphism shows that B_0 is structurally identical to its image $A_0 = f_0(B_0)$. One would therefore like to effect an "exchange", by replacing A_0 with B_0, so as to obtain an actual extension of B_0 with the desired properties. A difficulty arises, however, because B_0 may contain elements that also occur in A in a structurally different and conflicting way. This obstacle may be overcome by first replacing the elements of $A - A_0$ with new elements having nothing to do with B_0, and then effecting the exchange. The result is a Boolean algebra B that contains B_0 as a subalgebra and that is isomorphic to A via a mapping that extends f_0. The assertion that all of this works out as expected is called the *exchange principle*.

Here are the details. Let A_1 be the set of elements that are in A but not in A_0. Choose a set (any set) B_1 with the same number of elements as A_1 and with no elements in common with B_0. The assumption that B_1 and A_1

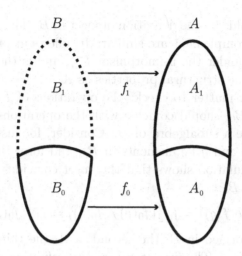

have the same size means that there is a bijection f_1 from B_1 to A_1. (See the diagram.) Take B to be the union of the sets B_0 and B_1, and define a mapping f from B to A by

$$f(p) = \begin{cases} f_0(p) & \text{if } p \in B_0, \\ f_1(p) & \text{if } p \in B_1. \end{cases}$$

Then f is a well-defined bijection from B to A; the easy proof depends on the facts that f_0 and f_1 are bijections, and the sets B_0 and B_1 are disjoint, as are the sets A_0 and A_1. Turn B into a Boolean algebra by defining operations of meet, join, and complement that are the counterparts, under f^{-1}, of the operations of meet, join, and complement in A. More precisely, to form the meet, join, and complements of two elements p and q in B, translate these elements to A using the mapping f, form the meet, join, and complements of the translations in A, and then translate the results back to B using f^{-1}:

$$p \wedge q = f^{-1}(f(p) \wedge f(q)),$$
$$p \vee q = f^{-1}(f(p) \vee f(q)),$$
$$p' = f^{-1}(f(p)').$$

The operations on the right sides of the equations are those of A, while the ones on the left are the operations that are being defined on B. Under these definitions, f automatically becomes a Boolean isomorphism from B to A. For instance, f preserves meet because

$$f(p \wedge q) = f(f^{-1}(f(p) \wedge f(q))) = f(p) \wedge f(q).$$

The first equality holds by the definition of meet in B. The arguments that f preserves join and complement are similar. It follows at once that B is the inverse image of A under the isomorphism f^{-1}, and is therefore a Boolean algebra with the same structural properties as A.

There is another matter to check: the operations of B, when restricted to the elements of B_0, should coincide with the operations of B_0. In other words, B_0 should be a subalgebra of B. Consider, for instance, the operation of meet. Let p and q be elements in B_0, and form their meet in B; a straightforward calculation shows that this meet coincides with the meet of the two elements in B_0:

$$p \wedge q = f^{-1}(f(p) \wedge f(q)) = f_0^{-1}(f_0(p) \wedge f_0(q)) = f_0^{-1}(f_0(p \wedge q)) = p \wedge q.$$

(The first meet is formed in B, the second in A, the third in A_0, and the fourth and fifth in B_0.) The first step uses the definition of meet in B; the second step uses the fact that on elements of B_0, the mapping f coincides with f_0, and on elements of A_0, the mapping f^{-1} coincides with f_0^{-1}; and the third step uses the isomorphism properties of f_0. Join and complement are handled in a similar fashion.

The function f maps B isomorphically to A. Its restriction to the subalgebra B_0 coincides with f_0, and therefore maps B_0 isomorphically to A_0. The discussion of the exchange principle is now complete.

An isomorphism between Boolean algebras preserves every infinite supremum and infimum that happens to exist, but in general a mere homomorphism will not do so. A homomorphism f is called *complete* in case it preserves all suprema (and, consequently, all infima) that happen to exist. This means that if $\{p_i\}$ is a family of elements in the domain of f with supremum p, then the family $\{f(p_i)\}$ has a supremum and that supremum is equal to $f(p)$. There is an interesting connection between complete monomorphisms and regular subalgebras: a monomorphism is complete if and only if its range is a regular subalgebra. For later use, here is a more precise formulation.

Lemma 1. *A Boolean monomorphism f from B into A is complete if and only if the image of B under f is a regular subalgebra of A.*

Proof. The image of B under the monomorphism f is certainly a subalgebra of A; denote it by C. It is to be shown that C is a regular subalgebra of A just in case f is complete.

Assume first that C is a regular subalgebra of A. To prove that f is complete, consider an arbitrary family $\{p_i\}$ of elements in B with a supremum p.

The mapping f is an isomorphism from B to C, so $f(p)$ is certainly the supremum of $\{f(p_i)\}$ in C. It follows that $f(p)$ is also the supremum of $\{f(p_i)\}$ in A, since C is a regular subalgebra of A.

For the reverse implication, suppose f is complete. To prove that C is a regular subalgebra of A, consider a family $\{q_i\}$ of elements in C with a supremum q in C; it is to be shown that q is also the supremum of the family in A. Since f is one-to-one, there are uniquely determined elements p_i and p in B such that

$$f(p_i) = q_i \qquad \text{and} \qquad f(p) = q.$$

The element p is the supremum of the family $\{p_i\}$ in B, because f is an isomorphism from B to C. The completeness of f ensures that $f(p)$ is the supremum of $\{f(p_i)\}$ in A, and this directly implies the desired conclusion.

Exercises

1. Let f be the mapping from the Boolean algebra of finite and cofinite subsets of an infinite set into 2 that takes each finite set to 0 and each cofinite set to 1. Verify that f is a Boolean homomorphism.

2. Let f be a Boolean homomorphism from B into A, and suppose C is a Boolean subalgebra of B. Prove that the restriction of f to C is a homomorphism from C into A.

3. Let f be a mapping from a Boolean algebra B onto an algebraic structure A with binary operations $\wedge$ and $\vee$, and a unary operation $'$. Complete the proof that if f satisfies the identities (1)–(3), then A must be a Boolean algebra with zero element $f(0)$ and unit $f(1)$.

4. Prove that if f is a Boolean homomorphism, then $f(1) = 1$.

5. Prove that a degenerate Boolean algebra cannot be mapped homomorphically into a non-degenerate one.

6. Prove that a mapping between Boolean algebras that preserves join and complement is a Boolean homomorphism.

7. Prove that a mapping between Boolean algebras that preserves the Sheffer stroke is a Boolean homomorphism.

8. Is every ring homomorphism between Boolean algebras a Boolean homomorphism? What if it preserves 1?

9. If a mapping f between Boolean algebras preserves 0, 1, $\wedge$, and $\vee$, is it necessarily a Boolean homomorphism?

10. If a mapping f between Boolean algebras preserves order, is it necessarily a Boolean homomorphism?

11. If a bijection f between Boolean algebras satisfies the *order-preserving equivalence*

$$p \leq q \qquad \text{if and only if} \qquad f(p) \leq f(q)$$

for all p and q in its domain, is f necessarily a Boolean isomorphism?

12. A *dual isomorphism* between Boolean algebras is a bijection f between the algebras that satisfies the following conditions for all p and q in its domain:

$$f(p \wedge q) = f(p) \vee f(q), \qquad f(p \vee q) = f(p) \wedge f(q), \qquad f(p') = f(p)'.$$

Prove that a bijection f is a dual isomorphism if and only if it satisfies the *order-reversing equivalence*

$$p \leq q \qquad \text{if and only if} \qquad f(p) \geq f(q)$$

for all p and q in its domain.

13. Suppose f is a bijection between two lattices. Show that the following conditions on f are equivalent:

   (a) $f(p \wedge q) = f(p) \wedge f(q)$ for all p and q in the domain of f;
   (b) $f(p \vee q) = f(p) \vee f(q)$ for all p and q in the domain of f;
   (c) $p \leq q$ if and only if $f(p) \leq f(q)$ for all p and q in the domain of f.

A bijection satisfying one of these three conditions is called a (*lattice*) *isomorphism*.

14. Suppose f is a bijection between two lattices. Show that the following conditions on f are equivalent:

   (a) $f(p \wedge q) = f(p) \vee f(q)$ for all p and q in the domain of f;
   (b) $f(p \vee q) = f(p) \wedge f(q)$ for all p and q in the domain of f;

   (c) $p \leq q$ if and only if $f(p) \geq f(q)$ for all p and q in the domain
   of f.

   A bijection satisfying one of these three conditions is called a *dual*
   (*lattice*) *isomorphism*.

15. If A, B, and C are lattices, and if f and g are (lattice) isomorphisms
    from B to A, and from C to B, respectively, then the composition $f \circ g$ is
    an isomorphism from C to A. What can be said when f and g are dual
    isomorphisms (Exercise 14)? What if one of them is an isomorphism
    and the other a dual isomorphism?

16. Suppose that both f and g are A-valued homomorphisms on B. Define
    a mapping $f \vee g$ from B into A by

$$(f \vee g)(p) = f(p) \vee g(p).$$

    Is $f \vee g$ a homomorphism? What about $f + g$ (defined similarly)?

17. If A is a relativization of a Boolean algebra B, prove that a subset of A
    has a supremum in B if and only if it has a supremum in A, and if these
    suprema exist, then they are equal. Conclude that a relativization of a
    complete Boolean algebra is necessarily complete.

18. If A is a relativization of a Boolean algebra B, prove that an element
    from A is an atom in A if and only if it is an atom in B.

19. The notion of a relativization of a Boolean algebra can be somewhat
    extended. If B is a subalgebra of a Boolean algebra A, and if p_0 is
    an element of A that is not necessarily in B, define the notion of the
    relativization of B to p_0. Show that this relativization is a homomorphic
    image of B and a subalgebra of the relativization of A to p_0.

20. Prove that a Boolean isomorphism maps every atom in the domain to
    an atom in the range. Is the same true of a Boolean monomorphism?

21. Prove that the Boolean algebra of periodic sets of integers of period
    three is isomorphic to the Boolean algebra of subsets of $\{0, 1, 2\}$.

22. If sets X and Y have the same number of elements, prove that the field
    of finite and cofinite subsets of X is isomorphic to the field of finite and
    cofinite subsets of Y.

23. Let A be the interval algebra of the real numbers, and let E be the subset of A consisting of the intervals $[n, n+1)$, where n ranges over the integers. Prove that the subalgebra of A generated by E is isomorphic to the field of finite and cofinite sets of integers. (See Exercise 11.14.)

24. Let A be the field of finite and cofinite sets of integers, and B the subfield of A consisting of the finite sets of even integers and their complements. Are A and B isomorphic?

25. Are the field of all sets of real numbers and the Boolean algebra of regular open sets of real numbers isomorphic?

26. Consider a Boolean homomorphism f from B into A. Prove that if B_0 is a subalgebra of B, then the image

$$f(B_0) = \{f(p) : p \in B_0\}$$

is a subalgebra of A. Prove also that if A_0 is a subalgebra of A, then the inverse image

$$f^{-1}(A_0) = \{p \in B : f(p) \in A_0\}$$

is a subalgebra of B. Conclude that if f is an epimorphism, then a subset A_0 of A is a subalgebra if and only if there is a subalgebra B_0 of B such that

$$A_0 = f(B_0).$$

27. Consider a Boolean isomorphism f from B into A, an arbitrary element q_0 in B, and the image element $p_0 = f(q_0)$ in A. Prove that the appropriate restriction of f maps the relativization $B(q_0)$ isomorphically to the relativization $A(p_0)$.

28. Prove that if the range of an A-valued homomorphism f includes a set of generators of A, then f is an epimorphism.

29. If f is a Boolean epimorphism from B to A, and if E is a set of generators of B, prove that $\{f(p) : p \in E\}$ is a set of generators of A.

30. Prove that if E generates B, and if f and g are A-valued homomorphisms on B such that $f(p) = g(p)$ whenever p is in E, then $f = g$. What if B is the complete algebra generated by E, and f and g are complete homomorphisms?

31. Prove that if E generates B, and if p_0 is an arbitrary element of B, then the set

$$F = \{q \wedge p_0 : q \in E\}$$

generates the relativization of B to p_0.

32. Prove that a necessary and sufficient condition for an arbitrary mapping f from a set B to a set A, and an arbitrary mapping g from A to B, to be bijections and inverses of one another is that the compositions $f \circ g$ and $g \circ f$ be the identity functions on A and B respectively.

33. If f is a Boolean homomorphism from B to A, and g a Boolean homomorphism from C to B, prove that the composition $f \circ g$ is a Boolean homomorphism from C to A.

34. Let f be a Boolean homomorphism. If a family $\{p_i\}$ of elements in the domain of f has a supremum p, show that $f(p)$ is an upper bound for the family $\{f(p_i)\}$. Conclude that if $\bigvee_i f(p_i)$ exists, then

$$\bigvee_i f(p_i) \le f(p).$$

35. Prove that a Boolean isomorphism preserves all suprema and infima that happen to exist.

36. Prove that a complete homomorphism preserves all infima that exist.

37. If a Boolean homomorphism preserves all infima that happen to exist, prove that the homomorphism must be complete.

38. Prove that the following condition is necessary and sufficient for a Boolean homomorphism f to be complete: whenever a family $\{p_i\}$ of elements in the domain of f has the infimum 0, then the family $\{f(p_i)\}$ in the range of f has the infimum 0. (Compare this exercise with Exercise 11.22.)

39. Formulate and prove the dual to Exercise 38.

40. (Harder.) Give an example of an incomplete homomorphism between complete Boolean algebras. Can such an example be a monomorphism? An epimorphism?

41. Prove that a subalgebra B of a Boolean algebra A is regular if and only if the identity mapping of B into A is a complete homomorphism.

42. Prove that if a subalgebra B of a complete Boolean algebra A happens to be complete (considered as an algebra in its own right), then a necessary and sufficient condition that B be a complete subalgebra of A is that the identity mapping of B into A be a complete homomorphism.

43. (Harder.) Show that if a complete homomorphism has a complete domain, then its range is a regular subalgebra. What if the domain is not complete?

44. (Harder.) Can Lemma 1 be generalized to arbitrary homomorphisms? In other words, is the range of a homomorphism a regular subalgebra if and only if the homomorphism is a complete homomorphism?

Chapter 13

Extensions of Homomorphisms

A Boolean homomorphism f is called an *extension* of a Boolean homomorphism g if the domain of g is a subalgebra of the domain of f, and if

$$f(p) = g(p)$$

for every p in the domain of g. If f is an extension of every member of a family of homomorphisms, then f is said to be a *common extension* of the family. A family of Boolean homomorphisms does not, in general, have a common extension; there is, however, a special case when such an extension does exist. Call a family $\{f_i\}$ of A-valued homomorphisms *directed* if any two homomorphisms f_i and f_j in the family have a common extension f_k in the family.

Lemma 1. *A directed family of A-valued homomorphisms always has a common extension to an A-valued homomorphism. If the homomorphisms in the family are one-to-one, then so is the common extension.*

Proof. Let $\{f_i\}$ be a directed family of A-valued homomorphisms. The domains of these homomorphisms form a directed family of Boolean algebras, by the assumption that any two homomorphisms in the family $\{f_i\}$ have a common extension in the family. The union B of the domains is a Boolean algebra, and each domain is a subalgebra of B (see Exercise 11.14). Define a mapping f on B by

$$f(p) = f_i(p)$$

S. Givant, P. Halmos, *Introduction to Boolean Algebras*,
Undergraduate Texts in Mathematics, DOI: 10.1007/978-0-387-68436-9_13,
© Springer Science+Business Media, LLC 2009

whenever p is in the domain of f_i. The mapping f is well defined in the sense that it does not depend on the particular choice of the homomorphism f_i. Indeed, suppose p is also in the domain of f_j. The two homomorphisms f_i and f_j have a common extension f_k, by assumption, so

$$f_i(p) = f_k(p) = f_j(p).$$

It is easy to verify that f is a homomorphism from B into A. If p and q are elements in B, then p is in the domain of some f_i, and q is in the domain of some f_j. The two mappings have a common extension f_k, by assumption; the elements p and q, together with their join, are in the domain of the common extension, and

$$f(p \vee q) = f_k(p \vee q) = f_k(p) \vee f_k(q) = f(p) \vee f(q),$$

by the definition of f and the homomorphism properties of f_k. The verification that f preserves complement is similar, but easier (there is no need to pass to a common extension f_k):

$$f(p') = f_i(p') = f_i(p)' = f(p)'.$$

An analogous argument shows that f is one-to-one whenever each mapping f_i is one-to-one.

The lemma applies, in particular, to families of homomorphisms that are linearly ordered by the relation of being an extension. Such families are called chains. More precisely, a family $\{f_i\}$ of A-valued homomorphisms is called a *(homomorphism) chain* if for any two members f_i and f_j of the family, one of them is an extension of the other.

The action of a homomorphism on a Boolean algebra is completely determined by its action on a generating set, as the following lemma shows.

Lemma 2. *If two A-valued homomorphisms on a Boolean algebra B agree on the elements of a generating set, they agree on all of B.*

Proof. Suppose B is generated by a set E. Let f and g be A-valued homomorphisms on B that agree on E; in other words,

$$f(p) = g(p)$$

for every element p in E. Define C to be the set of elements in B on which f and g agree. The set E is included in C, by assumption. Furthermore, if p and q belong to C, then so do $p \vee q$ and p', since

$$f(p \vee q) = f(p) \vee f(q) = g(p) \vee g(q) = g(p \vee q),$$

and

$$f(p') = f(p)' = g(p)' = g(p'),$$

by the definition of C and the homomorphism properties of f and g. Conclusion: C is a subalgebra of B that includes E. The algebra B is, by assumption, the smallest subalgebra of itself that includes E, so B must coincide with C. It follows that f and g agree on all of B.

The preceding lemma suggests the problem of finding necessary and sufficient conditions for an A-valued function on a set of generators of a Boolean algebra to have an extension to an A-valued homomorphism. The solution to the problem is usually called the *homomorphism extension criterion*, and is due to Roman Sikorski [63]. (It generalizes an earlier result of Kuratowski and Posament, given in Corollary 1 below.) The formulation of the criterion requires some notation that was introduced before. For each element i of a Boolean algebra, write

$$p(i,j) = \begin{cases} i & \text{if } j = 1, \\ i' & \text{if } j = 0. \end{cases}$$

Theorem 4. *A mapping g from a generating set E of a Boolean algebra B into a Boolean algebra A can be extended to a homomorphism from B into A just in case for every 2-valued function a on a finite subset F of E,*

$$\bigwedge_{i \in F} p(i, a(i)) = 0 \qquad \text{implies} \qquad \bigwedge_{i \in F} p(g(i), a(i)) = 0.$$

Proof. Assume that a mapping g from E into A satisfies the extension criterion formulated in the theorem. It must be shown that g can be extended to an A-valued homomorphism on B. Consider, first, the case when E is finite. The algebra B is then finite, its atoms are the non-zero elements of the form

$$p_a = \bigwedge_{i \in E} p(i, a(i)),$$

every atom of B can be written in exactly one way as such a meet, and every element in B is the join of a uniquely determined set of atoms, by Theorem 2 (p. 81). Completely analogous remarks apply to the subalgebra C of A generated by the set

$$g(E) = \{g(q) : q \in E\}.$$

This generating set is finite, so C is finite, its atoms are the non-zero elements of the form

$$q_a = \bigwedge_{i \in E} p(g(i), a(i)),$$

every atom of C can be written in exactly one way as such a meet, and every element in C is the join of a uniquely determined set of atoms.

Let K be the set of 2-valued mappings a on E such that p_a is non-zero. Each element r in B can be written in the form

$$r = \bigvee_{a \in X} p_a$$

for a uniquely determined subset X of K, by Theorem 2. Define a function f from B into C by writing

$$f(r) = \bigvee_{a \in X} q_a.$$

In other words, f takes $\bigvee_{a \in X} p_a$ to $\bigvee_{a \in X} q_a$ for each subset X of K.

If

$$r = \bigvee_{a \in X} p_a \qquad \text{and} \qquad s = \bigvee_{a \in Y} p_a,$$

then

$$r \vee s = \bigvee_{a \in X \cup Y} p_a,$$

by Theorem 2 (see, in particular, equation (11.7)), and therefore

$$f(r \vee s) = \bigvee_{a \in X \cup Y} q_a = \left(\bigvee_{a \in X} q_a \right) \vee \left(\bigvee_{a \in Y} q_a \right) = f(r) \vee f(s).$$

This argument shows that f preserves join.

The argument that f preserves complement is similar, but it makes use of the extension criterion. Let X be a subset of K, and write X' for the complement of X with respect to K. It follows from Theorem 2 (see, in particular, equation (11.8)) that the two elements

$$\bigvee_{a \in X} p_a \qquad \text{and} \qquad \bigvee_{a \in X'} p_a$$

are complements of one another. An analogous argument shows that

$$\bigvee_{a \in X} q_a \qquad \text{and} \qquad \bigvee_{a \in X'} q_a$$

are complements of one another. Technically, the notation X' in the final join refers to the complement of X with respect to the set L of all 2-valued mappings a on E such that q_a is not zero. The key point, however, is that L is a subset of K, by the extension criterion, and $q_a = 0$ whenever a is not in L, by the definition of L. Consequently, the last two joins remain complements of one another when X' is interpreted as referring to the complement of X in K. If

$$r = \bigvee_{a \in X} p_a, \quad \text{then} \quad r' = \bigvee_{a \in X'} p_a,$$

and therefore

$$f(r') = \bigvee_{a \in X'} q_a = \left(\bigvee_{a \in X} q_a \right)' = f(r)',$$

by the preceding observations. In other words, f preserves complement. Conclusion: f is a homomorphism from B into C, and therefore into A.

It remains to demonstrate that f agrees with g on E. An element r in E can be written as

$$r = p(r, b(r)),$$

where b is the function from $\{r\}$ into 2 that maps r to 1. The proof of Theorem 2 (equation (11.9) with $F = \{r\}$) shows that

$$r = \bigvee \{p_a : a \in K \text{ and } a \text{ extends } b\} = \bigvee \{p_a : a \in K \text{ and } a(r) = 1\}.$$

Similarly, the element $g(r)$ is in $g(E)$ and

$$g(r) = p(g(r), b(r)).$$

An analogous argument shows that

$$g(r) = \bigvee \{q_a : a \in K \text{ and } a \text{ extends } b\} = \bigvee \{q_a : a \in K \text{ and } a(r) = 1\}.$$

(Technically, equation (11.9) yields the preceding identities for $g(r)$ with the set L — described in the preceding paragraph — in place of K. However, L is a subset of K, and $q_a = 0$ for a in $K - L$, so L may be replaced by K without affecting the validity of the identities.) These observations combine with the definition of f to give

$$f(r) = \bigvee \{q_a : a \in K \text{ and } a(r) = 1\} = g(r).$$

The proof that g can be extended to an A-valued homomorphism when the generating set E is finite is thus complete.

Consider, next, the case of an arbitrary generating set E. For each finite subset F of E, let B_F be the subalgebra of B generated by F, and let g_F be the restriction of the mapping g to F. Because g satisfies the extension criterion, its restriction g_F must satisfy a restricted extension criterion in which the set E is replaced by F. The first part of the proof shows that there is an A-valued homomorphism f_F on B_F that extends g_F. The family of homomorphisms

$$\{f_F : F \text{ is a finite subset of } E\}$$

is directed. Indeed, consider two finite subsets F and G of E. Their union

$$H = F \cup G$$

is also a finite subset of E. The subalgebra B_H includes B_F, because its generating set H includes the generating set F of B_F. In other words, the domain of f_H includes the domain of f_F. Both f_H and f_F agree with g on the elements of F, so f_F and the restriction of f_H to B_F must be equal, by Lemma 2. This just means that f_H extends f_F. The proof that f_H extends f_G is similar.

Every directed family of homomorphisms has a common extension, by Lemma 1. There is consequently a homomorphism f that extends each of the homomorphisms f_F. The domain of f is the union, over all finite subsets F of E, of the directed family of subalgebras B_F. This union is a subalgebra of B, by Lemma 11.1, and it includes the generating set E, because it includes each finite subset of E. The union must therefore coincide with B. Conclusion: f is an A-valued homomorphism on B that agrees with g on each finite subset of E, and therefore agrees with g on all of E.

The converse implication of the theorem is easier to establish. Suppose a mapping g from E into A can be extended to an A-valued homomorphism f on B. It is to be shown that g necessarily satisfies the extension criterion. A simple example should suffice to illustrate the general argument. Suppose F consists of three elements in E, say r, s, and t, and suppose a is the 2-valued function on F that assigns the value 1 to s, and the value 0 to both r and t. Then

$$\bigwedge_{i \in F} p(i, a(i)) = r' \wedge s \wedge t'$$

and

$$\bigwedge_{i \in F} p(g(i), a(i)) = g(r)' \wedge g(s) \wedge g(t)'.$$

The homomorphism properties of f, and the fact that f extends g, imply

$$f(r' \wedge s \wedge t') = f(r)' \wedge f(s) \wedge f(t)' = g(r)' \wedge g(s) \wedge g(t)'.$$

If, therefore,

$$r' \wedge s \wedge t' = 0, \qquad \text{then} \qquad g(r)' \wedge g(s) \wedge g(t)' = 0,$$

since $f(0) = 0$.

In the case of an arbitrary finite subset F of E, and an arbitrary 2-valued function a on F, write

$$p_a = \bigwedge_{i \in F} p(i, a(i)) \qquad \text{and} \qquad q_a = \bigwedge_{i \in F} p(g(i), a(i)).$$

An argument similar to the one in the preceding paragraph shows that

$$f(p_a) = q_a.$$

When $p_a = 0$, the homomorphism properties of f imply $q_a = 0$, so that the extension criterion is satisfied.

A minor addition to the preceding argument yields a *monomorphism extension criterion* that, in its application to fields of sets, goes back to Kuratowski and Posament [39].

Corollary 1. *A mapping g from a generating set E of a Boolean algebra B into a Boolean algebra A can be extended to a monomorphism from B into A just in case, for every 2-valued function on a finite subset F of E,*

$$\bigwedge_{i \in F} p(i, a(i)) = 0 \qquad \text{if and only if} \qquad \bigwedge_{i \in F} p(g(i), a(i)) = 0.$$

Proof. Assume the monomorphism extension criterion is satisfied, and consider first the case when the generating set E is finite. Recall from the preceding proof that K is the set of 2-valued functions a on E such that

$$p_a = \bigwedge_{i \in E} p(i, a(i))$$

is not zero. Every element in B can be written in one and only one way as a join of elements (and actually atoms) p_a with a from K, by the definition of K. The monomorphism criterion implies that K is also the set of functions a such that

$$q_a = \bigwedge_{i \in E} p(g(i), a(i))$$

is not zero, and consequently that every element in the subalgebra of A generated by $g(E)$ can be written in one and only one way as a join of elements (and actually atoms) q_a with a in K. In other words, for subsets X and Y of K,

$$\bigvee_{a \in X} q_a = \bigvee_{a \in Y} q_a \qquad \text{if and only if} \qquad X = Y,$$

$$\text{if and only if} \qquad \bigvee_{a \in X} p_a = \bigvee_{a \in Y} p_a.$$

The extension homomorphism f is defined to map $\bigvee_{a \in X} p_a$ to $\bigvee_{a \in X} q_a$, so it must be one-to-one.

Consider now the case of an arbitrary generating set E. The preceding argument shows that for each finite subset F of E, the extension homomorphism f_F from B_F (the subalgebra of B generated by F) into A is one-to-one. Since the common extension of a directed family of one-to-one homomorphisms is one-to-one, by Lemma 1, the common extension f of the directed family of monomorphisms $\{f_F\}$ must be one-to-one.

To prove the converse direction of the corollary, assume f is a monomorphism from B into A that extends the mapping g. We saw in the preceding proof that $f(p_a) = q_a$ for each 2-valued function a on a finite subset of E. Because $f(0) = 0$, it follows from the one-to-oneness of f that $p_a = 0$ if and only if $q_a = 0$. In other words, the monomorphism extension criterion is satisfied.

Suppose C is a Boolean subalgebra of B. It is natural to look for conditions under which a homomorphism g from C into A can be extended to a homomorphism from B into A. One such condition is very easy to formulate: it suffices that the Boolean algebra A be complete. To prove this assertion it is helpful to establish first a special case. Call B a *one-step (Boolean) extension* of C if B is generated by $C \cup \{r\}$ for some element r. (The case when r is in C is not excluded, and in this case $B = C$.)

Lemma 3. *A Boolean homomorphism into a complete Boolean algebra can be extended to any one-step extension of its domain.*

Proof. Let g be a homomorphism from a Boolean algebra C into a complete Boolean algebra A, and suppose B is a one-step extension of C generated, say, by $C \cup \{r\}$. The goal is to find an element s in A such that the homomorphism extension criterion is satisfied when s is taken for $g(r)$ and $E = C \cup \{r\}$. The set C is closed under complements and meets, so the extension criterion can be formulated more simply: for all elements p in C,

$$p \wedge r = 0 \quad \text{implies} \quad g(p) \wedge s = 0$$

and

$$p \wedge r' = 0 \quad \text{implies} \quad g(p) \wedge s' = 0.$$

The first implication says that $r \leq p'$ implies $s \leq g(p')$; because C is closed under complement, this is equivalent to saying that $r \leq q$ implies $s \leq g(q)$. (Recall, in this connection, that $g(p)' = g(p')$, by the homomorphism properties of g.) The second implication says that $p \leq r$ implies $g(p) \leq s$. We are therefore looking for an element s in A such that

(1) $$p \leq r \leq q \quad \text{implies} \quad g(p) \leq s \leq g(q)$$

for all p and q in C.

Write

$$P = \{g(p) : p \in C \text{ and } p \leq r\} \quad \text{and} \quad Q = \{g(q) : q \in C \text{ and } r \leq q\}.$$

The supremum s_1 of P, and the infimum s_2 of Q, both exist in the complete algebra A. The homomorphism properties of g imply that $g(p) \leq g(q)$ whenever p and q are elements of C with $p \leq r \leq q$. Every element of Q is therefore an upper bound of P. The element s_1 is the least upper bound of P, so it is below every element of Q. In other words, s_1 is a lower bound of Q. The greatest lower bound of Q is s_2, so $s_1 \leq s_2$. Take s to be any element in A satisfying $s_1 \leq s \leq s_2$. Then

$$g(p) \leq s_1 \leq s \leq s_2 \leq g(q)$$

whenever p and q are elements in C satisfying the hypothesis of (1).

The desired element s has been found. The discussion in the first paragraph of the proof shows that the homomorphism extension criterion is satisfied. Theorem 4 now guarantees the existence of a homomorphism f from B into A that agrees with g on C and maps r to s.

Each choice of an element s in A satisfying condition (1) determines a homomorphism from B into A that extends g and maps r to s. Different choices for s lead to different extensions, so the extension homomorphism f is in general not unique.

The next theorem, due to Sikorski [58], is usually called the *homomorphism extension theorem*. It says that homomorphisms into complete Boolean algebras can always be extended.

Theorem 5. *A Boolean homomorphism into a complete Boolean algebra can be extended to any Boolean extension of its domain.*

Proof. Let g be a homomorphism from a Boolean algebra C into a complete Boolean algebra A, and let B be a Boolean extension of C. Enumerate the elements of B in a (possibly) transfinite sequence $\{p_i\}_{i<\alpha}$ indexed by the set of ordinals less than a given ordinal number α. Define a corresponding transfinite sequence $\{f_i\}_{i\leq\alpha}$ of A-valued homomorphisms with the following properties: (1) $f_0 = g$; (2) f_j is an extension of f_i whenever $j \geq i$; (3) p_i is in the domain of f_{i+1}. We shall write B_i for the domain of f_i.

The definition of the sequence of homomorphisms proceeds by transfinite induction on ordinals. The base case is completely determined by condition (1): put $f_0 = g$; then f_0 is an A-valued homomorphism, by assumption, and conditions (2) and (3) hold vacuously. For the induction step, consider an ordinal $k \leq \alpha$, and suppose A-valued homomorphisms f_i have been defined for $i < k$ so that the family $\{f_i\}_{i<k}$ satisfies conditions (1)–(3). When k is a successor ordinal, say $k = i + 1$, take B_k to be the one-step extension of B_i generated by $B_i \cup \{p_i\}$. The homomorphism f_i from B_i into A can be extended to a homomorphism f_k from B_k into A, by Lemma 3. When k is a limit ordinal, invoke Lemma 1 (for chains) to obtain an A-valued homomorphism f_k that is a common extension of the homomorphisms f_i for $i < k$. The domain of f_k is the union of family of domains $\{B_i\}_{i<k}$ (this union is a subalgebra of B, by Lemma 11.1); f_k is defined at each element p in its domain by

$$f_k(p) = f_i(p)$$

whenever p is in B_i. This completes the construction of the family $\{f_i\}_{i\leq\alpha}$.

The algebra B coincides with the domain of f_α. Indeed, if p is any element of B, then $p = p_i$ for some ordinal i, so p is in the domain of f_{i+1}, by condition (3). Consequently, p is in the domain of f_α, by condition (2). The mapping f_α is a homomorphism of B_α into A, by construction. It extends each homomorphism f_i, by condition (2), so it extends g, by condition (1).

Exercises

1. Prove that the common extension of a directed family of one-to-one A-valued homomorphisms is one-to-one.

2. Prove directly, without using Lemma 1, that a chain of A-valued homomorphisms has a common extension to an A-valued homomorphism.

3. Suppose the set E in Theorem 4 consists of just two elements, say u and v. Use the definition of the function f in the proof of the theorem to show directly that

$$f(u) = g(u), \quad f(u \wedge v) = g(u) \wedge g(v), \quad f(u \vee v) = g(u) \vee g(v),$$
$$f(v) = g(v), \quad f(u \wedge v') = g(u) \wedge g(v)', \quad f(u \vee v') = g(u) \vee g(v)',$$
$$f(u') = g(u)', \quad f(u' \wedge v) = g(u)' \wedge g(v), \quad f(u' \vee v) = g(u)' \vee g(v),$$
$$f(v') = g(v)', \quad f(u' \wedge v') = g(u)' \wedge g(v)', \quad f(u' \vee v') = g(u)' \vee g(v)'.$$

4. Let g be a mapping from a subset E of a Boolean algebra B into a Boolean algebra A. Prove that when the set E is finite, the homomorphism extension criterion,

$$\bigwedge_{i \in F} p(i, a(i)) = 0 \quad \text{implies} \quad \bigwedge_{i \in F} p(g(i), a(i)) = 0$$

for all finite subsets F of E, is equivalent to a restricted version in which the implication is assumed to hold only for the set $F = E$:

$$\bigwedge_{i \in E} p(i, a(i)) = 0 \quad \text{implies} \quad \bigwedge_{i \in E} p(g(i), a(i)) = 0.$$

5. Let g be a mapping from a subset E of a Boolean algebra B into a Boolean algebra A. Prove that when the set E is finite, the monomorphism extension criterion,

$$\bigwedge_{i \in F} p(i, a(i)) = 0 \quad \text{if and only if} \quad \bigwedge_{i \in F} p(g(i), a(i)) = 0$$

for all finite subsets F of E, is equivalent to a restricted version in which the equivalence is assumed to hold only for the set $F = E$:

$$\bigwedge_{i \in E} p(i, a(i)) = 0 \quad \text{if and only if} \quad \bigwedge_{i \in E} p(g(i), a(i)) = 0.$$

6. Let $E = \{p_1, p_2, p_3, \dots\}$ be a countably infinite subset of a Boolean algebra B, and g a mapping from E into a Boolean algebra A. Write

$$E_n = \{p_1, p_2, \dots, p_n\}$$

for each positive integer n. Prove that the homomorphism extension criterion is equivalent to a restricted version in which the equivalence is assumed to hold only for the sets $F = E_n$:

$$\bigwedge_{i \in E_n} p(i, a(i)) = 0 \qquad \text{implies} \qquad \bigwedge_{i \in E_n} p(g(i), a(i)) = 0$$

for all positive integers n.

7. Formulate and prove the analogue of Exercise 6 for the monomorphism extension criterion.

8. (Harder.) Construct the homomorphism f in the proof of Lemma 3 directly, without using the homomorphism extension criterion and without making implicit use of the directed family construction from Theorem 4.

Chapter 14

Atoms

The most natural field of subsets of a set is the field of all its subsets. Does that field have a simple algebraic characterization? The answer is yes; the purpose of this chapter is to exhibit such a characterization.

An *atom* of a Boolean algebra is an element that has no non-trivial proper subelements. Better: q is an atom if $q \neq 0$ and if there are only two elements p such that $p \leq q$, namely 0 and q. A typical example of an atom is a singleton in a field of sets. In the Boolean algebra 2^X, the atoms are the characteristic functions of singletons — the functions that map a single element of X to 1 and the remaining elements of X to 0.

There are a number of characterizations of atoms. Here are some of the more useful ones.

Lemma 1. *The following conditions on an element q in a Boolean algebra are equivalent:*

(1) *q is an atom;*

(2) *for every element p, either $q \leq p$ or $q \wedge p = 0$, but not both;*

(3) *for every element p, either $q \leq p$ or $q \leq p'$, but not both;*

(4) *$q \neq 0$, and if q is below a join $p \vee r$, then $q \leq p$ or $q \leq r$;*

(5) *$q \neq 0$, and if q is below the supremum of a family $\{p_i\}$, then q is below p_i for some i.*

Proof. The proofs of most implications are automatic. To see, for example, that (2) implies (5), argue by contraposition. Let p be the supremum of a

S. Givant, P. Halmos, *Introduction to Boolean Algebras*,
Undergraduate Texts in Mathematics, DOI: 10.1007/978-0-387-68436-9_14,
© Springer Science+Business Media, LLC 2009

family $\{p_i\}$, and suppose q is not below any element p_i. Then $q \wedge p_i = 0$ for each index i, by condition (2), so $q \wedge p = 0$, by Lemma 8.3. In other words, either $q = 0$ or q is not below p.

An element in a Boolean algebra may not be the supremum of a set of atoms, but if it is, that set is uniquely determined.

Lemma 2. *If an element p in a Boolean algebra is the supremum of a set of atoms E, then E is the set of all atoms below p.*

Proof. Assume p is the supremum of a set of atoms E. Certainly, every element in E is below p, by the definition of a supremum. If r is an arbitrary atom below p, then r is below some atom q in E, by Lemma 1. It follows from the minimality of atoms that $r = q$, and therefore that r is in E.

A Boolean algebra is said to be *atomic* if every non-zero element dominates at least one atom. A Boolean algebra is *atomless* if it has no atoms. (Note that these two concepts are not just the negations of one another.) A field of sets is usually (but not always) atomic: the field of all subsets, or the finite–cofinite algebra, of a set are obvious examples. A counterexample is the interval algebra of the real numbers; it is atomless. The regular open algebra of a topological space X is quite likely to be atomless; the absence of separation axioms and the presence of isolated points is likely to introduce atoms (see, for example, Exercise 29.31).

The next lemma essentially goes back to Tarski [71].

Lemma 3. *The following conditions on a Boolean algebra are equivalent.*

(1) *The algebra is atomic.*

(2) *Every element is the supremum of the atoms it dominates.*

(3) *The unit is the supremum of the set of all atoms.*

Proof. The statement in (2) is intended to convey the information that the supremum in question always exists (without any assumption of completeness). Observe also that even the zero element does not have to be excluded from the statement. Now for the proof that (1) implies (2): begin with the trivial comment that each element p is an upper bound of the set, say E, of the atoms that it dominates. It is to be proved that if r is an arbitrary upper bound of E, then $p \leq r$. Assume that, on the contrary, $p - r \neq 0$. It follows from the assumption of atomicity that there exists an atom q with $q \leq p - r$.

Since $p - r \leq p$, the atom q belongs to E, and is therefore below r. Consequently,

$$q \leq (p - r) \wedge r = p \wedge r' \wedge r = 0,$$

and this contradicts the fact that q is an atom.

The implication from (2) to (3) is trivial. To prove that (3) implies (1), let E be the set of all atoms, and p an arbitrary non-zero element, of the Boolean algebra. Then

$$p = p \wedge 1 = p \wedge \bigvee E = \bigvee \{p \wedge q : q \in E\},$$

by Lemma 8.3. Because the element p is not zero, the preceding equalities show that there is at least one atom q in E for which $p \wedge q \neq 0$. For such an atom q, we must have $p \wedge q = q$, by the minimality of atoms, and therefore $q \leq p$.

The axioms of Boolean algebra were selected in order to capture the basic properties of fields of sets. Do they fulfill this task? There are a number of ways to answer this question. One way is to answer another question, the so-called *representation problem*: is every Boolean algebra isomorphic to a field of sets? An isomorphism from a Boolean algebra A to a field of sets is called a *representation* of A. Equivalently, a representation of A (over a set X) is an embedding of A into the field $\mathcal{P}(X)$. A representation that preserves all existing suprema as unions (and hence all existing infima as intersections) is said to be *complete*. In other words, a representation f is complete if whenever $\{p_i\}$ is a family of elements with a supremum p, then

$$f(p) = \bigcup_i f(p_i).$$

In the case when f represents A over a set X, the preceding condition is equivalent to saying that f is a complete monomorphism of A into $\mathcal{P}(X)$.

We shall see later that the representation problem has a positive solution. For now, we prove a special representation theorem (essentially due to Tarski [71]) for atomic Boolean algebras.

Theorem 6. *Let A be an atomic Boolean algebra, and X its set of atoms. The correspondence*

$$p \rightarrow \{q \in X : q \leq p\}$$

is a complete representation of A over X.

Proof. Let f be the mapping on A that takes each element p to the set of atoms below p. It is easy to see that f is one-to-one: if $f(p) = f(r)$, then p and r dominate the same set of atoms, say E, and consequently

$$p = \bigvee E = r,$$

by Lemma 3.

The proof that f preserves arbitrary suprema is only slightly more involved. Let $\{p_i\}$ be a family of elements in A with supremum p. It is to be shown that $f(p)$ is the union of the sets $f(p_i)$. For an arbitrary atom q in A,

$$
\begin{array}{lll}
q \in f(p) & \text{if and only if} & q \leq p, \\
& \text{if and only if} & q \leq p_i \quad \text{for some } i, \\
& \text{if and only if} & q \in f(p_i) \quad \text{for some } i, \\
& \text{if and only if} & q \in \bigcup_i f(p_i).
\end{array}
$$

The first and third equivalences use the definition of f, the second equivalence uses Lemma 1, and the last equivalence uses the definition of the union of a family of sets. Since $f(p)$ and $\bigcup_i f(p_i)$ are sets of atoms, it follows that

$$f(p) = \bigcup_i f(p_i),$$

as desired.

The preceding argument establishes, in particular, that f preserves the join of any two elements. A similar argument shows that f preserves complement: for any atom q in A,

$$
\begin{array}{lll}
q \in f(p') & \text{if and only if} & q \leq p', \\
& \text{if and only if} & q \nleq p, \\
& \text{if and only if} & q \notin f(p), \\
& \text{if and only if} & q \in f(p)'.
\end{array}
$$

The first and third equivalences use the definition of f, the second equivalence uses Lemma 1, and the last equivalence uses the definition of the complement of a set. Since $f(p')$ and $f(p)'$ are sets of atoms, it follows that

$$f(p') = f(p)'.$$

The preceding theorem can be used to give an algebraic characterization (due to Tarski [71]) of the Boolean algebras that are isomorphic to a field of all subsets of some set.

Corollary 1. *A necessary and sufficient condition that a Boolean algebra A be isomorphic to the field of all subset of some set is that A be complete and atomic.*

Proof. The necessity of the conditions is obvious: every field of all subsets of a set is complete and atomic, and so is every isomorphic copy of such a field. Suppose now that A is a complete and atomic Boolean algebra, and let X be the set of all atoms of A. The mapping f that takes each element in A to the set of atoms it dominates is a complete monomorphism of A into $\mathcal{P}(X)$, by the preceding theorem. It remains to check that f maps A onto $\mathcal{P}(X)$. If E is an arbitrary subset of X, then the assumed completeness of A implies that E has a supremum p in A. Since E is a set of atoms with supremum p, it must be the set of all atoms below p, by Lemma 2. Consequently, $f(p) = E$, by the definition of f.

A closer examination of the proof of Theorem 6 and its corollary reveals that only two properties of the Boolean algebra A and the field $\mathcal{P}(X)$ are used to establish their isomorphism: they are complete, and they have the same number of atoms. Thus, a more general theorem is actually true.

Corollary 2. *Two complete, atomic Boolean algebras with the same number of atoms are isomorphic. In fact, every bijection between the sets of atoms extends to an isomorphism between the algebras.*

Proof. Let A and B be complete and atomic Boolean algebras with sets of atoms X and Y respectively. Then A is isomorphic to the field $\mathcal{P}(X)$ via the mapping f that takes each element of A to the set of atoms it dominates. Similarly, B is isomorphic to the field $\mathcal{P}(Y)$ via the mapping g that takes each element of B to the set of atoms it dominates. Suppose that the two algebras have the same number of atoms, so that X and Y have the same cardinality. Let ϕ be any bijection from X to Y, and take h to be the isomorphism from $\mathcal{P}(Y)$ to $\mathcal{P}(X)$ induced by ϕ (see Chapter 12):

$$h(Q) = \phi^{-1}(Q) = \{p \in X : \phi(p) \in Q\}$$

for each subset Q of Y. The composition

$$k = g^{-1} \circ h^{-1} \circ f$$

is an isomorphism from A to B:

$$A \xrightarrow{\ f\ } \mathcal{P}(X) \xleftarrow{\ h\ } \mathcal{P}(Y) \xleftarrow{\ g\ } B.$$

For each atom q of A,

$$k(q) = g^{-1}(h^{-1}(f(q))) = g^{-1}(h^{-1}(\{q\})) = g^{-1}(\{\phi(q)\}) = \phi(q),$$

by the definition of f, h, and g. Therefore, k extends the bijection ϕ.

It is occasionally useful to have a more direct formulation of the definition of the isomorphism k from the preceding proof. The isomorphism f maps each element p in A to the set P of atoms below p, and the isomorphism g maps each element q in B to the set Q of atoms below q. The inverse isomorphism g^{-1} therefore maps the set of atoms Q in B to its supremum:

$$g^{-1}(Q) = \bigvee Q = q.$$

The isomorphism h^{-1} maps the set P of atoms in A to the corresponding set in B, under the bijection ϕ:

$$h^{-1}(P) = \phi(P) = \{\phi(r) : r \in P\} = \{\phi(r) : r \text{ is an atom in } A \text{ and } r \leq p\}.$$

These observations lead easily to an explicit formula for $k(p)$:

$$k(p) = g^{-1}(h^{-1}(f(p))) = g^{-1}(h^{-1}(P)) = g^{-1}(\phi(P))$$
$$= \bigvee \phi(P) = \bigvee \{\phi(r) : r \text{ is an atom in } A \text{ and } r \leq p\}.$$

In words, k maps each element p to the supremum of the images, under the bijection ϕ, of the atoms below p.

In general, an infinite Boolean algebra is neither complete, nor completely distributive. An exception is provided by the field of all subsets of an infinite set, which has both properties. In fact, these two properties actually characterize the Boolean algebras that are isomorphic to the field of all subsets of some set. (The characterization is due to Tarski [71].)

Theorem 7. *A Boolean algebra is isomorphic to the field of all subsets of a set if and only if it is complete and completely distributive.*

Proof. The field of all subsets of a set is complete and completely distributive, so every isomorphic copy of such a field has the same properties. To prove the converse, let A be a complete and completely distributive Boolean algebra. By Corollary 1, it suffices to prove that A is atomic. The proof is a kind of infinitary version of part of the proof of Theorem 2.

Take I to be A, and write

$$p(i,j) = \begin{cases} i & \text{if } j = 1, \\ i' & \text{if } j = 0, \end{cases}$$

for each i in I. Since

$$\bigvee_{j \in 2} p(i,j) = p(i,1) \vee p(i,0) = i \vee i' = 1$$

for every i, it follows that

$$\bigwedge_{i \in I} \bigvee_{j \in 2} p(i,j) = 1,$$

and consequently that

$$\bigvee_{a \in 2^I} \bigwedge_{i \in I} p(i, a(i)) = 1,$$

by the assumed complete distributivity (8.3) of A. The proof will be completed by demonstrating that each non-zero element of the form $\bigwedge_{i \in I} p(i, a(i))$ is an atom; the preceding equation then implies that the unit is a sum of atoms, so A is atomic, by Lemma 3.

Suppose, accordingly, that

$$q = \bigwedge_{i \in I} p(i, a(i)) \neq 0.$$

Notice that q is a well-determined element, by the assumption that A is complete. Let r be an arbitrary element of A. If $a(r) = 1$, then

$$p(r, a(r)) = p(r, 1) = r$$

and therefore $q \leq r$. If $a(r) = 0$, then

$$p(r, a(r)) = p(r, 0) = r'$$

and therefore $q \leq r'$. Both inequalities cannot hold simultaneously, since q is not zero. It now follows from Lemma 1 that q is an atom.

Theorem 7 and Corollary 1 together yield the following conclusion (due to Lindenbaum and Tarski; see [71]).

Corollary 3. *A complete Boolean algebra is atomic if and only if it is completely distributive.*

A subalgebra of an atomic Boolean algebra need not be atomic. For example, the interval algebra of the real numbers is a subfield of the field of all sets of reals numbers, but it is not atomic. This situation cannot arise when the subalgebra in question is a regular subalgebra (as was observed by Hirsch and Hodkinson [26]).

Lemma 4. *A regular subalgebra of an atomic Boolean algebra is atomic.*

Proof. Let A be an atomic Boolean algebra, and B a regular subalgebra of A. It is to be shown that for every non-zero element p in B, there is an atom in B below p. The algebra A is atomic, by assumption, so it contains an atom q that is below p. Write

$$E = \{r \in B : q \leq r \text{ (in } A)\}.$$

The set E contains p, since $q \leq p$. Furthermore, for each element r in B, exactly one of r and r' is in E; indeed, q is an atom, so it is below exactly one of r and r', by Lemma 1. Finally, and most importantly, the set E has a non-zero lower bound in B. To see this, assume to the contrary that no such lower bound exists. The element 0 is clearly a lower bound of E, so it must be the greatest lower bound of E in B. Consequently, 0 is the greatest lower bound of E in A, by the assumption that B is a regular subalgebra of A. This contradicts the fact that q is a non-zero lower bound of E in A.

Let s be a non-zero lower bound of E in B. For each element r of B, exactly one of r and r' is in E, and therefore exactly one of these two elements is above s, by the definition of E. Thus, s is an atom of B, by Lemma 1. Also, s is below p, since p is in E.

Theorem 6 guarantees that an atomic Boolean algebra — complete or not — always has a complete representation. The converse (due to Hirsch and Hodkinson [25]) is also true.

Theorem 8. *A Boolean algebra with a complete representation is necessarily atomic.*

Proof. Let A be a Boolean algebra, and suppose f is a complete representation of A, say over the set X. This means that f is a complete monomorphism of A into the field $\mathcal{P}(X)$. Write B for the range of A under f. The assumption that f is complete implies that B is a regular subalgebra of $\mathcal{P}(X)$, by Lemma 12.1. The field $\mathcal{P}(X)$ is obviously atomic, so the regular subalgebra B must be atomic, by the preceding lemma. Since A is isomorphic to B via the mapping f, it must also be atomic.

Does every Boolean algebra have a complete representation? The preceding theorem implies that the answer is, in general, negative. An algebra with such a representation would have to be atomic, and as we have already seen, there are many Boolean algebras that are very far from being atomic. In fact, the Boolean algebra of regular open subsets of $\mathbb{R}^n$ and the interval algebra of the real line are both atomless.

The notion of an atom was first introduced by Schröder in §47 of [52], under the name "Individuum". The term "atom" appears to be due to Tarski [71].

Exercises

1. What are the atoms of the field of sets of integers of period four (see Chapter 5, and in particular Exercise 5.3)? Show directly that every element in this field is a join of atoms.

2. Write out a complete proof of Lemma 1.

3. Prove that a field of all subsets of a set is complete and completely distributive.

4. Prove that a Boolean algebra is atomless if and only if for each non-zero element q, there is a non-zero element p that is strictly below q (in the sense that $p \leq q$ and $p \neq q$).

5. Prove directly (without using Corollary 3 or the results of Chapter 11) that every finite Boolean algebra is atomic. (This theorem is due to Huntington [28]. Since a finite algebra is obviously complete and completely distributive, Corollary 3 would yield the desired conclusion at once. The conclusion is too elementary, however, to deserve such a relatively high-powered treatment.)

6. Prove directly (without using either Theorem 6 and its corollaries, or the results of Chapter 11) that the total number of elements in every finite Boolean algebra is a power of 2, and that two finite Boolean algebras with the same number of elements must be isomorphic. (This theorem is due to Huntington [28].)

7. (Harder.) Prove that if p is a non-zero element of an atomic Boolean algebra A, then there exists a 2-valued homomorphism f on A such that $f(p) = 1$.

8. (Harder.) Characterize the topological spaces whose regular open algebras are (1) atomic, (2) atomless.

9. (Harder.) Does the set of all atoms in a Boolean algebra always have a supremum?

10. (Harder.) Show that the hypothesis of completeness in Corollary 3 is superfluous. In other words, show that an arbitrary Boolean algebra is atomic if and only if it is completely distributive. (This improvement is pointed out in Horn and Tarski [27], which refers to the proof of Theorem 7 given in Birkhoff [7].)

11. Use Exercise 12 to give another proof of Theorem 8. Conclude that the following conditions are equivalent in a Boolean algebra A: (1) A is atomic; (2) A is completely distributive; (3) A is completely representable.

Chapter 15

Finite Boolean Algebras

The observations of the previous chapter yield a complete description (due to Huntington [28]) of all finite Boolean algebras. In formulating this description it is helpful to use the von Neumann definition of the natural number n as the set $\{0, 1, \ldots, n-1\}$.

Lemma 1. *A finite Boolean algebra is atomic.*

Proof. It is to be shown that every non-zero element p is above an atom. If p itself is an atom, we are done. If not, then there must be a non-zero element p_1 strictly below p. If p_1 is an atom, then again we are done. If not, there must be a non-zero element p_2 strictly below p_1, and so on. Eventually this process must lead to an atom below p; otherwise, the Boolean algebra would have an infinite, strictly descending chain of elements, contradicting the assumption that the algebra is finite.

The number of atoms in a finite Boolean algebra uniquely determines the isomorphism type of the algebra.

Corollary 1. *Every finite Boolean algebra A is isomorphic to the field $\mathcal{P}(n)$, or, equivalently, to the Boolean algebra 2^n, for some non-negative integer n. In fact, n is the number of atoms in A.*

Proof. A finite Boolean algebra A is atomic, by the preceding lemma, and it is obviously complete. Let n be the number of atoms in A. The field $\mathcal{P}(n)$ is also atomic with the same number of atoms — its atoms are the singletons $\{0\}, \ldots, \{n-1\}$ — and it is also complete. Two complete, atomic Boolean algebras with the same number of atoms are isomorphic, by Corollary 14.2. Consequently, A and $\mathcal{P}(n)$ are isomorphic; in fact, any bijection

S. Givant, P. Halmos, *Introduction to Boolean Algebras*,   127
Undergraduate Texts in Mathematics, DOI: 10.1007/978-0-387-68436-9_15,
© Springer Science+Business Media, LLC 2009

between the sets of atoms extends to an isomorphism between the two algebras. The assertion that A and 2^n are isomorphic follows from the observation (Chapter 3) that $\mathcal{P}(n)$ and 2^n are isomorphic via the mapping that takes each subset of n to its characteristic function.

How many Boolean algebras are there of a given finite size? If the size is not a power of two, the answer is zero, by the preceding corollary. If the size is a power of two, the answer is one, provided that isomorphic Boolean algebras are treated as being the same. Certainly, there is a Boolean algebra of size 2^n, namely the Boolean algebra 2^n. That there is only one such algebra, up to isomorphic copies, follows from the next corollary.

Corollary 2. *Two finite Boolean algebras with the same number of elements are isomorphic.*

Proof. Consider two finite Boolean algebras A and B with the same number of elements. Both algebras are atomic, by Lemma 1; say A has m atoms and B has n atoms. It follows from Corollary 1 that A is isomorphic to the Boolean algebra 2^m, and B is isomorphic to the Boolean algebra 2^n. The two algebras are assumed to have the same size, so we must have $m = n$. Consequently, A and B are isomorphic to the same Boolean algebra, so they are isomorphic to each other.

Simply put, the finite Boolean algebras are, up to isomorphic copies, precisely the algebras $2^0, 2^1, 2^2, 2^3, \ldots$. These algebras can be thought of as forming a subalgebra chain, since the algebra 2^m can be embedded into the algebra 2^n whenever $m \leq n$. The proof of this last observation is perhaps easiest to grasp if one considers the fields $\mathcal{P}(m)$ and $\mathcal{P}(n)$ instead of the corresponding powers of 2. Let ϕ be the mapping from n to m defined by

$$\phi(i) = \begin{cases} i & \text{if} \quad 0 \leq i < m, \\ m - 1 & \text{if} \quad m \leq i < n. \end{cases}$$

Then ϕ induces a homomorphism from $\mathcal{P}(m)$ to $\mathcal{P}(n)$, namely the function f determined by

$$f(P) = \phi^{-1}(P) = \begin{cases} P & \text{if} \quad m - 1 \notin P, \\ P \cup \{m, \ldots, n - 1\} & \text{if} \quad m - 1 \in P, \end{cases}$$

for each set P in $\mathcal{P}(m)$ (see Chapter 12). Notice that

$$f(P) \cap m = P.$$

Consequently, if $f(P) = f(Q)$, then

$$P = f(P) \cap m = f(Q) \cap m = Q,$$

which proves that f is one-to-one, and hence a monomorphism.

Axioms (2.11)–(2.20) were intended to axiomatize the laws true in fields of sets. It is natural to ask whether they actually accomplish this task. In other words, is an identity that is true in all fields of sets necessarily derivable from the axioms? The answer is affirmative; in fact, the axioms of Boolean algebra are strong enough to imply all the laws that are true in the two-element Boolean algebra.

Theorem 9. *A Boolean identity is derivable from axioms* (2.11)–(2.20) *if and only if the identity is true in the Boolean algebra* 2.

Proof. Before starting the proof proper, it will be helpful to make some preliminary observations.

A Boolean term is an expression built up from variables and the constant symbols 0 and 1 using the symbols $\vee$, $\wedge$, and $'$ that denote the operations of join, meet, and complement, and using parentheses. Every Boolean term is provably equivalent to a term ρ of the form

(1) $(\rho_{11} \vee \cdots \vee \rho_{1n_1}) \wedge (\rho_{21} \vee \cdots \vee \rho_{2n_2}) \wedge \cdots \wedge (\rho_{m1} \vee \cdots \vee \rho_{mn_m}),$

where each term ρ_{ij} is either one of the constants 0, 1, or a variable, or the complement of a variable. In other words, for each Boolean term σ one can construct a Boolean term ρ of the form (1) such that the identity $\sigma = \rho$ is derivable from the axioms of Boolean algebra.

To construct ρ from σ, first apply the De Morgan laws (2.17) repeatedly to move all occurrences of the complement symbol in σ inward to the variables and constants; use the double complement law (2.15) to cancel two occurrences of complement that are next to each other; and use (2.11) to get rid of any occurrence of a complement symbol that is next to a constant. In this manner, we arrive at an intermediate term δ that is built up from the constant symbols 0 and 1, from variables, and from complements of variables, using only the symbols $\vee$ and $\wedge$ and parentheses (but not complement), and such that the equation $\sigma = \delta$ is derivable (from the Boolean axioms). Next, apply repeatedly the distributive laws (2.20) to δ (as well as the commutative laws (2.18), and associative laws (2.19)) to write all joins of meets as meets of joins, and in this way arrive at a term ρ of the form (1) such that the equation $\delta = \rho$ is derivable. It follows that the equation $\sigma = \rho$ is also derivable.

Consider now a term ρ of the form (1). Suppose first that in every subterm of ρ of the form

$$(2) \qquad\qquad \rho_{i1} \vee \cdots \vee \rho_{in_i}$$

either one of the terms ρ_{ij} is 1, or else one of the terms ρ_{ij} is a variable and another of the terms, say ρ_{ik}, is the complement of that same variable. Then each equation

$$\rho_{i1} \vee \cdots \vee \rho_{in_i} = 1$$

is derivable with the help of (2.12) and the complement laws (2.14); therefore, the equation $\rho = 1$ is derivable, using also the identity laws (2.13).

On the other hand, if in some subterm of ρ of the form (2) there is no occurrence of the constant 1, and there are also no terms ρ_{ij} and ρ_{ik} such that ρ_{ij} is a variable and ρ_{ik} is the complement of that variable, then it is possible to find an assignment of the values 0 and 1 to the variables of ρ that makes the given subterm (2), and hence also ρ, evaluate to 0 in the algebra 2; whenever ρ_{ij} is a variable of the given subterm, assign to that variable the value 0 everywhere in ρ, and whenever ρ_{ij} is the complement of a variable in the given subterm, assign to that variable the value 1 everywhere in ρ. Since a variable and its complement do not occur simultaneously in (2), by assumption, this prescription does not lead to conflicting assignments. Assign to all other variables occurring in ρ (the ones that do not occur in the given subterm) one of the values 0 and 1 (it does not matter which). Under this assignment, the given subterm (2) evaluates to 0 (in the algebra 2), by the laws (2.11) and (2.13). Consequently, ρ itself, under this assignment, evaluates to 0, by the laws (2.12).

It has been shown that, for every term ρ of the form (1), either the equation $\rho = 1$ is derivable from the Boolean axioms, or there is an assignment of the values 0 and 1 to its variables that makes ρ evaluate to 0 in the algebra 2. Suppose now that an arbitrary Boolean identity is given. It is an equation of the form $\sigma = \tau$, where σ and τ are Boolean terms. It is easy to show that $\sigma = \tau$ is provably equivalent to the identity

$$(\sigma \wedge \tau) \vee (\sigma' \wedge \tau') = 1$$

in the sense that each identity is derivable from the other one on the basis of the Boolean axioms. The term $(\sigma \wedge \tau) \vee (\sigma' \wedge \tau')$ is provably equivalent to a term ρ of the form (1), so the given identity $\sigma = \tau$ is provably equivalent to the identity $\rho = 1$. We have seen that the identity $\rho = 1$ is either derivable from the axioms or else there is an assignment of values 0 and 1 to its variables that makes ρ evaluate to 0, and consequently that makes the identity $\rho = 1$ fail, in the algebra 2. It follows that the original identity $\sigma = \tau$ is either provable from the axioms, or there is an assignment of values 0 and 1 to its variables that makes it fail in 2.

For an illustration of how the preceding theorem may be applied, consider the equation

$$p' \vee (q' \wedge (q' \vee p)) = (p \wedge q)'.$$

Straightforward computations, using the definitions of the operations of join, meet, and complement in the Boolean algebra 2, show that this equation is true in 2 for all possible values of p and q (see the table below); consequently, the equation must be derivable from axioms (2.11)–(2.20), by the theorem.

p	q	$p' \vee (q' \wedge (q' \vee p))$	$(p \wedge q)'$
1	1	0	0
1	0	1	1
0	1	1	1
0	0	1	1

Theorem 9 is a Boolean algebraic version of the completeness theorem for propositional logic, due to Post [50]. It implies that an identity is true in one non-degenerate Boolean algebra if and only if it is true in every other non-degenerate Boolean algebra; hence, the same identities are true of all non-degenerate Boolean algebras. (The degenerate algebra must be excluded from consideration, because every identity is trivially true in it.)

Corollary 3. *The same set of identities is true in every non-degenerate Boolean algebra.*

Proof. Let A be an arbitrary non-degenerate Boolean algebra. The corollary will follow if it can be shown that an identity is true in A if and only if it is true in 2. An identity is a universal assertion about the elements and operations of a Boolean algebra; if it holds in the algebra A, then it must hold in every subalgebra B, since the elements of B are among the elements of A, and the operations of B are restrictions of the operations of A. Since A is assumed to be non-degenerate, it includes (a copy of) 2 as a subalgebra. Consequently, every identity true in A is also true in 2. Conversely, if an identity is true in 2, then it is derivable from the Boolean axioms, by the previous theorem, and therefore it must be true in every Boolean algebra, including A.

Theorem 9 and its corollary yield the following surprising conclusion: if an identity is true in some (any) non-degenerate Boolean algebra — and in particular if it is true in some non-degenerate field of sets — then it is true in the Boolean algebra 2, and therefore it is derivable from axioms (2.11)–(2.20).

Exercises

1. Prove directly that 2^m is embeddable into 2^n. In other words, define an appropriate mapping from 2^m into 2^n, and prove that the mapping is an embedding.

2. For each Boolean term σ below, use the algorithm described in the proof of Theorem 9 to find a term ρ of the form (1) such that $\sigma = \rho$ is derivable from axioms (2.11)–(2.20). Show that in each case the term is provably equal to 1 or else there is an assignment of the values 0 and 1 to the variables so that the term evaluates to 0 in the Boolean algebra 2.

   (a) $(p \vee q) \wedge (r \vee 0 \vee q')'$.

   (b) $[((0' \vee p) \wedge q)' \vee (p' \vee 0)']'$.

   (c) $(p \wedge q \wedge r')' \vee (p' \wedge q) \vee (p \wedge r')$.

3. Determine whether the equation

$$[p \wedge (p \wedge p')] \vee [p' \wedge (p \wedge p')'] = p'$$

   is derivable from axioms (2.11)–(2.20) by checking the validity of the equation in the Boolean algebra 2.

4. (Harder.) Let m and n be natural numbers with $m \leq n$. Describe all of the embeddings of $\mathcal{P}(m)$ into $\mathcal{P}(n)$.

5. (Harder.) If a bijection f between two finite Boolean algebras preserves order in the sense that

$$p \leq q \qquad \text{implies} \qquad f(p) \leq f(q),$$

   is f necessarily an isomorphism? (Compare this exercise with Exercises 12.10 and 12.11.)

6. Show that the Boolean identities $\sigma = \tau$ and

$$(\sigma \wedge \tau) \vee (\sigma' \wedge \tau') = 1$$

are provably equivalent. In other words, show that the first identity is derivable from the second, and the second is derivable from the first, on the basis of the Boolean axioms.

7. Theorem 9 has a dual proof. First, one shows that every Boolean term is provably equivalent to a term ρ of the form

$$(\rho_{11} \wedge \cdots \wedge \rho_{1n_1}) \vee (\rho_{21} \wedge \cdots \wedge \rho_{2n_2}) \vee \cdots \vee (\rho_{m1} \wedge \cdots \wedge \rho_{mn_m}),$$

where each term ρ_{ij} is either one of the constants 0, 1, or a variable, or the complement of a variable. Second, one shows that if ρ has this form, then either the equation $\rho = 0$ is derivable from the Boolean axioms, or else there is an assignment of the values 0 and 1 to its variables that makes ρ evaluate to 1 in the algebra 2. Third, one shows that an arbitrary Boolean identity $\sigma = \tau$ is provably equivalent to the identity

$$(\sigma \wedge \tau') \vee (\sigma' \wedge \tau) = 0.$$

Finally, one argues that the identity $\sigma = \tau$ is either derivable from the axioms, or there is some assignment of values to its variables that makes it false in 2. Write out the details of this proof.

Chapter 16

Atomless Boolean Algebras

The discussion in Chapter 14 focused on atomic Boolean algebras. At the other extreme lie the atomless Boolean algebras, which have no atoms at all. The degenerate (one-element) Boolean algebra is vacuously atomless (and vacuously atomic); it has no atoms because it has no non-zero elements. Interval algebras provide examples of non-degenerate atomless Boolean algebras. For instance, the interval algebra of the real numbers is atomless, and so is its subalgebra consisting of the finite unions of left half-closed intervals with endpoints that are rational numbers (or $\pm\infty$). Notice that this last algebra is countable. Quite surprisingly, it is the only possible example of a countable atomless Boolean algebra that is not degenerate, at least up to isomorphic copies. The purpose of the present chapter is to present a proof (apparently due to Tarski — see [84], footnote 21) of this assertion.

The technique employed in the proof is a *back-and-forth argument* that finds applications in more advanced parts of the theory of Boolean algebras, and in other areas of mathematics as well. It goes back to Cantor [13], with a refinement by Huntington [29], and was used by them to prove that, up to isomorphic copies, the rational numbers are the only example of a denumerable dense linear order without endpoints.

Every atomless Boolean algebra with more than one element must be infinite. Indeed, the unit 1 is different from zero, so there is a non-zero element p_1 strictly below 1; otherwise, 1 would be an atom. Because p_1 is not zero, there must be a non-zero element p_2 strictly below p_1; otherwise, p_1 would be an atom. Continue in this fashion to produce an infinite, strictly decreasing sequence of elements $1 > p_1 > p_2 > \cdots$.

S. Givant, P. Halmos, *Introduction to Boolean Algebras*,
Undergraduate Texts in Mathematics, DOI: 10.1007/978-0-387-68436-9_16,
© Springer Science+Business Media, LLC 2009

Theorem 10. *Any two countable, atomless Boolean algebras with more than one element are isomorphic.*

Proof. Let A and B be countable, atomless Boolean algebras with more than one element. The assumption of countability implies that the elements of each algebra can be enumerated in a sequence indexed by the positive integers (or, equivalently, by any infinite subset of the positive integers). It will simplify the notation of the back-and-forth argument to enumerate the elements of A using even indices, and the elements of B using odd indices, say, $p_2, p_4, p_6, \ldots$ and $q_1, q_3, q_5, \ldots$.

A bit of auxiliary notation will be needed. As usual, for each element i of a Boolean algebra write

$$p(i,j) = \begin{cases} i & \text{if } j = 1, \\ i' & \text{if } j = 0. \end{cases}$$

Also, put

$$I_n = \{1, 2, \ldots, n\}$$

for every positive integer n.

The principal part of the proof involves the construction of elements p_n in A, for odd n, and elements q_n in B, for even n, such that the correspondence taking p_i to q_i for $i = 1, \ldots, n$ satisfies an equivalent version of the monomorphism extension criterion, namely

$$(1) \qquad \bigwedge_{i \in I_n} p(p_i, a(i)) = 0 \qquad \text{if and only if} \qquad \bigwedge_{i \in I_n} p(q_i, a(i)) = 0$$

for all positive integers n and all 2-valued functions a on I_n (see Exercise 13.7). Once this construction is accomplished, Corollary 13.1 may be applied to obtain an isomorphism g from the subalgebra of A generated by the family $\{p_n\}$ to the subalgebra of B generated by the family $\{q_n\}$ such that

$$g(p_n) = q_n$$

for each n. The former family includes, in particular, the elements p_n with even indices n, so it includes every element of A. Similarly, the latter family includes the elements q_n with odd indices n, so it includes every element of B. It follows that g is an isomorphism from A to B, as desired.

The construction of the elements p_n for odd n, and q_n for even n, uses a back-and-forth argument. Let g_0 be the isomorphism from the trivial subalgebra of A to the trivial subalgebra of B. For $n = 1$, the element q_1 is assumed to be in B, and an element p_1 in A is to be defined so that criterion (1) is satisfied. In this case, the criterion involves two conditions:

$$p_1 = 0 \quad \text{if and only if} \quad q_1 = 0,$$

and

$$p_1' = 0 \quad \text{if and only if} \quad q_1' = 0.$$

If q_1 is zero or the unit, take $p_1 = g_0^{-1}(q_1)$; the criterion is obviously satisfied because g_0 is an isomorphism. If q_1 is not zero or the unit, take p_1 to be any element of A different from zero and the unit; the criterion is vacuously satisfied, because neither p_1 nor q_1 is 0 or 1. The mapping that takes p_1 to q_1 can therefore be extended to an isomorphism g_1 from the subalgebra of A generated by p_1 to the subalgebra of B generated by q_1, by Corollary 13.1.

Next, suppose $n = 2$. The element p_2 in A is given, and an element q_2 in B must be chosen so that criterion (1) is satisfied. In the present case, the criterion involves four conditions:

$$
\begin{array}{lll}
p_1 \wedge p_2 = 0 & \text{if and only if} & q_1 \wedge q_2 = 0, \\
p_1 \wedge p_2' = 0 & \text{if and only if} & q_1 \wedge q_2' = 0, \\
p_1' \wedge p_2 = 0 & \text{if and only if} & q_1' \wedge q_2 = 0, \\
p_1' \wedge p_2' = 0 & \text{if and only if} & q_1' \wedge q_2' = 0.
\end{array}
$$

Choose an element $x \le q_1$ as follows: if the meet $p_1 \wedge p_2$ is generated (in A) by p_1, take $x = g_1(p_1 \wedge p_2)$; otherwise, take x to be any non-zero element strictly below q_1. Such a selection is always possible: if the meet is not generated by p_1, then the meet cannot be zero, and therefore p_1 cannot be zero; this implies (by the argument of the previous paragraph) that q_1 is not zero, so the assumption that B is atomless ensures the existence of the required element x. Choose an element $y \le q_1'$ in a completely analogous way: if $p_1' \wedge p_2$ is generated by p_1, take $y = g_1(p_1' \wedge p_2)$; otherwise, take y to be any non-zero element strictly below q_1'. Again, the assumption that B is atomless ensures the existence of such a y.

It is not difficult to check that

$$
\begin{array}{lll}
x = 0 & \text{if and only if} & p_1 \wedge p_2 = 0, \\
x = q_1 & \text{if and only if} & p_1 \wedge p_2' = 0, \\
y = 0 & \text{if and only if} & p_1' \wedge p_2 = 0, \\
y = q_1' & \text{if and only if} & p_1' \wedge p_2' = 0.
\end{array}
$$

The first equivalence follows readily from the isomorphism properties of g_1 and the definition of x. In more detail, if x is zero, then the definition of x implies that

$$g_1(p_1 \wedge p_2) = x = 0;$$

therefore $p_1 \wedge p_2$ is zero, by the isomorphism properties of g_1. On the other hand, if $p_1 \wedge p_2$ is zero, then it is certainly generated by p_1, and therefore

$$x = g_1(p_1 \wedge p_2) = g_1(0) = 0,$$

by the definition of x and the isomorphism properties of g_1.

The proof of the second equivalence is similar, but slightly more involved. If $x = q_1$, then x must have been defined by the clause $x = g_1(p_1 \wedge p_2)$; since $q_1 = g_1(p_1)$, it follows that

$$g_1(p_1) = q_1 = x = g_1(p_1 \wedge p_2),$$

and therefore that $p_1 = p_1 \wedge p_2$ (because g_1 is one-to-one). On the other hand, if this last equation holds, then $p_1 \wedge p_2$ is certainly generated by p_1, and therefore

$$x = g_1(p_1 \wedge p_2) = g_1(p_1) = q_1.$$

Thus,

$$x = q_1 \qquad \text{if and only if} \qquad p_1 \wedge p_2 = p_1.$$

The equation on the right is equivalent to the inequality $p_1 \leq p_2$, and hence to the equation

$$p_1 \wedge p_2' = 0.$$

The third and fourth equivalences are established in a completely analogous way.

Write $q_2 = x \vee y$, and observe that

$$q_1 \wedge q_2 = x, \qquad\qquad q_1 \wedge q_2' = q_1 \wedge x',$$
$$q_1' \wedge q_2 = y, \qquad\qquad q_1' \wedge q_2' = q_1' \wedge y'.$$

For instance,

$$q_1' \wedge q_2 = q_1' \wedge (x \vee y) = (q_1' \wedge x) \vee (q_1' \wedge y) = 0 \vee y = y,$$

since $x \leq q_1$ and $y \leq q_1'$. Similarly,

$$q_1' \wedge q_2' = q_1' \wedge (x \vee y)' = q_1' \wedge x' \wedge y' = q_1' \wedge y',$$

since $x \leq q_1$, and therefore $q_1' \leq x'$.

It is not difficult to verify (1), using the observations of the preceding paragraphs. For example,

$$p_1' \wedge p_2 = 0 \qquad \text{if and only if} \qquad y = 0,$$
$$\text{if and only if} \qquad q_1' \wedge q_2 = 0.$$

Similarly,

$$p_1' \wedge p_2' = 0 \qquad \text{if and only if} \qquad y = q_1',$$
$$\text{if and only if} \qquad q_1' \wedge y' = 0,$$
$$\text{if and only if} \qquad q_1' \wedge q_2' = 0.$$

The second equivalence holds because $y \le q_1'$.

The construction of the elements p_n for even n, and q_n for odd n, in the general case is very similar to the preceding construction. Assume, as the induction hypothesis, that the sequences

$$p_1, p_2, \ldots, p_{n-1} \qquad \text{and} \qquad q_1, q_2, \ldots, q_{n-1}$$

have been defined so that criterion (1) holds with $n - 1$ in place of n. Corollary 13.1 then implies the existence of an isomorphism g_{n-1} from the subalgebra of A generated by the first sequence to the subalgebra of B generated by the second sequence such that $g_{n-1}(p_i) = q_i$ for $i = 1, \ldots, n - 1$.

Suppose n is even. An element q_n in B must be selected so that (1) is satisfied. Let K be the set of 2-valued functions on I_{n-1}. Each 2-valued function a on I_n is the extension of a unique function b in K, namely the restriction of a to I_{n-1}. Conversely, each function in K has exactly two extensions to a 2-valued function on I_n; one extension maps n to 1, and the other maps n to 0. Criterion (1) may therefore be reformulated in terms of functions in K. Write

$$p_b = \bigwedge_{i \in I_{n-1}} p(p_i, b(i)) \qquad \text{and} \qquad q_b = \bigwedge_{i \in I_{n-1}} p(q_i, b(i)).$$

In terms of this notation, (1) says that

$$p_b \wedge p_n = 0 \qquad \text{if and only if} \qquad q_b \wedge q_n = 0,$$

and

$$p_b \wedge p_n' = 0 \qquad \text{if and only if} \qquad q_b \wedge q_n' = 0,$$

for each function b in K. Notice that

$$g_{n-1}(p_b) = q_b,$$

by the isomorphism properties of g_{n-1}.

Associate with each b in K an element $x_b \leq q_b$ as follows: if $p_b \wedge p_n$ is generated by p_b, write

$$x_b = g_{n-1}(p_b \wedge p_n);$$

otherwise, take for x_b any non-zero element strictly below q_b. Such a selection is always possible: if $p_b \wedge p_n$ is not zero, then neither is p_b; the element q_b is therefore also not zero (since the isomorphism g_{n-1} maps p_b to q_b), so the assumption that B is atomless ensures the existence of the required element x_b.

It is not difficult to show that

$$x_b = 0 \qquad \text{if and only if} \qquad p_b \wedge p_n = 0,$$

and

$$x_b = q_b \qquad \text{if and only if} \qquad p_b \wedge p_n' = 0.$$

The argument is nearly identical to the one given above in the case $n = 2$.
Write

$$q_n = \bigvee_{c \in K} x_c,$$

and observe that

$$q_b \wedge q_n = x_b \qquad \text{and} \qquad q_b \wedge q_n' = q_b \wedge x_b'.$$

To prove this, recall that the elements q_b and q_c are disjoint for distinct functions b and c in K. (The functions b and c differ on some index i, so q_b is below one of the elements q_i and q_i', while q_c is below the other.) Also, each element x_c is below q_c, and therefore

$$q_b \wedge x_c \leq q_b \wedge q_c = 0.$$

It follows that

$$q_b \wedge q_n = q_b \wedge \bigvee_{c \in K} x_c = \bigvee_{c \in K}(q_b \wedge x_c) = q_b \wedge x_b = x_b.$$

Similarly, q_b is disjoint from x_c, and therefore below x_c', for $c \neq b$. Consequently, $q_b \wedge x_c' = q_b$. It follows that

$$q_b \wedge q_n' = q_b \wedge \left(\bigvee_{c \in K} x_c \right)' = q_b \wedge \bigwedge_{c \in K} x_c' = \bigwedge_{c \in K}(q_b \wedge x_c') = q_b \wedge x_b'.$$

The verification of (1), using the observations of the preceding paragraphs, is routine. For instance,

$$p_b \wedge p_n = 0 \qquad \text{if and only if} \qquad x_b = 0,$$
$$\text{if and only if} \qquad q_b \wedge q_n = 0.$$

Similarly,

$$p_b \wedge p_n' = 0 \qquad \text{if and only if} \qquad x_b = q_b,$$
$$\text{if and only if} \qquad q_b \wedge x_b' = 0,$$
$$\text{if and only if} \qquad q_b \wedge q_n' = 0.$$

The second equivalence holds because $x_b \leq q_b$.

When n is even, an element p_n in A must be selected so that (1) holds. The argument that such an element exists is symmetric to the preceding argument, and is left to the reader. This completes the proof of the theorem.

Exercises

1. Prove that the interval algebra of the real numbers is atomless. What is its cardinality?

2. Prove that the interval algebra of the rational numbers is atomless and countable.

3. For each non-negative integer n, let A_n be the field of periodic sets of integers of period 2^n (see Chapter 5). Show that the union of the fields A_n is a countable, atomless Boolean algebra.

4. Prove that the regular open algebra of the space of real numbers is atomless. What is its cardinality?

5. Find two atomless Boolean algebras that are not isomorphic. Can they have the same cardinality?

6. The proof of Theorem 10, in the case $n = 2$, asserts that

$$y = q_1' \qquad \text{if and only if} \qquad p_1' \wedge p_2' = 0,$$

and that

$$q_1 \wedge q_2 = x \qquad \text{and} \qquad q_1 \wedge q_2' = q_1 \wedge x'.$$

Prove these assertions, and show that

$$p_1 \wedge p_2 = 0 \qquad \text{if and only if} \qquad q_1 \wedge q_2 = 0,$$

and

$$p_1 \wedge p_2' = 0 \quad \text{if and only if} \quad q_1 \wedge q_2' = 0.$$

7. Give the details of the construction of the element p_3 in the proof of Theorem 10, and the verification of criterion (1) for the case $n = 3$.

8. Give the details of the construction of the element p_n in the proof of Theorem 10, and the verification of criterion (1) for the case of an arbitrary positive odd integer n.

Chapter 17

Congruences and Quotients

Congruences on algebras are a way of gluing elements of the algebra together to form structurally similar, but simpler algebras. The prototypical example is that of a modular congruence on the ring of integers. Define two integers p and q to be congruent modulo a fixed positive integer n if they have the same remainder upon division by n, or, what amounts to the same thing, if their difference $p - q$ is divisible by n. The notation

$$p \equiv q \mod n$$

is usually used to express this relation.

Congruence modulo n is a binary relation on the set of integers that is reflexive,

$$p \equiv p \mod n \quad \text{for all integers } p,$$

symmetric,

$$\text{if} \quad p \equiv q \mod n, \quad \text{then} \quad q \equiv p \mod n,$$

and transitive,

$$\text{if} \quad p \equiv q \mod n \quad \text{and} \quad q \equiv r \mod n, \quad \text{then} \quad q \equiv r \mod n.$$

A convenient way to express these three properties is to say that congruence modulo n is an *equivalence relation* on the set of integers. The relation has two further properties that are quite important: it preserves addition and multiplication. In more detail, whenever

$$p \equiv r \mod n \quad \text{and} \quad q \equiv s \mod n,$$

we also have

S. Givant, P. Halmos, *Introduction to Boolean Algebras*,
Undergraduate Texts in Mathematics, DOI: 10.1007/978-0-387-68436-9_17,
© Springer Science+Business Media, LLC 2009

$$p + q \equiv r + s \mod n \quad \text{and} \quad p \cdot q \equiv r \cdot s \mod n.$$

Congruence modulo n partitions the set of integers into n mutually disjoint subsets called equivalence classes: two integers are put into the same equivalence class just in case they are congruent modulo n. For instance, congruence modulo 2 partitions the integers into two equivalence classes, the even integers and the odd integers. Congruence modulo 3 partitions the integers into three equivalence classes, namely the sets of integers whose remainder upon division by 3 is 0, 1, or 2 respectively. Write $[p]_n$ for the equivalence class of an integer p modulo n, so that

$$[p]_n = \{q : p \equiv q \mod n\}.$$

The preservation conditions make it possible to define operations of addition and multiplication on the set of equivalence classes: for any two equivalence classes $[p]_n$ and $[q]_n$, define their sum and product by

$$[p]_n + [q]_n = [p+q]_n \quad \text{and} \quad [p]_n \cdot [q]_n = [p \cdot q]_n.$$

(The operations on the right sides of these equations are addition and multiplication of integers.) The set of equivalence classes under these operations is easily seen to form a ring, the ring of integers modulo n. When $n = 2$, we get an isomorphic copy of the Boolean ring 2.

The preceding construction, suitably modified, works for any algebraic structure, and in particular for Boolean algebras. A *Boolean congruence* (*relation*) is defined to be an equivalence relation on a Boolean algebra B that preserves the operations of meet, join, and complement. In other words, it is a binary relation Θ on B that is *reflexive, symmetric,* and *transitive* in the sense that

(1) $\qquad p \equiv p \mod \Theta \quad$ for all integers p,

(2) $\quad$ if $\quad p \equiv q \mod \Theta$, $\quad$ then $\quad q \equiv p \mod \Theta$,

(3) $\quad$ if $\quad p \equiv q \mod \Theta \quad$ and $\quad q \equiv r \mod \Theta$, $\quad$ then $\quad q \equiv r \mod \Theta$,

and such that whenever

(4) $\qquad p \equiv r \mod \Theta \quad$ and $\quad q \equiv s \mod \Theta$,

we also have

(5) $\qquad p \wedge q \equiv r \wedge s \mod \Theta,$

(6) $\qquad p \vee q \equiv r \vee s \mod \Theta,$

(7) $\qquad p' \equiv r' \mod \Theta.$

Other Boolean operations such as $+$ and $\Rightarrow$ are definable in terms of meet, join, and complement, so they are also preserved by Boolean congruences. In other words, if Θ is a Boolean congruence on B, and if (4) holds, then

$$p + q \equiv r + s \quad \text{mod } \Theta \qquad \text{and} \qquad p \Rightarrow q \equiv r \Rightarrow s \quad \text{mod } \Theta.$$

The proofs are easy computations based on (5)–(7). For example, to show that $\Rightarrow$ is preserved, use first (7) and then (6) (with p and r replaced by p' and r') to arrive at

$$p' \vee q \equiv r' \vee s \quad \text{mod } \Theta,$$

which, in view of the definition of $\Rightarrow$ in (6.3), is just the desired result.

The important property in the preceding argument was the definability of the Boolean operations in terms of meet, join, and complement. If an equivalence relation Θ on a Boolean algebra preserves enough Boolean operations so that all others are definable in terms of them, then Θ is a Boolean congruence. For instance, if Θ preserves join and complement (that is, if it satisfies (6) and (7)), then Θ also preserves meet and is therefore a Boolean congruence.

The *equivalence classes* of a congruence Θ on a Boolean algebra B are the sets of the form

$$p/\Theta = \{q : p \equiv q \quad \text{mod } \Theta\}.$$

The properties of reflexivity, symmetry, and transitivity imply that

(8) $\qquad\qquad p/\Theta = q/\Theta \qquad \text{if and only if} \qquad p \equiv q \quad \text{mod } \Theta.$

An easy consequence of this observation is that two equivalence classes of Θ are always either equal or disjoint.

Preservation conditions (5)–(7) make it possible to define operations $\wedge$, $\vee$, and $'$ on the set of all equivalence classes of Θ in the following way:

(9) $\qquad\qquad\qquad (p/\Theta) \wedge (q/\Theta) = (p \wedge q)/\Theta,$

(10) $\qquad\qquad\qquad (p/\Theta) \vee (q/\Theta) = (p \vee q)/\Theta,$

(11) $\qquad\qquad\qquad\qquad (p/\Theta)' = (p')/\Theta.$

(The operations on the right sides of the equations are those of the Boolean algebra B.) To show that these operations are well defined, it must be checked that the definitions do not depend on the particular choice of the elements in the equivalence classes that are being used to define the operations. For instance, to verify that $\wedge$ is well defined, suppose

$$p/\Theta = r/\Theta \quad \text{and} \quad q/\Theta = s/\Theta.$$

In view of (8), these two equations, when translated into the language of congruences, say that the conditions in (4) hold. It follows that condition (5) also holds, by the definition of a Boolean congruence. The translation of (5) into the language of equivalence classes says, by (8), that

$$(p \wedge q)/\Theta = (r \wedge s)/\Theta.$$

Conclusion: in definition (9) it does not matter whether p or r is used as a representative of the first equivalence class, nor does it matter whether q or s is used as a representative of the second equivalence class; all choices yield the same result.

The set of all equivalence classes of Θ is denoted by B/Θ. Under the operations defined by (9)–(11), this set becomes a Boolean algebra, the so-called *quotient of B modulo* Θ. A direct verification of this fact is not difficult; the validity of axioms (2.11)–(2.20) in B must be checked. To verify the commutative law for meet, for example, let p and q be elements of B. Then

$$(p/\Theta) \wedge (q/\Theta) = (p \wedge q)/\Theta = (q \wedge p)/\Theta = (q/\Theta) \wedge (p/\Theta);$$

the first and last equalities follow from the definition of meet in the quotient, while the middle equality follows from the validity of the commutative law in B. The other axioms are verified in a similar fashion.

There is another, more efficient way of proving that the quotient B/Θ is a Boolean algebra: it suffices to show that B/Θ is a homomorphic image of B, by the remarks in the second paragraph of Chapter 12. Define a mapping f from B onto B/Θ by

$$f(p) = p/\Theta.$$

Simple computations show that f satisfies conditions (12.1) and (12.3), and is therefore an epimorphism:

$$f(p \wedge q) = (p \wedge q)/\Theta = (p/\Theta) \wedge (q/\Theta) = f(p) \wedge f(q)$$

and

$$f(p') = (p')/\Theta = (p/\Theta)' = f(p)'.$$

The mapping f is called the *canonical homomorphism*, or the *projection*, from B onto B/Θ.

Here are two examples of Boolean congruences. For the first, let p_0 be an arbitrary element of a Boolean algebra B. Define a binary relation Θ on B as follows: two elements p and q in B are congruent modulo Θ if

(12) $$p \wedge p_0 = q \wedge p_0.$$

It is easy to check that Θ is an equivalence relation on B. Indeed, equation (12) obviously holds when $p = q$, so Θ is reflexive. If (12) holds for p and q, then it holds with p and q interchanged, so Θ is symmetric. Finally, if p and q are congruent modulo Θ, and also q and r, then

$$p \wedge p_0 = q \wedge p_0 \qquad \text{and} \qquad q \wedge p_0 = r \wedge p_0,$$

and consequently

$$p \wedge p_0 = r \wedge p_0.$$

It follows that p and r are congruent modulo Θ, so that Θ is transitive.

To show that Θ preserves meet and complement, assume (4) holds. Then

$$p \wedge p_0 = r \wedge p_0 \qquad \text{and} \qquad q \wedge p_0 = s \wedge p_0.$$

An application of the idempotent law for meet gives

$$p \wedge q \wedge p_0 = p \wedge p_0 \wedge q \wedge p_0 = r \wedge p_0 \wedge s \wedge p_0 = r \wedge s \wedge p_0.$$

Thus (5) holds, by the definition of Θ. The elements $p \wedge p_0$ and $r \wedge p_0$ belong to the relativization $B(p_0)$. Their assumed equality implies the equality of their complements in the relativization; in other words,

$$p' \wedge p_0 = r' \wedge p_0.$$

Thus, (7) holds.

For the second example of a Boolean congruence, let B be a field of subsets of some set X. Define two sets P and Q in B to be congruent if they differ by at most finitely many elements, that is, if they contain exactly the same elements with at most finitely many exceptions. In still other words, P and Q are defined to be congruent if their symmetric difference $P + Q$ is finite. The binary relation Θ defined in this way is easily seen to be reflexive and symmetric. It is reflexive because the set $P + P$ is empty and therefore finite. It is symmetric because the sets $P + Q$ and $Q + P$ are equal (so one is finite just in case the other is). The transitivity of Θ follows from the inclusion

$$P + R \subseteq (P + Q) \cup (Q + R)$$

(Exercise 7.9(a)): if $P + Q$ and $Q + R$ are finite, then so is their union, and hence also $P + R$, by the inequality. Similarly, the preservation condition for join follows from the inclusion

$$(P \cup Q) + (R \cup S) \subseteq (P + R) \cup (Q + S)$$

(Exercise 7.9(b)): if $P + R$ and $Q + S$ are finite, then so is their union, and hence also $(P \cup Q) + (R \cup S)$, by the inequality. The preservation condition for complement follows at once from the equation

$$P' + Q' = P + Q$$

(Exercise 6.2(g)).

Exercises

1. Let Θ be an equivalence relation on a set B. Prove that

   $$p/\Theta = q/\Theta \qquad \text{if and only if} \qquad p \equiv q \mod \Theta.$$

   Conclude that two equivalence classes of Θ are either equal or disjoint.

2. Prove that congruence modulo n is a congruence relation on the set of integers, that is, it is an equivalence relation that preserves addition and multiplication.

3. Verify that the operations of addition and multiplication defined on the set of equivalence classes of the integers modulo n are in fact well defined.

4. Prove that the operations of join and complement defined in (10) and (11) on the set of equivalence classes of a Boolean congruence Θ are well defined.

5. A congruence on a Boolean ring is an equivalence relation on the ring that preserves (Boolean) addition and multiplication. Is every congruence on a Boolean algebra also a congruence on the associated Boolean ring?

6. Is every congruence on a Boolean ring also a congruence on the associated Boolean algebra?

7. Give a precise definition of the notion of a congruence relation on an arbitrary ring. Define operations of addition and multiplication on the set of equivalence classes of the congruence, and show that these operations are well defined. Prove that the set of equivalence classes under these operations is a ring.

8. Let Θ be a congruence on a ring R. Define the projection f from R to the quotient R/Θ, and prove that f is an epimorphism.

9. Let B be a field of subsets of a set X, and x_0 a fixed element of X. Define two sets P and Q in B to be congruent if either both sets contain x_0 or else neither set contains x_0. Prove that this is a congruence relation on B. Describe the elements and operations of the quotient algebra.

10. Let B be a field of subsets of a set X. Define two sets P and Q in B to be congruent if they differ on at most countably many elements, that is, if their symmetric difference

$$P + Q = (P \cap Q') \cup (P' \cap Q)$$

is countable. Prove that this relation is a congruence on B. If X is a countable set, what is the resulting quotient algebra?

11. Define two elements p and q in a Boolean algebra B to be congruent if their sum $p + q$ can be written as the join of finitely many atoms. Prove that the relation so defined is a congruence on B.

12. Define two elements p and q in Boolean algebra B to be congruent if there are no atoms below $p + q$. Prove that the relation so defined is a congruence on B.

Chapter 18

Ideals and Filters

A Boolean congruence Θ obviously determines each of its equivalence classes, and in particular it determines the equivalence class of 0, which is called the *kernel* of Θ. It is a happy state of affairs that, conversely, Θ is completely determined by its kernel via the equivalence

$$(1) \qquad p \equiv q \mod \Theta \qquad \text{if and only if} \qquad p + q \equiv 0 \mod \Theta.$$

In other words, to check whether two elements p and q are congruent, it is necessary and sufficient to check whether their Boolean sum is in the kernel. For the proof, suppose first that p and q are congruent (modulo Θ). The element q is congruent to itself, by reflexivity, and congruences preserve Boolean addition. Therefore, the sums $p + q$ and $q + q$ are congruent. Since

$$(2) \qquad\qquad\qquad q + q = 0,$$

the sum $p+q$ is congruent to 0, and is therefore in the kernel. For the reverse implication, suppose that $p + q$ is congruent to 0. Then $p + q + q$ and $0 + q$ are congruent; the first element is p, by (2), and the second is q.

Under what conditions is a subset of a Boolean algebra B the kernel of some congruence Θ on B? Three properties of the kernel of Θ are immediately evident:

$$(3) \qquad\qquad\qquad 0 \equiv 0 \mod \Theta,$$

(4) if $p \equiv 0 \mod \Theta$ and $q \equiv 0 \mod \Theta$, then $p \vee q \equiv 0 \mod \Theta$,

(5) if $p \equiv 0 \mod \Theta$ and $q \in B$, then $p \wedge q \equiv 0 \mod \Theta$.

Motivated by these properties, we make the following definition: a (*Boolean*) *ideal* in a Boolean algebra B is a subset M of B such that

S. Givant, P. Halmos, *Introduction to Boolean Algebras*,
Undergraduate Texts in Mathematics, DOI: 10.1007/978-0-387-68436-9_18,
© Springer Science+Business Media, LLC 2009

(6) $$0 \in M,$$

(7)   if $p \in M$ and $q \in M$,   then $p \vee q \in M$,

(8)   if $p \in M$ and $q \in B$,   then $p \wedge q \in M$.

Observe that condition (6) in the definition can be replaced by the superficially less restrictive condition that M be not empty, without changing the concept of ideal. Indeed, if M is not empty, say $p \in M$, and if M satisfies (8), then $p \wedge 0$ (that is, 0) is in M.

Two further properties of an ideal M are quite useful:

(9)   if $p \in M$ and $q \in M$,   then $p + q \in M$,

(10)   if $p \in M$ and $q \in B$,   then $p \cdot q \in M$.

Indeed, if p and q are in M, then so are $p \wedge q'$ and $p' \wedge q$, by condition (8). Therefore, the join $(p \wedge q') \vee (p' \wedge q)$ — that is to say, the sum $p + q$ — is in M, by condition (7). Equation (10) is an immediate consequence of condition (8), since multiplication in a Boolean algebra is defined to be meet.

The kernel of every congruence on B is an ideal, as is evident from properties (3)–(5). The converse is also true: every ideal M in B uniquely determines a congruence of which it is the kernel. The equivalence in (1) suggests how the congruence should be defined: it is the binary relation Θ on B determined by

(11) $p \equiv q \mod \Theta$ if and only if $p + q \in M$

for every pair of elements p and q in B. It is not difficult to show that Θ is a congruence. The reflexivity of Θ is a direct consequence of the identity (2) and condition (6). Symmetry is a consequence of the commutative law for Boolean addition: if $p \equiv q \mod \Theta$, then $p + q$ is in M; since

$$q + p = p + q,$$

it follows that $q + p$ is in M, and therefore (by (11)) $q \equiv p \mod \Theta$. To establish the transitivity of Θ, assume

$$p \equiv q \mod \Theta \quad \text{and} \quad q \equiv r \mod \Theta.$$

The sums $p + q$ and $q + r$ are then both in M, by (11), so the sum of these two sums is in M, by (9). Since

$$p + r = (p + q) + (q + r),$$

by (2), it may be concluded that $p \equiv r \mod \Theta$.

To verify that Θ preserves meet, assume

$$p \equiv r \mod \Theta \qquad \text{and} \qquad q \equiv s \mod \Theta.$$

This means that the sums $p+r$ and $q+s$ are in M. The products $(p+r) \cdot q$ and $(q+s) \cdot r$ are then both in M, by (10), and therefore the Boolean sum of these two products is in M, by (9). This sum is just $p \cdot q + r \cdot s$:

$$(p+r) \cdot q + (q+s) \cdot r = (p \cdot q + r \cdot q) + (q \cdot r + s \cdot r)$$
$$= p \cdot q + (r \cdot q + r \cdot q) + r \cdot s = p \cdot q + 0 + r \cdot s = p \cdot q + r \cdot s,$$

by the distributive law (1.9), the associative law (1.1), the commutative law (1.4), the identity law (1.5), and (2). Invoke (11) to arrive at

$$p \cdot q \equiv r \cdot s \mod \Theta.$$

In view of the definition of multiplication in a Boolean algebra (see (3.3)), it may be concluded that

$$p \wedge q \equiv r \wedge s \mod \Theta.$$

The proof that Θ preserves complement is similar, but simpler, and uses the identity

$$p' + q' = p + q$$

from Exercise 6.7(a). If $p \equiv q \mod \Theta$, then $p + q$ is in M. It follows from the given identity that $p' + q'$ is then in M, so

$$p' \equiv q' \mod \Theta.$$

The kernel of a congruence is the set of elements congruent to 0. The kernel of the congruence Θ defined in (11) is therefore the set of elements p such that $p + 0$ belongs to M. This set is of course just M. Conclusion: the kernel of Θ is M. In view of (1), every congruence is completely determined by its kernel, so Θ is the only congruence on B with kernel M. This completes the proof that every ideal in B uniquely determines a congruence of which it is the kernel.

The equivalence classes of any congruence Θ can be computed directly from the kernel M of the congruence. For an arbitrary element p of B, the equivalence class p/Θ coincides with the *coset*

$$p + M = \{p + r : r \in M\}.$$

The proof of this assertion amounts to checking that the sets in question have the same elements. If q is in B, then

$$q \in p/\Theta \qquad \text{if and only if} \qquad p \equiv q \mod \Theta,$$

$$\text{if and only if} \quad p + q \in M,$$
$$\text{if and only if} \quad q \in p + M.$$

The first step uses the definition of an equivalence class, and the second step uses (1), which is equivalent to (11). For the third step, observe that if $p + q$ belongs to M, and if $r = p + q$, then $p + r$ belongs to $p + M$; since

$$q = 0 + q = p + p + q = p + r,$$

it follows that q is in $p + M$. On the other hand, if q belongs to $p + M$, then $q = p + r$ for some element r in M; since

$$p + q = p + p + r = 0 + r = r,$$

it follows that $p + q$ belongs to M.

Definitions (17.9)–(17.11) of the Boolean algebraic operations of the quotient algebra B/Θ can be expressed in terms of cosets in the following manner:

(12) $$(p + M) \wedge (q + M) = (p \wedge q) + M,$$
(13) $$(p + M) \vee (q + M) = (p \vee q) + M,$$
(14) $$(p + M)' = (p') + M.$$

The zero and unit of the quotient algebra are the cosets $0 + M$ — which coincides with M — and $1 + M$. The Boolean algebra of cosets under the operations defined by (12)–(14) is identical to the Boolean algebra of equivalence classes of Θ. It is called the *quotient algebra of B modulo the ideal M*. It is helpful to adopt a notation for this quotient that is reminiscent of the notation used in the case of congruences. For that reason, we shall write B/M for the quotient, we shall usually write p/M for the coset $p + M$, and we shall write

$$p \equiv q \quad \mod M \quad \text{instead of} \quad p \equiv q \quad \mod \Theta.$$

The characterization given in (17.8) of when two equivalence classes modulo Θ are equal can also be rephrased in terms of the ideal M:

(15) $$p/M = q/M \quad \text{if and only if} \quad p \equiv q \quad \mod M,$$
$$\text{if and only if} \quad p + q \in M.$$

The second equivalence follows from (11).

The canonical homomorphism that maps each element of B to its equivalence class modulo Θ can also be expressed in terms of cosets; it is the function f from B to B/M defined by

$$f(p) = p/M.$$

The *kernel* of f is, by definition, the set of elements that are mapped to the zero element of B/M. It is easy to check that this set is just M:

$$f(p) = 0/M \quad \text{if and only if} \quad p/M = 0/M,$$
$$\text{if and only if} \quad p \equiv 0 \mod M,$$
$$\text{if and only if} \quad p \in M;$$

the first equality uses the definition of f, the second uses (17.8) (formulated in the notation introduced in the previous paragraph), and the third uses the fact that M is the kernel of the congruence determined by it. The kernel of the homomorphism f and the kernel of the congruence Θ are thus equal; they are both M.

A consequence of the preceding discussion is that in the study of Boolean algebras, congruences can be dispensed with entirely; they can be replaced by ideals, and congruence classes can be replaced by cosets — in fact, they are cosets. The quotient of a Boolean algebra modulo a congruence is identical to the quotient of the Boolean algebra modulo the ideal that is the kernel of the congruence.

A similar situation exists in the theory of rings. Motivated by the example of congruence modulo n on the ring of integers, we define a *congruence* on a ring R to be an equivalence relation Θ on R that preserves the ring operations of addition, multiplication, and formation of negatives (additive inverses). Operations of addition, multiplication, and formation of negatives are defined on the set of equivalence classes of Θ as follows: the sum and product of equivalence classes p/Θ and q/Θ are the equivalence classes $(p + q)/\Theta$ and $(p \cdot q)/\Theta$, and the negative of p/Θ is $(-p)/\Theta$. The preservation conditions ensure that these operations are well defined. The set of all equivalence classes, under the operations just defined, is a ring, the *quotient ring of R modulo Θ*. The projection mapping that takes each element of R to the corresponding equivalence class of Θ is an epimorphism from R to R/Θ.

An *ideal* in a ring is defined to be an arbitrary subset M of the ring that satisfies conditions (6), (9), (10), and

(16) $$\text{if } p \in M, \quad \text{then} \quad -p \in M,$$
(17) $$\text{if } p \in M \text{ and } q \in B, \quad \text{then} \quad q \cdot p \in M,$$

where $+$, $\cdot$, and $-$ are the ring operations of addition, multiplication, and formation of negatives, and 0 is the zero element of the ring. Condition (16) is needed for rings without unit; for rings with unit, it is a consequence of (10). Similarly, condition (17) is needed for non-commutative rings; for commutative rings, it is a consequence of (10). The *kernel* of a ring congruence — the equivalence class of 0 — is always an ideal. Conversely, an ideal M uniquely determines a ring congruence of which it is the kernel, namely the binary relation Θ determined by

$$p \equiv q \mod \Theta \qquad \text{if and only if} \qquad p - q \in M$$

(where $p - q$ is the sum of p and the additive inverse of q). The equivalence classes of Θ coincide with the cosets of M, the sets of the form $p + M$. The ring operations on the equivalence classes may be written as operations on cosets in the following way:

$$(p + M) + (q + M) = (p + q) + M \quad \text{and} \quad (p + M) \cdot (q + M) = (p \cdot q) + M.$$

Conclusion: in the study of rings, congruence relations can be, and almost always are, replaced by ideals, and one speaks of the quotient of a ring modulo an ideal instead of the quotient of a ring modulo a congruence.

Every Boolean algebra is (or, better, can be turned into) a Boolean ring, and conversely. What is the relationship between the corresponding notions of ideal? It turns out that ideals are ideals, or, to put it more precisely, a subset M of a Boolean algebra B is a Boolean ideal if and only if it is an ideal in the corresponding Boolean ring. Indeed, every Boolean ideal M is a ring ideal, since conditions (7) and (8) imply conditions (9) and (10). To prove the converse, suppose M is an ideal in the sense of ring theory. Condition (10) at once implies condition (8), since meet is defined as ring multiplication. To verify condition (7), let p and q be elements of M. The join of these two elements is defined by

$$p \vee q = p + q + p \cdot q$$

(see (3.4)), and the sum on the right side of this equation belongs to M, by conditions (9) and (10). Consequently, $p \vee q$ is in M.

The concept of a Boolean ideal can also be defined in order-theoretic terms, but the language of order does not have much to contribute to ideal theory. This much can be said: condition (8) can be replaced by

(18)   if $p \in M$ and $q \leq p$,   then $q \in M$,

without changing the concept of ideal. The proof is elementary.

Every example of a Boolean congruence (such as the ones in Chapter 17) gives rise to an example of an ideal, namely its kernel. Thus, if a congruence Θ on a Boolean algebra B is defined in terms of a fixed element p_0 in B by

$$p \equiv q \mod \Theta \qquad \text{if and only if} \qquad p \wedge p_0 = q \wedge p_0,$$

then the corresponding ideal consists of those elements p for which $p \wedge p_0 = 0$, or, equivalently, $p \leq p_0{}'$. If a congruence Θ is defined on the field of all subsets of a set X by

$$P \equiv Q \mod \Theta \qquad \text{if and only if} \qquad P + Q \text{ is finite,}$$

then the corresponding ideal consists of all finite sets in the field. More generally, the class of all those finite sets that happen to belong to some particular field is an ideal in that field. A similar generalization is available for each of the following two examples. The class of all countable sets is an ideal in the field of all subsets of an arbitrary set; and the class of all nowhere dense sets is an ideal in the field of all subsets of a topological space.

Every Boolean algebra B has a *trivial* ideal, namely the set $\{0\}$ consisting of 0 alone; all other ideals of B will be called *non-trivial*. Every Boolean algebra B has an *improper* ideal, namely B itself; all other ideals will be called *proper*. Observe that an ideal is proper if and only if it does not contain 1. (This follows at once from condition (18).)

The intersection of every family of ideals in a Boolean algebra B is again an ideal of B. (The intersection of the empty family is, by convention, the improper ideal B.) The proof consists in verifying that conditions (6)–(8) hold in the intersection. For instance, to verify condition (7), let p and q be elements in the intersection of a family of ideals. Every ideal in the family must contain both p and q, and therefore must also contain the join $p \vee q$ (by condition (7) applied to each ideal in the family). It follows that $p \vee q$ is in the intersection of the family. The other two conditions are verified in a completely analogous fashion.

One consequence of the preceding remark is that if E is an arbitrary subset of B, then the intersection of all those ideals that happen to include E is an ideal. (There is always at least one ideal that includes E, namely the improper ideal B.) That intersection, say M, is the smallest ideal in B that includes E; in other words, M is included in every ideal that includes E. The ideal M is called the ideal *generated* by E.

The definition just given is top-down and non-constructive; it does not describe the elements in the ideal generated by E. (An advantage of the definition is that with minimal changes, it applies to every algebraic structure in which there is a suitable analogue of the notion of an ideal.) Fortunately, it is possible to give a more explicit description (due to Stone [66]) of the elements of the ideal.

Theorem 11. *An element p of a Boolean algebra is in the ideal generated by a set E if and only if there is a finite subset F of E such that $p \leq \bigvee F$.*

Proof. Let M be the ideal generated by a subset E of a Boolean algebra B, and let N be the set of elements p in B such that p is below the join of some finite subset of E. It is to be proved that M and N are equal. Every join of a finite subset of E is certainly in M, by condition (7), and consequently every element below such a join is in M, by condition (18). Thus, N is a subset of M. To establish the reverse inclusion, it suffices to prove that N is an ideal. Since N obviously includes E, it then follows that N must include the smallest ideal that includes E, namely M.

The empty set is a finite subset of E, and its join is 0; hence, 0 belongs to N. If p and q are elements of N, then there are finite subsets F and G of E such that

$$p \le \bigvee F \qquad \text{and} \qquad q \le \bigvee G.$$

The set $H = F \cup G$ is a finite subset of E, and $p \vee q$ is below $\bigvee H$, by monotony; consequently, $p \vee q$ is in N. Finally, if p is an element of N, say p is below the join of a finite subset F of E, then any element q that is below p is also below $\bigvee F$, and therefore belongs to N, by the definition of N. Conclusion: N satisfies the three conditions (6), (7), and (18) that characterize ideals, so it is an ideal in B.

The ideal generated by the empty set is the smallest possible ideal of B, namely the trivial ideal $\{0\}$. An ideal generated by a singleton $\{p\}$ is called a *principal ideal*; it consists of all the subelements of p, by Theorem 11, and it is usually denoted by (p). Both the trivial ideal $\{0\}$ and the improper ideal B are principal; the former consists of the subelements of 0, and the latter of the subelements of 1. An ideal is said to be *finitely generated* if it is generated by a finite set of elements.

Corollary 1. *Every finitely generated ideal in a Boolean algebra is principal.*

Proof. Suppose E is a finite subset of a Boolean algebra B. An element p in B will be below the join of some finite subset of E just in case it is below the join $\bigvee E$. The ideal generated by E therefore coincides with the ideal generated by the single element $\bigvee E$, by the preceding theorem.

Theorem 11 can also be used to characterize when the set of atoms of a Boolean algebra generates a proper ideal.

Corollary 2. *The ideal generated by the set of all atoms in a Boolean algebra is a proper ideal if and only if the Boolean algebra is infinite.*

Proof. Let E be the set of all atoms in a Boolean algebra B, and let M be the ideal generated by E. The argument proceeds by contraposition. If B is finite, then it is atomic, by Lemma 15.1. In this case $1 = \bigvee E$, by Lemma 14.3. Because E is also finite, the unit 1 must be in M, so M is the improper ideal.

Now assume that M is improper. By Theorem 11, there must be a finite subset F of E such that $1 \leq \bigvee F$, and therefore $1 = \bigvee F$. In other words, the unit must be the supremum of a finite set F of atoms. This implies (Lemmas 14.3 and 14.2) that B is atomic and that F is the set of all of its atoms. An atomic Boolean algebra with finitely many atoms is perforce finite (see, for instance, Theorem 6, p. 119).

An important special case of Theorem 11, the *ideal extension lemma*, describes how to extend an ideal by adjoining a single element.

Lemma 1. *Let M be an ideal, and p_0 an element, in a Boolean algebra. The ideal generated by $M \cup \{p_0\}$ is the set*

$$N = \{p \vee q : p \leq p_0 \text{ and } q \in M\}.$$

Proof. Put $E = M \cup \{p_0\}$, and invoke Theorem 11. An element r of the Boolean algebra is in the ideal generated by E just in case there is a finite subset F of E such that $r \leq \bigvee F$. No generality is lost by assuming that the element p_0 is in F (adding this element to F can only make the join bigger). Also, the elements of F that are in M may be combined into a single element, since M is closed under join. Thus, r is in the ideal generated by E if and only if there is an element s in M such that $r \leq s \vee p_0$. This last inequality holds just in case

$$r = r \wedge (s \vee p_0) = (r \wedge s) \vee (r \wedge p_0).$$

The element $q = r \wedge s$ is again a member of M, and the element $p = r \wedge p_0$ is below p_0. Therefore, r is in the ideal generated by E just in case

$$r = q \vee p$$

for some q in M and some $p \leq p_0$.

The lemma yields a simple criterion for determining when the extension of a proper ideal by a new element is again a proper ideal.

Corollary 3. *Let M be an ideal, and p_0 an element, in a Boolean algebra. If p_0' is not in M, then the ideal generated by $M \cup \{p_0\}$ is proper.*

Proof. The argument proceeds by contraposition. Let N be the ideal generated by $M \cup \{p_0\}$, and assume N is improper. It is to be shown that p_0' belongs to M. Since N is improper, it certainly contains p_0'. There must therefore be elements $p \le p_0$ and q in M such that $p_0' = p \vee q$, by Lemma 1. A simple computation shows that $p_0' \le q$:

$$p_0' = p_0' \wedge p_0' = p_0' \wedge (p \vee q) = (p_0' \wedge p) \vee (p_0' \wedge q)$$
$$\le (p_0' \wedge p_0) \vee (p_0' \wedge q) = 0 \vee (p_0' \wedge q) = p_0' \wedge q \le q.$$

Therefore, p_0' is in M, by condition (18).

The concepts of subalgebra and homomorphism are in a certain obvious sense self-dual; the concept of ideal is not. The dual concept is defined as follows. A (*Boolean*) *filter* in a Boolean algebra B is a subset N of B such that

(19) $1 \in N,$

(20)   if $p \in N$ and $q \in N,$ then $p \wedge q \in N,$

(21)   if $p \in N$ and $q \in B,$ then $p \vee q \in N.$

Condition (19) can be replaced by the condition that N be non-empty. Condition (21) can be replaced by

(22)   if $p \in N$ and $p \le q,$ then $q \in N.$

Neither of these replacements will alter the concept being defined. The filter generated by a subset of B, and in particular a principal filter, are defined by an obvious dualization of the corresponding definitions for ideals. In more detail, the intersection of any family of filters in B is again a filter in B. The filter *generated* by a subset E of B is defined to be the intersection of the filters that include E. (There is always one such filter, namely B.) In other words, it is the smallest filter that includes E. A filter N is *principal* if it is generated by a single element p. It is not difficult to show that in this case

$$N = \{q \in B : p \le q\}.$$

The relation between filters and ideals is a very close one. The fact is that filters and ideals come in dual pairs. This means that there is a one-to-one correspondence that pairs each ideal to a filter, its dual, and by means of which every statement about ideals is immediately translatable to a statement about filters. The pairing is easy to describe. If M is an ideal, write $N = \{p : p' \in M\}$, and, in reverse, if N is a filter, write $M = \{p : p' \in N\}$. It is trivial to verify that this construction does indeed convert an ideal into a filter, and vice versa, since the conditions (6)–(8) and (19)–(21) are dual to one another. We shall have more to say about this subject in the next chapter.

Exercises

1. Let M be an ideal of a ring R, and let Θ be the binary relation on R determined by

$$p \equiv q \mod \Theta \qquad \text{if and only if} \qquad p - q \in M$$

(where $p - q$ is the sum of p and the additive inverse of q). Prove that Θ is a congruence relation on R with kernel M, and that the equivalence class p/Θ coincides with the coset

$$p + M = \{p + q : q \in M\}.$$

2. Prove that condition (8) in the definition of a (Boolean) ideal can equivalently be replaced by condition (18).

3. Prove that the class of finite sets of integers, the class of finite sets of even integers, and the class of finite sets of odd integers are all ideals in the Boolean algebra of all subsets of the set of integers.

4. Prove that in any field of sets, the countable sets form an ideal.

5. Prove that the nowhere dense sets form an ideal in the field of all subsets of a topological space.

6. Let E be a subset of a Boolean ring B with or without a unit. The *annihilator* of E is the set of all elements p in B such that $p \cdot r = 0$ for all r in E. Prove that the annihilator of a set is always a ring ideal. (This observation is due to Stone [66]. The notion of an annihilator of a set, and the proof that the annihilator of a set is an ideal, hold in the more general setting of commutative rings.)

7. Prove that the intersection of two ideals M and N in a Boolean ring (with or without a unit) is the trivial ideal if and only if N is included in the annihilator of M (Exercise 6).

8. A subset of a Boolean ring B (with or without a unit) is called *dense* if every non-zero element in B is above a non-zero element of the subset. Prove that an ideal M in B is dense if and only if M has a non-trivial

intersection with every non-trivial ideal in B. Conclude that M is dense if and only if its annihilator (Exercise 6) is the trivial ideal.

9. (Harder.) Prove that the class of all sets of measure zero is an ideal in the field of all measurable subsets of a measure space.

10. It has been shown that a subset of a Boolean algebra is a Boolean ideal if and only if it is a ring ideal. Prove an analogous result for Boolean rings without unit. In other words, prove that a subset M of a Boolean ring without unit satisfies conditions (6)–(8) if and only if it satisfies conditions (6), (9), and (10), where the meet and join operations in the ring are defined by equations

$$p \wedge q = p \cdot q \qquad \text{and} \qquad p \vee q = p + q + p \cdot q.$$

11. Complete the proof that the intersection of every family of ideals in a Boolean algebra B is again an ideal in B.

12. Show that Theorem 11 also holds for Boolean rings without unit.

13. In general, homomorphisms do not preserve all structural properties of a Boolean algebra. The following example demonstrates this in a dramatic way. Let X be the set of integers, and M the ideal of finite subsets of X. The field $\mathcal{P}(X)$ is atomic. Prove that the quotient $\mathcal{P}(X)/M$ is atomless. Conclude that an epimorphism need not preserve the property of an element being an atom, nor the property of an algebra being atomic.

14. Describe the ideal generated by the set of all atoms in a Boolean algebra.

15. Prove that a subset N of a Boolean algebra is a filter if and only if it is non-empty and satisfies conditions (20) and (22).

16. Prove that the intersection of a family of filters in a Boolean algebra is again a filter.

17. Formulate and prove the analogue of Theorem 11 for filters.

18. Prove that the following conditions on an ideal M are equivalent: (1) M is improper; (2) there is an element p such that p and p' are both in M; (3) 1 is in M.

19. A subset E of a Boolean algebra is said to have the *finite join property* if the unit is not the supremum (join) of any finite subset of E. Prove that the ideal generated by a set E is proper if and only if E has the finite join property.

20. Formulate and prove the dual of Exercise 19.

21. Show that a set E generates an ideal in a Boolean algebra if and only if the set $E' = \{p' : p \in E\}$ generates the dual filter.

22. Formulate and prove, for filters, the analogue of the assertion that every finitely generated ideal is principal.

23. Prove that the join of two elements p and q in a Boolean algebra belongs to an ideal M if and only if both p and q belong to M.

24. Formulate and prove, for filters, the analogue of Exercise 23.

25. Suppose B is a Boolean subalgebra of A. Prove that if M is an ideal in A, then the intersection $N = M \cap B$ is an ideal in B. Prove further that N is a proper ideal in B just in case M is a proper ideal in A.

26. Suppose f is a Boolean homomorphism from B to A. Prove that if M is an ideal in A, then its inverse image under f, the set

$$f^{-1}(M) = \{p \in B : f(p) \in M\},$$

is an ideal in B. Prove also that if f is an epimorphism, then the image under f of an ideal N in B, the set

$$f(N) = \{f(p) : p \in N\},$$

is an ideal in A.

27. The intersection of a family $\{M_i\}$ of ideals in a Boolean algebra B is an ideal. In fact, it is the largest ideal that is included in each M_i. In other words, under the partial ordering of inclusion, it is the *infimum* of the family $\{M_i\}$. Define an appropriate notion of the *supremum* of the family $\{M_i\}$ so that this supremum is again an ideal. Prove that under these notions of supremum and infimum, the set of all ideals of B is a complete distributive lattice. (See Chapter 7.)

28. Prove the following assertions about ideals and filters in a Boolean algebra B.

    (a) If M is an ideal, then the set $N = \{p' : p \in M\}$ is a filter. It is called the *dual filter* of M.

    (b) If N is a filter, then the set $M = \{p' : p \in N\}$ is an ideal. It is called the *dual ideal* of N.

    (c) If M is an ideal and N its dual filter, then the dual ideal of N is just M.

    (d) If N is a filter and M its dual ideal, then the dual filter of M is just N.

    (e) An ideal M_1 is included in an ideal M_2 if and only if the dual filter of M_1 is included in the dual filter of M_2.

    (f) The correspondence that takes each ideal to its dual filter maps the class of ideals in B bijectively to the class of filters in B, and is a lattice isomorphism.

29. Prove that if M is an ideal in a Boolean algebra and if N is its dual filter, then the set-theoretic union $M \cup N$ is a subalgebra.

30. If an ideal M in a Boolean algebra is countably generated, show that there must be a (countable) ascending chain

$$p_1 \leq p_2 \leq \cdots \leq p_n \leq \cdots,$$

of generators in M such that an element p of the Boolean algebra belongs to M if and only if $p \leq p_n$ for some positive integer n.

31. It was proved in this chapter that congruences can equivalently be replaced by ideals. The purpose of this exercise is to show that congruences can equivalently be replaced by filters.

    (a) Prove that the *cokernel* of a congruence — the equivalence class of 1 — is always a filter.

    (b) Prove that a congruence is uniquely determined by its cokernel.

    (c) Prove that every filter uniquely determines a congruence of which it is the cokernel.

    (d) Describe the equivalence classes of a congruence in terms of the "cosets" of its cokernel.

    (e) Describe the operations of the quotient algebra in terms of the "cosets" of the cokernel.

(f) Describe the canonical homomorphism in terms of the "cosets" of the cokernel.

32. Let B be a Boolean subalgebra of A, and p_0 an arbitrary element in A. Show that the set M of elements in B that are disjoint from p_0 is an ideal in B, and that the relativization of B to p_0,

$$B(p_0) = \{p \wedge p_0 : p \in B\}$$

(Exercise 12.19), is isomorphic to the quotient B/M via the correspondence defined by

$$f(p \wedge p_0) = p/M$$

for each p in B.

Chapter 19

Lattices of Ideals

The class of all ideals in a Boolean algebra B is partially ordered by the relation of set-theoretic inclusion — the relation of one ideal being a subset of another. Under this partial order, a family $\{M_i\}$ of ideals always has an infimum; it is just the intersection $\bigcap_i M_i$ of the ideals in the family. The family also has a supremum, but in general the supremum is not the union of the ideals in the family, since that union is rarely an ideal; rather, the supremum is the ideal generated by the union $\bigcup_i M_i$. In other words, the supremum of the family $\{M_i\}$ is the intersection of the class of those ideals in B that include every ideal M_i. (This class is not empty, because it always contains B.) All of this may be summarized by saying that the class of ideals of a Boolean algebra B is a complete lattice under the partial order of inclusion; the infimum of an arbitrary family of ideals is the intersection of the family, and the supremum is the ideal generated by the union of the family.

The special case when a family of ideals is empty is worth discussing for a moment, if only to avoid later confusion. The improper ideal B is vacuously included in every ideal in the empty family (there is no ideal in the empty family that does not include B). Clearly, then, B is the largest ideal that is included in every ideal in the empty family, so it is, by definition, the infimum of that family. Similarly, the trivial ideal $\{0\}$ vacuously includes every ideal in the empty family (there is no ideal in the empty family that is not included in $\{0\}$). Since $\{0\}$ is obviously the smallest ideal that includes every ideal in the empty family, it must be the supremum of that family.

It is worthwhile formulating a simple but quite useful alternative description of the join and meet of two ideals (which goes back to Stone [66]).

S. Givant, P. Halmos, *Introduction to Boolean Algebras*,   164
Undergraduate Texts in Mathematics, DOI: 10.1007/978-0-387-68436-9_19,
© Springer Science+Business Media, LLC 2009

Lemma 1. *If M and N are ideals in a Boolean algebra, then*

$$M \vee N = \{p \vee q : p \in M \text{ and } q \in N\}$$

and

$$M \wedge N = \{p \wedge q : p \in M \text{ and } q \in N\}.$$

Proof. To establish the first identity, write

$$L = \{p \vee q : p \in M \text{ and } q \in N\}.$$

It must be shown that

$$L = M \vee N.$$

If p and q are elements of M and N respectively, then both elements are in $M \vee N$, and therefore so is their join. Consequently, L is included in $M \vee N$, by the definition of L. To establish the reverse inclusion, suppose r is an element of $M \vee N$. There must then exist finite subsets E of M, and F of N, such that r is below the join of $E \cup F$, by Theorem 11 (p. 155). The ideals M and N are closed under finite joins, so the elements

$$s = \bigvee E \qquad \text{and} \qquad t = \bigvee F$$

belong to M and N respectively, and $r \leq s \vee t$. It follows that the elements

$$p = r \wedge s \qquad \text{and} \qquad q = r \wedge t$$

are also in M and N respectively, by condition (18.8). Since

$$r = r \wedge (s \vee t) = (r \wedge s) \vee (r \wedge t) = p \vee q,$$

the element r belongs to L, by the definition of L.

The proof of the second identity is similar, but easier. Write

$$L = \{p \wedge q : p \in M \text{ and } q \in N\}.$$

It must be shown that

$$L = M \wedge N.$$

An element r in $M \wedge N$ belongs to both M and N. The meet of r with itself — which is just r — is therefore in L, by the definition of L. It follows that $M \wedge N$ is included in L. For the reverse inclusion, consider an arbitrary element r in L; it has the form $r = p \wedge q$ for some p in M and some q in N. Since r is below each of p and q, it belongs to both M and N, by condition (18.18), and consequently it belongs to $M \wedge N$. Thus, each element of L is in $M \wedge N$. The proof of the lemma is complete.

The formulas in the preceding lemma assume a particularly perspicuous form when applied to principal ideals.

Corollary 1. *Let p and q be elements in a Boolean algebra. Then*

$$(p) \vee (q) = (p \vee q) \qquad and \qquad (p) \wedge (q) = (p \wedge q).$$

Proof. The previous lemma, applied to the principal ideals (p) and (q), asserts that $(p) \vee (q) = L$, where

(1) $$L = \{r \vee s : r \leq p \text{ and } s \leq q\}.$$

To obtain the first identity of the corollary, then, it suffices to check that $L = (p \vee q)$. The inclusion of L in $(p \vee q)$ is an immediate consequence of (1) and the monotony laws:

$$r \leq p \quad \text{and} \quad s \leq q \quad \text{implies} \quad r \vee s \leq p \vee q.$$

The reverse inclusion follows from the simple observation that $p \vee q$ is in L; since L is an ideal, every element below $p \vee q$ must also be in L.

The second identity is established in a completely analogous fashion.

The lattice of ideals is not only complete, it is also distributive in the sense that the laws (2.20) hold identically in it. The proof is a direct application of Lemma 1. Let L, M, and N be three ideals in a Boolean algebra. Then

(2) $$L \vee (M \wedge N) = \{p \vee (q \wedge s) : p \in L, q \in M, \text{ and } s \in N\}$$

and

(3) $$(L \vee M) \wedge (L \vee N) = \{(p \vee r) \wedge (q \vee s) : p, q \in L, r \in M, \text{ and } s \in N\},$$

by the lemma. Since

$$p \vee (q \wedge s) = (p \vee q) \wedge (p \vee s),$$

by the distributive laws for Boolean algebras, every element in the ideal (2) is also in the ideal (3). To demonstrate the reverse inclusion, let t be an element in the ideal (3), say,

$$t = (p \vee r) \wedge (q \vee s),$$

where p and q are in L, and r and s in M and N respectively. Apply the distributive and monotony laws to obtain

$$t = (p \vee r) \wedge (q \vee s) = (p \wedge q) \vee (p \wedge s) \vee (r \wedge q) \vee (r \wedge s)$$
$$\leq p \vee p \vee q \vee (r \wedge s) = (p \vee q) \vee (r \wedge s).$$

The join $p \vee q$ is in L, and the meet $r \wedge s$ is in $M \wedge N$, so the element $(p \vee q) \vee (r \wedge s)$ is in the ideal (2), by Lemma 1. Consequently, the element t is also in that ideal, by condition (18.18).

It has been demonstrated that

$$L \vee (M \wedge N) = (L \vee M) \wedge (L \vee N)$$

for all ideals L, M, and N of a Boolean algebra. The dual distributive law can be established in a similar fashion, or it can be derived directly from the preceding law. (Each version of the distributive law is derivable from its dual in all lattices; see Exercise 7.19.)

The ideals of a Boolean algebra form a complete, distributive lattice, but they do not, in general, form a Boolean algebra. To give an example, it is helpful to introduce some terminology. An ideal is *maximal* if it is a proper ideal that is not properly included in any other proper ideal. We shall see in the next chapter that an infinite Boolean algebra B always has at least one maximal ideal that is not principal. Assume this result for the moment. A "complement" of such an ideal M in the lattice of ideals of B would be an ideal N with the property that

$$M \wedge N = \{0\} \qquad \text{and} \qquad M \vee N = B.$$

Suppose the first equality holds. If q is any element in N, then $p \wedge q = 0$, and therefore $p \leq q'$, for every element p in M, by Lemma 1. In other words, the ideal M is included in the principal ideal (q'). The two ideals must be distinct, since M is not principal. This forces (q') to equal B, by the maximality of M. In other words, $q' = 1$, and therefore $q = 0$. What has been shown is that the meet $M \wedge N$ can be the trivial ideal only if N itself is trivial. In this case, of course, $M \vee N$ is M, not B. Conclusion: a maximal, non-principal ideal does not have a complement in the lattice of ideals.

This is not to say that no ideal has a complement; some do. For example, each principal ideal (p) has the complement (p'), since

$$(p) \vee (p') = (p \vee p') = (1) = B$$

and

$$(p) \wedge (p') = (p \wedge p') = (0) = \{0\},$$

by Corollary 1.

Let's return for a moment to the question of the supremum of a family of ideals. As was mentioned earlier, the union of a family of ideals usually fails to be an ideal. There is, however, an important exception. A family $\{M_i\}$ of ideals is said to be *directed* if any two ideals of the family are always included in some ideal of the family; in other words, whenever M_i and M_j are ideals in the family, there is an ideal M_k in the family such that

$$M_i \subseteq M_k \quad \text{and} \quad M_j \subseteq M_k.$$

Lemma 2. *The union of a non-empty directed family of ideals in a Boolean algebra is again an ideal. The ideal is proper if and only if each ideal in the family is proper.*

Proof. Let $\{M_i\}$ be a non-empty directed family of ideals in a Boolean algebra, and let M be the union of this family. It is to be shown that M satisfies the defining conditions (18.6)–(18.8) for ideals. The family contains at least one ideal, by assumption, and 0 is in that ideal. Therefore, 0 is in the union M. If p and q are elements in M, then there must be ideals M_i and M_j in the family that contain p and q respectively. Since the family is directed, one of its ideals, say M_k, includes both M_i and M_j. The elements p and q then both belong to M_k, so their join $p \vee q$ is in M_k as well. Consequently, this join is also in M. Finally, an element p in M must belong to one of the ideals M_i of the family. For each element q in the Boolean algebra, the meet $p \wedge q$ belongs to M_i and therefore also to M. Conclusion: M is an ideal.

The ideal M is improper just in case it contains 1. Since M is the union of a directed family of ideals, it contains 1 just in case some ideal in the family contains 1, that is to say, just in case some ideal in the family is improper. The second conclusion is now immediate: M is proper just in case every ideal in the family is proper.

The lemma applies, in particular, to non-empty families of ideals that are linearly ordered by inclusion in the sense that for any two members M_i and M_j of the family, either $M_i \subseteq M_j$ or $M_j \subseteq M_i$. Such families are called *chains*.

Everything that has been said about ideals can be repeated almost *verbatim* for filters. The class of all filters in a Boolean algebra is partially ordered by the relation of set-theoretic inclusion, and under this partial order the class becomes a complete lattice. The infimum of a family of filters is the intersection of the family, and the supremum is the filter generated by the union of the family (namely, the intersection of all filters that include each member in the family). The lattice is distributive, but in general it is not complemented; that is to say, it is not a Boolean algebra.

Actually, even more is true. The canonical mapping f that takes every ideal M of a Boolean algebra B to its dual filter $N = \{p' : p \in M\}$ is an isomorphism between the lattice of ideals and the lattice of filters in B, and its inverse is the mapping g that takes each filter N to its dual ideal $M = \{p' : p \in N\}$ (see Exercise 12.13). Indeed,

$$g(f(M)) = g(\{p' : p \in M\}) = \{p'' : p \in M\} = M$$

for every ideal M, and, similarly,

$$f(g(N)) = N$$

for every filter N. It follows from these equations that f maps the lattice of ideals one-to-one onto the lattice of filters, and that its inverse mapping is g. If an ideal M_1 is included in an ideal M_2, then obviously the filter $N_1 = \{p' : p \in M_1\}$ is included in the filter $N_2 = \{p' : p \in M_2\}$, and dually. In other words, the correspondence f preserves the partial ordering of the lattices in the strong sense that

$$M_1 \subseteq M_2 \quad \text{if and only if} \quad f(M_1) \subseteq f(M_2).$$

Conclusion: f is a lattice isomorphism.

The observation that the class of ideals in a Boolean algebra forms a complete distributive lattice was first made by Stone [66] and Tarski [74].

Exercises

1. Fill in the details of the following alternative proof of Lemma 1. The set

   $$L = \{p \vee q : p \in M \text{ and } q \in N\}$$

   is an ideal that includes M and N, and that is included in $M \vee N$; consequently, $L = M \vee N$. Similarly, the set

   $$K = \{p \wedge q : p \in M \text{ and } q \in N\}$$

   is an ideal that includes $M \wedge N$ and that is included in both M and N; consequently, $K = M \wedge N$.

2. Derive the second identity in Corollary 1.

3. Extend Lemma 1 to Boolean rings without unit.

4. Show by a direct argument that the ideals of a Boolean algebra satisfy the distributive law

   $$L \wedge (M \vee N) = (L \wedge M) \vee (L \wedge N).$$

5. Show that the mapping from a Boolean algebra B into the lattice of ideals of B that takes each element p to the principal ideal (p) is a lattice monomorphism (a one-to-one mapping that preserves the operations of join and meet).

6. Prove directly (without using Lemma 2) that the union of a non-empty chain of ideals in a Boolean algebra is again an ideal, and that the ideal is proper if and only if each ideal in the chain is proper.

7. Prove directly that the class of filters in a Boolean algebra is a complete lattice under the relation of set-theoretic inclusion.

8. Formulate and prove the analogue of Lemma 1 for filters.

9. Show by a direct argument that the lattice of filters is distributive.

10. Define what it means for a family of filters in a Boolean algebra to be *directed*, and prove that the union of a non-empty directed family of filters is again a filter.

11. Verify that the dual of a principal ideal is a principal filter, and conversely. Conclude that the canonical isomorphism between the lattice of ideals and the lattice of filters of a Boolean algebra maps the sublattice of principal ideals onto the sublattice of principal filters.

12. The class of all congruences on a Boolean algebra B is also a complete lattice under the partial ordering of set-theoretic inclusion. The infimum of any family of congruences on B is the intersection of the family; the supremum is the congruence generated by the union of the family, that is to say, it is the intersection of all those congruences on B that include every congruence in the given family. Prove that the correspondence that maps each ideal M in B to the congruence on B determined by M is a lattice isomorphism.

Chapter 20

Maximal Ideals

An ideal is *maximal* if it is a proper ideal that is not properly included in any other proper ideal. Equivalently, to say that M is a maximal ideal in B means that $M \neq B$, and, moreover, if N is an ideal such that $M \subseteq N$, then either $N = M$ or $N = B$. Thus, the maximal ideals in B are just the maximal elements in the lattice of ideals of B. Examples: the trivial ideal is maximal in 2; the ideals, in fields of sets, defined by the exclusion of one point are maximal (see Exercise 2).

Maximal ideals are characterized by a curious algebraic property.

Lemma 1. *An ideal M in a Boolean algebra B is maximal if and only if either p is in M or p' is in M, but not both, for each p in B.*

Proof. Assume first that for some p in B, neither p nor p' is in M. If N is the ideal generated by $M \cup \{p\}$, then N is a proper ideal, by Corollary 18.3, and it properly includes M because it contains p. Consequently, M is not maximal.

For the converse, assume that always either p or p' is in M, and suppose that N is an ideal properly including M; it is to be proved that $N = B$. Since $N \neq M$, there is an element p in N that does not belong to M. The assumption implies that the element p' is in M, and therefore also in N; since N is an ideal, it follows that $p \vee p'$ is in N. Therefore, N coincides with B.

There is another characterization of maximal ideals, due to Stone [66], that is quite useful. A Boolean ideal M is said to be *prime* if it is proper and if the presence of $p \wedge q$ in M always implies that at least one of p and q is in M.

S. Givant, P. Halmos, *Introduction to Boolean Algebras*,
Undergraduate Texts in Mathematics, DOI: 10.1007/978-0-387-68436-9_20,
© Springer Science+Business Media, LLC 2009

Corollary 1. *A Boolean ideal M is maximal if and only if it is prime.*

Proof. Assume that M is a maximal ideal, and argue by contraposition that it is prime. If neither p nor q is in M, then both p' and q' are in M, by Lemma 1, and consequently so is the join $p' \vee q'$. Since this join may be written as $(p \wedge q)'$, by the De Morgan laws, M cannot contain $p \wedge q$, by the lemma.

Now suppose M is a prime ideal. Since 0 is in M, and $0 = p \wedge p'$, at least one of the elements p and p' must belong to M, by the assumption that M is prime. Both of them cannot belong to M, for then M would contain the join $p \vee p'$, and would consequently be an improper ideal. It follows from Lemma 1 that M is maximal.

Corollary 2. *A principal ideal (p) is maximal if and only if p' is an atom.*

Proof. Maximality for a principal ideal (p) is equivalent to the conditions that $p \neq 1$ and that $q \leq p$ or $q' \leq p$ for every element q, by Lemma 1. Reformulated in terms of p', these conditions say that $p' \neq 0$ and that $p' \leq q'$ or $p' \leq q$ for every element q. This reformulation just expresses the fact that p' is an atom, by Lemma 14.1.

So far, we do not know that maximal ideals exist. The next theorem guarantees that they exist in abundance. It is usually called the *maximal ideal theorem*, and is due to Stone [66] and (for fields of sets) Tarski [70].

Theorem 12. *Every proper ideal in a Boolean algebra is included in a maximal ideal.*

Proof. Let M be a proper ideal in a Boolean algebra B. Enumerate the elements of B in a (possibly) transfinite sequence $\{p_i\}_{i<\alpha}$ indexed by the set of ordinals less than some ordinal α. Define a corresponding sequence $\{M_i\}_{i\leq\alpha}$ of proper ideals in B with the following properties: (1) $M_0 = M$; (2) $M_i \subseteq M_j$ whenever $i \leq j \leq \alpha$; (3) either p_i or p_i' is in M_{i+1} for each $i < \alpha$. The definition of the sequence proceeds by induction on ordinal numbers.

Put $M_0 = M$. Then M_0 is a proper ideal by assumption, and condition (1) is satisfied automatically, while conditions (2) and (3) (with α replaced by 0) hold vacuously for the family $\{M_i\}_{i\leq 0}$. For the induction step, consider an ordinal $k \leq \alpha$, and suppose proper ideals M_i have been defined for each ordinal $i < k$ so that the family $\{M_i\}_{i<k}$ satisfies conditions (1), (2), and (3) (with α replaced by k, with $j < k$ in condition (2), and with $i + 1 < k$ in condition (3)). When k is a successor ordinal, say $k = i + 1$, the definition of M_k splits into two cases. If either p_i or p_i' is in M_i, put $M_k = M_i$; otherwise, define M_k to be the ideal generated by the set $M_i \cup \{p_i\}$. The ideal M_k is proper, either by the induction hypothesis (in the first case) or by Corollary 18.3 (in the second case). When k is a limit ordinal, put

$$M_k = \bigcup_{i<k} M_i.$$

This union is a proper ideal, by Lemma 19.2. It is a simple matter to check that conditions (1)–(3) (with α replaced by k) hold for the family $\{M_i\}_{i \leq k}$.

The ideal M_α is the desired maximal extension of M. It is a proper ideal, by construction, and it extends M, by conditions (1) and (2). To verify maximality, consider an arbitrary element p in B. This element occurs somewhere in the enumeration of the elements of B, say $p = p_i$. Either p_i or p_i' is in M_{i+1}, by condition (3), so either p or p' is in M_α, by condition (2). Both elements cannot belong to M_α, for that would force M_α to be improper. The maximality of M_α now follows by Lemma 1.

The following somewhat sharper formulation of the maximal ideal theorem is frequently useful.

Corollary 3. *For every proper ideal M in a Boolean algebra B, and for every element p in B that does not belong M, there exists a maximal ideal that includes M and does not contain p.*

Proof. The ideal N generated by $M \cup \{p'\}$ is proper, by Corollary 18.3. Apply the maximal ideal theorem to obtain a maximal ideal that includes N, and therefore also includes M. Since that maximal ideal contains p', it does not contain p.

The maximal ideal theorem can be used to prove the existence, in every infinite Boolean algebra B, of maximal ideals that are not principal. (For fields of sets, this result goes back to Tarski [70].) Notice, first of all, that the unit of an infinite Boolean algebra cannot be written as a join of finitely many atoms; for otherwise the algebra would be atomic with finitely many atoms, by Lemmas 14.2 and 14.3, and would therefore be finite, by Theorem 6 (p. 119). Consider now the ideal M generated by the set of all atoms in B; it consists of those elements of B that can be written as joins of finite sets of atoms (Exercise 18.14). The unit cannot be written as such a join, so the ideal M is proper; consequently, M can be extended to a maximal ideal N, by the maximal ideal theorem. Since N is a proper ideal and contains every atom, it cannot contain the complement of any atom. Consequently, it cannot be a principal ideal, by the characterization of maximal principal ideals in Corollary 2.

Corollary 4. *Every infinite Boolean algebra has maximal ideals that are non-principal.*

The dual of the notion of a maximal ideal plays an important role in Boolean algebra and in other areas of mathematics. A *maximal filter*, or an *ultrafilter* as it is often called, is a proper filter that is not properly included in any other proper filter. Thanks to the isomorphism between the lattice of ideals and the lattice of filters of a Boolean algebra, we can immediately conclude from the maximal ideal theorem and its corollary that every proper filter can be extended to an ultrafilter, and that every infinite Boolean algebra has a non-principal ultrafilter.

Exercises

1. Prove that a subset M of a Boolean algebra B is a maximal ideal if and only if it contains 0, is closed under join, and contains exactly one of p and p', for every element p in B.

2. Let B be a field of subsets of a set X, and x_0 an element of X. Prove that the class of all sets in B that do not contain x_0 is a maximal ideal in B.

3. Prove that every ideal in a Boolean algebra is the intersection of the maximal ideals that include it. (This theorem is due to Stone [66] and Tarski — see [73] and [74].)

4. Suppose B is a Boolean subalgebra of A. If M is a maximal ideal in A, prove that the intersection $M \cap B$ is a maximal ideal in B.

5. Suppose f is a Boolean homomorphism from B into A. If M is a maximal ideal in A, prove that the inverse image of M under f, the set

$$N = \{p \in B : f(p) \in M\},$$

is a maximal ideal in B.

6. Use induction on the ordinals $k \leq \alpha$ to prove that the family $\{M_i\}_{i \leq k}$ of proper ideals defined in the proof of the maximal ideal theorem satisfies conditions (1)–(3) of that proof (with α replaced by k in conditions (2) and (3)).

7. Formulate and prove a characterization of ultrafilters that is analogous to Lemma 1.

8. Formulate and prove the dual of Corollary 1, for filters.

9. Formulate and prove the dual of Corollary 2, for filters. In other words, characterize when a principal filter is maximal.

10. Prove directly, without using the maximal ideal theorem, that every proper filter is contained in an ultrafilter.

11. Prove directly that every infinite Boolean algebra has a non-principal ultrafilter.

12. A subset E of a Boolean algebra is said to have the *finite meet property*, or the *finite intersection property* in the case of a field of sets, if the meet of every finite subset of E is non-zero. Prove that every subset with the finite meet property is included in an ultrafilter.

13. Formulate and prove the dual of Exercise 12.

14. Let B be the field of finite and cofinite subsets of the natural numbers. Describe the maximal ideals and the maximal filters in B.

15. (Harder.) Prove that if B is a proper Boolean subalgebra of A, then there is a maximal ideal in B that has at least two different extensions to a maximal ideal in A.

16. (Harder.) Give an example of an incomplete epimorphism between two complete Boolean algebras.

17. (Harder.) Suppose p is an element in a Boolean algebra A, and B is the relativization of A to p. Prove that every maximal ideal in B has a unique extension to a maximal ideal in A that does not contain p. Conversely, show that every maximal ideal in A that does not contain p is the extension of a uniquely determined maximal ideal in B.

18. Show that every ideal (and in particular every maximal ideal) in a Boolean algebra B is a Boolean ring (possibly without unit) under the ring operations of B.

19. (Harder.) Prove that every Boolean ring A (possibly without unit) is a maximal ideal in some Boolean algebra B such that the ring operations

of A coincide with the ring operations of B, restricted to A. To what extent is the extension B unique? (This result is due to Stone [66]. Together with Exercise 18, it shows that the study of Boolean rings — possibly without units — is essentially the same as the study of maximal ideals in Boolean algebras, or, equivalently, in Boolean rings with unit.)

20. Prove that every non-degenerate Boolean ring with unit can be obtained by adjoining a new unit to a Boolean ring that may or may not have a unit. (See Exercise 1.10.)

21. Prove that a Boolean ring without unit is necessarily infinite. (This theorem is due to Stone [66]. Compare Exercise 1.12.)

22. A maximal ideal M in a Boolean algebra B is a Boolean ring (possibly without a unit) under the restricted ring operations of B, by Exercise 18. Prove that every ring homomorphism from M into a Boolean algebra A — including the trivial homomorphism — can be extended in one and only one way to a Boolean homomorphism from B into A, and every Boolean homomorphism from B into A is the extension of a ring homomorphism from M into A.

23. (Harder.) An ideal in a commutative ring is said to be *prime* if it is proper and satisfies the following condition: for any elements p and q in the ring, if $p \cdot q$ is in the ideal then at least one of p and q is in the ideal. Prove that even when a Boolean ring does not have a unit, an ideal is prime if and only if it is maximal. (This theorem is due to Stone [66].)

24. Prove that an ideal M in a Boolean ring B (with or without a unit) is prime if and only if the quotient B/M is non-degenerate and has no zero-divisors. (This observation is due to Stone [66]. An element is a *zero-divisor* if it is non-zero and its product with some other non-zero element is zero; see Exercise 1.12.)

25. (Harder.) A version of Corollary 3 applies to Boolean rings without unit. Prove that if M is an ideal in a Boolean ring B without unit, and if p an element of B that is not in M, then there is a maximal ideal in B that includes M and does not contain p. (This theorem is due to Stone [66].)

26. Formulate and prove the analogue of Exercise 3 for Boolean rings without unit. (This result is due to Stone [66].)

27. (Harder.) A version of Corollary 3 also applies to distributive lattices. A subset M of a lattice B is called an *ideal* if it is non-empty and satisfies conditions (18.7) and (18.8). The ideal is said to be *prime* if it is proper (it does not equal B) and, for any elements p and q in B, the presence of $p \wedge q$ in M implies that either p or q must be in M. Prove that if M is an ideal in a distributive lattice B, and if p is an element of B that is not in M, then there is a prime ideal in B that includes M and does not contain p.

28. Formulate and prove the analogue of Exercise 3 for distributive lattices.

Chapter 21

Homomorphism and Isomorphism Theorems

The *kernel* of a homomorphism f from a Boolean algebra B to a Boolean algebra A is the set of those elements in B that f maps onto 0 in A. In symbols, the kernel M of f is defined by

$$M = f^{-1}(\{0\}) = \{p \in B : f(p) = 0\}.$$

The kernel of a homomorphism is always an ideal. The proof is a straight-forward verification that conditions (18.6)–(18.8) are satisfied. Suppose M is the kernel of f. Certainly 0 is in M, since $f(0) = 0$. If p and q are in M, then

$$f(p \vee q) = f(p) \vee f(q) = 0 \vee 0 = 0,$$

so that $p \vee q$ is in M. If p is in M, and if q is an arbitrary element of B, then

$$f(p \wedge q) = f(p) \wedge f(q) = 0 \wedge f(q) = 0,$$

so that $p \wedge q$ is in M.

Every example of a homomorphism (such as the ones we saw in Chapter 12) gives rise to an example of an ideal, namely its kernel. Thus, if f is the relativizing homomorphism defined by $f(p) = p \wedge p_0$ for every element p, then the corresponding ideal consists of all those elements p for which $p \wedge p_0 = 0$, or, equivalently, $p \leq p_0'$. In other words, it is the principal ideal (p_0'). If f is defined on a field of subsets of X so that $f(P)$ is the value of the character-istic function of P at some particular point x_0 of X, then the corresponding ideal consists of all those sets P in the field that do not contain x_0. If, finally, the homomorphism f is induced by a mapping ϕ from a set X into a set

S. Givant, P. Halmos, *Introduction to Boolean Algebras*,
Undergraduate Texts in Mathematics, DOI: 10.1007/978-0-387-68436-9_21,
© Springer Science+Business Media, LLC 2009

Y, then the corresponding ideal consists of all those subsets P of Y that are disjoint from the range of ϕ.

Here is a general and useful remark about homomorphisms and their kernels: a necessary and sufficient condition that a homomorphism be a monomorphism (one-to-one) is that its kernel be $\{0\}$. Proof of necessity: if f is one-to-one and $f(p) = 0$, then $f(p) = f(0)$, and therefore $p = 0$. Proof of sufficiency: if the kernel of f is $\{0\}$ and if $f(p) = f(q)$, then

$$f(p+q) = f(p) + f(q) = f(p) + f(p) = 0,$$

so that $p + q = 0$, and this means that $p = q$.

The definition of ideals was formulated so as to guarantee that the kernel of every homomorphism is an ideal. It is natural and important to raise the converse question: is every ideal the kernel of some homomorphism? The answer is easily seen to be yes: if M is an ideal of a Boolean algebra B, then the projection f of B onto the quotient B/M is an epimorphism with kernel M. This proves the following result, known as the *homomorphism theorem*.

Theorem 13. *Every ideal is the kernel of some epimorphism, namely the projection onto the corresponding quotient algebra.*

What do the homomorphic images of a Boolean algebra B look like? A quotient of B modulo an ideal is always a homomorphic image of B. The next theorem says that, up to isomorphism, these are the only homomorphic images of B. The result is usually called the *first isomorphism theorem*.

Theorem 14. *If f is a Boolean homomorphism from B onto A, and if M is the kernel of f, then B/M is isomorphic to A via the mapping*

$$p/M \rightarrow f(p).$$

Proof. Let p and q be elements of B. A short computation shows that

$$f(p) = f(q) \qquad \text{if and only if} \qquad p/M = q/M.$$

Indeed,

$$
\begin{aligned}
f(p) = f(q) \quad &\text{if and only if} \quad f(p) + f(q) = 0, \\
&\text{if and only if} \quad f(p+q) = 0, \\
&\text{if and only if} \quad p + q \in M, \\
&\text{if and only if} \quad p/M = q/M.
\end{aligned}
$$

The first equivalence uses the observation from (1.10) that every element is its own inverse under the operation of Boolean addition. The second equivalence uses the homomorphism properties of f, the third uses the definition of the kernel of f, and the fourth uses the characterization of congruence classes given in (18.15).

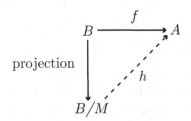

Let h be the mapping from B/M into A that takes each coset p/M to the image $f(p)$. (See the diagram above.) The preceding computation shows that h is well defined and maps B/M one-to-one into A; it maps B/M onto A because f maps B onto A. The proof that h preserves meet and complement makes use of the definitions of meet and complement in B/M, the definition of h, and the homomorphism properties of f:

$$h((p/M) \wedge (q/M)) = h((p \wedge q)/M) = f(p \wedge q)$$
$$= f(p) \wedge f(q) = h(p/M) \wedge h(q/M),$$

and

$$h((p/M)') = h(p'/M) = f(p') = f(p)' = h(p/M)'.$$

This completes the proof that h is an isomorphism from B onto A.

It is occasionally helpful to view the first isomorphism theorem as an assertion about factoring homomorphisms. From this perspective it says that every homomorphism on a Boolean algebra can be factored into the composition of a uniquely determined monomorphism and an appropriate projection.

Corollary 1. *Let f_0 be a Boolean homomorphism from B into A_0 with kernel M, and f the projection from B to B/M. There is a unique monomorphism g from B/M into A_0 such that $f_0 = g \circ f$.*

Proof. Let A be the image of B under f_0. The quotient B/M is isomorphic to A via an isomorphism g that maps each coset p/M to $f_0(p)$, by the first

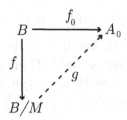

isomorphism theorem. (See the diagram above.) Obviously, g is a monomorphism of B/M into A_0. The composition $g \circ f$ coincides with f_0 because it maps each element p in B to the element

$$g(f(p)) = g(p/M) = f_0(p).$$

If h is any homomorphism from B/M into A_0 for which the composition $h \circ f$ coincides with f_0, then

$$h(p/M) = h(f(p)) = f_0(p)$$

for each coset p/M, by the definition of $f(p)$ and the assumed equality. Consequently, h coincides with g.

Associated with the isomorphism theorem there is a cluster of results of the universal algebraic kind, some of which we now proceed to state.

Suppose that M is a proper ideal in a Boolean algebra B. Write $[M, B]$ for the class, or *interval*, of ideals N in B such that $M \subseteq N \subseteq B$. This interval forms a sublattice, and in fact a complete sublattice, of the lattice of ideals in B: the infimum (the intersection) and supremum (the ideal generated by the union) of a family of ideals in the interval is again an ideal in the interval. It turns out that this sublattice is isomorphic to the lattice of ideals of the quotient B/M in a natural way.

Write $A = B/M$, and let f be the projection from B to A. The projection associates with every ideal N in the interval $[M, B]$ an ideal in A, namely the image set

$$P = f(N) = \{f(p) : p \in N\}.$$

The argument that P really is an ideal is straightforward. The zero element of A is the image, under f, of the zero element of B; since the zero of B is in the ideal N, the zero of A is in P. If r and s are elements in P, then they are images, under f, of elements p and q in N. The join $p \vee q$ is in N (because N is an ideal), and the image of this join under f is

$$f(p \lor q) = f(p) \lor f(q) = r \lor s;$$

consequently, $r \lor s$ is in P. If, finally, r and s are elements in P and A respectively, then they are images, under f, of elements p in N and q in B respectively (s is the image of an element q in B because f maps B onto, and not just into, A). The meet $p \land q$ is in the ideal N, and the image of this meet under f is

$$f(p \land q) = f(p) \land f(q) = r \land s;$$

consequently, $r \land s$ is in P. Thus, P satisfies the conditions for being an ideal in A. A common notation for this ideal is N/M (or sometimes $N + M$), because

$$P = f(N) = \{f(p) : p \in N\} = \{p/M : p \in N\} = \{p + M : p \in N\}.$$

Conclusion: the correspondence induced by f,

(1) $N \to f(N) = N/M,$

is a mapping from the sublattice $[M, B]$ into the lattice of ideals in A.

An argument analogous to the proceeding one shows that if P is an arbitrary ideal in A, then its inverse image

$$f^{-1}(P) = \{p \in B : f(p) \in P\}$$

is an ideal in B that includes M. In other words, it is an ideal in the sublattice $[M, B]$. What is its image under f? And what is the inverse image of $f(N)$ when N is an ideal in B? It is not difficult to check that

(2) $f(f^{-1}(P)) = P$ and $f^{-1}(f(N)) = N$

for every ideal P in A and every ideal N in B. Here is the proof of the second identity. Every element p in N is contained in $f^{-1}(f(p))$, and therefore in $f^{-1}(f(N))$, by the definition of the inverse image of a set under a function. Thus, N is included in $f^{-1}(f(N))$. To establish the reverse inclusion, consider an arbitrary element p in $f^{-1}(f(N))$. The image $f(p)$ belongs to $f(N)$, by the definition of the inverse image of $f(N)$ under f, so there must be an element q in N such that $f(p) = f(q)$. It follows from the homomorphism properties of f and from (1.10) that

$$f(p + q) = f(p) + f(q) = 0.$$

The Boolean sum $p + q$ therefore belongs to the kernel of f, which is M. The ideal N includes M, by assumption, so it contains $p + q$. The element q

also belongs to N, and ideals are closed under Boolean addition, by condition (18.9), so the sum $(p + q) + q$ must be in N. This last sum is just p; consequently, p is in N. A similar but simpler argument establishes the first identity.

The identities in (2) imply that the correspondence (1) maps the interval $[M, B]$ bijectively to the lattice of ideals in A. Indeed, if P is an ideal in A, then $N = f^{-1}(P)$ is an ideal in $[M, B]$, and P is the image of N under f, by the first identity in (2). In other words, correspondence (1) is onto. If N_1 and N_2 are two ideals in $[M, B]$ such that

$$f(N_1) = f(N_2),$$

then

$$N_1 = f^{-1}(f(N_1)) = f^{-1}(f(N_2)) = N_2,$$

by the second identity in (2). Consequently, correspondence (1) is one-to-one. A similar argument shows that correspondence (1) preserves the lattice ordering of inclusion. If $N_1 \subseteq N_2$, then obviously $f(N_1) \subseteq f(N_2)$. If, conversely, the latter inclusion holds, then

$$f^{-1}(f(N_1)) \subseteq f^{-1}(f(N_2)),$$

so that $N_1 \subseteq N_2$, by the second identity in (2). Because the correspondence (1) preserves the lattice ordering, it is a lattice isomorphism.

The formal statement of the preceding observations is usually called the *correspondence theorem.*

Theorem 15. *For every ideal M in a Boolean algebra B, the correspondence*

$$N \to N/M$$

is an isomorphism from the sublattice of ideals in B that include M to the lattice of ideals in B/M.

The relationship between the ideals of a quotient B/M and the ideals of B that extend M goes beyond what is expressed in the correspondence theorem. Quotients of B/M by quotient ideals N/M are in fact isomorphic to quotients of B. It is therefore unnecessary to consider quotients of quotient Boolean algebras, quotients of quotients of quotient Boolean algebras, and so on. Each such quotient essentially reduces to a quotient of the original Boolean algebra. A precise formulation of this fact is contained in the *second isomorphism theorem.*

Theorem 16. *Let M and N be ideals in a Boolean algebra B, with $M \subseteq N$. The quotient of B/M by the ideal N/M is isomorphic to the quotient B/N via the mapping*

$$(p/M)/(N/M) \to p/N.$$

Proof. Write

$$A = B/M \qquad \text{and} \qquad C = (B/M)/(N/M).$$

The projection f from B onto A, and the projection g from A onto C, are both epimorphisms, so the composition $h = g \circ f$ is an epimorphism from B onto C. The kernel of h is N, as can be verified in two steps: the kernel of g is the ideal N/M, by the observations following (18.12)–(18.14); and the set of elements of B that are mapped into N/M by f is just

$$f^{-1}(N/M) = f^{-1}(f(N)) = N,$$

by the second identity in (2).

The first isomorphism theorem, applied to the epimorphism h, says that the quotient B/N is isomorphic to C via the mapping that takes each coset p/N to $h(p)$. The proof is completed by observing that

$$h(p) = g(f(p)) = g(p/M) = (p/M)/(N/M).$$

The second isomorphism theorem may also be formulated as an assertion about "factoring" homomorphisms f_0. The corollary to the first isomorphism theorem assumes that the kernel of f_0 coincides with the ideal M, and concludes that f_0 can be factored into the composition of a uniquely determined monomorphism and the projection f from B to B/M. The next corollary assumes only that the kernel of f_0 includes the ideal M, and concludes that f_0 can be factored into the composition of a uniquely determined homomorphism and the projection f.

Corollary 2. *Let f_0 be a Boolean homomorphism from B into A_0, and suppose that its kernel includes the ideal M. There is then a unique homomorphism g from B/M into A_0 such that $f_0 = g \circ f$, where f is the projection from B to B/M.*

Proof. Let A be the image of B under f_0, and M_0 the kernel of f_0. It is assumed that $M \subseteq M_0$. The quotient B/M_0 is isomorphic to A via the function g_0 that maps each coset p/M_0 to $f_0(p)$, by the first isomorphism theorem. Of course, g_0 is a monomorphism of B/M_0 into A_0. The projection f of B onto B/M is an epimorphism, as is the projection g_2 of B/M onto $(B/M)/(M_0/M)$. The latter quotient is isomorphic to B/M_0 via the function g_1 that maps $(p/M)/(M_0/M)$ to p/M_0 for each p, by the second isomorphism theorem. The composition

$$g = g_0 \circ g_1 \circ g_2$$

is a homomorphism from B/M into A_0 with the property that $f_0 = g \circ f$ (see the diagram). Indeed, an easy computation shows that g maps each element p/M to $f_0(p)$:

$$g(p/M) = g_0(g_1(g_2(p/M))) = g_0(g_1((p/M)/(M_0/M))) = g_0(p/M_0) = f_0(p);$$

therefore, the composition $g \circ f$ maps each element p in B to $f_0(p)$.

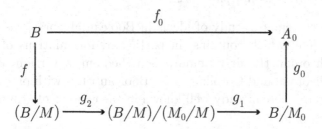

To prove the uniqueness of g, consider any homomorphism h from B/M into A_0 with the property that $f_0 = h \circ f$. The definition of f and the assumed equality imply that

$$h(p/M) = h(f(p)) = f_0(p)$$

for each coset p/M. Consequently, h coincides with g.

A Boolean algebra is called *simple* if it is not degenerate and has no non-trivial proper ideals. The underlying intuition is that a simple Boolean algebra B cannot be "simplified" by passing to a quotient B/M: each such quotient is either degenerate, or else isomorphic to B via the projection homomorphism. The former happens when M is improper ($M = B$) and the latter when M is trivial ($M = \{0\}$). Simplicity is a universal algebraic concept, but, as it turns out, in the context of Boolean algebras it is not a fruitful one. The reason is that there is just one simple algebra, namely 2. Clearly 2 is simple, since it has just two ideals: the trivial ideal $\{0\}$ and the improper ideal $\{0, 1\}$. Assume now that B is any simple Boolean algebra, and consider an arbitrary non-zero element p in B. The principal ideal generated by p is non-trivial (it contains p), and therefore must be improper, by the assumed simplicity of B. This can happen only if $p = 1$. Thus, any element in B different from 0 must equal 1, and consequently $B = 2$.

The correspondence between the ideals of a quotient algebra and the ideals of its "numerator" (formulated in the correspondence theorem) shows that the quotient algebra is simple if and only if its "denominator" (the ideal) is maximal. Indeed, a Boolean quotient B/M is not simple and not degenerate if and only if it has a proper, non-trivial ideal. By the correspondence theorem, this happens if and only if there is an ideal N in B that is between M and B, but different from both. Such an ideal N exists in B if and only if the ideal M is proper but not maximal, by the maximal ideal theorem.

Corollary 3. *An ideal in a non-degenerate Boolean algebra B is maximal if and only if B/M is isomorphic to 2.*

The first systematic study of ideals in Boolean algebras was carried out by Stone in [66], which contains, in particular, formulations of the homomorphism theorem, the first isomorphism theorem, a version of the correspondence theorem, and Corollary 3 for Boolean rings with or without unit. (These theorems were already well known in the context of groups and commutative rings.)

Exercises

1. The *cokernel* of a Boolean homomorphism f from B to A is the set of those elements in B that f maps to 1. Prove that f is one-to-one if and only if its cokernel is $\{1\}$.

2. Prove that the kernel of a Boolean homomorphism from B into A is a proper ideal in B if and only if A is not degenerate.

3. Prove that every non-degenerate Boolean algebra can be mapped homomorphically to 2.

4. Prove that distinct elements in a Boolean algebra can always be distinguished by a 2-valued homomorphism. More precisely, show that if p and q are distinct elements in a Boolean algebra B, then there is a 2-valued homomorphism f on B such that $f(p) \neq f(q)$.

5. If X is an infinite set, show that there is a homomorphism from $\mathcal{P}(X)$ to 2 that maps the finite subsets of X to 0 and the cofinite subsets to 1.

6. Let M be an ideal in a Boolean algebra B, and f the projection from B onto B/M. Show that if P is an ideal in B/M, then $f^{-1}(P)$ is an ideal in B, and

$$f(f^{-1}(P)) = P.$$

7. Formulate and prove the analogue of the correspondence theorem for filters. (See Exercise 18.31.)

8. Prove that if two epimorphisms on a Boolean algebra have the same kernel, then the image algebras are isomorphic.

9. Suppose M and N are ideals in a Boolean algebra B, and $M \subseteq N$. Set-theoretically, a coset p/M is the set of translations by p of elements of the ideal M:

$$p + M = \{p + q : q \in M\}.$$

The quotient ideal N/M in B/M is the class of cosets corresponding to elements of N:

$$N/M = \{p/M : p \in N\} = \{p + M : p \in N\}$$
$$= \{\{p + q : q \in M\} : p \in N\}.$$

Set-theoretically, what are the cosets p/N and $(p/M)/(N/M)$? Are they identical?

10. (Harder.) Prove that if B is a proper Boolean subalgebra of A, then there is a 2-valued homomorphism on B that can be extended in two different ways to a 2-valued homomorphism on A.

11. (Harder.) Formulate and prove an analogue of Corollary 3 for Boolean rings without a unit. (This result is due to Stone [66].)

Chapter 22

The Representation Theorem

The representation problem asks whether every Boolean algebra is isomorphic to a field of sets. In other words, given Boolean algebra A, does there always exist a set X such that A is isomorphic to a subalgebra of $\mathcal{P}(X)$? Each point x_0 in a set X can be used to define a 2-valued homomorphism on the field $\mathcal{P}(X)$: the homomorphism takes the value 1 on the subsets of X that contain x_0, and the value 0 on the subsets that do not contain x_0 (see Chapter 12). This comment suggests that if we start with a Boolean algebra A and seek to represent it as a field over some set X, a reasonable place to conduct the search for points suitable to make up X is among the 2-valued homomorphisms of A. The suggestion would be impractical if it turned out that A has no 2-valued homomorphisms. Our first result along these lines is that there is nothing to fear; there is always a plethora of 2-valued homomorphisms.

Lemma 1. *For every non-zero element p of every Boolean algebra A there is a 2-valued homomorphism x on A such that $x(p) = 1$.*

Proof. Let p be a non-zero element in a Boolean algebra A, and consider the principal ideal N generated by the complement p'. The elements in N are just the elements of A that are below p'. Since p is not 0, its complement is not 1, and therefore 1 is not in N. It follows that N is a proper ideal. Extend N to a maximal ideal M, by the maximal ideal theorem (p. 172), and observe that p does not belong to M, since M contains p' (Lemma 20.1). The quotient A/M is a two-element Boolean algebra, by Corollary 21.3. If z is the projection from A to A/M that maps each element q to the coset q/M, and if y is the (unique) isomorphism from A/M to 2 that maps $0/M$ to 0, and $1/M$ to 1, then the composition $x = y \circ z$ is the desired 2-valued homomorphism on A. Indeed,

S. Givant, P. Halmos, *Introduction to Boolean Algebras*,
Undergraduate Texts in Mathematics, DOI: 10.1007/978-0-387-68436-9_22,
© Springer Science+Business Media, LLC 2009

$$x(q) = y(z(q)) = y(q/M) = \begin{cases} 0 & \text{if } q \in M, \\ 1 & \text{if } q \notin M. \end{cases}$$

In particular, $x(p) = 1$, since p is not in M.

The following assertion (due to Stone [67]) is known as the (*Stone*) *representation theorem*, and is one of the most fundamental results about Boolean algebras.

Theorem 17. *Let X be the set of 2-valued homomorphisms on a Boolean algebra A. Then A is embeddable into $\mathcal{P}(X)$ via the mapping defined by*

$$f(p) = \{x \in X : x(p) = 1\}$$

for each p in A.

Proof. The verification that f is a homomorphism is purely mechanical. For instance, if p and q are elements in A, then

$$\begin{aligned} f(p \vee q) &= \{x \in X : x(p \vee q) = 1\} \\ &= \{x \in X : x(p) \vee x(q) = 1\} \\ &= \{x \in X : x(p) = 1 \text{ or } x(q) = 1\} \\ &= \{x \in X : x(p) = 1\} \cup \{x \in X : x(q) = 1\} \\ &= f(p) \cup f(q). \end{aligned}$$

The first and last equalities use the definition of f, the second uses the homomorphism properties of the mappings x in X, the third uses the definition of join in $\mathbf{2}$, and the fourth uses the definition of union. Similarly,

$$\begin{aligned} f(p') &= \{x \in X : x(p') = 1\} \\ &= \{x \in X : x(p)' = 1\} \\ &= \{x \in X : x(p) = 0\} \\ &= \{x \in X : x(p) = 1\}' \\ &= f(p)'. \end{aligned}$$

In order to demonstrate that f is one-to-one, it suffices to show that its kernel contains only 0. If $p \neq 0$, then there is a 2-valued homomorphism x on A such that $x(p) = 1$, by Lemma 1; consequently, the set $f(p)$ is not empty. Thus, each non-zero element p in A is mapped by f to a non-empty set, so p cannot be in the kernel of f.

The mapping f of the theorem is often called the *canonical embedding* of A.

Corollary 1. *Every Boolean algebra is isomorphic to a field of sets.*

There are at least two variations of the representation f in Theorem 17. Instead of the set X of 2-valued homomorphisms on A, one can use the set Y of kernels of these homomorphisms. The kernel of a 2-valued homomorphism on A is a maximal ideal in A, and conversely, every maximal ideal in A is the kernel of a 2-valued homomorphism on A (Corollary 21.3). The set Y is therefore just the collection of maximal ideals in A. The mapping g that one uses to embed A into $\mathcal{P}(Y)$ is defined by

$$g(p) = \{M \in Y : p \notin M\}$$

for p in A.

One can prove directly that g is a monomorphism. An alternative proof, using the monomorphism properties of the representation f in Theorem 17, goes as follows. The function ϕ that takes each 2-valued homomorphism on A to its kernel is a bijection of X to Y. This bijection induces an isomorphism k from $\mathcal{P}(Y)$ to $\mathcal{P}(X)$ that maps each set of maximal ideals in A to the corresponding set of 2-valued homomorphisms on A (see p. 94). The composite monomorphism $k^{-1} \circ f$ coincides with g, since

$$
\begin{aligned}
k^{-1}(f(p)) &= k^{-1}(\{x \in X : x(p) = 1\}) \\
&= k^{-1}(\{x \in X : x(p) \neq 0\}) \\
&= \{\phi(x) : x \in X \text{ and } x(p) \neq 0\} \\
&= \{M \in Y : p \notin M\} \\
&= g(p).
\end{aligned}
$$

The second variation of the representation f is the dual of the first. Instead of the set Y of maximal ideals in A, one uses the set Z of ultrafilters (maximal filters) in A. The corresponding embedding of A into $\mathcal{P}(Z)$ — call it h — is defined by

$$h(p) = \{N \in Z : p \in N\}$$

for all p in A. There are several psychological advantages to using the set of ultrafilters instead of the set of maximal ideals. First, the definition of the embedding h has a positive form, whereas the definition of g involves a negation ("$p \notin M$"). Second, there is a simple intuition underlying the representation when one uses ultrafilters. Recall that every atomic Boolean algebra is isomorphic to a field of subsets of the set of its atoms (Theorem 6, p. 119). To represent A, it therefore suffices to construct an extension of A in which there is an atom below each non-zero element. A set E of elements in A generates a proper filter if and only if E has the *finite meet property*, that is to say, if and only if the meet of any finite subset of E is not zero (Exercise 18.20). Ultrafilters are just the maximal subsets of A that have the finite meet property, and it is not difficult to prove that every subset of A with the finite meet property can be extended to an ultrafilter (Exercise 20.12). The intuition underlying the definition of h is that each ultrafilter in A should determine a unique atom q, and in fact q should be the infimum of the elements of N. The atoms of the field $\mathcal{P}(Z)$ are the singletons of ultrafilters N, and for each element p in A, the atom $\{N\}$ is below the set $h(p)$ in $\mathcal{P}(Z)$ just in case N is an element of $h(p)$, or, equivalently, just in case p is in N. If each element p in A is identified with its isomorphic image $h(p)$, then the atom $\{N\}$ can be thought of as the infimum, in $\mathcal{P}(Z)$, of the set of elements in N. We shall have more to say about this in the next chapter, in Lemma 23.1.

There is still another formulation of the representation theorem that is useful and that offers its own insights. The field of sets $\mathcal{P}(X)$ is isomorphic to the Boolean algebra 2^X via the mapping that takes each subset of X to its characteristic function (see Chapter 3). The composition of this isomorphism with the monomorphism f from Theorem 17 is therefore an embedding of A into 2^X. The embedding takes each element p in A to the characteristic function (on X) of the set of those 2-valued homomorphisms (on A) that map p to 1.

Corollary 2. *Every Boolean algebra is embeddable into a power of 2.*

Exercises

1. Let Y be the set of maximal ideals in a Boolean algebra A. Prove directly that the mapping g defined on A by

$$g(p) = \{M \in Y : p \notin M\}$$

   is an embedding of A into $\mathcal{P}(Y)$.

2. Let Z be the set of ultrafilters in a Boolean algebra A. Prove directly that the mapping h defined on A by

$$h(p) = \{N \in Z : p \in N\}$$

   is an embedding of A into $\mathcal{P}(Z)$.

3. Let $B = 2 \times 2 \times 2$. Describe the set Z of ultrafilters in B and the canonical embedding of B into $\mathcal{P}(Z)$.

4. (Harder.) A relativization of a Boolean algebra is not a subalgebra, but it constitutes a Boolean algebra in a natural way (see Chapter 12). Is that Boolean algebra necessarily isomorphic to a subalgebra of the whole algebra?

5. Is every complete Boolean algebra isomorphic to a complete field of sets?

6. Is every Boolean algebra isomorphic to a subalgebra of a complete algebra?

7. (Harder.) Prove that every distributive lattice is isomorphic to a lattice of sets under the operations of intersection and union. (This theorem is due to Birkhoff [5].)

Chapter 23

Canonical Extensions

If A is a Boolean algebra and if X is the set of 2-valued homomorphisms on A, then A is mapped isomorphically to a subalgebra of $\mathcal{P}(X)$ via the canonical embedding f that takes each element p in A to the set of 2-valued homomorphisms on A that map p to 1 (see Theorem 17, p. 189). The algebra $\mathcal{P}(X)$ can therefore be viewed as a Boolean extension of A. The purpose of this chapter is to characterize this extension algebraically.

The most obvious properties of $\mathcal{P}(X)$ are that it is complete and atomic, and that it contains an isomorphic copy of A as a subalgebra. In order to describe other properties in the most perspicuous way, it is convenient to identify each element p in A with its image $f(p)$ in $\mathcal{P}(X)$. In terms of this identification, another, less obvious property, can be formulated as follows: any two (distinct) atoms q and r in $\mathcal{P}(X)$ are *separated* by some element p in A in the sense that $q \leq p$ and $r \leq p'$. Indeed, the atoms must have the form $q = \{x\}$ and $r = \{y\}$ for some distinct 2-valued homomorphisms x and y on A. The distinctness of the homomorphisms implies the existence of an element p in A such that

$$x(p) = 1 \qquad \text{and} \qquad y(p) = 0.$$

It follows that x belongs to the set $f(p)$, and y to $f(p')$, by the definition of f. In other words,

$$q = \{x\} \subseteq f(p) \qquad \text{and} \qquad r = \{y\} \subseteq f(p').$$

When p is identified with $f(p)$, these inclusions say that $q \leq p$ and $r \leq p'$.

Yet another property of $\mathcal{P}(X)$ is its *compactness* with respect to A: if a subset E of A has, as its supremum in $\mathcal{P}(X)$, an element q in A, then a finite subset of E must already have q as its supremum (in A and in $\mathcal{P}(X)$). The

S. Givant, P. Halmos, *Introduction to Boolean Algebras*, 193
Undergraduate Texts in Mathematics, DOI: 10.1007/978-0-387-68436-9_23,
© Springer Science+Business Media, LLC 2009

proof of this assertion begins with the special case in which the supremum in question is 1, and the argument in that case proceeds by contraposition. Consider an arbitrary subset E of A, and suppose that no finite subset of E has 1 as its supremum (in A). It is to be shown that 1 is not the supremum of E in $\mathcal{P}(X)$. The ideal generated by E in A is the set

$$\{p \in A : p \le \bigvee F \text{ for some finite } F \subseteq E\}$$

(Theorem 11, p. 155), and it does not contain 1, by assumption. Consequently, this ideal is proper and can therefore be extended to a maximal ideal M in A, by the maximal ideal theorem. The maximality of M implies the existence of a 2-valued homomorphism x on A with kernel M (Theorem 14, p. 179, and Corollary 21.3). For each element p in M, we have $x(p) = 0$, so x cannot belong to $f(p)$. It follows that x does not belong to the union $\bigcup_{p \in M} f(p)$. On the other hand, x does belong to $f(1)$, because 1 is not in M and therefore $x(1) = 1$. These observations (and the fact that M includes E) show that

$$\bigcup_{p \in E} f(p) \subseteq \bigcup_{p \in M} f(p) \ne f(1).$$

Since suprema are unions in $\mathcal{P}(X)$, the set $f(1)$ cannot be the supremum of the set

$$f(E) = \{f(p) : p \in E\}$$

in $\mathcal{P}(X)$. When the elements of A are identified with their images under f, this conclusion says that 1 cannot be the supremum of the set E in $\mathcal{P}(X)$.

To prove the general case of the compactness property, consider an arbitrary subset E of A, and suppose that an element q from A is the supremum of E in $\mathcal{P}(X)$. The set $E \cup \{q'\}$ is also a subset of A (since q belongs to A), and its supremum in $\mathcal{P}(X)$ is 1, since

$$\bigvee (E \cup \{q'\}) = q' \vee \bigvee E = q' \vee q = 1;$$

consequently, there must be a finite subset F of E such that

$$1 = q' \vee \bigvee F,$$

by the observations of the previous paragraph. Form the meet of both sides of this equation with q, and use the fact that q is an upper bound of F, to conclude that

$$q = q \wedge 1 = q \wedge (q' \vee \bigvee F) = (q \wedge q') \vee (q \wedge \bigvee F) = 0 \vee \bigvee F = \bigvee F.$$

In other words, q is the supremum of a finite subset F of E.

The properties discussed above can be formulated in an abstract setting. An extension of a Boolean algebra A is said to have the *atom separation property* with respect to A if any two atoms q and r in the extension are separated by some element p in A in the sense that $q \le p$ and $r \le p'$. The extension is said to have the *compactness property* with respect to A provided that whenever a subset E of A has a supremum in the extension, and that supremum — say q — belongs to A, then some finite subset of E already has q as its supremum (in A and in the extension). The argument of the preceding paragraph shows that the general compactness property is equivalent to the special case in which $q = 1$. A complete and atomic Boolean extension of A that has the atom separation and the compactness properties is called a *canonical extension*, or a *perfect extension*, of A.

If A is an arbitrary Boolean algebra, and if f is the canonical embedding of A into $\mathcal{P}(X)$ (where X is the set of 2-valued homomorphisms on A), then the argument above shows that $\mathcal{P}(X)$ is a canonical extension of the subalgebra that is an isomorphic copy of A under f. An application of the exchange principle (Chapter 12) leads to the conclusion that A itself has a canonical extension. This proves the following *existence theorem for canonical extensions.*

Theorem 18. *Every Boolean algebra has a canonical extension.*

In particular, every Boolean algebra has a complete and atomic extension. Happily, a Boolean algebra has just one canonical extension, up to isomorphic copies. The key step in the proof of this assertion is formulated in the next lemma, and is closely related to the intuition underlying the proof of the representation theorem that was discussed at the end of Chapter 22.

Lemma 1. *If B is a canonical extension of a Boolean algebra A, then the distinct atoms in B are precisely the infima of the distinct ultrafilters in A.*

Proof. Every atom in B is the infimum of a uniquely determined ultrafilter in A. For the proof, consider an atom q in B. The principal filter generated by q in B, the set

$$\{p \in B : q \le p\},$$

is an ultrafilter, by the dual of Corollary 20.2, and consequently the intersection of this ultrafilter with A, the set

(1) $$N = \{p \in A : q \le p\},$$

is an ultrafilter in A, by the dual of Exercise 20.4. The infimum of the set N exists in B, by the assumed completeness of B; call it s. It is to be shown that $q = s$. The atom q is obviously a lower bound of N, and s is by definition the greatest lower bound of N; consequently, $q \leq s$. In order to establish equality, it suffices to show that q is the only atom below s, because every element in an atomic Boolean algebra — and in particular, s — is the supremum of the set of atoms that it dominates (Lemma 14.3). Consider any atom r in B that is different from q. There must be an element p in A such that $q \leq p$ and $r \leq p'$, by the atom separation property. The element p belongs to N, by (1), and therefore $s \leq p$, since s is the infimum of N. It follows that r and s are disjoint, since

$$s \wedge r \leq p \wedge p' = 0.$$

Thus, no atom different from q can be below s.

Consider now an arbitrary ultrafilter M in A. We shall show that the infimum of M in B — call it s — is an atom in B. The assumption that $s = 0$ leads to a contradiction. Indeed, some finite subset of M would then have infimum 0, by the dual of the compactness property. Since M is closed under finite meets, this would imply that 0 is in M, and therefore that M is an improper filter. But M is an ultrafilter, so it must be proper. Thus, $s \neq 0$. The algebra B is atomic, and s is not zero, so there is at least one atom q below s. The set N defined by (1) is an ultrafilter in A, and its infimum is q, by the observations of the first paragraph. Moreover, N includes M because every element in M is above s, by assumption, and therefore also above q. The assumed maximality of M now implies that $M = N$. Consequently, the infimum of M coincides with the infimum of N. In other words, $q = s$, so that s is an atom.

The correspondence that takes each atom q in B to the ultrafilter N in A defined by (1) is a one-to-one mapping, since the atom q uniquely determines, and is uniquely determined by, the ultrafilter N. The correspondence maps the set of atoms in B onto the set of ultrafilters in A, because the infimum of an arbitrary ultrafilter N in A is an atom q such that N is the ultrafilter defined by (1). Conclusion: the set of atoms in B is in bijective correspondence with the set of ultrafilters in A.

Two canonical extensions, say B and C, of a Boolean algebra A have the same number of atoms, because the number of atoms is equal to the number of ultrafilters in A, by Lemma 1. The two extensions are also complete, so any bijection between the sets of atoms extends to an isomorphism between B

and C, by Corollary 14.2. It follows that B and C are isomorphic. It is actually possible to construct an isomorphism that is the identity mapping on A, but some care must be exercised in selecting the bijection between the two sets of atoms. Here are the details.

Lemma 1 says that each ultrafilter in A uniquely determines an atom in B and an atom in C, namely the atom that is the infimum of the ultrafilter; conversely, each atom in each of the two algebras is uniquely determined by some ultrafilter in A. Let ϕ be the function that for each ultrafilter N in A, maps the atom in B determined by N (namely, the infimum of N in B) to the atom in C determined by N (namely, the infimum of N in C). The mapping ϕ is a bijection from the set of atoms in B to the set of atoms in C, by Lemma 1. The isomorphism f from B to C induced by this bijection is defined by

$$f(p) = \bigvee \{\phi(q) : q \text{ is an atom in } B \text{ and } q \le p\},$$

by the remarks following Corollary 14.2. What is the value of f on an element p in A? An atom q in B is below p just in case p belongs to the ultrafilter N determined by q in A (see (1) in the proof of Lemma 1). The image atom $\phi(q)$ in C determines the same ultrafilter N, by the definition of ϕ, so $\phi(q)$ is below p just in case p belongs to N. It follows that ϕ maps the set of atoms in B that are below p onto the set of atoms in C that are below p. Of course, p is the supremum of the set of atoms it dominates in each algebra, by Lemma 14.3. The definition of f therefore yields

$$f(p) = \bigvee \{\phi(q) : q \text{ is an atom in } B \text{ and } q \le p\}$$
$$= \bigvee \{r : r \text{ is an atom in } C \text{ and } r \le p\} = p.$$

In other words, f maps each element in A to itself. The following *uniqueness theorem for canonical extensions* has been proved.

Theorem 19. *Any two canonical extensions of a Boolean algebra A are isomorphic via a mapping that is the identity on A.*

The theorem may be viewed as a justification for the common practice of referring to *the* canonical extension of a Boolean algebra.

What is the size of the canonical extension of a Boolean algebra A, compared to the size of A? A finite Boolean algebra is its own canonical extension, so there is no increase in size. Suppose A has an infinite number m of elements. It can be shown that the number of ultrafilters in A is between m and 2^m. Each ultrafilter determines, and is determined by, a unique atom in the canonical extension, and each element in the canonical extension determines, and is determined by, a unique set of atoms. There are therefore as many elements in the canonical extension as there are subsets of the set of ultrafilters in A. Conclusion: the canonical extension has between 2^m and 2^{2^m} elements.

The algebraic characterization of the canonical extension discussed in this chapter is due to Jónsson and Tarski [32].

Exercises

1. Prove that a finite Boolean algebra is its own canonical extension.

2. Prove that the canonical extension of a Boolean algebra A possesses the following *dual compactness property*: if an element p in A is the infimum in B of a subset E of A, then p is the infimum of some finite subset of E.

3. (Harder.) Give an example of complete and atomic Boolean algebras A and B such that B is a subalgebra, but not a complete subalgebra, of A.

4. If two Boolean algebras are isomorphic via a mapping g, prove that their canonical extensions are isomorphic via a mapping that extends g.

5. Give an example of an incomplete monomorphism between complete Boolean algebras.

6. Prove that every homomorphism between Boolean algebras can be extended to a homomorphism between the corresponding canonical extensions.

7. (Harder.) Prove that, in fact, every homomorphism between Boolean algebras can be extended to a complete homomorphism between the corresponding canonical extensions. More precisely, let A and B be Boolean algebras, and A_1 and B_1 the corresponding canonical extensions. Take E to be the set of infima in B_1 of subsets of B:

$$E = \{r \in B_1 : r = \bigwedge F \text{ for some } F \subseteq B\}.$$

Prove that if g is a homomorphism from B to A, then the mapping f from B_1 to A_1 defined by

$$f(p) = \bigvee \left\{ \bigwedge \{g(s) : s \in B \text{ and } q \le s\} : q \in E \text{ and } q \le p \right\}$$

for p in B_1 is a complete homomorphism that extends g. Show further that if g is one-to-one or onto, then so is f. (This is a special case of a much more general theorem due to Jónsson and Tarski [32].)

8. Prove that if B is a Boolean subalgebra of A, then the canonical extension of B is (up to an isomorphism that is the identity on A) a complete subalgebra of the canonical extension of A, and in fact it is the complete subalgebra generated by B.

Chapter 24

Complete Homomorphisms and Complete Ideals

A homomorphism between Boolean algebras preserves suprema and infima of finite sets, but in general it will not preserve suprema and infima of infinite sets. A Boolean homomorphism f from B to A is said to be *complete* if it preserves all suprema that do exist. This means that if a family $\{p_i\}$ of elements in B has a supremum p, then the family $\{f(p_i)\}$ has a supremum in A, and that supremum is $f(p)$.

A complete homomorphism f automatically preserves all infima that happen to exist. For the proof, suppose a family $\{p_i\}$ of elements has an infimum, say p. The supremum of the family $\{p_i'\}$ is then p', because

$$p' = \Big(\bigwedge_i p_i\Big)' = \bigvee_i p_i',$$

by Lemma 8.1. The homomorphism f is assumed to preserve all existing suprema, so

$$f(p)' = f(p') = f\Big(\bigvee_i p_i'\Big) = \bigvee_i f(p_i') = \bigvee_i f(p_i)' = \Big(\bigwedge_i f(p_i)\Big)'.$$

Form the complements of the first and last terms to conclude that

$$f(p) = \bigwedge_i f(p_i).$$

There is a simple and useful criterion for completeness: a Boolean homomorphism f from B into A is complete if and only if whenever a family $\{p_i\}$ in B has the unit as its supremum, then the family $\{f(p_i)\}$ in A also has the

S. Givant, P. Halmos, *Introduction to Boolean Algebras*,
Undergraduate Texts in Mathematics, DOI: 10.1007/978-0-387-68436-9_24,
© Springer Science+Business Media, LLC 2009

unit as its supremum. The necessity of the condition is obvious; it follows directly from the definition of a complete homomorphism. To demonstrate the sufficiency of the condition, assume that f satisfies the condition, and consider an arbitrary family $\{p_i\}$ in B with supremum p. The element $f(p)$ is certainly an upper bound of the family $\{f(p_i)\}$ in A, since $p_i \leq p$, and therefore $f(p_i) \leq f(p)$, for each i. Also, the family obtained by adjoining p' to $\{p_i\}$ clearly has the supremum 1 in B, since

$$1 = p' \vee p = p' \vee \bigvee_i p_i.$$

The assumed condition on f therefore implies that the family obtained by adjoining $f(p)'$ to $\{f(p_i)\}$ has the supremum 1 in A:

$$1 = f(1) = f(p' \vee \bigvee_i p_i) = f(p') \vee \bigvee_i f(p_i) = f(p)' \vee \bigvee_i f(p_i).$$

If q is any upper bound of $\{f(p_i)\}$ in A, then $q' \wedge f(p_i) = 0$ for each i, and therefore

$$q' = q' \wedge 1 = q' \wedge (f(p)' \vee \bigvee_i f(p_i))$$

$$= (q' \wedge f(p)') \vee \bigvee_i (q' \wedge f(p_i)) = (q' \wedge f(p)') \vee 0 = q' \wedge f(p)'.$$

It follows that

$$q' \leq f(p)',$$

or, equivalently, that $f(p) \leq q$. Consequently, $f(p)$ is the least upper bound of the family $\{f(p_i)\}$.

We have already encountered several examples of complete homomorphisms. One is the relativizing homomorphism on a Boolean algebra induced by an element p_0: it maps each p in the algebra to the meet $p \wedge p_0$. Another is the homomorphism on a complete field of sets induced by a point x_0: it maps each set P in the field to 1 or 0 according as x_0 is, or is not, in P. Not all homomorphisms are complete, however. For instance, let X be the set of non-negative integers, and consider the field B consisting of the finite sets of positive integers and the complements of such sets in X, namely the cofinite subsets of X that contain the integer 0. The identity function f on B is certainly a monomorphism from B into the field A of all subsets of X, but it is not complete. To see this, write $P_i = \{i\}$ for each positive integer i. The family $\{P_i\}$ has X as its supremum in B, and $X - \{0\}$ as its supremum in A. Consequently, the monomorphism f does not map the supremum of $\{P_i\}$

in B to the supremum of $\{P_i\}$ in A. The field B is of course not a complete Boolean algebra. However, there even exist incomplete homomorphisms between complete Boolean algebras (see Exercise 23.5).

The kernels of homomorphisms on a Boolean algebra B are just the ideals in B, by the homomorphism theorem. Is there an analogous theorem for complete homomorphisms? The first step in answering this question is to introduce an appropriate modification of the notion of an ideal. Define a *complete ideal* in a Boolean algebra B to be a subset M of B such that

(1) $0 \in M,$

(2)   if $\{p_i\}$ is a family in M with a supremum p in B, then $p \in M$,

(3)   if $p \in M$ and $q \in B$, then $p \wedge q \in M$.

In other words, a complete ideal is an ideal that satisfies condition (2) for infinite families of elements.

It is not difficult to check that kernels of complete homomorphisms are complete ideals. Indeed, if M is the kernel of a complete homomorphism f, then M is certainly an ideal (Chapter 21), so it suffices to check that M also satisfies condition (2). Consider a family $\{p_i\}$ of elements in M, and suppose that the supremum of the family exists in B, say it is p. The completeness of f implies that

$$f(p) = f\left(\bigvee_i p_i\right) = \bigvee_i f(p_i) = 0;$$

therefore, p belongs to M.

It is natural to ask about the converse: is every complete ideal the kernel of a complete homomorphism? Consider a complete ideal M in a Boolean algebra B. The projection f of B onto the quotient B/M, which maps each element p to the coset p/M, has M as its kernel. The question just posed will be answered positively if it can be shown that f is complete as a homomorphism. Let $\{p_i\}$ be a family of elements in B with supremum 1. It is to be shown that $f(1)$ is the supremum of the family $\{f(p_i)\}$ in the quotient B/M. In other words, it is to be shown that $1/M$ is the supremum of the family $\{p_i/M\}$. Certainly, $1/M$ is an upper bound of the family: the inequality $p_i \leq 1$ implies that $p_i/M \leq 1/M$. Consider now any upper bound q/M of the family. Since $p_i/M \leq q/M$, by assumption, we have

$$(p_i \wedge q')/M = (p_i/M) \wedge (q/M)' = 0/M.$$

In other words, $p_i \wedge q'$ is in the ideal M for each i. The supremum of the family $\{p_i \wedge q'\}$ in B is q', because

$$q' = q' \wedge 1 = q' \wedge \bigvee_i p_i = \bigvee_i (q' \wedge p_i) = \bigvee_i (p_i \wedge q'),$$

by Lemma 8.3. It follows from the assumed completeness of M that M contains q'. In other words, $q'/M = 0/M$, and therefore $q/M = 1/M$. Conclusion: $1/M$ is the least upper bound of the family $\{p_i/M\}$.

The preceding argument proves the natural analogue, for complete ideals, of the homomorphism theorem.

Theorem 20. *Every complete ideal is the kernel of some complete epimorphism, namely the corresponding projection.*

Principal ideals are examples of complete ideals. Indeed, if M is the ideal generated by an element p in a Boolean algebra, and if E is any subset of M, then p is certainly an upper bound of E. Consequently, the supremum of E, if it exists, must be below p and hence in M.

The intersection of an arbitrary family of complete ideals in a Boolean algebra is again a complete ideal. For the proof, consider such a family $\{M_i\}$, and let M be its intersection. Certainly, M is an ideal, by the remarks in Chapter 18. To verify that M also satisfies condition (2), let $\{p_i\}$ be a family of elements in M, and suppose that the supremum of this family exists, say it is p. Each ideal M_i is assumed to be complete and to contain every element in the family $\{p_i\}$, so M_i must contain the supremum p of the family, by condition (2) (applied to M_i). Consequently, the intersection M also contains p.

It follows from the observations of the preceding paragraph that if E is an arbitrary subset of a Boolean algebra B, then the intersection of the complete ideals in B that include E is itself a complete ideal. (There is at least one complete ideal that includes E, namely the improper ideal B.) That intersection, say M, is the smallest complete ideal that includes E; in other words, every complete ideal that includes E also includes M. The ideal M is called the complete ideal *generated* by E. Warning: the complete ideal generated by a set E is not the same as the ideal generated by E; the latter is always included in the former, but the reverse inclusion generally fails.

The definition just given is non-constructive. It gives no idea of the elements of B that actually belong to the complete ideal generated by E. There is a description of these elements that is somewhat analogous to the description of the elements belonging to the ideal generated by E (Theorem 11, p. 155). To formulate it, we introduce some notation. Let E^d be the set of elements in B that are below some element of E,

(4) $E^d = \{p \in B : p \leq q \text{ for some } q \in E\}$.

The set E^d is called the *downward closure* of E (in B).

Lemma 1. *An element p of a Boolean algebra is in the complete ideal generated by a set E if and only if it is the supremum of some subset of E^d.*

Proof. Let M be the complete ideal generated by a set E in a Boolean algebra B, and let N be the set of elements in B that are suprema of subsets of E^d. It is to be proved that M and N are equal. The first step is verifying that N is a complete ideal that includes E. An element p in E is always the supremum of a subset of E^d, namely the subset $\{p\}$. Therefore, E is included in N. Clearly, N contains 0, since 0 is the supremum of the empty set. To verify condition (2), consider a family $\{p_i\}$ of elements in N, and suppose that this family has a supremum, say p, in B. Each p_i is the supremum of some subset F_i of E^d, by the definition of N. The union F of these subsets is itself a subset of E^d, and its supremum is p, since

$$p = \bigvee_i p_i = \bigvee_i \left(\bigvee_i F_i \right) = \bigvee_i F,$$

by Lemma 8.2. Therefore, p is in N, by the definition of N. To check condition (3), suppose p is in N and q in B. The definition of N implies that p is the supremum of a subset F of E^d. Each element r in F is below some element s in E, by definition (4). Since

$$r \wedge q \leq r \leq s,$$

the meet $r \wedge q$ must also belong to E^d. Thus, the set of these meets,

$$G = \{r \wedge q : r \in F\},$$

is a subset of E^d. The supremum of G is $p \wedge q$, because

$$p \wedge q = \left(\bigvee F \right) \wedge q = \bigvee \{r \wedge q : r \in F\} = \bigvee G,$$

by Lemma 8.3. It now follows from the definition of N that $p \wedge q$ belongs to N. Conclusion: N is a complete ideal that includes E.

The set M is the smallest complete ideal that includes E, and N is a complete ideal that includes E, so M must be included in N. On the other hand, every element of E^d certainly belongs to M, by condition (18.18). The completeness of M therefore implies that whenever a subset of E^d has a supremum in B, that supremum must be in M. It follows (from the definition of N) that N is included in M. Thus, $M = N$, as desired.

There is another description of the complete ideal generated by a set E in a Boolean algebra that is worth formulating. Recall that the set of upper bounds of E is, by definition, the set U of elements in the Boolean algebra that are above every element of E. The set of lower bounds of U is the set L of elements in the Boolean algebra that are below every element in U. It turns out that L is the complete ideal generated by E.

Lemma 2. *The complete ideal generated by a set E in a Boolean algebra is the set of lower bounds of the set of upper bounds of E.*

Proof. Let M be the complete ideal generated by a set E in a Boolean algebra B, let U be the set of upper bounds of E (in B), and let L be the set of lower bounds of U. It is to be shown that M and L coincide. The first step is verifying that L is a complete ideal that includes E. Each element in E is a lower bound of U, by the definition of U, so E is included in L, by the definition of L. It is equally obvious that 0 is in L: the set U is not empty (it contains 1), and 0 is below every element in U. To verify that L satisfies condition (2), consider an arbitrary family $\{p_i\}$ of elements in L, and suppose the family has a supremum p in B. The elements of U are upper bounds for the set L, by the definition of L, so they are upper bounds of the family $\{p_i\}$. It follows that they are all above the least upper bound p. Hence, p is in L, by the definition of L. Condition (3) is also easy to check. If p is in L and q in B, then p is below every element of U, and consequently so is $p \wedge q$. It follows that $p \wedge q$ is in L, by the definition of L.

The set M is, by assumption, the smallest complete ideal that includes E. It has just been shown that L is also a complete ideal that includes E. Consequently, M is included in L. To establish the reverse inclusion, consider an arbitrary element p in L, and let F be the set of elements in E^d that are below p. It will be shown that p is the supremum of F in B. It then follows from the previous lemma that p is in M, and therefore that L is included in M.

The element p is certainly an upper bound of F, by the definition of F. Consider any other upper bound of F, say q. It must be proved that $p \leq q$. The first step is to prove that every element of E is below $q \vee p'$. For each r in E, the meet $r \wedge p$ is below p and belongs to E^d, by (4), so it is in F, by the definition of F. Every element of F is below q, by assumption, so $r \wedge p \leq q$. A straightforward computation yields

$$r = r \wedge 1 = r \wedge (p \vee p') = (r \wedge p) \vee (r \wedge p') \leq q \vee p'.$$

It has been shown that $q \vee p'$ is an upper bound of E, so this join belongs to the set U of upper bounds of E. The element p belongs to the set L of lower bounds of U, by assumption, so $p \leq q \vee p'$ and consequently

$$p = p \wedge (q \vee p') = (p \wedge q) \vee (p \wedge p') = (p \wedge q) \vee 0 = p \wedge q.$$

In other words, $p \leq q$, as was to be shown.

There is a close connection between complete ideals and the "cuts" that play a crucial role in Dedekind's classical construction of the real numbers from the rational numbers (see [15]). A *Dedekind cut* in the set of rational numbers is a pair (P, Q) of non-empty sets that partition the rational numbers (every rational number is in exactly one of the two sets) and such that every number in P is less than every number in Q. The set P has the characteristic property that it is *downward closed*: if p is in P, and if q is less than p, then q is in P. Similarly, the set Q is *upward closed*: if p is in Q, and if q is greater than p, then q is in Q. The set Q can of course be reconstructed from the set P; it is just the complement of P in the set of rational numbers. Thus, one could define a Dedekind cut to be simply a non-empty, downward closed set of rational numbers that is different from the set of all rational numbers.

Consider now a subset E of a Boolean algebra B. The set U of upper bounds of E is upward closed, and the set L of lower bounds of U is downward closed, and the two sets L and U have at most one element in common. The pair (L, U) is therefore a kind of Dedekind cut in the partial ordering of B. (The fact that L and U may have one element in common — namely, the supremum of the set E, if it exists — is of no real significance.) The set U can, of course, be reconstructed from the set L — it is just the set of upper bounds of L — so one can consider L itself to be a Dedekind cut in the partial ordering of B. In view of the preceding lemma, we may conclude that complete ideals are the analogues, for Boolean algebras, of Dedekind cuts of rational numbers.

The class of complete ideals in a Boolean algebra B forms a complete lattice. The infimum of any family $\{M_i\}$ of complete ideals in B is just the intersection of the family. The supremum of $\{M_i\}$ is the complete ideal generated by the union of the ideals in the family, or, in different words, it is the intersection of the complete ideals that include each M_i. (There is always one such complete ideal, namely the improper ideal B.)

The lattice of complete ideals in B is not a sublattice of the lattice of all ideals of B. The (binary) operation of meet is the same in both lattices, since the meet of two complete ideals is just their intersection. The operation of join, however, is not the same. In the lattice of all ideals, the join of two complete ideals M and N is the intersection of all ideals that include $M \cup N$; in the lattice of complete ideals it is the intersection of all complete ideals that include $M \cup N$.

The difference can be illustrated by an example. Let B be the Boolean algebra of finite and cofinite subsets of integers, let M be the ideal of all finite sets of even integers, and let N the ideal of all finite sets of odd integers. It is easy to check that these two ideals are complete for a trivial reason: no infinite subset of M or N has a supremum in B. Indeed, suppose E is an infinite subset of M, and consider any upper bound P of E in B. The union of E is an infinite set of even integers, so the set P must be infinite, and hence cofinite. This means that it contains infinitely many odd integers; removing any one of them produces a proper subset of P that is still an upper bound of E in B. The proof that N is complete is analogous.

The join of the ideals M and N in the lattice of all ideals of B is the class of sets of the form $P \cup Q$, where P is a finite set of even integers and Q a finite set of odd integers, by Lemma 19.1. Every finite set of integers can be written in this form, so the join of the two ideals is really just the ideal of all finite sets of integers. On the other hand, the join of M and N in the lattice of complete ideals of B is the improper ideal B. To see this, write $P_i = \{i\}$ for each integer i. Every set P_i is in M or in N, and is therefore in the complete ideal L generated by $M \cup N$. The supremum of the family $\{P_i\}$ is the unit of B, the set of all integers. This supremum must be in L, by the definition of a complete ideal, so L contains the unit element. This forces L to coincide with B (the ideal generated by the unit).

The same example shows that the identity

(5) $$M \vee N = \{p \vee q : p \in M \text{ and } q \in N\}$$

fails to be true in the lattice of complete ideals of a Boolean algebra. The identity does hold, however, when the two complete ideals are principal, because

$$(p) \vee (q) = (p \vee q),$$

by Corollary 19.2, and the principal ideal $(p \vee q)$ is complete.

The lattice of complete ideals is not only complete (as a lattice), it is also distributive. The proof of this assertion is necessarily different from the proof of the analogous result for the lattice of all ideals, since the identity (5) no longer holds. Consider three complete ideals L, M, and N in a Boolean algebra. It is to be shown that in the lattice of complete ideals

(6) $L \wedge (M \vee N) = (L \wedge M) \vee (L \wedge N)$

and

(7) $L \vee (M \wedge N) = (L \vee M) \wedge (L \vee N).$

Begin with the proof of (6). The complete ideals $L \wedge M$ and $L \wedge N$ are certainly included in the left side of (6), since M and N are included in $M \vee N$. It follows that the complete ideal generated by the union of $L \wedge M$ and $L \wedge N$ is also included in the left side of (6). In other words, the ideal on the right side of (6) is included in the ideal on the left side. To establish the reverse inclusion, consider an arbitrary element p in the left-hand ideal. Since p is in the complete ideal generated by $M \cup N$, and since M and N coincide with their downward closures, by condition (18.18), there must be subsets E_1 of M, and E_2 of N, such that p is the supremum of $E_1 \cup E_2$, by Lemma 1. The element p is also in L, by assumption, so the sets E_1 and E_2 are included in L, by condition (18.18). It follows that E_1 is a subset of $L \wedge M$, and E_2 a subset of $L \wedge N$. The union $E_1 \cup E_2$ is therefore a subset of the right-hand ideal. The supremum p of this union must also be in the right-hand ideal, because that ideal is complete.

The dual distributive law (7) can be established by a similar argument. Alternatively, it can be derived directly from (6), because each of the two distributive laws for lattices is derivable from its dual. (See Exercise 7.19.)

There is another, quite surprising difference between the lattice of ideals and the lattice of complete ideals of a Boolean algebra. In the lattice of ideals, certain ideals may fail to have a complement. In the lattice of complete ideals, this never happens: every complete ideal has a complement. For the proof, consider a complete ideal M in a Boolean algebra B. The *annihilator* of M is the set N defined by

$$N = \{p \in B : p \wedge q = 0 \text{ for all } q \in M\}.$$

(See Exercise 18.6.) It is easy to check that N is a complete ideal in B. For instance, to verify condition (2), consider any family $\{p_i\}$ of elements in N that has a supremum p in B. For each element q in M,

$$p \wedge q = \left(\bigvee_i p_i \right) \wedge q = \bigvee_i (p_i \wedge q) = 0,$$

by Lemma 8.3. Consequently, the supremum p also belongs to the annihilator N.

The claim is that N is the complement of M in the sense that

(8) $M \wedge N = \{0\}$ and $M \vee N = B.$

The first identity is almost immediate: if p is in M and simultaneously in N, then $p \wedge p = 0$, by the definition of N, and therefore $p = 0$, by the idempotent law for meet. The second identity in (8) is equivalent to the assertion that the unit 1 belongs to the complete ideal generated by the set $M \cup N$. The unit will be in this ideal if and only if it is the supremum of the set $M \cup N$, by Lemma 1. Obviously, the unit is an upper bound of this union. Consider any other upper bound p of the union. Each element q in M is below p, and therefore $p' \wedge q = 0$. In other words, p' belongs to the annihilator N. Since p is also an upper bound of N, we conclude that $p' \leq p$, which can happen only if $p = 1$. It follows that the unit is the least upper bound of the set $M \cup N$, as desired.

The remarks in the preceding paragraphs lead to the conclusion (due to Stone [66], [68] and Tarski [75]) that the class of complete ideals in a Boolean algebra is itself a complete Boolean algebra.

Theorem 21. *The class of all complete ideals in a Boolean algebra B is itself a complete Boolean algebra with respect to the distinguished Boolean elements and operations defined by*

(1) $0 = \{0\},$
(2) $1 = B,$
(3) $M \wedge N = M \cap N,$
(4) $M \vee N = \bigcap \{L : L$ *is a complete ideal in B and $M \cup N \subseteq L\},$*
(5) $M' = \{p \in B : p \wedge q = 0$ *for all $q \in M\}.$*

The infimum and the supremum of a family of complete ideals are, respectively, the intersection of the family and the complete ideal generated by the union of the family.

Proof. The proof of the theorem amounts to showing that the identity laws (2.13), the complement laws (2.14), the commutative laws (2.18), and the distributive laws (2.20) are all valid in the algebra of complete ideals (see Exercise 2.2). The identity and commutative laws follow immediately from definitions (1)–(4). The complement and distributive laws were verified above.

Every principal ideal in a Boolean algebra is complete. What kind of structure, if any, does the class of these ideals possess? Here is one answer to this question.

Corollary 1. *The principal ideals in a Boolean algebra B form a regular subalgebra of the Boolean algebra of all complete ideals in B. The supremum of a family of principal ideals $\{(p_i)\}$ exists in the subalgebra just in case the supremum of the family of elements $\{p_i\}$ exists in B; in fact,*

$$\bigvee_i (p_i) = (p) \qquad \text{if and only if} \qquad p = \bigvee_i p_i.$$

Proof. Let A be the Boolean algebra of all complete ideals in B, and C the class of all principal ideals in B. Certainly, the zero ideal (0) is in C. Suppose p and q are elements in B. The principal ideals generated by these elements satisfy the equations

$$(p) \vee (q) = (p \vee q), \qquad (p) \wedge (q) = (p \wedge q), \qquad (p)' = (p')$$

in the lattice of all ideals in B, by Corollary 19.1 and the subsequent remarks. (The operations on the left sides of these equations are performed in the lattice of ideals, whereas those on the right are performed in B.) All the ideals in these equations are principal, so the equations also hold in A. (Keep in mind, though, that the operations of join in A and in the lattice of all ideals are in general not the same.) The class C is thus a subalgebra of A.

Turn now to the proof of the final assertion of the corollary, and assume first that the supremum of a family $\{(p_i)\}$ of principal ideals exists in C, say it is (p). Each ideal (p_i) is included in (p), so $p_i \leq p$. In other words, p is an upper bound (in B) of the family of elements $\{p_i\}$. If q is any other upper bound of this family of elements, then $p_i \leq q$, and therefore (p_i) must be included in (q), for each i. The ideal (p) is assumed to be the supremum of the given family of ideals, so (p) must be included in (q). It follows that $p \leq q$. This proves that p is the supremum of the family $\{p_i\}$ in B.

To prove the converse, assume that the supremum of a family $\{p_i\}$ of elements in B exists, say it is p. Then $p_i \leq p$ for each i, by the definition of supremum, so each ideal (p_i) is included in (p). In other words, (p) is an upper bound of the family of ideals $\{(p_i)\}$ in C. If (q) is any other upper bound in C of this family of ideals, then (p_i) is included in (q), and therefore $p_i \leq q$, for each i. The element p is assumed to be the supremum of the family $\{p_i\}$, so $p \leq q$. It follows that (p) is included in (q). Thus, (p) is the supremum of the family $\{(p_i)\}$ in C.

It remains to show that C is a regular subalgebra of A. Consider an arbitrary family $\{(p_i)\}$ of ideals in C, and suppose it has a supremum (p) in C. It must be shown that (p) is also the supremum in A of the given family of ideals. It is certainly an upper bound in A, since the ideal (p_i) is included in (p) for each i. To see that (p) is the least upper bound in A, consider any other complete ideal M that is an upper bound of $\{(p_i)\}$ in A. Each of the ideals (p_i) is included in M, by assumption, so the elements p_i all belong to M. The element p is the supremum of the family $\{p_i\}$ in B, by the observations of the second paragraph of the proof. Because the ideal M is assumed to be complete, the element p must also belong to M, by the definition of a complete ideal. It follows that (p) is included in M. Thus, (p) is the supremum of the family $\{(p_i)\}$ in A.

We close this chapter with a warning. There are other notions of a complete ideal that exist in the literature. For instance, some authors define an ideal to be complete if each subset of the ideal has a supremum and that supremum is in the ideal. This stronger notion of a complete ideal does not play a large role. The reason is that every such ideal is principal. Proof: if M is such an ideal, then $\bigvee M$ is in M.

Exercises

1. Prove that the relativizing homomorphism induced on a Boolean algebra by an element is complete.

2. Prove that the homomorphism induced on a complete field of sets by a point is complete.

3. Define the notion of a complete filter, and prove that an ideal is complete if and only if its dual filter is complete.

4. Prove that the class of complete filters in a Boolean algebra B is itself a complete Boolean algebra with respect to the distinguished Boolean elements and operations defined by

    (1) $$0 = \{1\},$$
    (2) $$1 = B,$$
    (3) $$M \wedge N = M \cap N,$$
    (4) $$M \vee N = \bigcap\{L : L \text{ is a complete filter in } B \text{ and } M \cup N \subseteq L\},$$
    (5) $$M' = \{p \in B : p \vee q = 1 \text{ for all } q \in M\},$$

    for all complete filters M and N in B. Show further that the Boolean algebras of complete ideals and of complete filters in B are isomorphic via the mapping that takes each complete ideal to its dual filter.

5. Prove that every complete filter is the cokernel of some complete epimorphism.

6. Prove that the complete ideal generated by a set E and the complete filter generated by the set $E' = \{p' : p \in E\}$ are the duals of one another.

7. Formulate and prove the analogue of Lemma 1 for complete filters.

8. Formulate and prove the analogue of Lemma 2 for complete filters.

9. If a Boolean homomorphism preserves all infima that happen to exist, prove that the homomorphism is complete.

10. Prove that a Boolean homomorphism f from B into A is complete if and only if whenever 0 is the infimum of a family $\{p_i\}$ in B, then 0 is the infimum of the family $\{f(p_i)\}$ in A.

11. Prove that a complete ideal in a complete Boolean algebra is closed under joins of arbitrary subsets of the ideal. Conclude that such an ideal is always principal.

12. Prove that the quotient of a complete Boolean algebra by a complete ideal is complete.

13. Prove that in the Boolean algebra of finite and cofinite sets of integers, the ideal of all finite sets of integers is not complete.

14. Prove that the collection of finite sets of integers of the form $3m$ for some integer m (in other words, the finite sets of multiples of 3) is a complete ideal in the Boolean algebra B of finite and cofinite sets of integers. Show that the same is true for the collection of finite sets of integers of the form $3m + 1$, and also for the collection of finite sets of integers of the form $3m + 2$. What is the join of these three ideals in the lattice of all ideals of B? What is the join of the three ideals in the lattice of all complete ideals of B? Generalize this example.

15. Define the *annihilator* of an arbitrary subset E (not necessarily an ideal) of a Boolean algebra to be the set of elements p such that $p \wedge q = 0$ for all q in E. Prove that the annihilator of E is a complete ideal.

16. Prove that the annihilator of a subset E of a Boolean algebra (Exercise 15) coincides with the set of complements of upper bounds of E.

17. Prove that the annihilator of a subset E of a Boolean algebra (Exercise 15) coincides with the annihilator of the complete ideal generated by E.

18. Prove that the annihilator of an ideal M is the largest ideal N with the property that

$$M \cap N = \{0\}.$$

19. Prove that an ideal M is complete if and only if it is the annihilator of the annihilator of M. In other words, in the notation of Theorem 21, prove that M is complete if and only if $M = M''$.

20. Prove that the complete ideal generated by a subset E in a Boolean algebra is just E'', the annihilator of the annihilator of E.

21. Give a direct proof that the distributive law (7) holds in the lattice of complete ideals of a Boolean algebra.

22. Verify directly, without appealing to Theorem 21, that the Boolean axioms (2.11), (2.12), (2.15), and (2.17) hold in the lattice of complete ideals of a Boolean algebra.

23. Define a correspondence f from a Boolean algebra B into its Boolean algebra of complete ideals by $f(p) = (p)$. In other words, $f(p)$ is defined to be the complete ideal generated by p. Prove that f is a complete Boolean monomorphism.

Chapter 25

Completions

The Stone representation theorem implies that every Boolean algebra is a subalgebra of a complete Boolean algebra, namely its canonical extension. One advantage of this extension is that it is atomic. A fundamental drawback is that all of the infinite joins that exist in the original algebra are changed in the passage to the canonical extension. More precisely, if an infinite subset E of a Boolean algebra A has a supremum p in A, and if p is not already the supremum of a finite subset of E, then the supremum of E in the canonical extension of A is definitely not p, by the compactness property. For many purposes, therefore, the canonical extension is not good enough.

What is needed is a complete extension in which missing suprema are "filled in", while the existing suprema are all left intact. A *completion* of a Boolean algebra A is a Boolean algebra B with the following properties: (1) A is a subalgebra of B; (2) every subset of A has a supremum in B; (3) every element in B is the supremum (in B) of some subset of A. Condition (3) is equivalent to the assertion that A is a *dense* subset of B in the sense that every non-zero element in B is above a non-zero element in A. One direction of this equivalence is obvious: if condition (3) holds, then every non-zero element of B, being the supremum of a subset of A, must be above some non-zero element of A. To prove the reverse direction of the equivalence, assume A is dense in B. Consider an arbitrary element p in B, and let E be the set of all elements in A that are below p. We shall show that p is the supremum of E in B. Clearly, p is an upper bound of E in B. Assume, for contradiction, that q is a strictly smaller upper bound of E in B. Then $p-q$ is non-zero, and hence (by density) is above a non-zero element r in A. The definition of E implies that r is in E, since $r \leq p$. The element q is an upper

S. Givant, P. Halmos, *Introduction to Boolean Algebras*,
Undergraduate Texts in Mathematics, DOI: 10.1007/978-0-387-68436-9_25,
© Springer Science+Business Media, LLC 2009

bound of E, so $r \leq q$. This last inequality contradicts the fact that $r \leq q'$. Conclusion: a Boolean algebra B is a completion of a Boolean algebra A if and only if A is a dense subalgebra of B and every subset of A has a supremum in B.

The argument just given shows that every element in a completion of a Boolean algebra A is the supremum of the set of all elements in A that are below it. A similar argument shows that a dense subalgebra of an arbitrary Boolean algebra is automatically a regular subalgebra. In other words, all infinite suprema that exist in the subalgebra are left intact. Suppose, indeed, that A is a dense subalgebra of B, and let E be a subset of A that has a supremum in A, say p. Certainly, p is an upper bound of E in B. If q were a strictly smaller upper bound of E in B, then $p - q$ would be a non-zero element of B, and therefore above a non-zero element r of A, by density. The difference $p - r$ would then be an upper bound of E in A that is strictly smaller than p, contradicting the assumption that p is the supremum of E in A.

We have yet to see that a completion B of a Boolean algebra A is in fact complete. Consider an arbitrary family $\{p_i\}$ of elements in B. Each p_i is the supremum (in B) of some subset E_i of A, by condition (3). The union $E = \bigcup_i E_i$ is a subset of A, and therefore has a supremum p in B, by condition (2). The generalized associative law formulated in Exercise 8.6 implies that p is the supremum of the family $\{p_i\}$, since

$$p = \bigvee E = \bigvee_i \left(\bigvee E_i \right) = \bigvee_i p_i.$$

A consequence of this observation is that a Boolean algebra B is a completion of A if and only if B is complete and includes A as a dense subalgebra.

It is not obvious that there are any completions at all, but fortunately every Boolean algebra does have a completion, and even more fortunately, that completion is unique, up to isomorphic copies. We first prove the *existence theorem for completions* (discovered independently by MacNeille [43] and Tarski [75] — see footnote 21 in [75]). Recall from Chapter 24 that an ideal M in a Boolean algebra A is said to be complete provided that whenever the supremum of a set of elements in M exists in A, that supremum belongs to M. The class of all complete ideals in A is a complete Boolean algebra: the meet of an arbitrary family of ideals is the intersection of the ideals in the family, the join of the family is the complete ideal generated by the union of the ideals in the family, and the complement of a complete ideal is the annihilator of the ideal (see Theorem 21, p. 209). It turns out that this algebra of complete ideals is a completion of A, provided that one identifies the principal ideals with the elements of A.

Theorem 22. *Every Boolean algebra A has a completion, namely (an isomorphic copy of) the Boolean algebra of complete ideals in A.*

Proof. Let A be a Boolean algebra, and B the class of complete ideals in A. Then B is a complete Boolean algebra, by Theorem 21. For each element p in A, the principal ideal (p) generated by p is a complete ideal, and therefore an element of B (see the remark following Theorem 20, p. 203). Define a mapping f from A into B by

$$f(p) = (p).$$

We shall show that f is a complete embedding of A into B. If p and q are elements of A, then

$$f(p \vee q) = (p \vee q) = (p) \vee (q) = f(p) \vee f(q)$$

and

$$f(p') = (p') = (p)' = f(p)'.$$

(Compare the displayed equations in the proof of Corollary 24.1, or see Corollary 19.1 and the remarks preceding Lemma 19.2.) Consequently, f is a homomorphism. If $f(p) = f(q)$, that is, if $(p) = (q)$, then $p = q$, since the generator of a principal ideal is the largest element in the ideal. Therefore, f is one-to-one. The range of f is the set of all principal ideals in A. This set is a regular subalgebra of B, by Corollary 24.1. It follows from Lemma 12.1 that f is a complete monomorphism from A into B.

Every non-zero complete ideal M in A obviously includes a non-zero principal ideal: if an element p in M is not zero, then (p) is a non-zero ideal included in M. The range of f — the image $f(A)$ — is therefore a dense subalgebra of B. Combine this observation with those of the preceding paragraphs to conclude that $f(A)$ is a (regular) dense subalgebra of the complete algebra B. In other words, B is a completion of $f(A)$. The exchange principle (see Chapter 12) now ensures that A itself has a completion.

A Boolean algebra A has many *complete extensions*, complete Boolean algebras that contain A as a subalgebra. For instance, when A is infinite, the canonical extension of A, the canonical extension of the canonical extension of A, and so on, are all distinct complete extensions of A. (They increase in size at an exponential rate, at the very least; see the remarks at the end of Chapter 23.) A completion of A distinguishes itself from other complete extensions of A by its *minimality*. There are several possible interpretations of this assertion, and they all turn out to be true.

Lemma 1. *Suppose B is a completion of a Boolean algebra A. The only embedding of B into itself that is the identity on A is the identity automorphism of B.*

Proof. Let f be an embedding of B into itself that is the identity mapping on A. For an arbitrary element q in B, the set of elements in A that are below q coincides with the set of elements in A that are below $f(q)$. Indeed, if an element p in A is below q, then

$$p = f(p) \leq f(q),$$

by the homomorphism properties of f and the fact that f is the identity mapping on A. Conversely, if $p \leq f(q)$, then

$$p = f^{-1}(p) \leq f^{-1}(f(q)) = q.$$

Since every element in B is the supremum of the set of elements in A that it dominates, by condition (2) in the definition of a completion, it now follows that $q = f(q)$. In other words, f is the identity mapping on B.

The next theorem, due independently to MacNeille [43] and Tarski [75], says that a completion of A is minimal in the sense that, up to isomorphism, it is a subalgebra of every other complete extension of A.

Theorem 23. *A completion of a Boolean algebra A can be embedded into any complete extension of A via a mapping that is the identity on A.*

Proof. Let B be a completion of A, and consider any complete extension C of A. The identity function f on A is a monomorphism of A into C. It can be extended to a homomorphism g from B into C, by the homomorphism extension theorem (Theorem 5, p. 114). To prove that g is one-to-one, it suffices to check that its kernel is trivial. Suppose $g(q) = 0$. Every element in B below q is then also mapped to 0, by the homomorphism properties of g. In particular, every element p in A with $p \leq q$ is mapped to 0 by g. But such elements p are mapped to themselves, since g extends the identity function on A. Conclusion: zero is the only element in A below q. Since q is the supremum (in B) of a set of elements in A, it follows that q is zero.

Another interpretation of minimality, given in the next corollary, says that a completion of A is a smallest complete extension of A.

Corollary 1. *A Boolean algebra B is a completion of a Boolean algebra A if and only if B is a complete extension of A, and no complete extension of A is a proper subalgebra of B.*

Proof. Suppose, first, that B is a completion of A. Consider any subalgebra C of B that is a complete extension of A. There is an embedding g of B into C that is the identity mapping on A, by the preceding theorem. Since C is a subalgebra of B, the mapping g may be viewed as an embedding of B into itself. Lemma 1 implies that g must be the identity mapping on B. In particular, the range of g is B, so B is also a subalgebra of C. Consequently, $B = C$. Conclusion: no complete extension of A is a proper subalgebra of B.

Consider now a complete extension B of A, and assume that no proper subalgebra of B is a complete extension of A. A completion C of A exists, by the existence theorem for completions, and there is an embedding g of C into B that is the identity mapping on A, by Theorem 23. The image $g(C)$ is a complete extension of A that is a subalgebra of B, so the assumption about B implies that $g(C) = B$. In other words, B is the isomorphic image of a completion of A via a mapping that is the identity on A. It follows that B must also be a completion of A.

We are ready to prove the *uniqueness theorem for completions*. It is an easy consequence of preceding observations.

Theorem 24. *Any two completions of a Boolean algebra A are isomorphic via a mapping that is the identity on A.*

Proof. Suppose B and C are completions of A. Then B can be embedded into C via a mapping g that is the identity on A, by Theorem 23. The image $g(B)$ is a complete extension of A that is a subalgebra of C, so it is equal to C by the preceding corollary. Thus, g is an isomorphism from B to C that is the identity on A.

The isomorphism g in the preceding proof is constructed indirectly, via the homomorphism extension theorem (see the proof of Theorem 23). It is useful to have a direct construction of g. Each element q in B is the supremum of a subset of A, namely the set E of all elements in A that are below q. Since g is an isomorphism that is the identity on A, it must map q to the supremum of E in C. Thus, g is the correspondence from B to C that for each subset E of A, maps the supremum of E in B to the supremum of E

in C. This argument also shows that g is the only isomorphism from B to C that is the identity on A.

The uniqueness theorem provides a justification for the common practice of referring to *the* completion of a Boolean algebra.

Dedekind [15] constructed the real numbers as a kind of order completion of the rational numbers, using subsets of rational numbers called Dedekind cuts (see the remarks following the proof of Lemma 24.2). MacNeille [43] extended Dedekind's methods to construct completions of partial orderings, and in particular completions of Boolean algebras. For this reason, the completion of a Boolean algebra A is sometimes called the *MacNeille completion* of A or even the *Dedekind–MacNeille completion* of A.

Exercises

1. Prove that every complete Boolean algebra is its own completion. Conclude that every finite Boolean algebra is its own completion.

2. Suppose B is the completion of a Boolean algebra A. Prove that an element in B is an atom if and only if it is already an atom in A.

3. Prove that the completion of a Boolean algebra A is atomic if and only if A is atomic. (This theorem is due to Tarski [75].) Conclude that if A is atomic, then its completion is isomorphic to the field of all subsets of the set of atoms of A.

4. Let A be the field of finite and cofinite subsets of an infinite set X. Prove that $\mathcal{P}(X)$ is the completion of A.

5. Prove that two atomic Boolean algebras with the same number of atoms have isomorphic completions.

6. Prove that the completion of an atomless algebra is atomless.

7. Prove that there is at most one isomorphism between two completions of a Boolean algebra A that is the identity on A.

8. Suppose B and C are both completions of a Boolean algebra A. Let g be the correspondence from B to C that, for each subset E of A, maps the supremum of E in B to the supremum of E in C. Prove directly that g is a well-defined isomorphism from B to C that is the identity mapping on A.

9. The completion of a Boolean algebra A has the property that every element is the supremum of the set of elements in A that it dominates. Does any other complete extension of A have this property?

10. Prove that every homomorphism between Boolean algebras can be extended to a homomorphism between the corresponding completions.

11. Show that not every homomorphism between Boolean algebras can be extended to a complete homomorphism between the corresponding completions. (Compare this with Exercise 23.7.)

12. Prove that every complete homomorphism between Boolean algebras can be extended to a complete homomorphism between the corresponding completions. More precisely, let A and B be Boolean algebras, and A_1 and B_1 the corresponding completions. Prove that if g is a homomorphism from B into A, then the mapping f from B_1 to A_1 defined by
$$f(p) = \bigvee \{g(s) : s \in B \text{ and } s \le p\}$$
for p in B_1 is a complete homomorphism that extends g. Show further that if g is one-to-one or onto, then so is f. (This is a special case of a much more general theorem due to Monk [44].)

13. If A and B are Boolean algebras, and if B is a regular subalgebra of A, prove that the completion of B is (up to an isomorphism that is the identity on A) a complete subalgebra of the completion of A, and in fact it is the complete subalgebra generated by B.

Chapter 26

Products of Algebras

A familiar way of making one new structure out of two old ones is to form their Cartesian product and, in case the structure involves some algebraic operations, to define the requisite operations coordinatewise. Boolean algebras furnish an instance of this procedure. The (*direct*) *product* of two Boolean algebras B and C is the algebra

$$A = B \times C$$

whose universe, the *Cartesian product* of the sets B and C, consists of the pairs (p, q) with p in B and q in C. The meet and join of two pairs in A is formed coordinatewise:

$$(p, q) \wedge (r, s) = (p \wedge r, q \wedge s) \qquad \text{and} \qquad (p, q) \vee (r, s) = (p \vee r, q \vee s),$$

where $p \wedge r$ and $p \vee r$ are the meet and join of p and r in B, while $q \wedge s$ and $q \vee s$ are the meet and join of q and s in C. The complement of a pair in A is likewise formed coordinatewise:

$$(p, q)' = (p', q'),$$

where p' and q' are the complements of p and q in B and C respectively. Under these operations, the product A is a Boolean algebra with zero $(0, 0)$ and unit $(1, 1)$. In fact, it is easy to verify that the Boolean axioms (2.11)–(2.20) hold in A. For instance, here is the proof that the commutative law for join holds in A:

$$(p, q) \vee (r, s) = (p \vee r, q \vee s) = (r \vee p, s \vee q) = (r, s) \vee (p, q).$$

The first and last equalities hold by the definition of join in A, while the middle equality holds because the commutative law for join is valid in the

S. Givant, P. Halmos, *Introduction to Boolean Algebras*,
Undergraduate Texts in Mathematics, DOI: 10.1007/978-0-387-68436-9_26,
© Springer Science+Business Media, LLC 2009

Boolean algebras B and C. The algebras B and C are called the *factors* of the product A.

The algebra 2×2 furnishes a concrete example of a product. Its universe consists of the four ordered pairs $(0,0)$, $(0,1)$, $(1,0)$, and $(1,1)$. Its operations, defined coordinatewise in terms of the operations of 2, are given in the following tables:

$\wedge$	$(0,0)$	$(0,1)$	$(1,0)$	$(1,1)$
$(0,0)$	$(0,0)$	$(0,0)$	$(0,0)$	$(0,0)$
$(0,1)$	$(0,0)$	$(0,1)$	$(0,0)$	$(0,1)$
$(1,0)$	$(0,0)$	$(0,0)$	$(1,0)$	$(1,0)$
$(1,1)$	$(0,0)$	$(0,1)$	$(1,0)$	$(1,1)$

,

$\vee$	$(0,0)$	$(0,1)$	$(1,0)$	$(1,1)$
$(0,0)$	$(0,0)$	$(0,1)$	$(1,0)$	$(1,1)$
$(0,1)$	$(0,1)$	$(0,1)$	$(1,1)$	$(1,1)$
$(1,0)$	$(1,0)$	$(1,1)$	$(1,0)$	$(1,1)$
$(1,1)$	$(1,1)$	$(1,1)$	$(1,1)$	$(1,1)$

,

$'$	$(0,0)$
$(0,0)$	$(1,1)$
$(0,1)$	$(1,0)$
$(1,0)$	$(0,1)$
$(1,1)$	$(0,0)$

.

In case B and C are fields of subsets of disjoint sets Y and Z respectively, their product A represents itself naturally as a field of subsets of the union $X = Y \cup Z$. The proof is based on some simple observations about the field of all subsets of X, and depends essentially on the assumption that Y and Z are disjoint. Every subset S of X can be written in one and only one way as a union $S = P \cup Q$ of a subset P of Y and a subset Q of Z. Indeed,

$$P = S \cap Y \qquad \text{and} \qquad Q = S \cap Z.$$

Furthermore, if S_1 and S_2 are subsets of X, say

$$S_1 = P_1 \cup Q_1 \qquad \text{and} \qquad S_2 = P_2 \cup Q_2,$$

then

(1) $$S_1 \cap S_2 = (P_1 \cap P_2) \cup (Q_1 \cap Q_2),$$
(2) $$S_1 \cup S_2 = (P_1 \cup P_2) \cup (Q_1 \cup Q_2),$$
(3) $$S_1' = P_1' \cup Q_1'.$$

The representation f of the product A as a field of subsets of X maps each pair (P, Q) in A to the union $P \cup Q$. Since every subset of X can be written in only one way as such a union, the mapping f is one-to-one. In more detail, if

$$f((P_1, Q_1)) = f((P_2, Q_2)),$$

then

$$P_1 \cup Q_1 = P_2 \cup Q_2;$$

intersect both sides of this equation with Y to obtain $P_1 = P_2$, and intersect both sides with Z to obtain $Q_1 = Q_2$. The proof that f preserves meet, join, and complement depends on the identities (1)–(3). For instance,

$$\begin{aligned}
f((P_1, Q_1) \wedge (P_2, Q_2)) &= f((P_1 \cap P_2, Q_1 \cap Q_2)) \\
&= (P_1 \cap P_2) \cup (Q_1 \cap Q_2) \\
&= (P_1 \cup Q_1) \cap (P_2 \cup Q_2) \\
&= f((P_1, Q_1)) \cap f((P_2, Q_2)).
\end{aligned}$$

The first equality uses the definition of meet in A, the second and fourth use the definition of the representation f, and the third equality uses (1). It follows that f preserves meet. The arguments for join and complement are similar.

It is natural to try to extend the preceding idea to more general classes of Boolean algebras. A Boolean algebra D is called an *internal product* of two Boolean algebras B and C if it includes B and C as subsets and has the following properties. First, every element s in D can be written in exactly one way as a join $s = p \vee q$ of elements p in B and q in C. Second, the operations of D obey the following identities for all elements p_1 and p_2 in B, and q_1 and q_2 in C:

(4) $$(p_1 \vee q_1) \wedge (p_2 \vee q_2) = (p_1 \wedge p_2) \vee (q_1 \wedge q_2),$$
(5) $$(p_1 \vee q_1) \vee (p_2 \vee q_2) = (p_1 \vee p_2) \vee (q_1 \vee q_2),$$
(6) $$(p_1 \vee q_1)' = p_1' \vee q_1',$$

where the meet $p_1 \wedge p_2$, the join $p_1 \vee p_2$, and the complement p_1' on the right sides of the equations are formed in B, while the meet $q_1 \wedge q_2$, the join $q_1 \vee q_2$, and the complement q_1' are formed in C, and all other operations are performed in D. The algebras B and C are called the (*internal*) *factors* of D.

There is a canonical isomorphism from the product A of two Boolean algebras B and C to an internal product D of the two algebras: it is the function f defined by

$$f((p,q)) = p \vee q.$$

The function is one-to-one because every element of D can be written in at most one way as the join of an element of B and an element of C: if $f((p_1, q_1)) = f((p_2, q_2))$, then $p_1 \vee q_1 = p_2 \vee q_2$, and therefore $p_1 = q_1$ and $p_2 = q_2$. The function maps A onto D because every element of D can be written in at least one way as a join of elements of B and C: if s is in D, then there are elements p in B and q in C such that $s = p \vee q$ and therefore

$$f((p,q)) = p \vee q = s.$$

Finally, the function preserves the operations of meet, join, and complement because of conditions (4)–(6). For instance,

$$\begin{aligned}
f((p_1, q_1) \wedge (p_2, q_2)) &= f((p_1 \wedge p_2, q_1 \wedge q_2)) \\
&= (p_1 \wedge p_2) \vee (q_1 \wedge q_2) \\
&= (p_1 \vee q_1) \wedge (p_2 \vee q_2) \\
&= f((p_1, q_1)) \wedge f((p_2, q_2)).
\end{aligned}$$

The first equality uses the definition of meet in A, the second and fourth equalities use the definition of the mapping f, and the third equality uses (4). It follows that f preserves meet. The arguments for join and complement are similar.

The relativizations of the product $A = B \times C$ to the elements $(1, 0)$ and $(0, 1)$ are the Boolean algebras

$$B_0 = B \times \{0\} = \{(p, 0) : p \in B\} \quad \text{and} \quad C_0 = \{0\} \times C = \{(0, q) : q \in C\}.$$

The canonical isomorphism f from A to the internal product D maps B_0 isomorphically to B, and C_0 isomorphically to C; in fact, f maps the pair $(p, 0)$ to the element p, and the pair $(0, q)$ to the element q. A number of properties of D can be deduced immediately from this observation. The algebras A, B_0, and C_0 have the same zero element, namely the pair $(0, 0)$, so D, B, and C must all have the same zero. The algebras B_0 and C_0 are disjoint, except for the common zero, so the algebras B and C are disjoint, except for the common zero. The units of B_0 and C_0 — the pairs $(1, 0)$ and $(0, 1)$ — are the complements of one another in A, so the units of B and C are complements of one another in A. The relativization of A to the unit of B_0 is just B_0, so the relativization of D to the unit of B is just B, and similarly, the relativization of D to the unit of C is just C.

Two internal products D_1 and D_2 of Boolean algebras B and C are always isomorphic via a mapping that is the identity on B and on C. In fact, if f_1 and f_2 are the canonical isomorphisms from the product $A = B \times C$ to D_1 and D_2 respectively, then the composition

$$g = f_2 \circ f_1^{-1}$$

maps D_1 isomorphically to D_2 (see the diagram). Furthermore, g maps each element p in B to itself, since

$$g(p) = f_2(f_1^{-1}(p)) = f_2((p,0)) = p.$$

Similarly, g maps each element in C to itself. These observations justify speaking of *the* internal product of B and C. We shall denote it by $B \otimes C$.

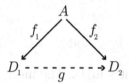

When does the internal product of two Boolean algebras B and C exist? Certainly, the two algebras must be disjoint, except for a common zero element. As it turns out, this is the only condition that is needed. For the proof, suppose B and C have the same zero element and are otherwise disjoint. Let A be the direct product of the two algebras, and let B_0 and C_0 be the relativizations of A defined above. Take h_1 to be the isomorphism from B to B_0, and h_2 the isomorphism from C to C_0, defined by

$$h_1(p) = (p,0) \qquad \text{and} \qquad h_2(q) = (0,q).$$

Notice that these two isomorphisms agree on the common zero element of B and C, and they map the rest of B and C to disjoint sets. An argument very similar to the exchange principle allows us to exchange B_0 for B, and C_0 for C, provided that the elements of A that are not in B_0 or in C_0 are first replaced by new elements that do not occur in B or C. The result is an algebra D that is the internal product of B and C.

The restriction of the internal product construction to pairs of Boolean algebras that have only zero in common is not severe. Given any pair of

Boolean algebras, one can always pass to a pair of isomorphic algebras that have zero, and no other element, in common.

An *internal decomposition* of a Boolean algebra D is a pair of Boolean algebras B and C such that

$$D = B \otimes C.$$

There is a very close connection between the internal decompositions of D and the relativizations of D. Recall (from Chapter 12) that the relativization of D to an element r in D is the set

$$D(r) = \{p \wedge r : p \in D\} = \{p : p \in D \text{ and } p \leq r\}$$

under the join and meet operations of D, restricted to $D(r)$; the complement of an element p in the relativization is defined to be $p' \wedge r$.

Lemma 1. *A Boolean algebra D is the internal product of the relativizations $D(r)$ and $D(r')$ for each element r in D.*

Proof. Write

$$B = D(r) \qquad \text{and} \qquad C = D(r').$$

Consider an arbitrary element s in D. The meets

$$p = s \wedge r \qquad \text{and} \qquad q = s \wedge r'$$

are in B and C respectively, and

$$s = s \wedge 1 = s \wedge (r \vee r') = (s \wedge r) \vee (s \wedge r') = p \vee q.$$

If p_1 and q_1 are any other elements of B and C such that $s = p_1 \vee q_1$, then

$$p = s \wedge r = (p_1 \vee q_1) \wedge r = (p_1 \wedge r) \vee (q_1 \wedge r) = p_1 \vee 0 = p_1,$$

and, similarly, $q = q_1$. The fourth inequality holds because p_1 is below r, while q_1 is below r' and therefore disjoint from r. The argument just given shows that every element of D can be written in exactly one way as the join of an element in B and an element in C.

It remains to verify identities (4)–(6) in D. Suppose p_1 and p_2 are elements in B, and q_1 and q_2 elements in C. Then

$$
\begin{aligned}
(p_1 \vee q_1) \wedge (p_2 \vee q_2) &= [(p_1 \wedge p_2) \vee (q_1 \wedge p_2)] \vee [(p_1 \wedge q_2) \vee (q_1 \wedge q_2)] \\
&= [(p_1 \wedge p_2) \vee 0] \vee [0 \vee (q_1 \wedge q_2)] \\
&= (p_1 \wedge p_2) \vee (q_1 \wedge q_2).
\end{aligned}
$$

The first equality uses the distributive law from Corollary 8.2, and the third uses the identity law for join. The second equality holds because p_1 and p_2 are below r, and therefore disjoint from q_1 and q_2 (which are below r'). The verifications of identities (5) and (6) are similar.

There is a trivial instance of the preceding lemma that is worth pointing out, namely when r is 0 or 1. The lemma then asserts that D is the internal product of the degenerate algebra $\{0\} = D(0)$ and $D = D(1)$ itself.

The lemma describes one method for decomposing a Boolean algebra into the internal product of two factors. As it turns out, there are no other possibilities.

Corollary 1. *A Boolean algebra D is the internal product of Boolean algebras B and C if and only if there is an element r in D such that*

$$B = D(r) \qquad and \qquad C = D(r').$$

Proof. If there is an element r in D for which the preceding equations hold, then D is certainly the internal product of B and C, by the previous lemma. Suppose, conversely, that D is the internal product of the two algebras B and C. The units of B and C are the complements of one another in D, and the relativizations of D to these units are just B and C, by the remarks preceding Lemma 1. Thus, if r is the unit of B, then the given equations hold.

Products play an important role in the study of algebraic structures. If a complicated algebra — a ring or a group, for example — can be written as the product of more basic factor algebras, then the analysis of the complicated algebra reduces to the analysis of these factors. The next corollary provides an example of this phenomenon. It asserts that every complete Boolean algebra is the internal product of a complete, atomic Boolean algebra and a complete, atomless Boolean algebra. Complete, atomic Boolean algebras are isomorphic to fields of all subsets of some set, by Corollary 14.1, so these algebras are in some sense fairly well understood. The analysis of complete Boolean algebras therefore reduces to the analysis of complete, atomless Boolean algebras.

Corollary 2. *Every complete Boolean algebra is the internal product of a complete, atomic Boolean algebra and a complete, atomless Boolean algebra.*

Proof. Let D be a complete Boolean algebra. The supremum r of the set of atoms exists in D, by the assumption that D is complete. Write

$$B = D(r) \qquad \text{and} \qquad C = D(r').$$

Then D is the internal product of B and C, by Lemma 1. It remains to show that B is complete and atomic, and that C is complete and atomless.

The atoms of B coincide with the atoms of D, by the definition of r, and the unit of B — the element r — is the supremum of the set of atoms (in B as well as in D); consequently, B is atomic, by Lemma 14.3. To show that B is complete, consider an arbitrary family $\{p_i\}$ of elements in B. The supremum p of this family certainly exists in D, by the assumption that D is complete. Each element p_i is in B, and is therefore below the unit r of B. In other words, r is an upper bound of the family $\{p_i\}$ (in B and in D). Since p is the least upper bound of this family in D, it follows that $p \leq r$ and hence that p is in B. Thus, p is the supremum of the family $\{p_i\}$ in B.

It is easy to check that an atom in C must also be an atom in D. Since every atom in D is below r, it follows that the algebra C must be atomless. The proof that C is complete is similar to the proof that B is complete.

From the point of view of products, the most basic algebras are those that cannot be decomposed further by means of products. Of course, every Boolean algebra is isomorphic to the product of itself and the degenerate (one-element) Boolean algebra. Such *trivial* decompositions are totally uninteresting. A Boolean algebra is said to be *directly indecomposable* if it is not degenerate and not isomorphic to the product of two non-degenerate Boolean algebras. As it turns out, there is just one directly indecomposable Boolean algebra (up to isomorphic copies), namely 2. (This observation is due to Stone [66].) An elementary cardinality argument shows that 2 is directly indecomposable. (The number 2 cannot be written as the product of two numbers both of which are greater than 1.) On the other hand, a Boolean algebra with more than two elements cannot be directly indecomposable. Indeed, such an algebra, say D, must contain an element r that is different from 0 and 1. The relativizations of D to r and to r' each have at least two elements, and

$$D = D(r) \otimes D(r'),$$

by Lemma 1; consequently, D is the product of two non-degenerate Boolean algebras.

We have looked at the product of two Boolean algebras from an external perspective, as a Cartesian product, and from an internal perspective, as an internal product. There is yet another perspective, a functional one that comes from category theory. If A is the (direct) product of B and C, then there are natural epimorphisms from A to the factor algebras, namely the (left and right) *projections* f_B and f_C defined by

$$f_B((p,q)) = p \quad \text{and} \quad f_C((p,q)) = q.$$

The verification that these mappings are epimorphisms is a simple exercise involving the definition of A. For instance, to verify that f_B preserves join and complement, consider two elements r_1 and r_2 in A, say

$$r_1 = (p_1, q_1) \quad \text{and} \quad r_2 = (p_2, q_2).$$

Then

$$r_1 \vee r_2 = (p_1 \vee p_2, q_1 \vee q_2) \quad \text{and} \quad r_1' = (p_1', q_1'),$$

so that

$$f_B(r_1) = p_1, \quad f_B(r_2) = p_2, \quad f_B(r_1 \vee r_2) = p_1 \vee p_2, \quad f_B(r_1') = p_1',$$

and therefore

$$f_B(r_1 \vee r_2) = f_B(r_1) \vee f_B(r_2) \quad \text{and} \quad f_B(r_1') = f_B(r_1)'.$$

The product A and the pair of projections (f_B, f_C) satisfy the following *lifting condition*: if D is any Boolean algebra, and if g_B and g_C are any homomorphisms from D to B and C, then there is a unique homomorphism g from D to A such that

(7) $$f_B \circ g = g_B \quad \text{and} \quad f_C \circ g = g_C$$

(see the diagram). The existence of g is straightforward to prove. Write

(8) $$g(p) = (g_B(p), g_C(p)).$$

The homomorphism properties of g_B and g_C imply that g is a homomorphism. For instance, g preserves meet because

$$g(p \wedge q) = (g_B(p \wedge q), g_C(p \wedge q)) = (g_B(p) \wedge g_B(q), g_C(p) \wedge g_C(q))$$
$$= (g_B(p), g_C(p)) \wedge (g_B(q), g_C(q)) = g(p) \wedge g(q);$$

the first and last equalities use the definition of g, the second equality uses the homomorphism properties of g_B and g_C, and the third equality uses the

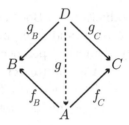

definition of meet in A. The identities in (7) follow at once from (8) and the definitions of the projections f_B and f_C; for instance,

$$f_B(g(p)) = f_B((g_B(p), g_C(p))) = g_B(p).$$

The uniqueness of g is equally easy to establish. Consider an arbitrary homomorphism h from D into A that satisfies (7) (with the function g replaced by h). If h maps the element p in D to the pair (r, s) in A, then

$$r = f_B((r, s)) = f_B(h(p)) = g_B(p)$$

and

$$s = f_C((r, s)) = f_C(h(p)) = g_C(p),$$

by the definitions of the projections and condition (7), so that

$$h(p) = (r, s) = (g_B(p), g_C(p)).$$

Consequently, h coincides with the homomorphism g defined in (8).

The Boolean algebra A and the pair of mappings (f_B, f_C) are uniquely determined, up to isomorphic copies, by the lifting condition: if a Boolean algebra D and a pair of homomorphisms (g_B, g_C) also satisfy the lifting condition, then there is an isomorphism g from D to A such that (7) holds. Indeed, the algebra A and the pair of projections (f_B, f_C) satisfy the lifting condition (in particular, with respect to the algebra D and the pair of mappings (g_B, g_C)), by the observations of the preceding paragraph, so there is a unique homomorphism g from D into A with the properties (7). It is assumed that the algebra D and the pair of mappings (g_B, g_C) also satisfy the lifting condition (in particular, with respect to the algebra A and the pair of mappings (f_B, f_C)), so there is a unique homomorphism f from A into D with the properties

(9) $g_B \circ f = f_B$ and $g_C \circ f = f_C.$

Equations (7) and (9) combine to yield

(10) $$g_B \circ f \circ g = g_B, \qquad g_C \circ f \circ g = g_C,$$

and

(11) $$f_B \circ g \circ f = f_B, \qquad f_C \circ g \circ f = f_C.$$

Since the algebra D and the pair of mappings (g_B, g_C) satisfy the lifting condition (in particular, with respect to themselves), there must be a unique homomorphism h from D into itself with the properties

$$g_B \circ h = g_B \qquad \text{and} \qquad g_C \circ h = g_C.$$

These equations are obviously satisfied if h is the identity automorphism on D, and they are also satisfied if h is the composition $f \circ g$, by (10). The assumed uniqueness of the homomorphism h implies that $f \circ g$ must be the identity automorphism on D. A similar argument, using (11), shows that $g \circ f$ is the identity automorphism on A. It follows that g and f are bijections and inverses of one another, so that g is an isomorphism from D to A with properties (7), as desired (see Exercise 12.32 or the section on bijections in Appendix A).

Almost everything that has been said so far can be generalized. By the (direct) product of a family $\{A_i\}_{i \in I}$ of Boolean algebras we shall understand their Cartesian product

$$A = \prod_{i \in I} A_i,$$

construed as a Boolean algebra with respect to the coordinatewise operations. The universe of the product consists of the functions p with domain I such that $p(i)$ — or p_i as we shall usually write — is an element of A_i for each index i. The meet and join of two functions p and q in A are the functions $p \wedge q$ and $p \vee q$ on I defined by

$$(p \wedge q)_i = p_i \wedge q_i \qquad \text{and} \qquad (p \vee q)_i = p_i \vee q_i,$$

while the complement of p is the function p' on I defined by

$$(p')_i = p'_i.$$

The right sides of these equations are computed in the Boolean algebra A_i for each i. The zero and unit of the product are the functions 0 and 1 on I defined by

$$0_i = 0 \qquad \text{and} \qquad 1_i = 1,$$

where the elements on the right sides of these equations are the zero and unit of A_i for each i. The algebras A_i are the *factors* of the product A.

The verification that the product of a family of Boolean algebras is again a Boolean algebra is quite similar to the verification of the analogous result for the product of two Boolean algebras, but the details have a superficially different appearance. For instance, to verify the commutative axiom (2.18) for join, consider two elements p and q in the product. Both $p \vee q$ and $q \vee p$ are functions on the index set I, so they will be equal just in case they agree on each index i. A simple computation based on the definition of join in the product and the commutative law in A_i yields

$$(p \vee q)_i = p_i \vee q_i = q_i \vee p_i = (q \vee p)_i.$$

The index set I is allowed to be empty. In this case there is just one function with domain I, namely the empty function, so the product of the family is the degenerate (one-element) Boolean algebra.

We shall indicate the products of finite and infinite sequences of Boolean algebras by such obvious and customary modifications of the symbolism as $\prod_{i=1}^{n} A_i$ and $\prod_{i=1}^{\infty} A_i$. When all the factors are equal to the same Boolean algebra B, the product $\prod_{i \in I} A_i$ is called a *power* of B, or, more precisely, the *I*th *power* of B, and is usually written as B^I. For instance, if $A_i = 2$ for each i, then $\prod_{i \in I} A_i$ is just the Boolean algebra 2^I discussed in Chapter 3 (where the symbol X was used instead of I). This power of 2 is isomorphic to the field $\mathcal{P}(I)$ of all subsets of I, as was shown in Chapter 3.

If each member of a family $\{A_i\}$ of Boolean algebras is a field of subsets of a set X_i, and if the sets X_i are mutually disjoint, then the product $A = \prod_i A_i$ is naturally represented as a field of subsets of the union $X = \bigcup_i X_i$ via the mapping that assigns to each element P in A the subset $\bigcup_i P_i$ of X. (Recall that P is a function on I, and P_i is a subset of X_i for each i.) Actually, the product A is isomorphic to a field of sets even when the sets X_i are not mutually disjoint. In this case, however, a modification in the argument is required. The set X must be taken to be, not the union of the sets X_i, but rather the union of disjoint copies of the sets X_i. For instance, put

$$X = \bigcup_i \{(x, i) : x \in X_i\}.$$

(The whole point of considering ordered pairs here is to force disjointness by means of the second coordinate.) The natural monomorphism from A into $\mathcal{P}(X)$ is the mapping that takes each element P of A to the set

$$\bigcup_i \{(x,i) : x \in P_i\}.$$

The representation of a product of fields of sets as a field of sets is a special case of a more general internal product construction. A Boolean algebra D is called the *internal product* of a family $\{A_i\}$ of Boolean algebras provided that it includes each set A_i as a subset, and has the following properties. First, if p_i is an element of A_i for each i, then the supremum of the family $\{p_i\}$ exists in D. Second, every element s in D can be written in one and only one way as a supremum $s = \bigvee_i p_i$, where p_i belongs to A_i for each i. Third, the operations of D satisfy the following identities whenever $\{p_i\}$ and $\{q_i\}$ are families of elements such that p_i and q_i are in A_i for each i:

$$(12) \qquad \left(\bigvee_i p_i\right) \wedge \left(\bigvee_i q_i\right) = \bigvee_i (p_i \wedge q_i),$$

$$(13) \qquad \left(\bigvee_i p_i\right) \vee \left(\bigvee_i q_i\right) = \bigvee_i (p_i \vee q_i),$$

$$(14) \qquad \left(\bigvee_i p_i\right)' = \bigvee_i (p_i'),$$

where the meet $p_i \wedge q_i$, the join $p_i \vee q_i$, and the complement p_i' on the right sides of equations (12)–(14) are formed in the Boolean algebra A_i for each i. The algebras A_i are the (internal) factors of D.

There is a canonical isomorphism f from the (direct) product A of a family $\{A_i\}$ of Boolean algebras to an internal product D of the family $\{A_i\}$ that is defined by

$$f(p) = \bigvee_i p_i$$

for each element p in A. The mapping is well defined because the supremum on the right exists in D for each element p in A, by the first condition in the definition of an internal product. The proof that the mapping is a bijection and preserves the Boolean operations is completely analogous to the proof in the case of the internal product of two algebras.

An internal product of a family of Boolean algebras exists if and only if the algebras in the family are pairwise disjoint, except for a common zero element. Moreover, two internal products of the family are always isomorphic via a mapping that is the identity on each of the factors; this justifies speaking of *the* internal product of the family. The proofs of these observations are again completely analogous to the proofs in the case of two algebras.

There is a close connection between the internal decompositions of a Boolean algebra D and the relativizations of D induced by partitions of the unit. A family $\{r_i\}$ of elements in D is called a *partition* of the unit provided that the elements of the family are pairwise disjoint — that is, $r_i \wedge r_j = 0$ for $i \neq j$ — and the supremum of the family is 1. We shall say that such a partition has the *supremum property* if for every family $\{p_i\}$ of elements in D satisfying $p_i \leq r_i$ for each i, the supremum of $\{p_i\}$ exists in D.

Lemma 2. *Let D be a Boolean algebra, and $\{r_i\}$ a partition of the unit with the supremum property. Then D is the internal product of the family of relativizations $\{D(r_i)\}$.*

Proof. Let $\{r_i\}$ be a family of elements in D satisfying the hypotheses of the lemma, and write $A_i = D(r_i)$. It is to be shown that D satisfies the defining conditions for being the internal product of the family $\{A_i\}$. If $\{p_i\}$ is a family of elements in D with p_i in A_i for each i, then $p_i \leq r_i$ for each i, by the definition of A_i, and therefore the supremum of the family exists in D, by the supremum property.

Consider, next, an arbitrary element s in D, and write

$$p_i = s \wedge r_i$$

for each i. Then p_i is in A_i, by the definition of A_i, and

$$\bigvee_i p_i = \bigvee_i (s \wedge r_i) = s \wedge \bigvee_i r_i = s \wedge 1 = s.$$

If $\{q_i\}$ is any other family with q_i in A_i for each i, and such that $s = \bigvee_i q_i$, then

$$p_j = p_j \wedge r_j = \bigvee_i (p_i \wedge r_j) = \left(\bigvee_i p_i \right) \wedge r_j = s \wedge r_j$$

$$= \left(\bigvee_i q_i \right) \wedge r_j = \bigvee_i (q_i \wedge r_j) = q_j \wedge r_j = q_j.$$

The first and last equalities hold because p_j and q_j are below r_j. The second and seventh equalities holds because r_i and r_j are disjoint for distinct indices i and j, and the elements p_i and q_i are below r_i; consequently,

$$p_i \wedge r_j = q_i \wedge r_j \leq r_i \wedge r_j = 0.$$

The third and sixth equalities hold by the distributive law in Lemma 8.3. This argument shows that every element in D can be written in one and only one way as the supremum of a family $\{p_i\}$, with p_i in A_i for each i.

The validity of (12) follows easily from the distributive law formulated in Corollary 8.2:

$$\left(\bigvee_i p_i\right) \wedge \left(\bigvee_j q_j\right) = \bigvee_{ij} (p_i \wedge q_j) = \bigvee_i (p_i \wedge q_i).$$

For the last step, observe that

$$p_i \wedge q_j \le r_i \wedge r_j = 0$$

when $i \ne j$. The validity of (13) is an immediate consequence of the generalized associative law in Lemma 8.2. To verify (14), let p_i be an element of the Boolean algebra A_i for each i, and let p_i' denote its complement in A_i. Then

$$p_i \wedge p_i' = 0 \qquad \text{and} \qquad p_i \vee p_i' = r_i.$$

The suprema $\bigvee_i p_i$ and $\bigvee_i p_i'$ exist, by the supremum property. Moreover,

$$\left(\bigvee_i p_i\right) \wedge \left(\bigvee_i p_i'\right) = \bigvee_i (p_i \wedge p_i') = \bigvee_i 0 = 0$$

and

$$\left(\bigvee_i p_i\right) \vee \left(\bigvee_i p_i'\right) = \bigvee_i (p_i \vee p_i') = \bigvee_i r_i = 1,$$

by (12) and (13). The preceding equations show that $\bigvee_i p_i'$ is the complement of $\bigvee_i p_i$, by Lemma 6.2. In other words, (14) holds. The proof of the lemma is complete.

The lemma describes one method for decomposing a Boolean algebra into the internal product of a family of factors. As it turns out, there are no other possibilities.

Corollary 3. *A Boolean algebra D is the internal product of a family of Boolean algebras $\{A_i\}$ if and only if there is a partition $\{r_i\}$ of the unit in D that has the supremum property and such that $A_i = D(r_i)$ for each i.*

Proof. If a partition of the unit of D with the stated properties exists, then D is the internal product of the family of corresponding relativizations, by the preceding lemma.

To prove the converse, assume that D is the internal product of a family $\{A_i\}$ of Boolean algebras, and write r_i for the unit of A_i. It must be shown that the family $\{r_i\}$ has the stated properties. Let A be the (direct) product of the family, and for each i, let s_i be the element in A defined by

$$s_i(k) = \begin{cases} r_i & \text{if } k = i, \\ 0 & \text{if } k \neq i, \end{cases}$$

for k in I. The elements s_i are pairwise disjoint. Indeed, if $i \neq j$, then for each index k at least one of $s_i(k)$ and $s_j(k)$ is zero, so that the meet

$$s_i(k) \wedge s_j(k)$$

is zero in A_k. In other words, $s_i \wedge s_j$ assumes the value zero at each argument k, so it is the zero element of A. Similarly, the supremum of the family $\{s_i\}$ is the unit of A. For the proof, suppose that q is any upper bound of this family in A. The inequality $s_i \leq q$ implies that $s_i(k) \leq q_k$ for each index k, and in particular

$$r_i = s_i(i) \leq q_i.$$

Thus, q_i must be the unit of A_i for each i, and therefore q is the unit of A, by the definition of a (direct) product. Conclusion: $\{s_i\}$ is a partition of the unit in A.

The definition of the product implies that the family $\{s_i\}$ has the supremum property in A. Indeed, suppose, for each i, that p_i is an element in A below s_i. Then

$$p_i(k) \leq s_i(k)$$

for each k. In particular, $p_i(k) = 0$ for each $k \neq i$, by the definition of s_i. The supremum of the family $\{p_i\}$ in A is therefore the element p in A defined by

$$p(i) = p_i(i)$$

for each i.

Let f be the canonical isomorphism from A to D. Recall that f maps each element q in A to the supremum $\bigvee_k q(k)$ in D. In particular,

$$f(s_i) = \bigvee_k s_i(k) = s_i(i) = r_i.$$

Since $\{s_i\}$ is a partition of the unit in A with the supremum property, it follows that the image of this family under f, namely $\{r_i\}$, must be a partition of the unit in D with the supremum property. Furthermore, f must map the set of elements in A below s_i bijectively to the set of elements in D below r_i, so that the image of the relativization $A(s_i)$ under f is $D(r_i)$. On the other hand, the elements in $A(s_i)$ are just the elements q in A that are below s_i; in other words, they are the functions q on the index set such that $q(i)$ belongs to A_i and $q(k) = 0$ for $k \neq i$, by the definition of s_i. The image of $A(s_i)$ under f is therefore just A_i, by the definition of f, so that

$$A_i = f(A(s_i)) = D(r_i).$$

The proof of the corollary is complete.

If A is the product of a family $\{A_i\}$ of Boolean algebras, then for each i there is a natural epimorphism from A to A_i, namely the *projection* f_i defined by $f_i(p) = p_i$. If, moreover, D is an arbitrary Boolean algebra, and if, for

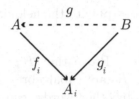

each i, there is a homomorphism g_i from D to A_i, then there is a unique homomorphism g from D to A such that $f_i \circ g = g_i$ for all i (see the diagram). In fact, g takes each element p in D to the element in A whose ith coordinate is $g_i(p)$, that is,

$$g(p) = q \qquad \text{if and only if} \qquad q_i = g_i(p)$$

for each i.

As in the case of the products of two algebras, one can describe this situation by saying that the product A and the family of projections $\{f_i\}$ satisfy the lifting condition. Moreover, A and the family $\{f_i\}$ are uniquely determined to within isomorphism by the lifting condition: if a Boolean algebra D and a family of homomorphisms $\{g_i\}$ also satisfy the lifting condition, then there is an isomorphism g from D to A such that $f_i \circ g = g_i$. The proof is similar to the proof in the case of the product of two algebras.

Exercises

1. Prove that the Boolean algebra 2×2 is isomorphic to the field of all subsets of a two-element set.

2. If Y and Z are disjoint sets, prove that every subset S of their union can be written in one and only one way as the union of a subset of Y with a subset of Z, and verify equations (1)–(3).

3. Suppose B and C are fields of subsets of disjoint sets Y and Z. Prove that the function f defined on the product $A = B \times C$ by

$$f((P,Q)) = P \cup Q$$

preserves join and complement. Prove, further, that if B and C are the fields of all subsets of Y and Z respectively, then f maps A onto the field of all subsets of $Y \cup Z$.

4. Suppose Y and Z are disjoint sets, and Z is finite. If B is the field of finite and cofinite subsets of Y, and C the field of all subsets of Z, prove that $B \times C$ is isomorphic to the field of finite and cofinite subsets of the set $Y \cup Z$.

5. Suppose $A = A_1 \times A_2$. Prove that $B = B_1 \times B_2$ is a subalgebra of A whenever B_1 and B_2 are subalgebras of A_1 and A_2 respectively. Is the converse true? In other words, can every subalgebra of A be decomposed into the product of a subalgebra of A_1 and a subalgebra of A_2?

6. Suppose $A = B \times C$. If M and N are ideals in B and C respectively, prove that $L = M \times N$ is an ideal in A. Can every ideal in A be written as the product of an ideal in B with an ideal in C?

7. Characterize the maximal ideals in a product $A = B \times C$.

8. Let B be the field of finite and cofinite sets of natural numbers. Describe the maximal ideals in $B \times B$.

9. Let f_1 be a (Boolean) homomorphism from B_1 into A_1, and f_2 a homomorphism from B_2 into A_2. Define a mapping f from $B = B_1 \times B_2$ into $A = A_1 \times A_2$ by

$$f((p_1, p_2)) = (f_1(p_1), f_2(p_2))$$

for all p_1 in B_1 and p_2 in B_2. Prove that f is a homomorphism. Prove further that f is one-to-one, or onto, if and only if f_1 and f_2 are both one-to-one, or both onto. Conclude that if B_1 and B_2 are isomorphic to A_1 and A_2 respectively, then B is isomorphic to A.

10. Can every Boolean homomorphism from a product $B_1 \times B_2$ into a product $A_1 \times A_2$ be decomposed as in Exercise 9? In other words, for each such homomorphism f, do there always exist homomorphisms f_1 from B_1 into A_1, and f_2 from B_2 into A_2, such that

$$f((p_1, p_2)) = (f_1(p_1), f_2(p_2))$$

for all p_1 in B_1 and p_2 in B_2?

11. Prove directly (without using Lemma 1) that if $A = B \times C$, then A is the internal product of the algebras

$$B_0 = B \times \{0\} \quad \text{and} \quad C_0 = \{0\} \times C.$$

12. Let D be the internal product, and A the (direct) product of two Boolean algebras B and C. Prove that the mapping f from A to D defined by

$$f((p, q)) = p \vee q$$

preserves join and complement.

13. Suppose B and C are Boolean algebras that are disjoint except for a common zero element. Give a careful proof, along the lines of the proof of the exchange principle (see Chapter 11), that the internal product of B and C exists.

14. Let B be a Boolean subalgebra of A, and r an element of A. Prove that the subalgebra of A generated by $B \cup \{r\}$ is just the internal product of the relativizations $B(r)$ and $B(r')$, formed in A. (See Exercise 12.19.)

15. Formulate and prove a version of Exercise 9 for internal products.

16. Prove that the set of atoms in the internal product of two Boolean algebras B and C is the union of the set of atoms in B with the set of atoms in C. Conclude that the internal product is atomic if and only if each of the two factors is atomic. Draw a similar conclusion for the direct product of B and C, and describe the set of atoms in this direct product.

17. Prove that the internal product of two Boolean algebras is complete if and only if each of the factor algebras is complete. Conclude that the same is true of the direct product of the two algebras.

18. If the supremum of the set of all atoms in a Boolean algebra exists, prove that the algebra can be decomposed into the internal product of an atomic Boolean algebra and an atomless Boolean algebra.

19. Prove that two countably infinite Boolean algebras with finitely many atoms are isomorphic if and only if they have the same number of atoms.

20. Are two countably infinite Boolean algebras with infinitely many atoms necessarily isomorphic?

21. Show, for every finite Boolean algebra A, that $A \times 2$ and A are not isomorphic.

22. Find an infinite Boolean algebra A such that $A \times 2$ and A are isomorphic. Can A be countable?

23. Find an infinite Boolean algebra A such that $A \times 2$ and A are not isomorphic. Can A be countable?

24. Verify that the product of a family of Boolean algebras is a Boolean algebra.

25. Suppose D is the internal product of a family $\{A_i\}$ of Boolean algebras. Show that the set of atoms in D is just the union of the sets of atoms in the individual factors A_i. Conclude that D is atomic if and only if each factor is atomic. Draw a similar conclusion for the direct product of the family $\{A_i\}$, and describe the set of atoms in this direct product. (This generalizes Exericise 16.)

26. Let $\{A_i\}$ be a family of Boolean algebras, and A its product. For each element q in A and each index i, write q_i for the ith coordinate of q. Prove that an element p in A is the supremum of a set E in A if and only if p_i is the supremum of the set

$$E_i = \{q_i : q \in E\}$$

in A_i for each i.

27. Prove that the product of a family of Boolean algebras is complete if and only if each of the factors is complete. (This generalizes Exercise 17.)

28. Consider two Boolean products

$$A = \prod_i A_i \qquad \text{and} \qquad B = \prod_i B_i$$

with the same index set, and suppose f_i is a Boolean homomorphism from B_i into A_i for each i. Define a mapping f from B into A as follows: if u is in A, and p in B, then

$$f(p) = u \qquad \text{if and only if} \qquad u_i = f_i(p_i)$$

for each i. Prove that f is a Boolean homomorphism. Prove further that f is one-to-one, or onto, if and only if every homomorphism f_i is one-to-one, or onto. Conclude that if B_i is isomorphic to A_i for each i, then B is isomorphic to A.

29. Suppose each algebra A_i in a family of Boolean algebras is a field of subsets of a set X_i. Suppose further that the sets X_i are mutually disjoint, and $X = \bigcup_i X_i$. Prove that the product $A = \prod_i A_i$ can be embedded into the field $\mathcal{P}(X)$ via the function that maps each element P in the product to the union $\bigcup_i P_i$.

30. Suppose D is the internal product, and A the (direct) product, of a family $\{A_i\}$ of Boolean algebras. Prove that the mapping f from A to D defined by

$$f(p) = \bigvee_i p_i$$

for each p in A is an isomorphism from A to D.

31. Prove that an internal product of a family of Boolean algebras exists if and only if the algebras in the family are mutually disjoint, except for a common zero element.

32. Prove that two internal products of a family of Boolean algebras are isomorphic via a mapping that is the identity on each of the factors.

33. Formulate and prove the analogue of Exercise 28 for internal products.

34. Prove that if $A = \prod_i A_i$ is a Boolean product, then for each index i, the projection from A to A_i is an epimorphism.

35. Show that a Boolean product $A = \prod_i A_i$ and the associated family $\{f_i\}$ of projections satisfy the lifting condition. In other words, if D is an arbitrary Boolean algebra and if g_i a homomorphism from D into A_i for each i, then there is a unique homomorphism g from D to A such that $f_i \circ g = g_i$ for each i.

36. Prove that the lifting condition characterizes the product $A = \prod_i A_i$ and the family $\{f_i\}$ of projections, up to isomorphic copies.

37. A product $A = \prod_i A_i$ includes two subalgebras, each of which might deserve some consideration as a kind of weak product of the family $\{A_i\}$.

One subalgebra — call it B — consists of those elements p for which p_i is in 2 for all but a finite set of indices i; the other, smaller, subalgebra — call it C — consists of those elements p for which either $p_i = 0$ for all but a finite set of indices i or else $p_i = 1$ for all but a finite set of indices i. Prove that B and C are subalgebras of A. Give an example for which all three algebras are distinct.

38. If D is the internal product of an infinite family $\{A_i\}$ of Boolean algebras, what subalgebra of D does the union $E = \bigcup_i A_i$ of the factors generate?

39. Exercise 37 can be generalized in several different ways when the factor algebras are all equal, that is, when $A = D^I$ for some Boolean algebra D and some set I. One subalgebra (of A) — call it B — consists of those elements p that have a finite range, that is, the set $\{p_i : i \in I\}$ is finite. Another, smaller, subalgebra — call it C — consists of those elements p that are constant on a cofinite subset of I, that is, there exists an element u in D such that $p_i = u$ for all but a finite set of indices i. Prove that B and C are subalgebras of A. Give other generalizations of Exercise 37 when the factor algebras are all equal.

40. Prove that if A is the product of two Boolean algebras B and C, then the canonical extension of A is the product of the canonical extensions of B and C.

41. Extend Exercise 40 to products of finite families of Boolean algebras.

42. Given an example to show that the result in Exercise 40 cannot be extended to infinite families of Boolean algebras.

43. Prove that if A is the product of an arbitrary family $\{A_i\}$ of Boolean algebras, then the completion of A is the product of the completions of the factor algebras A_i. (Contrast this result with those in Exercises 40 and 42.)

Chapter 27

Isomorphisms of Factors

To what extent do the laws of multiplication and exponentiation, familiar from the arithmetic of positive integers, carry over to products and powers of Boolean algebras? Many interesting problems arise from this question. For instance, if two positive integers divide one another, they must be equal. Does a form of this law hold for Boolean algebras? If, in other words, two Boolean algebras A and D are factors of one another, may it be concluded that A and D are isomorphic? The two algebras are factors of one another provided that

$$D = A \times B \qquad \text{and} \qquad A = D \times C$$

for some Boolean algebras B and C. (For typographical convenience we shall use the sign of equality in this and related contexts to denote isomorphism.) The question may therefore be reformulated in the following equivalent way: does $A = A \times B \times C$ imply $A = A \times B$? What if we restrict the question by assuming $B = C$? What if we restrict the question still further by assuming $B = C = 2$?

For grammatical convenience, these four problems may be expressed as statements rather than questions; the task is then to decide which statements are true and which ones false.

(1)   If $D = A \times B$ and $A = D \times C$, then $A = D$.

(2)   If $A = A \times B \times C$, then $A = A \times B$.

(3)   If $A = A \times B \times B$, then $A = A \times B$.

(4)   If $A = A \times 2 \times 2$, then $A = A \times 2$.

S. Givant, P. Halmos, *Introduction to Boolean Algebras*,
Undergraduate Texts in Mathematics, DOI: 10.1007/978-0-387-68436-9_27,
© Springer Science+Business Media, LLC 2009

As we have just seen, (1) and (2) are equivalent, (2) implies (3), and (3) implies (4)). It was Tarski who first raised the problem of determining which of these assertions are true; he proved in [76] and [77] that (1) — and therefore each of the four assertions — holds when the algebras in question possess a certain degree of completeness. In general, however, the answers to the questions are negative, for Hanf gave an example in [24] to show that (4) is false. The main purpose of this chapter is to present Tarski's theorem and Hanf's example.

To prove (1), it must be assumed that the algebras A and D are *countably complete*, or *σ-complete*, in the sense that the supremum and infimum of every countably infinite subset of A exist. We shall have much more to say about such algebras in Chapter 29. The proof of the next lemma (due to Sikorski [56] and Tarski [76], [77]) contains the heart of the argument; it is a Boolean-algebraic analogue of the proof of the Schröder–Bernstein theorem from set theory (see Appendix A, p. 463, and Exercises 1–4).

Lemma 1. *If a σ-complete Boolean algebra A is isomorphic to one of its relativizations $A(p)$, then it is isomorphic to every relativization $A(q)$ with $q \geq p$.*

Proof. Let f be an isomorphism from A to one of its relativizations $A(p)$. Notice that f maps A into itself. Given an arbitrary element $q \geq p$, it therefore makes sense to define two sequences $\{p_n\}$ and $\{q_n\}$ by induction on n as follows:

$$p_1 = 1 \quad \text{and} \quad p_{n+1} = f(p_n),$$
$$q_1 = q \quad \text{and} \quad q_{n+1} = f(q_n).$$

In other words, p_{n+1} and q_{n+1} are the result of applying n times the mapping f to the elements 1 and q respectively. The definitions of these two sequences, and the fact that f maps the unit of A to the unit of $A(p)$, imply that

$$p_2 = f(p_1) = f(1) = p \quad \text{and} \quad p \leq q = q_1 \leq 1 = p_1.$$

In particular,

$$p_2 \leq q_1 \leq p_1.$$

Apply the isomorphism f to each of these elements to obtain

$$f(p_2) \leq f(q_1) \leq f(p_1),$$

or, in other words,

$$p_3 \leq q_2 \leq p_2.$$

Iterate this process repeatedly to arrive at

$$p_{n+1} \le q_n \le p_n.$$

The inequalities combine to give

$$1 = p_1 \ge q_1 \ge p_2 \ge q_2 \ge p_3 \ge q_3 \ge \cdots.$$

The infimum r of this sequence exists, by the assumption of σ-completeness, and the preceding inequalities imply that

$$r = \bigwedge_n p_n = \bigwedge_n q_n.$$

The elements

$$p_1 \wedge q_1', \qquad q_1 \wedge p_2', \qquad p_2 \wedge q_2', \qquad q_2 \wedge p_3', \qquad \ldots,$$

together with r, form a countable partition of the unit 1 (see the diagram and see also Exercise 8.27), and this partition has the supremum property, by the assumption of σ-completeness. Consequently, A is the internal product of the corresponding sequence of relativizations, by Lemma 26.2. In more detail, if B_1 and C are the internal products of the families

$$\{A(p_n \wedge q_n') : n = 1, 2, 3, \ldots\} \qquad \text{and} \qquad \{A(q_n \wedge p_{n+1}') : n = 1, 2, 3, \ldots\},$$

respectively, then

$$A = B_1 \otimes C \otimes A(r).$$

Similarly, the elements

$$q_1 \wedge p_2', \qquad p_2 \wedge q_2', \qquad q_2 \wedge p_3', \qquad p_3 \wedge q_3', \qquad \ldots,$$

together with r, form a countable partition of q with the supremum property (see the diagram). Consequently, $A(q)$ is the internal product of the corresponding relativizations. In other words, if B_2 is the internal product of the family

$$\{A(p_n \wedge q_n') : n = 2, 3, 4, \ldots\},$$

then

$$A(q) = B_2 \otimes C \otimes A(r).$$

The isomorphism f maps the element $p_n \wedge q_n'$ to the element $p_{n+1} \wedge q_{n+1}'$, so it maps the relativization $A(p_n \wedge q_n')$ isomorphically to the relativization $A(p_{n+1} \wedge q_{n+1}')$. It therefore maps the internal product B_1 isomorphically to the internal product B_2. The identity function g on A obviously maps the

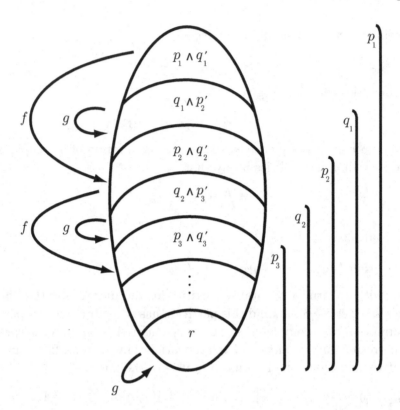

internal product $C \otimes A(r)$ isomorphically to itself. The function h defined on A by

$$h(s \vee t) = f(s) \vee g(t),$$

for every s in B_1 and every t in $C \otimes A(r)$, is the desired isomorphism from A onto $A(q)$ (see Exercises 26.9 and 26.15).

Tarski's theorem follows readily from the lemma.

Theorem 25. *If two σ-complete Boolean algebras are factors of one another, then they are isomorphic.*

Proof. Suppose two σ-complete Boolean algebras A and D are factors of one another. Since A is a factor of D, there must be a Boolean algebra B such that

$$D = A \times B.$$

The remarks in Chapter 26, in particular Corollary 26.1, imply the existence of an element p_0 in D such that A is isomorphic to $D(p_0)$ (and B to $D(p_0')$). Similarly, the assumption that D is a factor of A implies the existence of an element q in A such that D is isomorphic to $A(q)$.

Let f be an isomorphism from A to $D(p_0)$, and g an isomorphism from D to $A(q)$, and write $p = g(p_0)$. The appropriate restriction of g maps $D(p_0)$ isomorphically to $A(p)$ (Exercise 12.27), and therefore the composition $g \circ f$ maps A isomorphically to $A(p)$, since

$$g(f(1)) = g(p_0) = p.$$

The function g maps the unit of D to the element q, and therefore maps p_0 to an element below q. In other words, $p \leq q$. Invoke Lemma 1 to obtain an isomorphism h from A to $A(q)$. The composition $g^{-1} \circ h$ is the desired isomorphism from A to D.

We now present Hanf's example of two Boolean algebras that are factors of one another, but not isomorphic. The exposition is strongly influenced by several inspiring conversations with Dana Scott. Let $\{a_n\}$ and $\{b_n\}$ be two countable sets, disjoint from each other, and let X be their union. A bijection θ of X is defined by writing

$$\theta(a_n) = b_n \quad \text{and} \quad \theta(b_n) = a_n,$$

for $n = 1, 2, 3, \ldots$. A subset P of X is *invariant* under θ if $\theta(P) = P$. In other words, the set P is invariant provided it contains a_n whenever it contains b_n, and it contains b_n whenever it contains a_n. Every set I of positive integers gives rise to, or *induces*, a uniquely determined invariant set, namely the set

$$\{a_n : n \in I\} \cup \{b_n : n \in I\}$$

of elements in X with indices in I. Conversely, every invariant set P is induced by a uniquely determined set of positive integers, namely the set of indices of elements that occur in P.

The class of all those subsets of X that are invariant under θ is a complete field of subsets of X. The field is atomic, and its atoms are the couples (unordered pairs) $\{a_n, b_n\}$ for $n = 1, 2, 3, \ldots$.

Every subset R of X can obviously be written in a unique way as a union

$$R = P \cup Q$$

of an invariant set P and a set Q that is disjoint from P and that includes no non-empty invariant subset. (In other words, for no positive integer n are both a_n and b_n in Q.) Call P the *invariant part*, and Q the *variant part*, of R. A subset of X is said to be *almost invariant* if its variant part is finite, or, equivalently, if it differs from an invariant set by a finite set. Every triple (I, J, K) of sets of positive integers, with J and K finite and all three sets mutually disjoint, gives rise to an almost invariant set, namely

$$R = \{a_n : n \in I \cup J\} \cup \{b_n : n \in I \cup K\};$$

moreover, every almost invariant set can be written in this form for a unique triple (I, J, K) satisfying the given conditions. The invariant part of R is the invariant set

$$P = \{a_n : n \in I\} \cup \{b_n : n \in I\}$$

induced by I, and the variant part of R is the set

$$Q = \{a_n : n \in J\} \cup \{b_n : n \in K\}.$$

The class A of all almost invariant sets is a field. This field is atomic, and its atoms are the singletons of X, that is, the singletons $\{a_n\}$ and $\{b_n\}$. Note that every infinite almost invariant set (that is, every infinite set in A) includes an infinite invariant subset.

Lemma 2. *The relativization of A to any infinite invariant set is isomorphic to A.*

Proof. Suppose Y is the invariant set induced by an infinite set I of positive integers. Let ψ be any bijection from I to the set of all positive integers, and let ϕ be the corresponding bijection from Y to X:

$$\phi(a_n) = a_{\psi(n)} \qquad \text{and} \qquad \phi(b_n) = b_{\psi(n)}$$

for n in I. The bijection ϕ induces an isomorphism f from the field $\mathcal{P}(X)$ to the field $\mathcal{P}(Y)$ that is defined by

$$f(P) = \phi^{-1}(P)$$

for each subset P of X (see Chapter 12, p. 94). The isomorphism clearly maps invariant subsets of X to invariant subsets of Y, and finite subsets of X to finite sets of Y. It therefore maps almost invariant subsets of X to almost invariant subsets of Y. In other words, it maps A to the relativization of A to Y.

Corollary 1. *The algebra A is isomorphic to $A \times A$, and also to $A \times 2^k$ for every positive even integer k.*

Proof. Consider the invariant set Y induced by the set of positive even integers. Its complement Y' in X is the invariant set induced by the set of positive odd integers. The relativization of A to Y and the relativization of A to Y' — call them B and C respectively — are both isomorphic to A, by the preceding lemma, so that

$$B \times C = A \times A$$

(see Exercise 26.9). On the other hand, the algebra A is the internal product of the two relativizations, by Lemma 26.1, so that

$$A = B \times C.$$

It follows that A is isomorphic to $A \times A$.

The proof of the second assertion is similar. Suppose $k = 2n$, where n is a positive integer. Take Y to be the invariant set induced by the set of integers greater than n. The complement Y' is an invariant set with exactly k elements. Every subset of Y' is finite, and therefore almost invariant. Consequently, the relativization of A to Y' — call it C — coincides with the field of all subsets of Y'. It follows that C is isomorphic to 2^k (see Chapter 3). On the other hand, the relativization of A to Y — call it B — is isomorphic to A, by the preceding lemma. Therefore,

$$B \times C = A \times 2^k.$$

Since A is the internal product of the two relativizations, by Lemma 26.1, we may conclude that A is isomorphic to $A \times 2^k$.

The corollary shows that the Boolean algebras A and $A \times 2 \times 2$ are isomorphic. It remains to prove that A is not isomorphic to $A \times 2$. The algebra $A \times 2$ can be described, up to isomorphism, as follows. Let c be a new point not in X. The Boolean algebra 2 is isomorphic to the field $\mathcal{P}(\{c\})$, and therefore $A \times 2$ is isomorphic to the internal product of A and $\mathcal{P}(\{c\})$. This internal product consists of two classes of sets: the sets in A and the sets in A with the element c adjoined. We may therefore think of $A \times 2$ itself as consisting of the sets P and the sets $P \cup \{c\}$, where P ranges over the almost invariant subsets of X. Under this conception, the atoms of $A \times 2$ are the singletons of the elements in the set $X \cup \{c\}$. For notational convenience, we identify these singletons with the elements themselves, that is, we treat a_n, b_n, and c as the atoms of $A \times 2$.

The algebra $A \times 2$ has an automorphism of period two that leaves exactly one atom fixed. In other words, there is an automorphism g of $A \times 2$ such that $g \circ g$ is the identity automorphism, and such that exactly one atom of $A \times 2$ is mapped to itself by g. Indeed, just extend θ to a bijection of the set $X \cup \{c\}$ by requiring that the element c be mapped to itself, and take g to be an appropriate restriction of the automorphism of $\mathcal{P}(X \cup \{c\})$ induced by θ (see Chapter 12, p. 94). On the atoms of $A \times 2$, we have

$$g(a_n) = \theta(a_n) = b_n, \qquad g(b_n) = \theta(b_n) = a_n, \qquad g(c) = c,$$

and on arbitrary sets P in $A \times 2$ we have

$$g(P) = \{g(p) : p \in P\} = \{\theta(p) : p \in P\}.$$

Notice that $g \circ g$ is the identity mapping on $X \cup \{c\}$:

$$g(g(a_n)) = g(b_n) = a_n, \quad g(g(b_n)) = g(a_n) = b_n, \quad g(g(c)) = g(c) = c.$$

Consequently, $g \circ g$ is the identity on all of $A \times 2$:

$$g(g(P)) = \{g(g(p)) : p \in P\} = \{p : p \in P\} = P.$$

Also, the only atom left fixed by g is c.

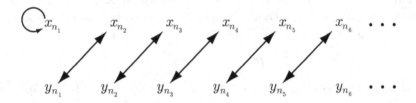

To prove that A is not isomorphic to $A \times 2$, we shall show that A has no such automorphism. Assume that, on the contrary, A has an automorphism g with period two that leaves exactly one atom fixed. Some terminology will be helpful: call the atoms a_n and b_n *associates* of one another. The atom left fixed by g is either a_{n_1} or b_{n_1} for some index n_1; denote it by x_{n_1}. The associate of x_{n_1} — call it y_{n_1} — cannot be mapped to itself, since g fixes no atom different from x_{n_1}, and y_{n_1} cannot be mapped to x_{n_1}, since x_{n_1} is mapped to itself. Consequently, y_{n_1} must be mapped to an atom with an index n_2 different from n_1 — either a_{n_2} or b_{n_2}. Call this atom x_{n_2}. Its associate y_{n_2} cannot be mapped to any of x_{n_1}, y_{n_1}, and x_{n_2}, and it cannot be mapped to itself, so it must be mapped to an atom with an index n_3 different from n_1 and n_2 — either a_{n_3} or b_{n_3}. Call this atom x_{n_3}. By applying this argument repeatedly, we obtain two infinite sequences $\{x_{n_k}\}$ and $\{y_{n_k}\}$ with two properties. First, the atoms x_{n_k} and y_{n_k} are associates, that is,

$$\{x_{n_k}, y_{n_k}\} = \{a_{n_k}, b_{n_k}\},$$

and the indices n_k are mutually distinct for distinct k. Second,

$$g(x_{n_1}) = x_{n_1}, \qquad g(y_{n_k}) = x_{n_{k+1}}, \qquad g(x_{n_{k+1}}) = y_{n_k}$$

for $k = 1, 2, 3, \ldots$ (see the diagram above). Let I be the set of indices n_k such that k has remainder 2 when divided by 3,

$$I = \{n_2, n_5, n_8, n_{11}, \ldots\},$$

and let P be the invariant set induced by I,

$$P = \{a_{n_2}, b_{n_2}, a_{n_5}, b_{n_5}, a_{n_8}, b_{n_8}, \ldots\} = \{x_{n_2}, y_{n_2}, x_{n_5}, y_{n_5}, x_{n_8}, y_{n_8}, \ldots\}.$$

The set P, being invariant, is in A. Its image under the automorphism g is the set

$$Q = \{y_{n_1}, x_{n_3}, y_{n_4}, x_{n_6}, y_{n_7}, x_{n_9}, \ldots\},$$

and this set is clearly not in A; no two elements of Q have the same index, so Q is an infinite set with no invariant subset. Conclusion: A cannot be closed under g, in contradiction to the assumption that g is an automorphism of A. This contradiction proves that the assumption of the existence of g is untenable.

A concrete illustration of the above construction may serve to elucidate the argument. Suppose the atom x_{n_k} is always a_{n_k}. Then y_{n_k} is always b_{n_k}. The argument constructs the infinite sequence $\{a_{n_k}\}$ to have the properties

$$g(a_{n_1}) = a_{n_1}, \qquad g(b_{n_k}) = a_{n_{k+1}}, \qquad g(a_{n_{k+1}}) = b_{n_k}$$

for $k = 1, 2, 3, \ldots$ (see the diagram below).

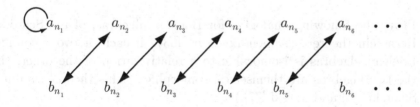

The image of the invariant set

$$P = \{a_{n_2}, b_{n_2}, a_{n_5}, b_{n_5}, a_{n_8}, b_{n_8}, \ldots\}$$

under g is the set

$$Q = \{b_{n_1}, a_{n_3}, b_{n_4}, a_{n_6}, b_{n_7}, a_{n_9}, \dots\}.$$

The set Q cannot be in A, since no two of its elements have the same index.

Hanf's counterexample to (4) is a large algebra (it has the power of the continuum); is there a countable one? The answer is negative, for Vaught proved that (4) holds whenever A is countable (see [24] and Exercise 30). Tarski asked whether there are countable counterexamples to (3). Such a counterexample was eventually devised by Hanf (see Chapter 45).

The *square root law* for positive integers says that squares of distinct positive integers are distinct. Is the same true of Boolean algebras? In other words, does $A^2 = B^2$ imply $A = B$? Hanf's example was used by Tarski to show that this question (also due to Tarski) has a negative answer (see [24] and Exercise 21). Is the implication true of countable Boolean algebras? Again, the answer is negative (see Chapter 45).

Here is a final example of a well-known problem regarding powers of Boolean algebras. It is not difficult to prove that

$$\mathcal{P}(X)^n = \mathcal{P}(X)$$

when X is an infinite set and n an arbitrary positive integer (see Exercise 14). Does $A^3 = A$ imply $A^2 = A$ for arbitrary Boolean algebras A? What if A is countable? The answer in both cases is no. An uncountable counterexample can be constructed using Hanf's counterexample (see Exercise 22). The question for countable algebras A was known as *Tarski's cube problem*, and was open for many years. It was finally solved negatively by Ketonen [34], using a difficult structure theorem for countable Boolean algebras.

Exercises

1. Prove the following analogue, for Boolean algebras, of the Schröder–Bernstein theorem (Appendix A, p. 463): if each of two σ-complete Boolean algebras is isomorphic to a relativization of the other, then the two algebras are themselves isomorphic. (This theorem is due to Sikorski [56] and Tarski [77].)

2. Show that the analogue of the Schröder–Bernstein theorem for Boolean algebras (Exercise 1) implies Theorem 25.

3. Show that, conversely, Theorem 25 implies the analogue of the Schröder–Bernstein theorem for Boolean algebras (Exercise 1).

4. Derive the Schröder–Bernstein theorem for sets (Appendix A, p. 463) from Theorem 25.

5. Show that the associative law

$$A \times (B \times C) = (A \times B) \times C$$

holds for products. (Interpret the equal sign in this and related contexts below to mean isomorphism.)

6. Verify the commutative law

$$A \times B = B \times A$$

for products.

7. Let X and Y be sets with the same number of elements, and B an arbitrary Boolean algebra. Prove that $B^X = B^Y$.

8. The law for multiplying exponential terms with the same base, in the context of Boolean algebras, asserts that if X and Y are disjoint sets, and if B is an arbitrary Boolean algebra, then

$$B^X \times B^Y = B^{X \cup Y}.$$

Does this law hold in general?

9. The law for raising an exponential term to a power, in the context of Boolean algebras, asserts that if X and Y are arbitrary sets, and if B is an arbitrary Boolean algebra, then

$$(B^X)^Y = B^{X \times Y},$$

where $X \times Y$ denotes the ordinary Cartesian product of sets,

$$X \times Y = \{(x, y) : x \in X \text{ and } y \in Y\}.$$

Does this law hold in general?

10. The law for multiplying exponential terms with the same exponent, in the context of Boolean algebras, asserts that if X is an arbitrary set, and if B and C are arbitrary Boolean algebras, then

$$B^X \times C^X = (B \times C)^X.$$

Does this law hold in general?

11. The *cancellation law* for products of Boolean algebras asserts that

$$A \times B = A \times C \qquad \text{implies} \qquad B = C.$$

Does this law hold in general?

12. Is the cancellation law for products (Exercise 11) true when the algebras are all countable?

13. Is the cancellation law for products (Exercise 11) true when the algebras are all finite?

14. Prove that if X is an infinite set, then

$$\mathcal{P}(X)^n = \mathcal{P}(X)$$

for every positive integer n.

15. Does $A = A \times B \times C$ imply $A = A \times B$ when A is finite?

16. Prove that the class of invariant sets (under the bijection θ defined in the chapter) is a complete field of sets.

17. Prove that the class of almost invariant sets (defined in the chapter) is a field of sets.

18. Prove that the field of almost invariant sets is not a complete Boolean algebra.

19. Prove that the mapping g defined in the chapter is an automorphism of $A \times 2$.

20. For a positive integer n, the nth *root law* for Boolean algebras asserts that

$$A^n = B^n \qquad \text{implies} \qquad A = B.$$

Does this law hold for finite Boolean algebras? Does it hold for σ-complete Boolean algebras? (The answer to this last question is due to Tarski [76], [77].)

21. Prove that the square root law (Exercise 20) fails for infinite Boolean algebras. In other words, find two Boolean algebras A and B such that

$$A \times A = B \times B,$$

but $A \neq B$. (This result is due to Tarski; see [24].)

22. Find a Boolean algebra B such that $B = B \times B \times B$, but $B \neq B \times B$.

23. (Harder.) Find a Boolean algebra A such that $A = A \times 2 \times 2 \times 2$, but $A \neq A \times 2$ and $A \neq A \times 2 \times 2$. (This example and its generalizations — see Exercise 25 below — are due to Hanf [24].)

24. Find Boolean algebras A_1 and A_2 such that $A_1 \times A_2 = A_1 \times A_2 \times 2$ but $A_1 \neq A_1 \times 2$ and $A_2 \neq A_2 \times 2$.

25. (Harder.) Find a Boolean algebra A such that $A = A \times 2 \times 2 \times 2 \times 2$, but $A \neq A \times 2$, and $A \neq A \times 2 \times 2$, and $A \neq A \times 2 \times 2 \times 2$.

26. Prove that if $A \times 2 = A$, then A must have infinitely many atoms.

27. Formulate and prove a generalization of Exercise 26.

28. (Harder.) Prove that if A is a countable Boolean algebra with infinitely many atoms, then $A = A \times 2$. (This theorem is due to Vaught; see [24].)

29. Prove that if A is a countable Boolean algebra with infinitely many atoms, and if B is a finite Boolean algebra, then $A = A \times B$. (This theorem is due to Vaught; see [24].)

30. Prove that if A is a countable Boolean algebra, and if B and C are finite Boolean algebras such that $A = A \times B \times C$, then $A = A \times B$. (This theorem is due to Vaught; see [24].)

Chapter 28

Free Algebras

The elements of every subset of every Boolean algebra satisfy various algebraic conditions (such as the distributive laws, for example) just by virtue of belonging to the same Boolean algebra. If the elements of some particular set E satisfy no conditions except these necessary universal ones, it is natural to describe E by some such word as "free." A crude but suggestive way to express the fact that the elements of E satisfy no special conditions is to say that the elements of E can be transferred to an arbitrary Boolean algebra in a completely arbitrary way with no danger of encountering a contradiction. In what follows we shall make these heuristic considerations precise. We shall restrict attention to sets that generate the entire algebra; from the practical point of view the loss of generality involved in doing so is negligible.

A set E of generators of a Boolean algebra B is called *free* if every mapping from E to an arbitrary Boolean algebra A can be extended to an A-valued homomorphism on B. In more detail: E is free in case for every Boolean algebra A and for every mapping g from E into A there exists an A-valued homomorphism f on B such that $f(p) = g(p)$ for every p in E. Equivalent expressions: "E freely generates B", or even "B is free on E". A Boolean algebra is called *free* if it has a free set of generators.

The definition is conveniently summarized by the subjoined diagram. The diagram is to be interpreted as follows. The arrow h is the identity mapping from E to B, expressing the fact that E is a subset of B. The arrow g is an arbitrary mapping from E to an arbitrary algebra A. The arrow f is, of course, the homomorphic extension required by the definition; it is dotted to indicate that it comes last, as a construction based on h and g. It is understood that the diagram is "commutative" in the sense that

S. Givant, P. Halmos, *Introduction to Boolean Algebras*,
Undergraduate Texts in Mathematics, DOI: 10.1007/978-0-387-68436-9_28,

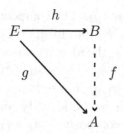

$$(f \circ h)(p) = g(p)$$

for every p in E.

The arrow diagram does not express the fact that E generates B. The most useful way in which that fact affects the mappings under consideration is to guarantee uniqueness: there can be only one A-valued homomorphism f on B that agrees with g on E, by Lemma 13.2. One way of expressing this latter fact is to say that f is uniquely determined by g and h.

There is another and even more important uniqueness assertion that can be made here. If B_1 and B_2 are Boolean algebras, free on subsets E_1 and E_2, respectively, and if E_1 and E_2 have the same number of elements, then B_1 and B_2 are isomorphic, via an isomorphism that interchanges E_1 and E_2. This says, roughly speaking, that B is uniquely determined (to within isomorphism) by the number of elements in E. It is therefore legitimate to speak of *the* free Boolean algebra on m generators, for any cardinal number m.

The proof is summarized by the following diagrams. Here g_1 is a bijec-

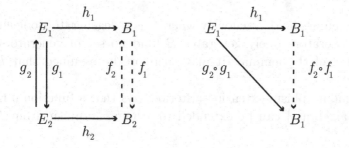

tion from E_1 to E_2, and g_2 is its inverse. The arrows h_1 and h_2 are the identity mappings from E_1 into B_1, and from E_2 into B_2, respectively. The assumption of free generation guarantees the existence of a homomorphism f_1 from B_1 into B_2 that extends g_1, and also the existence of a homomorphism f_2

from B_2 into B_1 that extends g_2. The composition $f_2 \circ f_1$ is an endomorphism of B_1 (a homomorphism of B_1 into itself) that extends $g_2 \circ g_1$. Since $g_2 \circ g_1$ is the identity mapping on E_1, the identity automorphism of B_1 is also an endomorphism that extends $g_2 \circ g_1$. The assumption that E_1 freely generates B_1 means, in particular, that $g_2 \circ g_1$ has only one extension to an endomorphism of B_1. Conclusion: $f_2 \circ f_1$ is the identity automorphism on B_1. A similar argument shows that the composition $f_1 \circ f_2$ is the identity automorphism on B_2. It follows that f_1 is a bijection mapping B_1 isomorphically to B_2, and f_2 is its inverse (see Exercise 12.32 or the section on bijections in Appendix A).

It is quite useful to have an intrinsic characterization of a set of free generators, one that is formulated in terms of the elements of the set alone, and not in terms of homomorphisms that extend mappings on the set. The next lemma gives such a characterization. The proof of the lemma is based on the homomorphism extension criterion formulated in Theorem 4 (p. 107). For each i in B write

$$p(i,j) = \begin{cases} i & \text{if } j = 1, \\ i' & \text{if } j = 0. \end{cases}$$

Lemma 1. *A necessary and sufficient condition for a set E of generators of a Boolean algebra B to be free is that whenever a is a 2-valued function on some finite subset F of E, then*

$$\bigwedge_{i \in F} p(i, a(i)) \neq 0.$$

Proof. The lemma holds vacuously when B is a degenerate Boolean algebra: no set of generators freely generates B, and no set of generators satisfies the condition of the lemma. It may therefore be assumed that B is non-degenerate.

The homomorphism extension criterion says that a function g from E to a Boolean algebra A can be extended to a homomorphism from B to A if and only if

(1) $\bigwedge_{i \in F} p(i, a(i)) = 0$ implies $\bigwedge_{i \in F} p(g(i), a(i)) = 0$

for every finite subset F of E and every 2-valued function a on F. If the condition of the lemma is satisfied, then the antecedent of (1) is never true, independently of the choice of a, and therefore the entire implication in (1) is always true, independently of the choice of a and g. It follows from Theorem 4 that every function g from E to a Boolean algebra A can be extended to a homomorphism from B into A, so that E freely generates B.

Assume now that E freely generates B. To show that the condition formulated in the lemma holds, consider an arbitrary 2-valued function a on a finite subset F of E. Let g be any 2-valued function on E that agrees with a on F; for concreteness, g may be taken to be the function defined by

$$g(i) = \begin{cases} a(i) & \text{if } i \in F, \\ 0 & \text{if } i \in E - F. \end{cases}$$

A direct computation shows that if i is in F, then

$$p(g(i), a(i)) = 1.$$

In more detail, if $a(i) = 1$, then

$$p(g(i), a(i)) = p(g(i), 1) = g(i) = a(i) = 1,$$

and if $a(i) = 0$, then

$$p(g(i), a(i)) = p(g(i), 0) = g(i)' = a(i)' = 0' = 1.$$

Consequently,

$$(2) \qquad \bigwedge_{i \in F} p(g(i), a(i)) = 1.$$

The assumption that E freely generates B implies that g can be extended to a 2-valued homomorphism on B. The implication in (1) must therefore be satisfied for g, by Theorem 4. The consequent of the implication is different from zero, by (2), so the antecedent must be different from zero, as desired.

There is one big gap in what we have seen so far of the theory of freely generated algebras. We may know all about uniqueness, but we know nothing about existence. The main thing to be known here is that for each cardinal number there actually exists a Boolean algebra that is free on a set having exactly that many elements. Let I be an arbitrary set of a given cardinality, and write $S = 2^I$. The elements of S are functions from I into 2, that is, they are functions x with arguments i in I and values $x(i) = x_i$ that are either 0 or 1. Consider the (direct) power 2^S of the two-element Boolean algebra. With each index i in I, there is a naturally associated *projection*, the function p_i from S into 2 defined by

$$p_i(x) = x_i$$

for each x in S. The *existence theorem for free Boolean algebras* asserts that the set of these projections freely generates a subalgebra of 2^S.

Theorem 26. *For each set I, the Boolean subalgebra of 2^{2^I} generated by the set of projections $E = \{p_i : i \in I\}$ is freely generated by E.*

Proof. Let B be the Boolean subalgebra of 2^{2^I} generated by the set E. To prove that E freely generates B, it suffices to show that the condition formulated in Lemma 1 is satisfied. In verifying the condition, it simplifies notation to consider 2-valued functions on finite subsets of I instead of on finite subsets of E. (A similar remark applies in other, related contexts below.)

Suppose a is a 2-valued function on a finite subset F of I. Let x be any element in 2^I that extends a (for instance, take $x_i = 0$ when i is in $I - F$). A straightforward computation shows that

$$p(p_i, a(i))(x) = 1$$

for each i in F. Indeed, if $a(i) = 1$, then

$$p(p_i, a(i))(x) = p(p_i, 1)(x) = p_i(x) = x(i) = a(i) = 1,$$

and if $a(i) = 0$, then

$$p(p_i, a(i))(x) = p(p_i, 0)(x) = p_i'(x) = p_i(x)' = x(i)' = a(i)' = 0' = 1.$$

Write

$$p_a = \bigwedge_{i \in F} p(p_i, a(i)).$$

The preceding argument shows that

$$p_a(x) = \bigwedge_{i \in F} p(p_i, a(i))(x) = 1;$$

in particular, p_a is not the zero element of B (the function on 2^I with constant value 0). Consequently, the set E satisfies the hypotheses of Lemma 1, so the subalgebra generated by E is in fact freely generated by E.

A description of the finitely generated free algebras can be obtained rather easily from the preceding existence theorem. Assume, in accordance with von Neumann's definition of the natural numbers, that each natural number m coincides with the set of its predecessors, $m = \{0, 1, \ldots, m - 1\}$ (see the section on natural and ordinal numbers in Appendix A).

Corollary 1. *For every natural number* m, *the Boolean algebra* 2^{2^m} *is free on* m *generators.*

Proof. Write $S = 2^m$ and $A = 2^S$. The set

$$E = \{p_i : i \in m\}$$

of the projections p_i from S to 2 has cardinality m and freely generates a subalgebra of A, by Theorem 26. It must be shown that the generated subalgebra coincides with A.

For each a in S, write

$$p_a = \bigwedge_{i \in m} p(p_i, a(i)),$$

where $p(p_i, a(i))$ is either p_i or p_i', according as $a(i)$ is 1 or 0. (This infimum exists because m is finite.) The element p_a is obviously generated by E, because it is a (finite) meet of elements and complements of elements from E. An easy computation shows that

(1) $\qquad\qquad p_a(x) = 1 \qquad$ if and only if $\qquad x = a,$

for each x in S. Indeed, $p_a(x)$ is the meet (in the Boolean algebra 2) of the elements $p(p_i, a(i))(x)$, and is therefore 1 just in case $p(p_i, a(i))(x) = 1$ for each i in m. The definition of p implies that

$$p(p_i, a(i))(x) = p_i(x) = x_i \qquad \text{or} \qquad p(p_i, a(i))(x) = p_i'(x) = x_i',$$

according as $a(i)$ is 1 or 0. Consequently, for $p_a(x)$ to be 1, we must have $x_i = 1$ when $a(i) = 1$, and $x_i = 0$ when $a(i) = 0$. This proves (1).

For every subset X of S, put

$$p_X = \bigvee_{a \in X} p_a.$$

(This supremum exists because X is finite.) Again, it is clear that p_X is generated by the elements p_a, and therefore by the set E. It follows from (1) that for each x in S,

$$p_X(x) = 1 \qquad \text{if and only if} \qquad x \in X.$$

An arbitrary element q in A is a 2-valued function on S, and is therefore completely determined by the set of those x in S for which $q(x) = 1$. If that set is X, then

$$q(x) = 1 \quad \text{if and only if} \qquad x \in X,$$
$$\qquad\quad \text{if and only if} \qquad p_X(x) = 1,$$

so that $q = p_X$. Consequently, every element q in A coincides with p_X for some subset X of S, and therefore q is generated by E, as was to be shown.

It is unreasonable to expect the corollary to be true for infinite cardinal numbers. When m is infinite, the free algebra on m generators has cardinality m, while 2^{2^m} has cardinality greater than m.

A combinatorial proof of the existence of free Boolean algebras is also available. One of its main virtues is that it shows how Boolean algebras (and, in particular, free Boolean algebras) arise in considerations of logic. We shall sketch a bare outline of this proof. (The construction dates back to Huntington [30] and Tarski [72].)

A general theory of the usual sentential connectives — conjunction (and), disjunction (or), negation (not), implication (if then), etc. — should be applicable to every conceivable collection of sentences. This implies that its basic constituents (generators) should be as unrestricted (free) as possible. Suppose now that we want to construct a theory equipped to deal with, say, at least m sentences simultaneously, where m is a cardinal number. The thing to do then is to take a set E of cardinality m — the set of propositional variables — and to consider all the formal expressions obtained by combining the elements of E and the sentential connectives in an intelligent manner. Ultimately the elements of E are to be replaced (or, at any rate replaceable) by sentences. All this can be done, and, incidentally, it is important that in the doing of it the cardinal number m should be allowed to be infinite. Even if a mathematician or logician wishes to consider only finite combinations of sentences, it seems both practically and theoretically undesirable to place a fixed upper bound on the number of sentences that may be combined. The only way to make one theory elastic enough to deal with all finite combinations is to provide it with an infinite supply of things that it may combine.

To achieve the desired end, a logician will usually begin by selecting enough sentential connectives so that all others are definable in terms of them; we know, for instance, that $\vee$ (or) and $'$ (not) will do. Next, given the set E, the logician will proceed to form all finite sequences whose terms are the selected connectives, or elements of E, or parentheses, put together in the usual and obvious manner. Precisely speaking, the admissible sequences consist of the one-term sequences whose term belongs to E; the sequences obtained by inserting $\vee$ between two others already admitted and enclosing the result in parentheses; the sequences obtained by following an already

admitted sequence by $'$ and enclosing the result in parentheses; and no others. The reason for the insistence on parentheses is caution. The distinction
between $(p \vee (q'))$ and $((p \vee q)')$ is obvious, whereas the customary decision
that $p \vee q'$ means the former and not the latter is the result of quite an
arbitrary and frequently unformulated convention. One other word of su-
percaution deserves mention: it must be assumed that neither the selected
connectives nor the parentheses that are used occur as elements of E.

If the sequences so obtained are to form a part of a general theory of
sentences, it is clear that certain identifications will have to be made. The
sequence $(p \vee q)$ is different from $(q \vee p)$, but if p and q are sentences, then
"p or q" and "q or p" are, in some sense, the same sentence. The customary
way to specify the identifications that sound logical intuition and practice
demand is first to define a special class of admissible sequences (called *tautologies*) and then to say that two admissible sequences are to be identified
just in case a certain easily describable combination of them is a tautology.
The procedure is similar to the formation of quotient Boolean algebras: first
we select an ideal and then we say that two elements of the given Boolean
algebra are congruent modulo the selected ideal just in case their Boolean
sum belongs to the ideal.

To define the set of tautologies we first define certain quite natural abbreviations, then, using these, we describe some tautologies, and finally we
obtain all tautologies by describing a simple operation that makes new tautologies out of old. The abbreviations are these: if S and T are admissible
sequences, we write $S \wedge T$ for $((S') \vee (T(T')))'$, we write $S \Rightarrow T$ for $(S') \vee T$,
and we write $S \Leftrightarrow T$ for $(S \Rightarrow T) \wedge (T \Rightarrow S)$. The initial set of tautologies
consists of all the sequences of one of the four forms

$$((S \vee S) \Rightarrow S),$$
$$(S \Rightarrow (S \vee T)),$$
$$((S \vee T) \Rightarrow (T \vee S)),$$
$$((S \Rightarrow T) \Rightarrow ((R \Rightarrow S) \Rightarrow (R \Rightarrow T))),$$

where R, S, and T are admissible sequences. (Each sequence of each of these
forms is called an *axiom*.) The way to make new tautologies out of old is this:
if S is a tautology and if $(S \Rightarrow T)$ is a tautology, then T is a tautology. (This
operation is a *rule of inference*, namely, in classical terms, *modus ponens*.) A
tautology is, by definition, a sequence that is either an axiom or obtainable
from the axioms by a finite number of applications of modus ponens.

Two sequences S and T are called *logically equivalent* in case $(S \Leftrightarrow T)$ is a tautology.

The structure outlined in this way, that is, the structure consisting of the set of all admissible sequences, the subset of tautologies, and the relation of logical equivalence, is known as the *propositional calculus*. The connection between the propositional calculus (based, as above, on a set of power m, say) and the theory of Boolean algebras is this: logical equivalence is an equivalence relation, the set of equivalence classes has in a natural way the structure of a Boolean algebra, and, in fact, that Boolean algebra is freely generated by m generators.

The involved construction of the propositional calculus outlined above is similar to, but definitely not identical with, a well-known construction of free groups (via "words" and equivalence classes). That familiar construction could also be adapted to the construction of free Boolean algebras; the result would be about equally painful with what we have already seen. It is unimportant but amusing to know that the cross-fertilization between the two theories is complete: the "axiom–rule" approach can be adapted to the construction of free groups.

In a subsequent chapter, we shall see yet another proof of the existence of free Boolean algebras. It is more economical than the proofs discussed above, because it is based on some powerful techniques that will be introduced later. The insights provided by the later proof are different from those provided by the proofs in this chapter, and are less algebraic in nature.

Infinite free Boolean algebras possess a number of interesting properties. Here is one example.

Corollary 2. *An infinite free Boolean algebra is atomless.*

Proof. A finitely generated Boolean algebra must be finite, by Corollary 11.2, so an infinite free Boolean algebra has an infinite set of free generators. Recall from the proof of Theorem 26 that if I is an infinite set with m elements, then the free Boolean algebra with m generators is (up to isomorphic copies) the subalgebra B of 2^{2^I} that is generated by the set of projections

$$E = \{p_i : i \in I\}.$$

The algebra B is the union of the directed family of subalgebras B_F generated by the finite subsets of projections

$$E_F = \{p_i : i \in F\},$$

where F ranges over the finite subsets of I (Corollary 11.1). Each algebra B_F, being finitely generated, is finite and therefore atomic; its atoms are the elements of the form

$$p_a = \bigwedge_{i \in F} p(p_i, a(i)),$$

where a is a function from F into 2; see Theorem 2, p. 81. (The elements p_a are non-zero, by Lemma 1.) To prove the corollary, it suffices to show that below each atom p_a of B_F there are two disjoint, non-zero elements in B (which of course do not belong to B_F); for then p_a cannot be an atom in B. Let i_0 be any index in I that is not in F. The function a has two extensions, say b and c, to functions from $F \cup \{i_0\}$ into 2; they are determined by requiring

$$b(i_0) = 1 \quad \text{and} \quad c(i_0) = 0.$$

The elements

$$p_b = \bigwedge_{i \in F \cup \{i_0\}} p(p_i, b(i)) \quad \text{and} \quad p_c = \bigwedge_{i \in F \cup \{i_0\}} p(p_i, c(i))$$

in B are obviously below p_a, since b and c extend a, and they are non-zero by Lemma 1. They are disjoint because

$$p_b \wedge p_c \le p(p_{i_0}, b(i_0)) \wedge p(p_{i_0}, c(i_0)) = p(p_{i_0}, 1) \wedge p(p_{i_0}, 0) = p_{i_0} \wedge p'_{i_0} = 0.$$

One surprising consequence of the preceding corollary and Theorem 10 (p. 134) is that the field of finite unions of left half-closed intervals of rational numbers is a free Boolean algebra on $\aleph_0$ (free) generators. The reader might find intriguing the problem of giving an explicit set of free generators for this algebra. (See Exercise 13.)

The early part of the theory of free Boolean algebras extends with no profound conceptual change to the class of complete algebras. The definition reads just as before except that all the Boolean algebras that enter into it, and all the homomorphisms also, are now required to be complete. The uniqueness theorems are proved just as before. The situation of the principal existence theorem, however, is startlingly different. Both Gaifman [17] and Hales [21] proved that for each cardinal number m there exists a countably generated complete Boolean algebra with m or more elements. (Generation is to be interpreted here in the sense appropriate to the class of complete Boolean algebras: the complete subalgebra generated by a set E is the intersection of all complete subalgebras that include E.) This result implies that the class of complete Boolean algebras does not contain a free algebra B on $\aleph_0$ generators, for then every countably generated algebra in the class, being a complete homomorphic image of B, would have to have cardinality at most that of B.

Exercises

1. What is the Boolean algebra freely generated by the empty set?

2. Prove that a generating set of a degenerate Boolean algebra is never free and never satisfies the condition of Lemma 1.

3. For which finite sets X is the algebra $\mathcal{P}(X)$ free? Examine especially sets X of cardinality 1, 2, 3, and 4.

4. If X is an infinite set, can $\mathcal{P}(X)$ be free?

5. Find a set of free generators for the algebra $\mathcal{P}(X)$ when $X = \{0, 1, 2, 3\}$. Do the same when the set is $X = \{0, 1, 2, 3, 4, 5, 6, 7\}$. Can these results be generalized?

6. (Harder.) Give a direct proof of the existence theorem for free Boolean algebras, without using Lemma 1 and the homomorphism extension criterion (Theorem 4, p. 107).

7. Theorem 26 follows rather easily from Lemma 1, as its proof shows. Prove that, conversely, Lemma 1 follows rather easily from Theorem 26.

8. If E is a set of free generators of a Boolean algebra, prove that every subset of E is a set of free generators of the subalgebra it generates.

9. Is every subalgebra of a free Boolean algebra free?

10. Is every infinite subalgebra of an infinite free Boolean algebra free?

11. Prove that every Boolean algebra is isomorphic to a quotient of a free one.

12. Prove that the following conditions on a subset E of a Boolean algebra B are equivalent: (1) E is a set of generators of B; (2) every 2-valued mapping on E has at most one extension to a 2-valued homomorphism on B; (3) every mapping from E into an arbitrary Boolean algebra A has at most one extension to an A-valued homomorphism on B.

13. (Harder.) Prove, without using Theorem 10, that a countable atomless Boolean algebra with more than one element is free. Conclude that all countable atomless Boolean algebras with more than one element are isomorphic.

Chapter 29

Boolean σ-algebras

Between Boolean algebras and complete Boolean algebras there is room for many intermediate concepts. The most important one is that of a Boolean σ-*algebra*; this means, by definition, a Boolean algebra in which every countable set has a supremum (and therefore, of course, an infimum). Similarly, a field of sets is a σ-*field* if it is closed under the formation of countable unions (and therefore under the formation of countable intersections).

It is a routine matter to imitate the entire algebraic theory developed so far for the two extremes (Boolean algebras and complete algebras) in the intermediate case of σ-algebras. Thus, a σ-*subalgebra* of a σ-algebra A is a subalgebra B of A that is closed under the formation of countable suprema (and hence under the formation of countable infima); more precisely, whenever p is the supremum in A of a countable family of elements in B, then p belongs to B. A σ-subalgebra of a σ-field of sets is called a σ-*subfield*. The intersection of a family of σ-subalgebras (of A) is of course itself a σ-subalgebra. (The intersection of the empty family of σ-subalgebras is, by convention, the improper σ-subalgebra A.) The σ-subalgebra generated by a subset E is, by definition, the intersection of the family of σ-subalgebras that include E as a subset. (Notice that this family is not empty, since it always contains A.) When more clarity is needed, E may be called a set of σ-generators.

Continuing in the same spirit, we define a σ-*homomorphism* as a homomorphism that preserves all the countable suprema (that is, the suprema of all the countable sets) that happen to exist. A *free σ-algebra* is defined the same way as a free Boolean algebra except that all the algebras and homomorphisms that enter the definition are now required to be σ-algebras and σ-homomorphisms. (The problem of the existence of σ-algebras free on sets

S. Givant, P. Halmos, *Introduction to Boolean Algebras,*
Undergraduate Texts in Mathematics, DOI: 10.1007/978-0-387-68436-9_29,
© Springer Science+Business Media, LLC 2009

of generators of arbitrary cardinality will be attacked later.)

A σ-*ideal* is, by definition, an ideal closed under the formation of countable suprema. The kernel of a σ-homomorphism on a σ-algebra is a σ-ideal. Indeed, if M is the kernel of a σ-homomorphism f, then M is certainly an ideal. To prove that it is a σ-ideal, consider a sequence $\{q_n\}$ of elements in M, and let q be the supremum of the sequence in B. Then $f(q_n) = 0$ for every n, by the definition of a kernel, and consequently

$$f(q) = \bigvee_n f(q_n) = 0,$$

by the definition of a σ-homomorphism. It follows that q is in M.

Conversely, every σ-ideal is the kernel of a σ-epimorphism. For the proof, suppose that B is a σ-algebra and M a σ-ideal in B. Form the quotient $A = B/M$, let f be the projection of B onto A, and recall that M is the kernel of f. We shall prove that A is a σ-algebra and f is a σ-homomorphism. The two assertions can be treated simultaneously by proving that if $\{q_n\}$ is a sequence of elements in B with supremum q, then the sequence $\{f(q_n)\}$ has a supremum in A, and, in fact,

$$\bigvee_n f(q_n) = f(q).$$

Write

$$f(q_n) = p_n \quad \text{and} \quad f(q) = p.$$

The inequality $q_n \leq q$ implies that $p_n \leq p$, for each n, by the homomorphism properties of f. It is to be proved that if $p_n \leq s$ for all n, then $p \leq s$. Let t be an element of B such that $f(t) = s$. (The element t exists because f maps B onto A.) Since

$$f(q_n) = p_n \leq s = f(t)$$

for all n, we have

$$f(q_n - t) = f(q_n) - f(t) = 0.$$

In other words, $q_n - t$ is in the kernel M. Because M is a σ-ideal, by assumption, the join $\bigvee_n (q_n - t)$ must also be in M. The equalities

$$\bigvee_n (q_n - t) = \left(\bigvee_n q_n \right) - t = q - t$$

(Exercise 8.25(c)) therefore imply that $q - t$ belongs to the kernel M. Consequently,

$$0 = f(q - t) = f(q) - f(t),$$

and therefore $f(q) \leq f(t)$. In other words, $p \leq s$, as promised. The following theorem (which dates back to Tarski [75]) has been proved. It contains, in particular, the analogue for σ-algebras of the homomorphism theorem (see Chapter 21).

Theorem 27. *The quotient of a σ-algebra by a σ-ideal is a σ-algebra. The corresponding projection is a σ-homomorphism, and its kernel is the given σ-ideal.*

The simplest way to be a σ-algebra is to be complete. There are other ways. The countable–cocountable algebra of every set is a σ-algebra that is not complete, unless the underlying set is countable. (Observe, by the way, that the class of all countable sets in this algebra is a non-trivial maximal ideal.) The most famous and useful incomplete σ-algebras arise in topological spaces. A *Borel set* in a topological space X is, by definition, a set belonging to the σ-field generated by the class of all open sets (or, equivalently, by the class of all closed sets). For instance, countable unions of closed sets and countable intersections of open sets are Borel; they are usually called *F_σ-sets* and *G_δ-sets* (or simply F_σ's and G_δ's) respectively. Here are some concrete examples. In the Euclidean space $\mathbb{R}^n$, every set consisting of just one point is closed, and the whole space with just one point removed is open, so every countable set is an F_σ, and every cocountable set is a G_δ. In the space $\mathbb{R}$, the open interval (a, b) is the union of the closed intervals $[a + 1/n, b - 1/n]$ for $n = 1, 2, \ldots$, and the closed interval $[a, b]$ is the intersection of the open intervals $(a - 1/n, b + 1/n)$ for $n = 1, 2, \ldots$, so every open interval in $\mathbb{R}$ is an F_σ, and every closed interval is a G_δ. Countable intersections of F_σ-sets and countable unions of G_δ-sets are also Borel; they are usually called *$F_{\sigma\delta}$-sets* and *$G_{\delta\sigma}$-sets* respectively. One can continue in this fashion to define *$F_{\sigma\delta\sigma}$-sets*, *$G_{\delta\sigma\delta}$-sets*, and so on, and all of these sets are Borel.

There is also an interesting σ-ideal that can be defined in topological terms. A subset of a topological space is called *nowhere dense* if the closure of its interior is empty, and it is called *meager* if it is the union of countably many nowhere dense sets (see Chapter 9). In any topological space, the empty set is meager (it is nowhere dense), the intersection of an arbitrary set with a meager set P is meager (a subset of a meager set is meager, by Exercise 9.27), and the union of a sequence of meager sets is meager (Exercise 9.28). This argument shows that the class of all meager subsets of a topological space X is a σ-ideal in $\mathcal{P}(X)$. Similarly, the class of all meager Borel sets is a σ-ideal in the σ-field of all Borel sets.

Compact topological spaces play a very important role in the theory of Boolean algebras. To define them, it is helpful to introduce an auxiliary notion. An *open cover* of a topological space is a family of open sets whose union is the whole space. A topological space is said to be *compact* if every open cover has a finite subcover; in other words, whenever the space is equal to the union of a family $\{U_i\}$ of open sets, it is already equal to the union of some finite subfamily $\{U_{i_1}, U_{i_2}, \ldots, U_{i_n}\}$. The contrapositive of this definition runs as follows: if $\{U_i\}$ is a family of open sets, and if no finite subfamily covers the whole space, then $\{U_i\}$ cannot cover the whole space.

There is a very useful characterization of compactness. A family of sets in a topological space has the *finite intersection property* if the intersection of every finite subfamily is non-empty. It turns out that a topological space is compact if and only if each family of closed sets with the finite intersection property has a non-empty intersection. For the proof, suppose that X is a compact space, and let $\{F_i\}$ be a family of closed sets with the finite intersection property. Then $\{F_i'\}$ is a family of open sets such that no finite subfamily covers X, by the De Morgan laws (2.7). The (contrapositive of the) definition of compactness implies that the union of $\{F_i'\}$ is not all of X; consequently, the intersection of the family $\{F_i\}$ is not empty, by the infinite De Morgan laws (8.1). The reverse implication is established by a completely analogous argument.

A subset Q of a topological space X is said to be compact if it is a compact space under the inherited topology. In other words, Q is compact if each open cover of Q (each family of open sets in X whose union includes Q) has a finite subcover. A closed and bounded subset of $\mathbb{R}^n$ is always compact; this is just the content of the celebrated Heine–Borel theorem. In particular, every finite interval $[a, b]$ in the space of real numbers is compact. On the other hand, it is easy to check that the set of all real numbers is not compact. For instance, the family of intervals $\{(n, n + 2)\}$, where n ranges over the integers, is an open cover of $\mathbb{R}$, but it has no finite subcover.

A topological space is said to be *Hausdorff* if any two points can be *separated* by open sets. This means that for any two points x and y in the space, there are disjoint open sets U and V such that x is in U and y is in V. The Euclidean spaces $\mathbb{R}^n$ are Hausdorff: if x and y are two points in $\mathbb{R}^n$, and if δ is the distance between these two points, then the open balls of radius $\delta/2$ centered at x and at y are disjoint and contain x and y respectively.

Every closed set in a compact space is compact. Proof: if Q is closed, and if $\{U_i\}$ is an open cover of Q, then the family $\{Q'\} \cup \{U_i\}$ is an open cover of the whole space. There must be a finite subcover of the whole space, say

$$\{Q', U_{i_1}, U_{i_2}, \ldots, U_{i_n}\},$$

so $\{U_{i_1}, U_{i_2}, \ldots, U_{i_n}\}$ must cover Q. The converse is not in general true, but it is true for Hausdorff spaces.

Lemma 1. *If Q is a compact set in a Hausdorff space, and if x is a point of the space that is not in Q, then x and Q can be separated by open sets. Consequently, each compact set is closed.*

Proof. Each point y in Q can be separated from x by open sets, because the space is Hausdorff. In other words, there are disjoint open sets V_y and W_y such that x is in V_y, and y is in W_y. The family $\{W_y\}_{y \in Q}$ is an open cover of Q, so it must have a finite subcover $\{W_{y_1}, W_{y_2}, \ldots, W_{y_n}\}$. The sets

$$V = \bigcap_{i=1}^{n} V_{y_i} \qquad \text{and} \qquad W = \bigcup_{i=1}^{n} W_{y_i}$$

are both open. (A finite intersection of open sets is open, as is an arbitrary union of open sets.) They are also disjoint: if a point belongs to V, then it belongs to each set V_{y_i}; consequently it cannot belong to any of the sets W_{y_i} (the sets V_{y_i} and W_{y_i} being disjoint), and therefore it cannot belong to W. Finally, x belongs to V (because x is in each set V_{y_i}), and Q is included in W (because the sets W_{y_i} form a finite subcover of Q).

One consequence of the observations of the preceding paragraph is that for every point x not in Q, there is an open set U_x (the set V above) that contains x and is disjoint from Q. The union of the family $\{U_x\}_{x \in Q'}$ is therefore an open set that is disjoint from Q and contains every point in Q'. Consequently, this union must coincide with Q'. Conclusion: Q' is open, and therefore Q is closed.

Corollary 1. *For every open set U in a compact Hausdorff space, and every point x in U, there is an open set V containing x such that the closure V^- is included in U.*

Proof. Suppose U is an open set, and x a point in U. The complement U' is closed, by definition, and therefore compact, since the whole space is assumed to be compact. Apply the lemma to obtain disjoint open sets V and W such that x belongs to V, and U' is included in W. The set W' is then closed, and

$$V \subseteq W' \subseteq U.$$

The closure of V must therefore be included in the closed set W', and hence also in U.

Corollary 2. *Any two points in a compact Hausdorff space are separated by open sets with disjoint closures.*

Proof. Given two points x and y in a compact Hausdorff space, there exist (by definition) disjoint open sets U and W such that x is in U and y in W. The complement U' is closed and includes W, so it includes the closure W^-. There is an open set V containing x such that V^- is included in U, by Corollary 1. The open sets V and W contain x and y respectively, and have disjoint closures, since V^- is included in U, and W^- in U'.

The following celebrated result, known as the *Baire category theorem* (see Baire [2]), is needed on most occasions when meager sets occur. It says that no non-empty open set can be meager in a compact Hausdorff space.

Theorem 28. *A meager open set in a compact Hausdorff space is empty.*

Proof. Suppose that U is a non-empty open set and that $\{S_n\}$ is a sequence of nowhere dense sets. We shall show that U contains at least one point that does not belong to any S_n. It follows that U cannot equal the union of the sets S_n. In other words, U cannot be meager.

Write $U_0 = U$. Let V_1 be a non-empty open set with the property that $V_1^- \subseteq U_0$; such a set exists by Corollary 1. The closed set S_1^- has an empty interior, by the assumption that S_1 is nowhere dense. Its complement, the open set $S_1^{-\prime}$, therefore has a non-empty intersection with every non-empty open set. In particular, it has a non-empty intersection with V_1. The set

$$U_1 = V_1 \cap S_1^{-\prime}$$

is a non-empty open subset of V_1 (it is open because it is the intersection of two open sets), and $U_1 \cap S_1 = \varnothing$, because U_1 is disjoint from S_1^-.

Repeat this argument with U_1 in place of U_0 to obtain a non-empty open set V_2 such that $V_2^- \subseteq U_1$, and to conclude that the set

$$U_2 = V_2 \cap S_2^{-\prime}$$

is a non-empty open subset of V_2 with the property that $U_2 \cap S_2 = \varnothing$. Continue this argument inductively: for each positive integer n, there is a non-empty open set V_n such that $V_n^- \subseteq U_{n-1}$, and the set

$$U_n = V_n \cap S_n^{-\prime}$$

is a non-empty open subset of V_n with the property that $U_n \cap S_n = \varnothing$.

The last equation implies that the intersection $\bigcap_k U_k$ is disjoint from each set S_n. The inclusions

$$U_n \subseteq V_n \subseteq V_n^- \subseteq U_{n-1}$$

imply that

$$\bigcap_k U_k = \bigcap_k V_k^-.$$

The family of non-empty closed sets $\{V_k^-\}$ is decreasing, and therefore has the finite intersection property:

$$V_1^- \cap V_2^- \cap \cdots \cap V_n^- = V_n^- \neq \varnothing.$$

The topological space is assumed to be compact, so the family $\{V_k^-\}$ has a non-empty intersection. This means that $\bigcap_k U_k$ is non-empty. Any point in this intersection has the desired properties: it belongs to $U = U_0$, and it does not belong to any of the sets S_n.

The ideal of meager sets makes contact with an earlier construction in a somewhat surprising way: Borel sets are almost regular open sets in the sense that each Borel set differs symmetrically from a uniquely determined regular open set by a meager set. It is helpful to formulate this result another way, using the notion of congruence modulo an ideal that was discussed in Chapter 18. Recall that two elements p and q of a Boolean algebra are defined to be congruent modulo an ideal M if the Boolean sum $p + q$ is in M, and in this case we write $p \equiv q \mod M$, or just $p \equiv q$ when the intended ideal M is clearly understood.

Lemma 2. *Every Borel set in a compact Hausdorff space is congruent to a unique regular open set modulo the σ-ideal of meager Borel sets.*

Proof. A subset S of a compact Hausdorff space X is said to have the *Baire property* if it is congruent to some open set, where "congruent" in the course of this proof means congruent modulo the σ-ideal M of meager Borel sets in X. The first step in the proof is to show that the class of sets with the Baire property is a σ-field that includes the open sets. Clearly every open set U has the Baire property: $U \equiv U$, since $U + U = \varnothing$ and the empty set is meager. Suppose $\{S_n\}$ is a sequence of sets with the Baire property, say $\{U_n\}$ is a sequence of open sets such that $S_n \equiv U_n$ for each n. The sums $S_n + U_n$ are in M, by the definition of congruence, so their union is also in M, because M is a σ-ideal. The sum of the unions $\bigcup_n S_n$ and $\bigcup_n U_n$ is included in the union of the sums,

$$\left(\bigcup_n S_n\right) + \left(\bigcup_n U_n\right) \subseteq \bigcup_n (S_n + U_n),$$

by Exercise 8.26, so it, too, belongs to M, by (18.18). It follows that

$$\left(\bigcup_n S_n\right) \equiv \left(\bigcup_n U_n\right).$$

Since the union of the family $\{U_n\}$ of open sets is open, the set $\bigcup_n S_n$ has the Baire property.

It remains to prove that the complement of a set with the Baire property also has the Baire property. Notice, first of all, that every open set U is congruent to its closure: $U \equiv U^-$. Indeed, the sum $U + U^-$ coincides with the difference $U^- - U$ (since $U - U^- = \varnothing$), which in turn is included in the boundary of U; consequently, the sum is nowhere dense, and hence meager, by Lemma 10.5 and Exercise 9.25. Suppose now that $S \equiv U$, where U is open. Since $U \equiv U^-$, it follows that $S \equiv U^-$. Boolean congruences preserve complementation, so $S' \equiv U^\perp$. This shows that the set S' has the Baire property, since $U^\perp$ is open (it is the complement of the closed set U^-).

It has been shown that the class of sets with the Baire property is a σ-field that includes all open sets. The class of Borel sets is, by definition, the smallest σ-field that includes the open sets. Conclusion: every Borel set has the Baire property.

The next step is to prove that every open set U is congruent to a regular open set, and in fact to $U^{\perp\perp}$. The assumption that U is open implies

$$U \subseteq U^{\perp\perp} \subseteq U^{\perp\perp -} = U^-.$$

The first inclusion is a consequence of Lemma 10.2. For the last equality, observe that

$$U^{-\prime} = U^\perp = U^{\perp\perp\perp} = U^{\perp\perp -\prime},$$

by the definition of $\perp$ and Lemma 10.3; form the complements of the first and last terms to obtain the desired equality. The difference $U^- - U$ is included in the boundary of U, which is nowhere dense and therefore meager. The preceding string of inclusions implies that the difference $U^{\perp\perp} - U$ is also included in $U^- - U$, and is therefore also meager. The sum $U + U^{\perp\perp}$ coincides with $U^{\perp\perp} - U$ (because $U - U^{\perp\perp} = \varnothing$), so it, too, is meager. In other words, $U \equiv U^{\perp\perp}$.

Every Borel set is congruent to an open set, and every open set is congruent to a regular open set, so every Borel set is congruent to a regular open set. To demonstrate that this regular open set is uniquely determined, it suffices to prove that two congruent regular open sets are in fact equal. Suppose U and V are regular open sets, and $U \equiv V$. Since

$$U \equiv U^- \qquad \text{and} \qquad V \equiv V^-,$$

it follows from the transitivity of $\equiv$ that

$$U^- \equiv V \qquad \text{and} \qquad U \equiv V^-.$$

Both $U^- + V$ and $U + V^-$ are therefore meager sets, and hence so are their subsets $V - U^-$ and $U - V^-$. Since these subsets are also open, the Baire category theorem implies that they are empty. In other words, $V \subseteq U^-$ and $U \subseteq V^-$. These two inclusions imply that $U^- = V^-$; consequently,

$$U = U^{\perp\perp} = U^{-\prime-\prime} = V^{-\prime-\prime} = V^{\perp\perp} = V.$$

The proof of the lemma is complete.

In view of the preceding lemma, the function f that takes each Borel set S to the regular open subset $f(S)$ such that $S \equiv f(S)$ is a well-defined mapping from the σ-field B of Borel sets into the complete Boolean algebra A of regular open sets. The function maps B onto A because every regular open set S is Borel, and $f(S) = S$. We have seen above that if $S \equiv U$, where U is open, then $S' \equiv U^\perp$; in particular,

$$f(S') = f(S)^\perp.$$

We have also seen that if $S_n \equiv U_n$ for $n = 1, 2, \ldots$, where again the U_n's are open, then

$$\bigcup_n S_n \equiv \bigcup_n U_n \equiv \left(\bigcup_n U_n \right)^{\perp\perp};$$

in particular,

$$f\left(\bigcup_n S_n \right) = \left(\bigcup_n f(S_n) \right)^{\perp\perp}.$$

These two assertions mean just that f is a σ-homomorphism, by the definition of complement and join in A (see Theorem 1, p. 66). The kernel of f is the class of Borel sets that are congruent to the empty set modulo the σ-ideal M, and this is just M itself. The mapping f is thus a σ-homomorphism from B onto A with kernel M, so that A is isomorphic to B/M. We summarize what has been accomplished in the following theorem.

Theorem 29. *Suppose B is the σ-field of Borel sets, and M the σ-ideal of meager Borel sets, in a compact Hausdorff space X. The correspondence f that takes each Borel set S to the regular open set $f(S)$ determined by*

$$S \equiv f(S) \quad \mathrm{mod}\ M$$

is a σ-homomorphism from B onto the complete algebra A of all regular open sets in X. The kernel of f is M, so that A is isomorphic to B/M.

One surprising aspect of the theorem is that the quotient of a σ-algebra by a σ-ideal, which is necessarily a σ-algebra itself, turns out to be a complete algebra (since A is complete). This is a special dividend; it is not to be expected in every case. (The fact that the quotient B/M in the preceding theorem is a complete Boolean algebra was first stated in [75] by Tarski, who refers to earlier work of Szpilrajn-Marczewski; it was independently observed by Birkhoff and Ulam in [6].)

It is tempting, but not particularly profitable, to define classes of Boolean algebras depending on other cardinal numbers the same way as σ-algebras depend on $\aleph_0$. The situation is analogous to the various generalizations of compactness depending on cardinal numbers. The questions undeniably exist, the answers are sometimes easy and sometimes not, and the answers are sometimes the same as for the ungeneralized concepts and sometimes not. In all cases, however, and in Boolean algebra as well as in topology, the generalized theory has much more the flavor of cardinal number theory than of the subject proper. The interested reader should have no trouble in reconstructing the basic theory. The problem is, given an infinite cardinal m, to define and to study m-algebras, m-fields, m-subalgebras, m-subfields, m-homomorphisms, free m-algebras, m-ideals, m-filters, etc. For complicated historical reasons the symbol $\aleph_0$ is always replaced by σ in such contexts, so that, for instance, $\aleph_0$-algebras are the same as the σ-algebras that constituted the main subject of this chapter.

Exercises

1. Prove that a Boolean algebra is a σ-algebra if and only if every countable set has an infimum.

2. Prove that a subalgebra B of a σ-algebra A is a σ-subalgebra if and only if the infimum (in A) of every countable set of elements in B belongs to B.

3. Show that the intersection of a family of σ-subalgebras (of a given σ-algebra) is always a σ-subalgebra.

4. Give some examples of incomplete σ-fields.

5. A family $\{B_i\}_{i \in I}$ of Boolean subalgebras is said to be *countably directed* if for every countable subset J of indices, there is an index i in I such that B_j is a subalgebra of B_i for each j in J. Prove that the union of a countably directed family of σ-subalgebras (of a given σ-algebra) is a σ-subalgebra.

6. Prove that if A is a σ-algebra, and if p is an element of the σ-subalgebra generated by a subset E of A, then E has a countable subset D such that p belongs to the σ-subalgebra generated by D.

7. Show that a homomorphism on a Boolean algebra is a σ-homomorphism if and only if it preserves all countable infima (that is, the infima of all countable sets) that happen to exist.

8. Define what it means for a family of A-valued homomorphisms to be countably directed. Prove that a countably directed family of A-valued σ-homomorphisms always has a common extension to an A-valued σ-homomorphism. If the homomorphisms in the family are one-to-one, show that the common extension is also one-to-one.

9. Formulate and prove the analogue of Exercise 12.26 for σ-homomorphisms between σ-algebras.

10. Formulate and prove the analogue of Exercise 12.29 for σ-epimorphisms between σ-algebras.

11. Prove that if two A-valued σ-homomorphisms on a σ-algebra B agree on the elements of a set of σ-generators of B, then they agree on all of B.

12. Formulate and prove the analogue for σ-algebras of the first isomorphism theorem.

13. Give a precise definition of the notion of a set E freely σ-generating a σ-algebra B. Prove that if two σ-algebras B_1 and B_2 are freely σ-generated by sets E_1 and E_2 of the same cardinality, then B_1 and B_2 are isomorphic via a mapping that interchanges E_1 and E_2.

14. Suppose A, B, and C are σ-algebras. If f is a σ-homomorphism from B to A, and g a σ-homomorphism from C to B, prove that the composition $f \circ g$ is a σ-homomorphism from C to A.

15. Prove that the intersection of a family of σ-ideals (in a given σ-algebra) is always a σ-ideal.

16. Define the notion of a σ-filter, and prove that the cokernel of a σ-homomorphism is a σ-filter.

17. If A is the field of all subsets of an infinite set X, prove that the σ-subfield of A generated by the set of singleton (that is, one-element) subsets of X coincides with the field of countable and cocountable subsets of X.

18. Prove that the class of all countable sets in the σ-field of countable and cocountable subsets of an uncountable set is a σ-ideal that is maximal (as an ideal, and therefore as a σ-ideal).

19. Prove that the class of F_σ-sets is closed under finite intersections and under countable unions. Formulate and prove an analogous result for G_δ-sets.

20. What kind of set is the complement of an F_σ?

21. (Harder.) Prove that in a metric space every closed set is a G_δ and every open set is an F_σ.

22. (Harder.) Prove that there are continuum many Borel sets of real numbers.

23. Complete the proof that a topological space is compact if and only if each family of closed sets with the finite intersection property has a non-empty intersection.

24. Prove that a topological space is compact just in case it satisfies the following condition: if the intersection of a family of closed sets is included in an open set, then the intersection of some finite subfamily of the closed sets is included in the open set.

25. Show that the family of intervals $(-n, n)$, where n ranges over the positive integers, is an open cover of $\mathbb{R}$ that has no finite subcover.

26. Show that the two-dimensional Euclidean space $\mathbb{R}^2$ is not compact.

27. Prove that in an arbitrary topological space, a closed subset of a compact set is compact.

28. Suppose Y is a subspace of a topological space X, and P is a subset of Y. Prove that P is compact in Y if and only if it is compact in X.

29. Suppose an infinite set X is endowed with the discrete topology (Chapter 9). Is the resulting space compact? Is it Hausdorff? What are the answers to these questions when X is endowed with the cofinite topology?

30. Show that in a Hausdorff space, singletons of points are closed sets.

31. Show that the atoms of the regular open algebra of a Hausdorff space are precisely the singletons of isolated points. (A point x is called *isolated* if $\{x\}$ is an open set.)

32. Show that a subset of a Hausdorff space, under the inherited topology, is a Hausdorff space.

33. Prove that in a compact Hausdorff space, two disjoint closed sets can always be separated by (disjoint) open sets.

34. (Harder.) Prove that a linearly ordered set endowed with the order topology (Exercise 9.33) is always a Hausdorff space. Prove further that the space is compact if and only if the ordering is complete in the sense that every subset has a supremum (see Exercise 7.23).

35. Define the notion of a *σ-regular subalgebra*, in analogy with the notion of a regular subalgebra that was introduced in Chapter 11. Investigate whether the results of Exercises 11.22 and 11.25 extend to this concept.

36. (Harder.) Is every set with the Baire property a Borel set?

37. (Harder.) Can the ideal of meager sets be maximal?

38. (Harder.) A topological space is called *locally compact* if for every point x, there is a compact set whose interior contains x. Prove the Baire category theorem for locally compact Hausdorff spaces.

39. Is the homomorphism f described in Theorem 29 complete?

40. Formulate and prove a version of Theorem 11 (p. 155) that applies to σ-ideals in σ-algebras.

41. Formulate and prove a version of Exercise 18.26 that applies to σ-ideals in σ-algebras.

42. Generalize Exercise 18.32 to σ-algebras.

43. Given a σ-algebra A generated by a set E, define by (transfinite) induction a transfinite sequence $\{E_i\}$, indexed by the set of countable ordinals, as follows: (1) $E_0 = E$; (2) if k is a countable successor ordinal, say $k = i + 1$, then E_k is the set of suprema (in A) of countable sets of elements and complements of elements from E_i; (3) if k is a countable limit ordinal, then $E_k = \bigcup_{i<k} E_i$. Prove that the sequence is increasing ($i \leq j$ implies $E_i \subseteq E_j$) and that its union is A. (The natural generalization of Theorem 3, p. 82, to σ-algebras fails because the infinite distributive laws may fail. This exercise provides an alternative that is adequate for many purposes.)

44. Formulate and prove the analogue of Exercise 12.31 for σ-algebras.

Chapter 30

The Countable Chain Condition

The algebraic behavior of the regular open algebra of a topological space reflects, at least in part, the topological properties of the space. One particular topological property, namely the possession of a countable base, has important algebraic repercussions. A *base* for a topology is a class S of open sets such that every open set in the topology is a union of sets in S. The space is said to have a *countable base* if at least one of its bases is countable. Here are some examples. The class of open intervals (a, b) constitutes a base for the Euclidean topology of the real numbers, and so does the class of open intervals with rational endpoints. The latter base is countable, so $\mathbb{R}$ (under the Euclidean topology) has a countable base. More generally, the open balls form a base for the Euclidean topology of $\mathbb{R}^n$, and so do the open balls for which the radii and the coordinates of the centers are rational numbers. Therefore, $\mathbb{R}^n$ has a countable base.

A Boolean algebra A is said to satisfy the *countable chain condition* if every disjoint set of non-zero elements of A is countable. (Recall that two elements p and q of a Boolean algebra are disjoint if $p \wedge q = 0$; a set E is *disjoint* if every two distinct elements of E are disjoint.) The regular open algebra of a space with a countable base does satisfy the countable chain condition. Proof: select a countable base, and, given a disjoint class of non-empty regular open sets, find in each one a non-empty set of the base. An algebra satisfying the countable chain condition is sometimes called *countably decomposable*.

Lemma 1. *A Boolean algebra A satisfies the countable chain condition if*

S. Givant, P. Halmos, *Introduction to Boolean Algebras*,
Undergraduate Texts in Mathematics, DOI: 10.1007/978-0-387-68436-9_30,
© Springer Science+Business Media, LLC 2009

*and only if every subset E of A has a countable subset D such that D and E
have the same set of upper bounds.*

Proof. We begin with a preliminary observation: if E is a disjoint set of non-zero elements in A, then no proper subset of E can have the same set of upper bounds as E. For the proof, consider a proper subset D of E, and let p be an element in E that is not in D. Since p is disjoint from each element in D, its complement p' is an upper bound for D ($p \wedge q = 0$ implies $q \leq p'$); but p' is certainly not an upper bound for E, since p is not zero, and therefore not below p'.

Assume now that the condition formulated in the lemma is satisfied, and suppose that E is a disjoint set of non-zero elements of A. Let D be a countable subset of E with the same set of upper bounds. Then $D = E$, by the remarks of the preceding paragraph, so E is countable.

To prove the converse, assume that the countable chain condition is satisfied, and consider an arbitrary subset E of A. Let M be the ideal generated by E; the elements of M are just those elements of A that are dominated by the supremum of some finite subset of E, by Theorem 11. It follows that M and E have the same set of upper bounds: any upper bound of E is certainly an upper bound of the set of suprema of finite subsets of E, and must therefore be an upper bound of M; on the other hand, an upper bound of M is obviously an upper bound of E, since E is a subset of M.

Construct a maximal disjoint set of non-zero elements in M as follows. Let $\{p_i\}_{i<\alpha}$ be an enumeration of the non-zero elements of M, indexed by the set of ordinals less than some ordinal number α. Define a corresponding transfinite sequence $\{F_i\}_{i\leq\alpha}$ of disjoint sets of non-zero elements in M such that (1) $F_0 = \varnothing$; (2) $F_i \subseteq F_j$ whenever $i \leq j$; (3) the element p_i is in F_{i+1} just in case it is disjoint from every element in F_i. The definition of the sequence proceeds by transfinite induction on ordinal numbers.

Put $F_0 = \varnothing$. Obviously this set is disjoint, and condition (1) holds by definition, while conditions (2) and (3) hold vacuously. For the induction step, consider an ordinal $k \leq \alpha$, and suppose disjoint sets F_i have been defined for each ordinal $i < k$ so that the family $\{F_i\}_{i<k}$ satisfies conditions (1)–(3) (with $j < k$ in (2), and $i+1 < k$ in (3)). When k is a successor ordinal, say $k = i + 1$, the definition of F_k splits into two cases: if p_i is disjoint from every element of F_i, put $F_k = F_i \cup \{p_i\}$; otherwise, put $F_k = F_i$. Clearly, F_k is a disjoint set in this case, by its very definition and the induction hypothesis that F_i is disjoint. When k is a limit ordinal, put $F_k = \bigcup_{i<k} F_i$. To check that this union is disjoint, consider any two of its elements p and q. There

must be ordinals $i, j < k$ such that p is in F_i and q is in F_j. One of the two ordinals is below the other, say $i \leq j$. Then p and q are in F_j, by condition (2), so $p \wedge q = 0$, by the induction hypothesis that F_j is a disjoint set. Conditions (1)–(3) (with $j \leq k$ in (2), and $i + 1 \leq k$ in (3)) are easily seen to hold for the family $\{F_i\}_{i \leq k}$.

The desired maximal set is F_α. We have already seen that this set is disjoint and consists of non-zero elements of M. To verify maximality, consider an arbitrary non-zero element p in M; it occurs somewhere in the given enumeration of M, say $p = p_i$. If p_i is not in F_α, then it is not in F_{i+1}, by conditions (2) and (3), and consequently it must have a non-zero meet with some element q in F_i. Of course, q is in F_α, by condition (2), so that p_i, that is to say p, cannot be adjoined to F_α to obtain a larger disjoint set.

Write $F = F_\alpha$. Reasoning as in the first paragraph of the proof, we infer that F and M have the same set of upper bounds. Indeed, assume p is not an upper bound of M, with the goal of showing that p is not an upper bound of F. The assumption implies the existence of an element q in M that is not below p, so $q \wedge p' \neq 0$. The meet $q \wedge p'$ is in the ideal M. If it is also in F, then p cannot be an upper bound of F, since p is not above $q \wedge p'$. If $q \wedge p'$ is not in F, then there must be an element r in F such that $r \wedge q \wedge p' \neq 0$, by the maximality of F. In this case, $r \wedge q \wedge p'$ is not below p, so obviously r cannot be below p. It follows that p is not an upper bound of F. Conclusion: every upper bound of F is an upper bound of M. The reverse implication is obvious, since F is a subset of M.

The countable chain condition is assumed to hold, so the set F is countable. Each element p in F is in the ideal M generated by E, and is therefore dominated by the supremum of some finite subset F_p of E, by Theorem 11 (p. 155). Write

$$D = \bigcup_{p \in F} F_p.$$

The set D is a countable union of finite sets, so it is a countable subset of E. If q is an upper bound of D, then q is an upper bound of F, since

$$p \leq \bigvee F_p \leq q$$

for every p in F; therefore q is also an upper bound of M and E, since the sets F, M, and E all have the same upper bounds. On the other hand, every upper bound of E is an upper bound of D, since D is a subset of E. Thus, D and E have the same set of upper bounds. The proof of the lemma is complete.

The following consequence of Lemma 1 is due to Tarski [75].

Corollary 1. *A Boolean σ-algebra that satisfies the countable chain condition is complete.*

Proof. Every countable supremum is formable by definition; by Lemma 1 every conceivable supremum coincides with some countable one.

The countable chain condition got its name from its close relation to a condition in which ascending chains do explicitly occur. An *ascending well-ordered chain* in a Boolean algebra A is a family $\{p_i\}_{i<\alpha}$ of elements in A, indexed by the set of all ordinals less than a particular ordinal α, with the property that $p_i \leq p_j$ whenever $i \leq j < \alpha$. The chain is *strictly ascending* if $p_i \neq p_j$ whenever $i < j$, and the chain is called countable in case the set of indices is countable.

Lemma 2. *If a Boolean algebra A satisfies the countable chain condition, then every strictly ascending well-ordered chain in A is countable.*

Proof. Suppose that $\{p_i\}_{i<\alpha}$ is a strictly ascending well-ordered chain in A. Write

$$q_i = p_{i+1} - p_i$$

whenever $i + 1 < \alpha$, and let E be the set of q_i's. The cardinality of E is the same as that of α. The elements of E are distinct from 0, since $p_{i+1} \neq p_i$. If $i < j$ and $j + 1 < \alpha$, then $p_{i+1} \leq p_j$, and therefore

$$q_i \wedge q_j = (p_{i+1} \wedge p_i{'}) \wedge (p_{j+1} \wedge p_j{'}) \leq p_{i+1} \wedge p_j{'} = 0.$$

In other words, E is a disjoint set of non-zero elements and therefore countable; it follows that the given chain is countable.

In a Boolean σ-algebra the converse of Lemma 2 is also true.

Lemma 3. *If every strictly ascending well-ordered chain in a Boolean σ-algebra A is countable, then A satisfies the countable chain condition.*

Proof. If the conclusion is false, then there exists a disjoint set E of cardinality $\aleph_1$ (the first uncountable cardinal number) consisting of non-zero elements of A. Establish a one-to-one correspondence between E and the set of all ordinal numbers less than ω_1 (the first uncountable ordinal number). Let p_i be the element of E corresponding to i (where $i < \omega_1$). Since the number of predecessors of i is countable, it makes sense to write $q_i = \bigvee_{j<i} p_i$

for each i. Since $\{q_i\}$ is a strictly ascending well-ordered chain (strictness follows from the disjointness of E), the hypothesis of the lemma leads to the contradictory conclusion that ω_1 is countable.

Exercises

1. Prove that the open circles for which the radii and the coordinates of the centers are rational numbers form a countable base for the topology of $\mathbb{R}^2$. Conclude that the σ-field of Borel sets in $\mathbb{R}^2$ is countably generated (as a σ-algebra).

2. Formulate and prove an extension of Exercise 1 to the spaces $\mathbb{R}^n$.

3. Let $x = (x_1, x_2)$ be any point in $\mathbb{R}^2$, and ϵ any positive real number. The *open square* (in $\mathbb{R}^2$) with center x and side ϵ is defined to be the set of points

$$\{(y_1, y_2) : x_1 - \epsilon/2 < y_1 < x_1 + \epsilon/2 \text{ and } x_2 - \epsilon/2 < y_2 < x_2 + \epsilon/2\}.$$

   Prove that the open squares for which the sides and the coordinates of the centers are rational numbers form a countable base for the topology of $\mathbb{R}^2$. Conclude that the σ-field of Borel sets in $\mathbb{R}^2$ is generated (as a σ-algebra) by the class of these open squares.

4. Does an infinite set endowed with the discrete topology have a countable base?

5. Does an infinite set endowed with the cofinite topology have a countable base?

6. If a Boolean algebra satisfies the countable chain condition, must every subalgebra satisfy the countable chain condition?

7. Suppose a Boolean algebra B satisfies the countable chain condition. Prove that a homomorphism on B that preserves countable suprema must preserve arbitrary suprema.

8. Suppose X is a countable set. Prove that every subfield of $\mathcal{P}(X)$ satisfies the countable chain condition. What if X is uncountable?

9. Prove that if A is the field of finite and cofinite subsets of an arbitrary set, then every strictly ascending well-ordered chain in A is countable.

10. If the regular open algebra of a topological space satisfies the countable chain condition, does it follow that the space has a countable base?

11. Show that the converse of Lemma 2 is false.

12. (Harder.) Show that the countable chain condition is not preserved by homomorphisms. (The main idea behind the solution to this exercise goes back to Sierpiński [55].)

13. Prove that a Boolean algebra A satisfies the countable chain condition if and only if every subset E of A that has a supremum has a countable subset D such that D has a supremum, and in fact $\bigvee D = \bigvee E$.

Chapter 31

Measure Algebras

Intuitively speaking, a measure is an assignment of "magnitude" or "size" to a collection of objects, real or conceptual. The lengths of lines (straight or curved), the areas of surfaces, and the volumes of solids are all examples of measures. It is sometimes not possible to assign a measure — a "size" — to every conceivable object of a certain type, but the class of objects that can be measured is usually closed under such operations as union, intersection, and complementation: if we know the length of a subset of the unit interval, then we know the length of its complement (with respect to the unit interval); if we know the lengths of two such subsets, and if we know how the subsets are related to one another, then we should be able to determine the lengths of their union and intersection. In analysis it is often important to be able to compute the measure of the union of an infinite sequence of sets, if we know the measures of the individual sets and if we know how the sets are related to one another. These considerations lead naturally to the study of abstract measures on Boolean algebras and σ-algebras.

A *measure* on a Boolean algebra A is a non-negative real-valued function μ on A — a function from A into the set of non-negative real numbers — such that whenever $\{p_n\}$ is a disjoint sequence of elements of A with a supremum p in A, then $\mu(p) = \sum_n \mu(p_n)$. The principal condition that this definition imposes is called *countable additivity*, so that a measure can be described as a non-negative and countably additive function on a Boolean algebra.

The concept just defined is the most useful one of a large collection of related concepts. Sometimes the word "measure" is applied to countably additive functions whose values are arbitrary real numbers, or complex numbers, or elements of much more general algebraic structures. Sometimes the

S. Givant, P. Halmos, *Introduction to Boolean Algebras*,
Undergraduate Texts in Mathematics, DOI: 10.1007/978-0-387-68436-9_31,
© Springer Science+Business Media, LLC 2009

condition of countable additivity is relaxed to *finite additivity*: if $p_1, p_2, \ldots, p_n$ is any finite disjoint sequence of elements (in A) with supremum p, then

$$\mu(p) = \mu(p_1) + \mu(p_2) + \cdots + \mu(p_n).$$

Note that μ is finitely additive if and only if

$$\mu(p \vee q) = \mu(p) + \mu(q)$$

whenever p and q are disjoint. If ever we need to make use of such generalized concepts we shall refer to them by appropriately qualifying "measure". (Thus, for instance, we may speak of a complex-valued finitely additive measure.)

Examples of measures are easy to obtain. For a combinatorial example consider the field $\mathcal{P}(X)$ of all subsets of a finite set X and, for each P in $\mathcal{P}(X)$, define $\mu(P)$ to be the number of points in P. A degenerate example is the *zero measure*: it assigns the value 0 to every element of a Boolean algebra. Many examples occur in analysis; perhaps the simplest is Lebesgue measure on the algebra of Lebesgue measurable subsets of the closed unit interval $[0, 1]$. The precise details of the definition need not concern us here, but it is helpful to know some of the more important properties of Lebesgue measure. The measure is not defined on all subsets of the unit interval, but the class of subsets on which it is defined is a σ-algebra. The sets in this algebra are said to be (*Lebesgue*) *measurable*. Every subinterval of $[0, 1]$, whether open, closed, or half-open, is measurable, and its measure is just the length of the interval. Every open set in the unit interval is a countable (disjoint) union of open intervals, so every open set is measurable, and consequently every Borel set is measurable. The converse is not true; there exist measurable sets that are not Borel. Each singleton $\{a\}$ can be written in the form $[a, a]$, so the measure of a singleton is 0. Every countable set, finite or infinite, is the union of a sequence of singletons, and is therefore measurable; countable additivity implies that its measure is zero. In general, a subset P of the unit interval has measure zero if and only if for every $\epsilon > 0$, there is a sequence of intervals $\{I_n\}$ covering P (that is, the union of the intervals includes P) such that $\sum_n \mu(I_n) < \epsilon$.

There are some basic properties that all measures possess. Consider a measure μ on a Boolean algebra. The most basic property is that $\mu(0) = 0$. (The first "0" refers to the zero of the Boolean algebra, while the second refers to the real number zero.) The proof is easy: use finite additivity to write

$$\mu(0) = \mu(0 \vee 0) = \mu(0) + \mu(0),$$

and then use the cancellation law for real numbers to conclude that $\mu(0) = 0$. The *subtraction property* says that

$$\mu(p - q) = \mu(p) - \mu(p \wedge q);$$

it is a direct consequence of finite additivity and the fact that p is the join of the disjoint elements $p - q$ and $p \wedge q$:

$$\mu(p) = \mu((p - q) \vee (p \wedge q)) = \mu(p - q) + \mu(p \wedge q).$$

The third property is *monotony*: if $q \leq p$, then $\mu(q) \leq \mu(p)$. It is a direct consequence of the subtraction property, since $q \leq p$ implies that $p \wedge q = q$ and therefore

$$\mu(p - q) = \mu(p) - \mu(q);$$

the term on the left side is a non-negative number, so the desired inequality follows.

The *join property* says that

$$\mu(p \vee q) = \mu(p) + \mu(q) - \mu(p \wedge q).$$

For the proof, observe that the elements $p - q$, $q - p$, and $p \wedge q$ are disjoint and join to $p \vee q$. Finite additivity therefore implies that

$$\mu(p \vee q) = \mu(p - q) + \mu(q - p) + \mu(p \wedge q).$$

Combine this identity with the subtraction property to arrive at the desired result. An analogous argument yields the *addition property*:

$$\mu(p + q) = \mu(p \vee q) - \mu(p \wedge q)$$

(where $p + q$ is the Boolean sum of p and q). The last property is sometimes called *countable subadditivity*: if $\{p_n\}$ is an arbitrary sequence of elements (not necessarily disjoint), and if that sequence has a supremum p, then

$$\mu(p) \leq \sum_n \mu(p_n).$$

For the proof, recall (Exercise 8.23) that the sequence $\{q_n\}$ defined by

$$q_n = p_n - (p_1 \vee p_2 \vee \cdots \vee p_{n-1})$$

is disjoint and has the same supremum as $\{p_n\}$. Notice that $q_n \leq p_n$, so that $\mu(q_n) \leq \mu(p_n)$. Countable additivity and monotony imply

$$\mu(p) = \sum_n \mu(q_n) \leq \sum_n \mu(p_n).$$

Notice that the derivations of most of the preceding properties require only finite additivity, but the derivation of countable subadditivity requires the full strength of countable additivity.

The preceding properties easily imply that if μ is a measure on a σ-algebra A, then the set of all elements of measure zero, that is, the set

$$M = \{p \in A : \mu(p) = 0\},$$

is a σ-ideal, and this ideal is proper if and only if μ is not the zero measure. The element 0 is in M because $\mu(0) = 0$. If p is in M, and if $q \leq p$, then

$$0 \leq \mu(q) \leq \mu(p) = 0,$$

by monotony, and consequently $\mu(q) = 0$; hence, q is in M. To demonstrate the closure of M under the formation of countable suprema, consider a sequence $\{p_n\}$ of elements in M. The supremum p of the sequence exists because A is a σ-algebra. Each element p_n has measure zero, by the definition of M, so

$$0 \leq \mu(p) \leq \sum_n \mu(p_n) = 0,$$

by countable subadditivity. It follows that $\mu(p) = 0$, and therefore that p is in M. If μ is the zero measure on A, then of course the unit 1 is in M, so the ideal is improper. If μ is not the zero measure, then $\mu(1) \neq 0$, and therefore 1 is not in M; in this case the ideal is proper.

A sufficient condition for two elements p and q in a Boolean algebra to have the same measure is that the sum $p + q$ be in the ideal of elements of measure zero. Indeed, if $p+q$ is in that ideal, then $\mu(p+q) = 0$, by definition. The differences $p - q$ and $q - p$ are both below $p + q$, so

$$\mu(p - q) = \mu(q - p) = 0.$$

The subtraction property implies that

$$\mu(p) = \mu(p - q) + \mu(p \wedge q) = 0 + \mu(p \wedge q) = \mu(p \wedge q),$$

and, similarly, that

$$\mu(q) = \mu(q \wedge p).$$

Consequently, $\mu(p) = \mu(q)$, as desired.

A measure μ is *normalized* if $\mu(1) = 1$. (Again, the occurrence of 1 on the left side of the equation denotes the unit of the Boolean algebra A, while the occurrence on the right side denotes the real number.) Every non-zero measure on a Boolean algebra can be normalized as follows. The value

$\mu(1) = t$ is not zero, since the measure is non-zero; define a real-valued function ν on A by

$$\nu(p) = (1/t) \cdot \mu(p)$$

for each p in A. It is a simple matter to check that ν is a normalized measure on A. The original measure μ can of course be regained from ν by writing

$$\mu(p) = t \cdot \nu(p)$$

for each p. For most purposes, then, it suffices to consider normalized measures.

A measure μ is *positive* if 0 is the only element at which μ takes the value 0. It takes a bit more work to turn a non-zero measure into a positive measure: one must identify two elements that differ (symmetrically) by an element of measure zero. From the measure-theoretic point of view, two such elements have the same properties, so there is no harm in making the identification. The formal way of doing this is to pass to the quotient algebra modulo the ideal of elements of measure zero.

Lemma 1. *Let ν be a normalized measure on a Boolean σ-algebra B, and let M be the σ-ideal of elements of measure zero. If $A = B/M$ and if f is the projection of B onto A, then there exists a unique measure μ on A such that*

$$\mu(f(q)) = \nu(q)$$

for all q in B; the measure μ is normalized and positive.

Proof. Given an element p in A, find q in B with $f(q) = p$ and write

$$\mu(p) = \nu(q).$$

The definition of μ is unambiguous; it does not depend on the choice of q. Indeed, if $f(q_1) = f(q_2)$, then

$$f(q_1 + q_2) = f(q_1) + f(q_2) = 0,$$

so that $q_1 + q_2$ is in M, and therefore $\nu(q_1) = \nu(q_2)$.

The quotient A is a σ-algebra, and the projection f from B to A is a σ-homomorphism, by Theorem 27 (p. 270). To prove that μ is countably additive, consider a disjoint sequence $\{p_n\}$ in A, say with supremum p, and let $\{q_n\}$ be a sequence in B such that $f(q_n) = p_n$. If q is the supremum of $\{q_n\}$ in B, then $f(q) = p$, because f is a σ-homomorphism. The sequence $\{q_n\}$ may not be disjoint, but it can be disjointed. More precisely, there exists a disjoint sequence $\{r_n\}$ with $f(r_n) = p_n$, obtained as follows:

$$r_1 = q_1,$$
$$r_2 = q_2 - q_1,$$
$$r_3 = q_3 - (q_1 \vee q_2),$$
$$r_4 = q_4 - (q_1 \vee q_2 \vee q_3),$$

$$\cdots$$

The sequence so defined is disjoint, and has the same supremum as $\{q_n\}$, namely q, by Exercise 8.23. A routine computation using the homomorphism properties of f shows that

$$f(r_n) = f(q_n - (q_1 \vee \cdots \vee q_{n-1})) = f(q_n) - (f(q_1) \vee \cdots \vee f(q_{n-1}))$$
$$= p_n - (p_1 \vee \cdots \vee p_{n-1}).$$

The elements $p_1, p_2, \ldots, p_{n-1}$ are disjoint from p_n, and hence so is their join. Consequently, $f(r_n) = p_n$. The countable additivity of μ is now an easy consequence of the corresponding property of ν:

$$\mu(p) = \mu(f(q)) = \nu(q) = \sum_n \nu(r_n) = \sum_n \mu(f(r_n)) = \sum_n \mu(p_n).$$

The second and fourth equalities use the definition of μ, while the third uses the countable additivity of ν.

It is obvious that μ is normalized: $f(1) = 1$, and therefore

$$\mu(1) = \nu(1) = 1.$$

To prove that μ is positive, suppose $\mu(p) = 0$ for some p in A. Let q be an element of B such that $f(q) = p$. Then $\nu(q) = 0$, by the definition of μ, so q is in the kernel M of f. It follows that $p = q/M$ is the zero element of A, as desired.

Lemma 1 says that under certain conditions measures can be transferred to quotient algebras. The reverse always works; a measure on a quotient can always be lifted to its numerator.

Lemma 2. *Let f be a Boolean σ-epimorphism from a σ-algebra B to a σ-algebra A, and let μ be a normalized measure on A. If*

$$\nu(q) = \mu(f(q))$$

for every q in B, then ν is a normalized measure on B. The kernel of f is included in the set of all those elements q of B for which $\nu(q) = 0$; the kernel coincides with that set if and only if the measure μ is positive.

Proof. Obviously, ν is a non-negative real-valued function. To check that it is countably additive, and therefore a measure on B, consider an arbitrary disjoint sequence $\{q_n\}$ of elements in B, say with supremum q. The family $\{f(q_n)\}$ is disjoint and has supremum $f(q)$ in A, because f is a σ-homomorphism. The definition of ν and the countable additivity of μ imply that

$$\nu(q) = \mu(f(q)) = \sum_n \mu(f(q_n)) = \sum_n \nu(q_n),$$

as desired. Finally, ν is normalized because μ is normalized:

$$\nu(1) = \mu(f(1)) = 1.$$

If q is in the kernel of f, then $\nu(q) = 0$, since

$$\nu(q) = \mu(f(q)) = \mu(0) = 0.$$

Assume that μ is a positive measure. If $\nu(q) = 0$, then q is in the kernel of f, since the equations

$$0 = \nu(q) = \mu(f(q))$$

and the positivity of μ imply that $f(q) = 0$. The kernel of f therefore coincides with the set of elements in B of measure zero (under ν). On the other hand, if μ is not positive, then there exists a non-zero element p in A such that $\mu(p) = 0$. Let q be an element in B such that $f(q) = p$. Then q is clearly not in the kernel of f, but

$$\nu(q) = \mu(f(q)) = \mu(p) = 0.$$

This shows that the kernel of f does not coincide with the set of elements in B of measure zero in this case.

It is sometimes useful to consider a measure as an intrinsic part of the Boolean algebra it is defined on. The appropriate definition is that of a *measure algebra*, defined as a Boolean σ-algebra A together with a positive, normalized measure μ on A. If A is not required to be a σ-algebra, but just a Boolean algebra, and if, correspondingly, μ is required to be only finitely additive, we may speak of a *finitely additive measure algebra*.

The theory of measure algebras has several points of contact, in both form and content, with the topological and algebraic results of the preceding two sections. Countability, for instance, enters through the essential countability properties of real numbers, as follows.

Lemma 3. *Every finitely additive measure algebra satisfies the countable chain condition.*

Proof. Consider a disjoint set E of non-zero elements in a Boolean algebra with a finitely additive measure μ. Define E_n to be the set of elements in E of measure at least $1/n$. The union of the sequence $\{E_n\}$ is E, because every element p in E is non-zero, and therefore has positive measure; if n is any positive integer such that $1/n < \mu(p)$, then p is in E_n. The set E_n has at most n elements. In fact, the assumption that $p_1, p_2, \ldots, p_{n+1}$ are distinct elements in E_n leads to the contradictory conclusion that

$$1 = \mu(1) \geq \mu(p_1 \vee p_2 \vee \cdots \vee p_{n+1})$$
$$= \mu(p_1) + \mu(p_2) + \cdots + \mu(p_{n+1}) \geq (n+1)/n > 1,$$

because the elements of E_n are disjoint, and μ is finitely additive. The argument just given shows that E is a countable union of finite sets, so it must be countable.

Corollary 1. *Every measure algebra is complete.*

Proof. Apply the preceding lemma and Corollary 30.1.

   The reduced Borel algebra (Borel sets modulo meager Borel sets) and the reduced measure algebra (Borel sets modulo Borel sets of measure zero) of the unit interval have much in common. Both algebras are obtained by reducing an incomplete σ-field modulo a σ-ideal; both algebras satisfy the countable chain condition and therefore (Corollary 30.1) both algebras are complete; and, incidentally, both algebras are atomless. (The proof of the last assertion is a trivial consequence of Theorem 29 (p. 277) for the reduced Borel algebra, since the algebra of regular open sets of the unit interval is atomless; for the reduced measure algebra it requires an elementary measure-theoretic argument.) No property of Boolean algebras that we have encountered so far is sharp enough to tell these two algebras apart; for all we know they are isomorphic. We conclude this chapter by showing that they are not: the reduced measure algebra has a non-zero measure, whereas the reduced Borel algebra does not. (This theorem is due to Birkhoff and Ulam; see [6].) It should be mentioned in passing that Borel sets modulo Borel sets of measure zero and Lebesgue measurable sets modulo Lebesgue measurable sets of measure zero are the same. This depends on the fact that every Lebesgue measurable set differs from some Borel set in a set of measure zero only.

Lemma 4. *Every measure on the reduced Borel algebra of the closed unit interval is identically zero.*

Proof. Let B be the σ-field of Borel sets in $[0, 1]$, and let M be the σ-ideal of meager sets in B. Write $A = B/M$, and let f be the projection of B onto A. If there is a non-zero measure μ on A, then we may assume, with no loss of generality, that μ is normalized. An application of Lemma 2 yields a normalized measure ν on B that vanishes on every meager Borel set.

We now show that there must be open intervals in B of arbitrarily small measure that contain any given rational number t (in $[0, 1]$). To see this, construct an infinite descending sequence $P_1 \supset P_2 \supset P_3 \supset \cdots$ of open intervals containing t such that the intersection of the intervals is $\{t\}$; for instance, P_{n+1} can be chosen so that its length is half that of P_n. Write

$$Q_n = P_n - P_{n+1} \quad \text{and} \quad Q_0 = \bigcap_n P_n = \{t\}.$$

The sets $Q_0, Q_1, Q_2, \ldots$ are mutually disjoint and their union is P_1. (See Exercise 8.27 for an abstract version of this construction.) The countable additivity of ν implies

$$\nu(P_1) = \sum_{n=0}^{\infty} \nu(Q_n).$$

In particular, the series $\sum_n \nu(Q_n)$ converges; so for any real number $\epsilon > 0$, there must be a positive integer k such that

$$\sum_{n=k}^{\infty} \nu(Q_n) < \epsilon.$$

The set Q_0 is nowhere dense (it contains only t), and hence it belongs to the ideal M. It is therefore mapped by the projection f to the zero element of A. The definition of ν implies that

$$\nu(Q_0) = \mu(f(Q_0)) = \mu(0) = 0.$$

The sequence of the sets $Q_0, Q_k, Q_{k+1}, Q_{k+2}, \ldots$ is disjoint, and its union is P_k, by Exercise 8.27. Invoke countable additivity again to conclude that

$$\nu(P_k) = \nu(Q_0) + \sum_{n=k}^{\infty} \nu(Q_n) = 0 + \sum_{n=k}^{\infty} \nu(Q_n) < \epsilon.$$

The set P_k is thus an open interval in B that contains t and that has measure less than ϵ.

Enumerate the rational numbers of the unit interval in an infinite sequence $\{t_n\}$. Given a positive real number ϵ, choose, for each positive integer n, an open interval U_n containing t_n such that

$$\nu(U_n) < \epsilon/2^n;$$

the intervals exist by the observations of the preceding paragraph. The union

$$U = \bigcup_n U_n$$

is an open set that contains all of the rational numbers. Its measure is less than ϵ, since

$$\nu(U) \leq \sum_{n=1}^{\infty} \nu(U_n) < \sum_{n=1}^{\infty} \epsilon/2^n = \epsilon,$$

by countable subadditivity. The complement of the open set U is a closed set that is nowhere dense. Indeed, every non-empty open set contains a rational number, and therefore has a non-empty intersection with U. Consequently, no non-empty open set can be included in U', so that the interior of U' is empty, as claimed.

It has been shown that for each $\epsilon > 0$, there is an open set in B of measure less than ϵ with a nowhere dense complement. For each positive integer n, let T_n be such a set of measure less than $1/n$. Write

$$T = \bigcap_n T_n \qquad \text{and} \qquad S = T' = \bigcup_n T_n'.$$

The sets S and T are both Borel. The measure of T is zero, since

$$\nu(T) \leq \nu(T_n) < 1/n$$

for every positive integer n. The set S, being a countable union of nowhere dense sets, is meager and hence an element of the ideal M. It is therefore mapped by the projection f to the zero element of A, so that

$$\nu(S) = \mu(f(S)) = \mu(0) = 0.$$

In other words, the unit interval is the disjoint union of two Borel sets, both of measure zero. This contradicts the fact that the measure ν is normalized: and let M be the σ-ideal of meager sets

$$\nu([0,1]) = \nu(S \cup T) = \nu(S) + \nu(T) = 0 + 0 = 0 \neq 1.$$

Exercises

1. Prove that a real-valued function μ on a Boolean algebra is finitely additive if and only if

$$\mu(p \vee q) = \mu(p) + \mu(q)$$

   whenever p and q are disjoint elements of the algebra.

2. Suppose μ is a measure on a Boolean algebra A. Show that for every positive real number t, the function ν defined on A by

$$\nu(p) = t \cdot \mu(p)$$

   is also a measure on A.

3. Formulate and prove the analogues of Lemmas 1 and 2 for finitely additive measures.

4. Suppose μ is a finitely additive measure on a Boolean algebra A. Prove that $\mu(p') = \mu(1) - \mu(p)$.

5. Suppose μ is a measure on a Boolean algebra A. Prove that if $\{p_n\}$ is an increasing sequence of elements with a supremum p, then

$$\lim_{n \to \infty} \mu(p_n) = \mu(p).$$

6. Formulate and prove a dual to Exercise 5.

7. Suppose μ is a finitely additive measure on a Boolean algebra A. Prove that for all elements p, q, and r in A,

$$\mu(q + r) = \mu((p \vee q) + (p \vee r)) + \mu((p \wedge q) + (p \wedge r)).$$

   (The first, second, and fourth occurrences of the symbol $+$ denote Boolean addition, while the third denotes addition of real numbers.)

8. (Harder.) Suppose A is a measure algebra with measure μ. Define a real-valued function d of two arguments on A by

$$d(p, q) = \mu(p + q).$$

   Prove that d is a metric on A, and that A is complete as a metric space. (This theorem goes back to Nikodym [48]. A metric space is said to be *complete* if every Cauchy sequence converges. A sequence of points $\{p_n\}$ in a metric space is *Cauchy* if for every $\epsilon > 0$, there is a positive integer n_0 such that $d(p_m, p_n) < \epsilon$ for $m, n \geq n_0$. The sequence is said to *converge* if there is a point p in the space with the property that for every $\epsilon > 0$, there is a positive integer n_0 such that $d(p, p_n) < \epsilon$ for $n \geq n_0$.)

9. Is the product of two measure algebras a measure algebra?

10. (Harder.) Suppose that $\{A_i\}$ is a family of non-degenerate Boolean algebras such that for each i, there exists a positive normalized measure on A_i. Under what conditions does it follow that there exists a positive normalized measure on $\prod_i A_i$?

Chapter 32

Boolean Spaces

There is a topological formulation of the Stone representation theorem (Theorem 17, p. 189) that establishes a fundamental connection between the class of Boolean algebras and a rather special class of topological spaces. The purpose of this chapter is to describe those spaces.

A *Boolean space* is a totally disconnected compact Hausdorff space. There are several possible definitions of total disconnectedness, but, as it turns out, they are all equivalent for compact Hausdorff spaces. The most convenient definition for our algebraic purposes is the one that demands that the clopen sets constitute a base. Explicitly: a Boolean space is a compact Hausdorff space with the property that every open set is the union of those simultaneously closed and open sets that it happens to include.

It is easy to see that the clopen subsets of a Boolean space X separate points. Indeed, for distinct points x and y in X, there must be an open set U that contains x but not y, because X is Hausdorff. The clopen sets form a base, so there is a clopen set P that contains x and is included in U. The set P and its complement are disjoint clopen sets that contain x and y respectively.

For Boolean spaces, as for every topological space, it is true that the class of all clopen sets is a field. The field of all clopen sets in a Boolean space X is called the *dual algebra* of X.

The simplest Boolean spaces are the finite discrete spaces. Recall that a space is *discrete* if every subset is open. Every subset of a discrete space is automatically closed (since its complement is open), and therefore clopen. The separation property holds trivially: two points x and y are separated by the disjoint open sets $\{x\}$ and $\{y\}$. When the space is finite, compactness

S. Givant, P. Halmos, *Introduction to Boolean Algebras*,
Undergraduate Texts in Mathematics, DOI: 10.1007/978-0-387-68436-9_32,
© Springer Science+Business Media, LLC 2009

also holds trivially, because there are only finitely many open sets. Since each subset of a finite discrete space X is clopen, the dual algebra of the space is the field of all subsets of X. Every finite Boolean algebra is isomorphic to the field of all subsets of some (necessarily finite) set, namely the set of its atoms (see Theorem 6, p. 119, or Corollary 15.1), so every finite Boolean algebra is isomorphic to the dual algebra of some finite Boolean space.

A less trivial collection of examples consists of the one-point compactifications of infinite discrete spaces. Explicitly, suppose a set X with a distinguished point x_0 is topologized as follows: a subset of X that does not contain the point $\{x_0\}$ is always open, and a subset that contains x_0 is open if and only if it is cofinite. It is easy to verify that the space X so defined is Boolean. For instance, a subset of X is clopen if and only if it is either a finite subset (of X) that does not contain $\{x_0\}$ or else a cofinite subset that contains x_0; indeed, a subset and its complement are both open just in case one of them (the one that contains x_0) is cofinite. The clopen sets form a base for the topology because every open set that contains x_0 is clopen, while every open set that does not contain x_0 is the union of its finite subsets. The singletons of points in $X - \{x_0\}$ are clopen, so it is easy to verify the separation property: two points x and y different from x_0 are separated by the clopen sets $\{x\}$ and $\{y\}$, while x and x_0 are separated by the clopen sets $\{x\}$ and $X - \{x\}$. An open cover of X must contain a cofinite set P (namely, any open set in the cover that contains x_0); the remaining (finite number of) points in $X - P$ can be covered by finitely many of the open sets in the cover. The dual algebra of X — the field of clopen sets — is isomorphic to the finite–cofinite algebra of $X - \{x_0\}$. In fact, the mapping f defined on the field of clopen sets by

$$f(P) = \begin{cases} P & \text{if } P \text{ is finite,} \\ P - \{x_0\} & \text{if } P \text{ is cofinite,} \end{cases}$$

is the desired isomorphism.

The set 2 is a Boolean algebra; from now on it will be convenient to construe it as a topological space as well, endowed with the discrete topology. For an arbitrary set I, the set 2^I of all functions from I into 2 (equivalently, the Cartesian product of copies of 2, one for each element of I) is a topological space under the *product topology* (due to Tychonoff [78], [79]), which we now define. Denote the value of a function x in 2^I at an element i of I by x_i. The product topology can be described succinctly by saying that the class of sets

(1) $\qquad \{x \in 2^I : x_i = 0\} \quad \text{and} \quad \{x \in 2^I : x_i = 1\},$

where i ranges over I, constitutes a *subbase* for the topology; this means that the collection of finite intersections of sets in this class is a base for the topology. In other words, every open set is the union of a family of finite intersections of subbase sets. In the sequel, the spaces 2^I endowed with the product topology will be called *Cantor spaces*.

The sets in (1) are complements of one another, so they are also closed and therefore clopen. Finite intersections of sets of type (1) are clopen (finite intersections of clopen sets are always clopen); we shall call them *basic* clopen sets because they constitute a base for the product topology. The open sets in the product topology are the subsets of 2^I that can be written as unions of basic clopen sets.

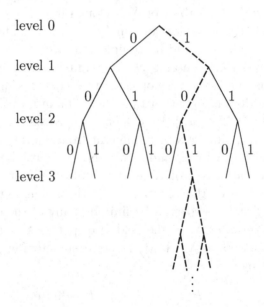

Basic clopen sets are easy to describe directly: they have the form

$$U_\delta = \{x \in 2^I : x_i = \delta_i \text{ for } i \in S\} = \bigcap_{i \in S}\{x \in 2^I : x_i = \delta_i\},$$

where S ranges over the finite subsets of I, and δ ranges over the functions from S into 2. For a concrete example, take I to be the set of non-negative integers, take $S = \{0, 1, 2\}$, and take δ to be the function that maps the integers 0 and 2 to Boolean unit 1, and the integer 1 to the Boolean zero 0. The corresponding clopen set is

$$U_\delta = \{x \in 2^I : x_0 = 1, \text{ and } x_1 = 0, \text{ and } x_2 = 1\}.$$

It can be visualized by picturing each function x in 2^I as an infinite branch in an infinite binary tree. At the ith level, the branch forks to the left if $x_i = 0$, and it forks to the right if $x_i = 1$. The clopen set U_δ consists of all functions in 2^I that begin by forking to the right, then fork to the left, and then fork to the right again (see the adjoining diagram).

Theorem 30. *Every Cantor space is a Boolean space.*

Proof. Parts of the theorem are quite easy to prove. The clopen sets form a base for the product topology of 2^I, by definition. The space is Hausdorff because distinct points y and z in 2^I differ at some argument i in I, and are therefore separated by the (disjoint) clopen sets

$$U_0 = \{x \in 2^I : x_i = 0\} \quad \text{and} \quad U_1 = \{x \in 2^I : x_i = 1\}.$$

The main task is to prove that the space is compact. Consider a family $\{F_j\}$ of closed sets with the finite intersection property, that is to say, with the property that every finite subfamily has a non-empty intersection. It is to be shown that the intersection of the entire family is non-empty (see Chapter 29). Each set F_j is the complement of an open set, and therefore must be the intersection of a class K_j of clopen sets (since every open set is the union of clopen sets). The class

$$K = \bigcup_j K_j$$

of all the clopen sets that are used to form the sets F_j has the finite intersection property, because the family $\{F_j\}$ has this property. Extend K to an ultrafilter N in the Boolean algebra of all clopen subsets of 2^I (Exercise 20.12).

Consider an element i in I, and let ϕ_i be the ith projection from 2^I to 2; thus, $\phi_i(x) = x_i$ for each x in 2^I. The image

$$\phi_i(U) = \{x_i : x \in U\}$$

of an arbitrary clopen set U in N is a subset of 2, and is therefore automatically clopen in the discrete topology of 2. The class of images

$$\{\phi_i(U) : U \in N\}$$

has the finite intersection property. To see this, let $U_1, \ldots, U_n$ be a finite sequence of sets in N. The intersection of this sequence is an element of N (N is a filter), and is therefore non-empty (N is proper); consequently, the image of the intersection under ϕ_i cannot empty. Since

$$\varnothing \neq \phi_i(U_1 \cap \cdots \cap U_n) \subseteq \phi_i(U_1) \cap \cdots \cap \phi_i(U_n),$$

it follows that the sequence $\phi_i(U_1), \ldots, \phi_i(U_n)$ of images has a non-empty intersection, as claimed. Invoke the compactness of the discrete space 2 to conclude that the intersection of the entire class of images is non-empty.

The observations of the preceding paragraph show that for each i in I, it is possible to choose an element y_i in the intersection

$$\bigcap_{U \in N} \phi_i(U).$$

These choices determine a function y in 2^I. We proceed to show that y is in every set of N. For each index i, the basic clopen set

$$V_i = \{x \in 2^I : x_i = y_i\}$$

has a non-empty intersection with every set in N, and therefore must itself belong to N. In more detail, suppose U is in N. Then y_i belongs to $\phi_i(U)$, by the choice of y_i. This means that there is an element z in U for which

$$\phi_i(z) = z_i = y_i.$$

The element z is also V_i, by the definition of V_i, so the intersection of V_i with U contains z and is therefore not empty. Since N is closed under finite intersections, it follows that any finite family of sets from the class $N \cup \{V_i\}$ has a non-empty intersection; in other words, $N \cup \{V_i\}$ has the finite intersection property. The maximality of the filter N now implies that V_i belongs to N.

Consider next an arbitrary basic clopen set U_δ containing y, where δ is some function from a finite subset S of I into 2. The assumption that y is in U_δ means that $\delta_i = y_i$ for each i in S. The sets V_i are all in N, so the finite intersections of these sets are also in N. Since

$$U_\delta = \bigcap_{i \in S} V_i,$$

it follows that U_δ is in N. Conclusion: every basic clopen set containing y is in N. This means, by the finite intersection property, that every basic clopen set containing y has a non-empty intersection with every set in N. Put somewhat differently, any given set in N has a non-empty intersection with every basic clopen set that contains y, and therefore with every open set that contains y. In other words, y belongs to the closure of every set in N. But the sets in N are all closed (and in fact clopen), so y belongs to every set in N, as claimed.

Every set F_j in the given family of closed sets is an intersection of sets in N, so y must belong to each F_j, and therefore it must belong to the intersection of the family $\{F_j\}$. It follows that this intersection is non-empty, as desired. The proof that 2^I is compact is complete.

The preceding theorem seems to have been first observed by Stone [67]. The main assertion of the theorem — that 2^I is compact — is a special case of the much more general theorem (due to Tychonoff) that the product of an arbitrary family of compact spaces is compact. The proof given above is a special case of Bourbaki's proof of Tychonoff's theorem.

The dual algebra of a Cantor space is not difficult to describe: it is the class A of all possible finite unions of basic clopen sets. For the proof, recall that a finite union of basic clopen sets is clopen (the clopen sets form a field). In particular, every set in A is clopen. To prove the converse, consider an arbitrary clopen set U. As an open set, U is the union of a family $\{U_i\}$ of basic clopen sets, by the definition of the product topology. As a closed subset of a compact space, U is compact (see Chapter 29), and is therefore covered by some finite subfamily of $\{U_i\}$. It follows that U is the union of finitely many basic clopen sets, so it belongs to A.

A field A of clopen sets in a topological space is called a *separating field* if any two points in the space can be separated by sets in A. This is equivalent to saying that for every pair of distinct points x and y in the space, there exists a set P in A such that x is in P and y is in P'. The following somewhat technical result is useful in the study of Boolean spaces.

Lemma 1. *If X is a compact space and if A is a separating field of clopen subsets of X, then X is a Boolean space and A is the field of all clopen subsets of X.*

Proof. The fact that A separates points clearly implies that X is Hausdorff. It also implies that A separates points and closed sets. This involves a standard compactness argument. Suppose, indeed, that F is a closed set and x is a point not in F. Notice that F, as a closed subset of a compact space, is compact. Separate each point y of F from x by a suitable set P_y in A that contains y but not x. The family $\{P_y\}_{y \in F}$ is an open cover of F. Compactness yields a finite cover of F by sets $P_{y_1}, \ldots, P_{y_n}$ in A, none of which contains x; their union is a set in A that separates F from x. (The union belongs to A because A is a field.)

The result of the preceding paragraph implies that A is a base for X, and this already implies that X is Boolean. In more detail, consider an arbitrary open subset U of X. For each point x in U, choose a clopen set Q_x in A that includes U' and does not contain x; such a set exists, by the observations of the preceding paragraph. The complement Q'_x is also in A (A is closed under complementation), it contains x, and it is included in U. Consequently, the union of the family $\{Q'_x\}_{x \in U}$ contains every point in U, and is therefore equal to U. This shows that U is a union of sets in A, so A is indeed a base for X.

It follows that every clopen set P in X is a finite union of sets in A, and is therefore itself a set in A. In more detail, P is a union of a family of sets in A because P is open and A is a base; P is the union of a finite subfamily because P is closed, and therefore compact. Since A is closed under the formation of finite unions, P must belong to A. It follows that A is the field of all clopen sets in X. The proof of the lemma is complete.

Corollary 1. *If a field of clopen subsets of a compact Hausdorff space is a base, then the space is Boolean and the field contains all clopen sets.*

Proof. Suppose a field A of clopen sets in a compact Hausdorff space X is a base for the topology of X. Distinct points x and y in X are separated by (disjoint) open sets U and V. There are sets U_0 and V_0 in A such that

$$x \in U_0 \subseteq U \qquad \text{and} \qquad y \in V_0 \subseteq V,$$

because A is a base. Since the sets U_0 and V_0 must be disjoint, A is a separating field. The desired conclusion now follows from the previous lemma.

Lemma 2. *Every closed subset Y of a Boolean space X is a Boolean space with respect to the topology it inherits from X. Every clopen set in Y is the intersection of Y with some clopen subset of X.*

Proof. The proof of the first assertion is straightforward. For instance, if the clopen sets form a base for the topology of X, then their intersections with Y form a base for the topology of Y. In more detail, if V is an open set in Y, then there is an open set U in X such that $V = U \cap Y$; consequently, if $U = \bigcup_i P_i$, where the sets P_i are clopen in X, then

$$V = \bigcup_i (P_i \cap Y),$$

and the sets $P_i \cap Y$ are clopen in Y. It is equally easy to check that Y is a compact Hausdorff space whenever X is (see Exercises 29.27 and 29.32).

If Q is clopen in Y, then it is open in Y, so there exists an open set U in X for which $Q = Y \cap U$. The set Q is also closed in Y, and Y is closed in X, so Q must be closed in X (Exercise 9.8), and therefore compact. The open set U is a union of clopen subsets of X, because X is a Boolean space. These clopen sets cover Q, so by compactness there exists a finite collection of clopen subsets of U whose union, say P, covers Q. Since

$$Q \subseteq P \subseteq U \quad \text{and} \quad Y \cap U = Q,$$

it follows that $Y \cap P = Q$. Thus, every clopen set in Y is the intersection with Y of a clopen set in X.

Exercises

1. Prove that the topology of a finite Hausdorff space must be discrete.

2. Prove that a discrete space is compact if and only if it is finite.

3. Show that a compact Hausdorff space has the discrete topology if and only if it is finite.

4. Suppose X is the one point compactification, with respect to a point x_0 in X, of an infinite discrete space $X - \{x_0\}$. Let f be the mapping defined on the field of clopen sets by

$$f(P) = \begin{cases} P & \text{if} \quad P \text{ is finite,} \\ P - \{x_0\} & \text{if} \quad P \text{ is cofinite.} \end{cases}$$

Prove that f is an isomorphism from the field of clopen subsets of X to the algebra of finite and cofinite subsets of $X - \{x_0\}$.

5. Prove that in a Boolean space, two disjoint closed sets are separated by a clopen set, that is to say, there is a clopen set that includes one of the closed sets and is disjoint from the other.

6. Prove that the class of open sets in a Cantor space is a topology.

7. Is the complement of a basic clopen set in a Cantor space always a basic clopen set?

8. Suppose the set $Y = \{0, 1\}$ is given the *Sierpiński topology*: the open sets are $\varnothing$, $\{0\}$, and Y. Let I be an arbitrary set, and write $X = Y^I$. A basic open set in X is defined to be a set of the form

$$\{x \in X : x_i = 0 \text{ for } i \in S\},$$

where S ranges over the finite subsets of I. Open sets are defined as arbitrary unions of basic open sets. Show that the class of open sets is a topology for X. (It is the product topology on X induced by the topology of Y.) Prove that the space X is compact under this topology. Is it Hausdorff?

9. Suppose the set $3 = \{0,1,2\}$ is given the discrete topology. For an arbitrary set I, define the product topology on the power 3^I. Prove directly that the resulting topological space is compact and Hausdorff. Conclude that the space is Boolean.

10. Suppose I is the set of positive integers. Define the *norm* or *absolute value* of an element x in 2^I to be the real number

$$|x| = \sum_{n=1}^{\infty} \frac{x(n)}{2^n}.$$

(The right side of this equation is to be understood as an infinite series of real numbers.) Prove that this norm satisfies the following *norm properties* for all x and y in 2^I.

(a) $|x| \geq 0$, and $|x| = 0$ if and only if x is the zero function (*strict positivity*).

(b) $|x + y| \leq |x| + |y|$ (*triangle inequality*),

where the sum $x + y$ is computed in the Boolean ring 2^I. Show also that every element in 2^I has norm at most 1.

11. Suppose I is the set of positive integers. Define a real-valued function d of two arguments on the set 2^I by

$$d(x,y) = |x + y| = \sum_{n=1}^{\infty} \frac{x(n) + y(n)}{2^n},$$

where $|x + y|$ is the norm (defined in Exercise 10) of the Boolean sum $x + y$. (The Boolean sum $x(n) + y(n)$ is computed, for each n, in the ring 2.) Prove that d is a metric, and show that the distance between any two points under this metric is at most 1.

12. (Harder.) The metric defined in Exercise 11 can be used to define a topology on 2^I. For each point x in 2^I and each real number ϵ with $0 < \epsilon \leq 1$, define the *open ball* of radius ϵ and center x to be the set

$$\{y \in 2^I : d(x, y) < \epsilon\}.$$

These open balls form the base for the *metric topology* on 2^I; the open sets are, by definition, the arbitrary unions of open balls. Prove that this topology coincides with the product topology on 2^I.

13. (Harder.) A topological space is called *separable* if it has a countable, dense subset. Show that if I is countable, then the Cantor space 2^I is separable.

14. (Harder.) Prove that a Boolean space with a countable base has a countable base consisting of clopen sets. Conclude that the field of clopen sets must be countable.

15. (Harder.) Let X be the set of all ordinal numbers up to and including some particular one. The set X is ordered (by magnitude), and, as such, has a natural topology, namely the one for which the open intervals constitute a base (see Exercise 9.33). Prove that X is a Boolean space.

16. Let X be the set of all ordinals up to and including the first infinite ordinal, and endow X with the order topology (see Exercise 15). Characterize the topology of X and describe its dual algebra.

17. (Harder.) Let X be the perimeter of a circle in the Cartesian plane. Order X as follows: (x_1, x_2) precedes (y_1, y_2) if $x_1 < y_1$ or (in case $x_1 = y_1$) $x_2 < y_2$. (This is known as the *lexicographic ordering*.) Endow X with the order topology (as defined in a similar situation in Exercise 9.33). Prove that X is a Boolean space whose dual algebra is the field of finite unions of half-closed intervals in the half-closed unit interval $[0, 1)$ (see Chapter 5).

18. (Harder.) A subset S of a topological space X is said to be *connected* if it cannot be split by open sets; more precisely, there do not exist open sets U and V in X such that $U \cap S$ and $V \cap S$ are non-empty and disjoint, and their union is S. A *component* of X is defined to be a maximal connected subset of X, that is to say, a connected subset that is not properly included in any other connected subset of X. Prove that a compact Hausdorff space is a Boolean space if and only if all its components are singletons. (Topological spaces with this last property are said to be *totally disconnected*.)

19. (Harder.) If I is an infinite set, of power m, say, what is the cardinal number of the dual algebra of the Cantor space 2^I?

20. Let $\{X_i\}_{i \in I}$ be a family of Boolean spaces, and let X be the Cartesian product of the family. The elements of X are functions x with domain I such that for each i in I, the value $x(i)$ — or x_i, as we shall usually write — is in X_i. The *product topology* on X is defined in the following way. The simplest open sets are those of the form

$$U = \{x \in X : x_i \in P\}$$

for some index i and some clopen subset P of X_i. Notice that the complement of such a set has the same form, since P' is clopen:

$$U' = \{x \in X : x_i \notin P\} = \{x \in X : x_i \in P'\}.$$

Sets of this form are therefore clopen in the product topology on X. Call them (temporarily) *simple clopen sets*. A *basic clopen set* is defined to be a finite intersection of simple clopen sets. In other words, a subset of X is a basic clopen set just in case it has the form

$$U_P = \{x \in X : x_i \in P_i \text{ for each } i \in S\}$$

for some finite subset S of I, and some function P with domain S such that P_i is a clopen subset of X_i for each i in S. The basic clopen sets form a base for the product topology on X in the sense that a subset of X is declared to be *open* in the product topology if it can be written as a union of basic clopen sets. (Notice that the simple clopen sets form a subbase for the product topology.)

(a) Prove that the class of open sets so defined really is a topology for X.

(b) Prove that for each index i and each open subset Q of X_i, the set

$$V = \{x \in X : x_i \in Q\}$$

is open in the product topology. Call such a set a *simple open set*.

(c) Define a *basic open set* to be a set of the form

$$V_Q = \{x \in X : x_i \in Q_i \text{ for each } i \in S\}$$

for some finite subset S of I, and some function Q with domain S such that Q_i is an open subset of X_i for each i in S. Define a subset of X to be open if it can be written as a union of basic open sets. Prove that the topology just defined coincides with the product topology defined above.

Note. For products of arbitrary families of topological spaces, the product topology must be defined as in (c), since the clopen sets do not in general form bases for the topologies of the component spaces.

21. (Harder.) Is the Cartesian product of a family of Boolean spaces a Boolean space with respect to the product topology defined in the preceding exercise?

Chapter 33

Continuous Functions

Continuous functions play a central role in topology, analogous to the role played by homomorphisms in algebra. The standard calculus definition of a continuous function cannot be extended to arbitrary topological spaces because there is no way of measuring the distance between points in such spaces. From a topological point of view, the most important property of the continuous functions one meets in calculus is that the inverse images of open sets under such mappings are always open. This property is taken as a definition in topology: a function ϕ mapping a topological space X into a topological space Y is called *continuous* if for every open set P in Y, the inverse image

$$\phi^{-1}(P) = \{x \in X : \phi(x) \in P\}$$

is an open set in X. If X and Y are metric spaces, this definition is equivalent to the usual ϵ, δ-definition from calculus, formulated in terms of metrics (Exercise 10).

The condition that the inverse image of every open set in Y be open in X is equivalent to the requirement that the inverse image of every closed set in Y be a closed set in X. The equivalence follows easily from the set-theoretic identity

$$(1) \qquad \phi^{-1}(Y - P) = X - \phi^{-1}(P),$$

which holds for any function ϕ mapping X into Y and any subset P of Y (see Exercise 1(c)). For the proof of the equivalence in one direction, suppose that the inverse image of every open set is open, and consider a closed subset P of Y. The complement $Y - P$ is open in Y, so its inverse image $\phi^{-1}(Y - P)$ is open in X, by the assumption. This inverse image is just the complement

S. Givant, P. Halmos, *Introduction to Boolean Algebras*,
Undergraduate Texts in Mathematics, DOI: 10.1007/978-0-387-68436-9_33,
© Springer Science+Business Media, LLC 2009

of $\phi^{-1}(P)$ in X, by (1), so the set $\phi^{-1}(P)$ must be closed in X. The same argument, with the words "open" and "closed" interchanged, establishes the reverse implication.

In order to show that a mapping ϕ from X to Y is continuous, it suffices to prove that the inverse image of every set in some subbase for the topology of Y is open in X.

Lemma 1. *Suppose ϕ is a function from a topological space X to a topological space Y. If for every set P in some subbase for Y, the inverse image $\phi^{-1}(P)$ is open in X, then ϕ is continuous.*

Proof. The class of finite intersections of sets in the subbase constitutes a base for Y, by the definition of a subbase. If Q belongs to this base, say

$$Q = P_1 \cap \cdots \cap P_n,$$

where the sets P_i are elements of the subbase, then

$$\phi^{-1}(Q) = \phi^{-1}(P_1 \cap \cdots \cap P_n) = \phi^{-1}(P_1) \cap \cdots \cap \phi^{-1}(P_n)$$

(see Exercise 1(b)). The sets $\phi^{-1}(P_i)$ are open by assumption, and a finite intersection of open sets is open; consequently, the set $\phi^{-1}(Q)$ is open.

Now consider an arbitrary open set U in Y. There must be a family $\{Q_i\}$ of base sets such that $U = \bigcup_i Q_i$. It is easy to check that

$$\phi^{-1}(U) = \bigcup_i \phi^{-1}(Q_i)$$

(see Exercise 5(a)). The sets $\phi^{-1}(Q_i)$ are open, by the argument of the preceding paragraph, so their union $\phi^{-1}(U)$ must be open as well. Thus, the inverse image of each open set is open, so ϕ is continuous.

Just as the composition of homomorphisms is a homomorphism, so too the composition of continuous functions is a continuous function. More explicitly, suppose X, Y, and Z are topological spaces, and ϕ a mapping from X to Y, and ψ a mapping from Y to Z. The composition $\psi \circ \phi$ is the mapping from X to Z whose value at each point x in X is given by

$$(\psi \circ \phi)(x) = \psi(\phi(x)).$$

For every subset P of Z,

$$(\psi \circ \phi)^{-1}(P) = \phi^{-1}(\psi^{-1}(P)).$$

If ϕ and ψ are continuous, and if P is open in Z, then $\psi^{-1}(P)$ is open in Y, and therefore $\phi^{-1}(\psi^{-1}(P))$ is open in X; consequently, the composition $\psi \circ \phi$ is continuous.

The class of 2-valued continuous functions on a Boolean space X plays a particularly important role in Boolean algebra. For each clopen subset P of X, the characteristic function of P is defined by

$$\phi(x) = \begin{cases} 1 & \text{if } x \in P, \\ 0 & \text{if } x \notin P. \end{cases}$$

The clopen sets $\{0\}$ and $\{1\}$ form a base for the discrete topology of 2, and the inverse images of these sets under ϕ are the clopen subsets

$$\phi^{-1}(\{1\}) = P \qquad \text{and} \qquad \phi^{-1}(\{0\}) = P'$$

of X. Consequently, ϕ is a continuous function from X to 2. More is true: every 2-valued continuous function on X is the characteristic function of some clopen set. Indeed, if ϕ maps X continuously into 2, then the inverse image of the clopen set $\{1\}$ under ϕ must be a clopen subset P of X. The elements of P and P' are mapped by ϕ to 1 and 0 respectively, so ϕ coincides with the characteristic function of P. We formulate these observations as a lemma.

Lemma 2. *The correspondence that takes each clopen subset of a Boolean space X to its characteristic function is a bijection from the class of clopen subsets of X to the class of 2-valued continuous functions on X.*

Continuous functions preserve a number of important topological properties, for example compactness.

Lemma 3. *The image of a compact set under a continuous mapping is compact.*

Proof. Let ϕ be a continuous mapping from a space X to a space Y, and let P be a compact subset of X. Consider an open cover $\{U_i\}$ of $\phi(P)$ in Y. The family $\{\phi^{-1}(U_i)\}$ of inverse images is an open cover of P, because ϕ is continuous. The assumed compactness of P implies the existence of a finite subcover $\phi^{-1}(U_{i_1}), \ldots, \phi^{-1}(U_{i_n})$ of P. The sets $U_{i_1}, \ldots, U_{i_n}$ must therefore cover $\phi(P)$.

Certain topological properties that are generally not preserved by continuous functions are preserved when the underlying spaces satisfy special hypotheses. Here is an example.

Corollary 1. *A continuous function from a compact space to a Hausdorff space maps closed sets to closed sets.*

Proof. Suppose ϕ is a continuous mapping from a compact space X to a Hausdorff space Y. If P is any closed subset of X, then P is necessarily compact; the image $\phi(P)$ is therefore compact in Y, by Lemma 3, and consequently it is closed in Y, by Lemma 29.1 and the assumption that Y is Hausdorff.

The conclusion of the corollary is not true when the domain space is not compact or the range space is not Hausdorff (see Exercise 14). The next lemma presents a somewhat specialized result that will be needed later.

Lemma 4. *If two continuous functions map a topological space into a Hausdorff space, then the set of points on which the functions agree is closed.*

Proof. Suppose ϕ and ψ are continuous functions from a topological space X into a Hausdorff space Y. It suffices to show that the set of points x in X such that $\phi(x) \neq \psi(x)$ is open. The complement of this set is then closed, and this complement is precisely the set of points on which ϕ and ψ agree.
If U and V are open subsets of Y, then the inverse images $\phi^{-1}(U)$ and $\psi^{-1}(V)$ are open subsets of X, because the mappings are assumed to be continuous; the intersection of the two inverse images,

$$S_{UV} = \phi^{-1}(U) \cap \psi^{-1}(V),$$

is therefore open. Let S be the union of the sets S_{UV}, where U, V range over the pairs of disjoint open sets in Y. Clearly, S is open in X. We shall show that S is just the set of points in X on which ϕ and ψ disagree.
Consider an arbitrary point x in X. If x belongs to S, then it belongs to one of the sets S_{UV}, and therefore $\phi(x)$ is in U, and $\psi(x)$ is in V. Since U and V are disjoint, the points $\phi(x)$ and $\psi(x)$ must be distinct. For the converse, suppose ϕ and ψ do not agree on x. The space Y is assumed to be Hausdorff, so there are disjoint open sets U and V in Y that contain $\phi(x)$ and $\psi(x)$ respectively. The element x is in each of the inverse images $\phi^{-1}(U)$ and $\psi^{-1}(V)$, so it is in S_{UV}, and therefore also in S.

Corollary 2. *If two continuous functions from a topological space X into a Hausdorff space agree on a dense subset of X, then they agree on all of X.*

Proof. Suppose two continuous functions ϕ and ψ from X into a Hausdorff space agree on a dense subset S of X. The set of all points on which ϕ and ψ

agree is closed in X, by the previous lemma, and it includes S by assumption, so it must include the closure of S. It therefore coincides with X.

As a simple application of the corollary, suppose a continuous mapping ϕ from a Boolean space X into itself is the identity function on a dense subset of X. The identity function on X is continuous, and it agrees with ϕ on the dense subset, so it agrees with ϕ on X. Conclusion: ϕ is the identity function on all of X.

The topological analogue of an isomorphism is a homeomorphism, a continuous bijection from one topological space to another, with a continuous inverse. More explicitly, a *homeomorphism* is a bijection ϕ from a space X to a space Y with the property that $\phi^{-1}(P)$ is an open subset of X if and only if P is an open subset of Y, or, equivalently, with the property that $\phi(P)$ is an open subset of Y if and only if P is an open subset of X. Thus, a homeomorphism carries the class of open subsets of its domain bijectively to the class of open subsets of its range. Two spaces are said to be *homeomorphic* if there exists a homeomorphism between them.

It is a simple observation that a bijection ϕ from a space X to a space Y maps open sets to open sets if and only if it maps closed sets to closed sets. Assume, for instance, that ϕ maps closed sets to closed sets. If P is open, then P' is closed, and therefore $\phi(P')$ is closed. It follows from the set-theoretic identity

$$(2) \qquad\qquad \phi(P') = \phi(P)'$$

(which is valid for all bijections — see Exercise 4) that $\phi(P)$ is open, as desired. The proof when ϕ maps open sets to open sets is completely analogous.

The criterion for being a homeomorphism may simplify when the underlying spaces satisfy some stronger conditions.

Lemma 5. *A bijection from a compact space to a Hausdorff space is a homeomorphism just in case it is continuous.*

Proof. A continuous bijection ϕ from a compact space X to a Hausdorff space Y maps closed subsets of X to closed subsets of Y, by Corollary 1. It therefore maps open sets to open sets, by the remark preceding the lemma. This just means that ϕ^{-1} is a continuous function.

Corollary 3. *A bijection between compact Hausdorff spaces is a homeomorphism just in case it maps open sets to open sets or, equivalently, just in case it maps closed sets to closed sets. A bijection from a Boolean space to* *a compact Hausdorff space is a homeomorphism just in case it maps clopen sets to clopen sets.*

Proof. Suppose ϕ is a bijection from a compact Hausdorff space X to a compact Hausdorff space Y. If ϕ maps open sets to open sets, then ϕ^{-1} is a continuous function from Y to X (see the preceding proof). It follows from the lemma that ϕ^{-1} is a homeomorphism, and consequently so is ϕ. If ϕ maps closed sets to closed sets, then it must also map open sets to open sets, so it is a homeomorphism, by the preceding observations. The final assertion of the corollary follows from the preceding argument and Lemma 1, since the clopen sets form a base for the topology of a Boolean space.

Two homeomorphic spaces have the same topological structure, just as isomorphic algebras have the same algebraic structure. This means that any property of spaces that is defined in terms of the topology will hold for one of the two spaces if and only if it holds for the other. For instance, if one of the spaces is compact, or Hausdorff, or Boolean, then so is the other. To demonstrate that two spaces are not homeomorphic, it suffices to exhibit a topological property that is possessed by one of the spaces, but not by the other. For example, suppose I is the set of positive integers. The Cantor space 2^I has the same cardinality as the one-dimensional Euclidean space of real numbers, so there are certainly bijections between the two spaces. They cannot be homeomorphic, however, because the Cantor space is compact, while the one-dimensional Euclidian space is not.

Examples of homeomorphisms are not hard to come by. If the topologies of two spaces are defined in a similar fashion in terms of cardinality, then any bijection between the spaces is a homeomorphism. For a concrete example, suppose two spaces of the same cardinality are endowed with the cofinite topology (or the discrete topology). The image of every cofinite set (or every singleton) under a bijection between the spaces is again a cofinite set (or a singleton), so the bijection and its inverse are continuous mappings. For another simple example, suppose Y is a discrete space with two points. For each set I, the product space Y^I is homeomorphic to the Cantor space 2^I; in fact, if ψ is any bijection from 2 to Y, then the mapping ϕ that takes each function x in 2^I to the function $\phi(x)$ in Y^I defined by

$$\phi(x)_i = \psi(x_i)$$

is a homeomorphism.

As a more sophisticated illustration of these ideas, we proceed to show that the Cantor set (to be defined in a moment) is homeomorphic to the

Cantor space 2^I when I is the set of positive integers. (This observation is due to Stone [67].) To construct the Cantor set, start with the unit interval, and define an infinite descending sequence of sets

$$K_0 \supseteq K_1 \supseteq K_2 \supseteq \cdots$$

as follows. The initial set is $K_0 = [0,1]$. The set K_1 is obtained by removing the open middle third interval $(\frac{1}{3}, \frac{2}{3})$ of K_0:

$$K_1 = [0, \tfrac{1}{3}] \cup [\tfrac{2}{3}, 1].$$

The set K_2 is obtained by removing the open middle thirds $(\frac{1}{9}, \frac{2}{9})$ and $(\frac{7}{9}, \frac{8}{9})$ of the two intervals that make up K_1:

$$K_2 = [0, \tfrac{1}{9}] \cup [\tfrac{2}{9}, \tfrac{3}{9}] \cup [\tfrac{6}{9}, \tfrac{7}{9}] \cup [\tfrac{8}{9}, 1].$$

Remove the open middle thirds from the four intervals that make up K_2 to

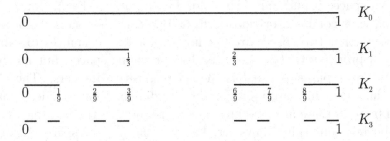

obtain K_3, and so on. (See the diagram.) The *Cantor* (*middle third*) *set* is defined to be the intersection of the sets K_n:

$$K = \bigcap_{n=0}^{\infty} K_n.$$

Each of the sets K_n is a finite union of closed intervals, and is therefore a closed subset of the real line. It follows that the intersection K is a closed subset of the real line; because K is included in the compact interval $[0,1]$, it is also compact. Thus, K is a compact Hausdorff space under the topology inherited from the real line.

To describe the real numbers in K, it is helpful to use the ternary analogue of the decimal representations of real numbers. Each number x in the interval $[0,1]$ may be written as an infinite series in the form

$$x = \frac{a_1}{3} + \frac{a_2}{3^2} + \frac{a_3}{3^3} + \cdots + \frac{a_n}{3^n} + \cdots,$$

where each a_n is one of the digits 0, 1, and 2. In analogy with the decimal representation of a real number, the *ternary representation* of x is defined to be

$$x = .a_1 a_2 a_3 \ldots a_n \ldots.$$

For instance, $1 = .22222\ldots$, since

$$\sum_{n=0}^{\infty} 1/3^n = 3/2, \quad \text{and therefore} \quad \sum_{n=1}^{\infty} 2/3^n = 2(3/2) - 2) = 1.$$

Similar calculations show that

$$2/3 = .200000\ldots = .122222\ldots, \quad 19/27 = .20100000\ldots = .20022222\ldots,$$
$$7/9 = .210000\ldots = .202222\ldots, \quad 104/243 = .10212000\ldots = .10211222\ldots.$$

As these examples indicate, every non-zero number x in the interval $[0,1]$ that has a finite ternary representation — a representation with only finitely many non-zero digits — also has an infinite ternary representation that is obtained from the finite representation by decreasing the rightmost non-zero digit by 1, and following it with a repeating infinite sequence of 2's. To obtain a unique representation of x in these cases, we make the following convention: if the rightmost non-zero digit in the finite representation of x is 2, use the finite representation; if the rightmost non-zero digit is 1, use the representation in which that occurrence of 1 is converted to 0, and followed by repeating 2's. This convention can also be formulated in the following way. The first digit a_1 in the ternary representation of x is determined by

$$a_1 = \begin{cases} 0 & \text{if } 0 \le x \le \frac{1}{3}, \\ 1 & \text{if } \frac{1}{3} < x < \frac{2}{3}, \\ 2 & \text{if } \frac{2}{3} \le x \le 1. \end{cases}$$

Thus, x is in K_1 if and only if $a_1 \ne 1$. The second digit a_2 in the ternary representation of x is determined by

$$a_2 = \begin{cases} 0 & \text{if } 0 \le x \le \frac{1}{9}, \text{ or } \frac{3}{9} < x \le \frac{4}{9}, \text{ or } \frac{6}{9} \le x \le \frac{7}{9}, \\ 1 & \text{if } \frac{1}{9} < x < \frac{2}{9}, \text{ or } \frac{4}{9} < x < \frac{5}{9}, \text{ or } \frac{7}{9} < x < \frac{8}{9}, \\ 2 & \text{if } \frac{2}{9} \le x \le \frac{3}{9}, \text{ or } \frac{5}{9} \le x < \frac{6}{9}, \text{ or } \frac{8}{9} \le x \le 1. \end{cases}$$

Thus, x is in K_2 if and only if $a_1 \ne 1$ and $a_2 \ne 1$. More generally, x is in K_n just in case the first n digits in its ternary expansion (under the convention made above) are all different from 1. It follows that x is in K if and only if the digit 1 does not occur in its ternary representation. (It is essentially this description of K that Cantor first published as a short note at the end of [12].)

Endow the set $\{0, 2\}$ with the discrete topology, and form the product space

$$X = \{0, 2\}^I.$$

Each point x in X may be viewed as a sequence of 0's and 2's. Define $\phi(x)$ to be the number in K whose ternary representation is given by x:

$$\phi(x) = .x_1\, x_2\, x_3 \ldots.$$

For example, if x has the value 2 at odd positive integers, and the value 0 at even positive integers, then

$$\phi(x) = .202020202\ldots = \tfrac{3}{4}.$$

The function ϕ so defined is a bijection from X to K. We shall show that ϕ is a homeomorphism.

Let ρ be any function from $\{1, 2, \ldots, n\}$ into $\{0, 2\}$, and consider the basic clopen subset

$$(3) \qquad U_\rho = \{x \in X : x_i = \rho_i \text{ for } i = 1, 2, \ldots, n\}$$

of X. The image $\phi(U_\rho)$ is the set of real numbers whose ternary representation begins with $.\rho_1 \rho_2 \ldots \rho_n$. If

$$a = .\rho_1 \rho_2 \ldots \rho_n 000\ldots \qquad \text{and} \qquad b = .\rho_1 \rho_2 \ldots \rho_n 222\ldots,$$

then obviously

$$(4) \qquad \phi(U_\rho) = [a, b] \cap K,$$

so that $\phi(U_\rho)$ is a closed subset of K.

An arbitrary basic clopen subset of X has the form

$$U_\delta = \{x \in X : x_i = \delta_i \text{ for } i \in S\}$$

for some finite subset S of I and some function δ from S into $\{0, 2\}$. Such a set can be written as a finite union of sets of the form (3): just take n to be the largest integer in S, and let $\rho_1, \ldots, \rho_m$ be a list of the finitely many functions from $\{1, \ldots, n\}$ into $\{0, 2\}$ that agree with δ on S (there are 2^k such functions, where k is the size of $\{1, \ldots, n\} - S$); then

$$U_\delta = U_{\rho_1} \cup \cdots \cup U_{\rho_m}.$$

The image of U_δ under ϕ can therefore be written as a finite union of closed sets of the form (4):

$$\phi(U_\delta) = \phi(U_{\rho_1}) \cup \cdots \cup \phi(U_{\rho_m}).$$

Since a finite union of closed sets is closed, the set $\phi(U_\delta)$ must be closed in K. In other words, the image under ϕ of an arbitrary basic clopen set in X is a closed set in K.

Every clopen set in X is a finite union of basic clopen sets; an argument similar to the one just given shows that the image under ϕ of an arbitrary clopen set in X is closed in K. Since the clopen sets form a field, the image under ϕ of every clopen set in X must also be open, and therefore clopen, in K. In more detail, if P is a clopen set in X, then its complement P' is also clopen, so that the images $\phi(P)$ and $\phi(P')$ must both be closed in K; since $\phi(P')$ coincides with $\phi(P)'$, by equation (2), it follows that the image $\phi(P)$ must also be open, and therefore clopen, in K. Conclusion: the bijection ϕ from X to the compact Hausdorff space K maps clopen sets to clopen sets, so it must be a homeomorphism, by Corollary 3.

The Cantor space 2^I is easily seen to be homeomorphic to X; in fact, the bijection from $\{0,1\}$ to $\{0,2\}$ that takes 1 to 2, and 0 to itself, induces a homeomorphism ψ from 2^I to X. The composition $\phi \circ \psi$ is the desired homeomorphism from the Cantor space 2^I to the Cantor set K.

Exercises

1. Let ϕ be a mapping from a set X into a set Y. Show that the following set-theoretic identities are valid for any subsets P and Q of Y.

   (a) $\phi^{-1}(P \cup Q) = \phi^{-1}(P) \cup \phi^{-1}(Q)$.

   (b) $\phi^{-1}(P \cap Q) = \phi^{-1}(P) \cap \phi^{-1}(Q)$.

   (c) $\phi^{-1}(Y - P) = X - \phi^{-1}(P)$.

2. If ϕ is any mapping from X into Y, and if P is any subset of X, and Q any subset of Y, show that

   $$Q \cap \phi(P) \neq \varnothing \quad \text{if and only if} \quad \phi^{-1}(Q) \cap P \neq \varnothing.$$

3. Let ϕ be a mapping from a set X into a set Y. Which of the following set-theoretic identities are valid for any subsets P and Q of X?

   (a) $\phi(P \cup Q) = \phi(P) \cup \phi(Q)$.

   (b) $\phi(P \cap Q) = \phi(P) \cap \phi(Q)$.

   (c) $\phi(X - P) = \phi(X) - \phi(P)$.

(d) $\phi(X - P) = Y - \phi(P)$.

4. Which of the set-theoretic identities in Exercise 3 holds when ϕ is one-to-one? When ϕ is a bijection?

5. Let ϕ be a mapping from a set X into a set Y, and let $\{P_i\}$ be an arbitrary family of subsets of Y. Verify the following set-theoretic identities.

   (a) $\phi^{-1}(\bigcup_i P_i) = \bigcup_i \phi^{-1}(P_i)$.
   (b) $\phi^{-1}(\bigcap_i P_i) = \bigcap_i \phi^{-1}(P_i)$.

6. Let ϕ be a mapping from a set X to a set Y, and let $\{P_i\}$ be an arbitrary family of subsets of X. Show that the identity

   (a) $\phi(\bigcup_i P_i) = \bigcup_i \phi(P_i)$

   always holds. Show that, in general, the identity

   (b) $\phi(\bigcap_i P_i) = \bigcap_i \phi(P_i)$

   fails, but it holds when ϕ is one-to-one.

7. Let X be the product of a family $\{X_i\}$ of Boolean spaces, under the product topology (Exercise 32.20), and let ϕ_i be the projection from X to X_i defined by $\phi_i(x) = x_i$ for each x in X. Prove that ϕ_i is continuous and maps open sets to open sets.

8. Prove that a real-valued function ϕ on a topological space X is continuous if and only if for every real number $\epsilon > 0$ and every point x_0 in X, there is a neighborhood U of x_0 such that

$$|\phi(x) - \phi(x_0)| < \epsilon$$

   for every x in U.

9. Prove that for functions ϕ from $\mathbb{R}$ to $\mathbb{R}$, the topological definition of continuity is equivalent to the standard ϵ, δ-definition from calculus: for every $\epsilon > 0$ and every point x_0 in $\mathbb{R}$, there is a $\delta > 0$ such that

$$|x - x_0| < \delta \qquad \text{implies} \qquad |\phi(x) - \phi(x_0)| < \epsilon.$$

10. Show that for metric spaces, the definition of continuity given in the text is equivalent to the standard ϵ, δ-definition from calculus (formulated in terms of the metrics).

11. A function ϕ from a metric space X to a metric space Y is said to be sequentially continuous provided that whenever a sequence $\{x_n\}$ converges to a point x in X, the corresponding sequence $\{\phi(x_n)\}$ converges to the point $\phi(x)$ in Y. Prove that the sequential continuity of ϕ implies its continuity.

12. Recall (Chapter 29) that a subset of a topological space is a G_δ (set) if it is a countable intersection of open sets, and an F_σ (set) if it is a countable union of closed sets. Is the inverse image of a G_δ under a continuous mapping always a G_δ? What about an F_σ?

13. Is the continuous image of a separable space separable? (A space is *separable* if it has a countable dense subset.)

14. Show that a continuous mapping from one topological space to another need not map closed sets to closed sets if the domain space is not compact or the range space is not Hausdorff, and this remains true even when the mapping is one-to-one.

15. (Harder.) Show that a continuous mapping from a compact Hausdorff space to a compact Hausdorff space need not map open sets to open sets, even when the mapping is one-to-one.

16. Suppose the set $Y = \{0, 1\}$ is given the Sierpiński topology: the open sets are $\varnothing$, $\{0\}$, and Y. Let I be some infinite set, and endow Y^I with the product topology (see Exercise 32.8). Prove that the Cantor space 2^I and the space Y^I are not homeomorphic.

17. Prove that the image of a connected set (Exercise 32.18) under a continuous mapping is connected.

18. (Harder.) Are the Euclidean spaces $\mathbb{R}$ and $\mathbb{R}^2$ homeomorphic?

19. Prove that any closed interval $[a, b]$ with $a < b$, under the topology inherited from $\mathbb{R}$, is homeomorphic to the interval $[0, 1]$.

20. Are the intervals $[0, 1]$ and $(0, 1)$ (under the inherited topology) homeomorphic? Is either of these intervals homeomorphic to $[0, 1)$?

21. Formulate and prove a topological version of the exchange principle discussed in Chapter 12.

22. Let W be a discrete space with two points. Prove that for any set I, the product spaces 2^I and W^I are homeomorphic. In fact, show that if ψ is any bijection from the discrete space $2 = \{0, 1\}$ to W, then the mapping ϕ that takes each function x in 2^I to the function $\phi(x)$ in W^I defined by

$$\phi(x)_i = \psi(x_i),$$

for i in I, is a homeomorphism from 2^I to W^I.

23. Prove that the Cantor set has Lebesgue measure zero (see Chapter 31, p. 289).

24. Prove that the Cantor set is nowhere dense.

25. Prove that the Cantor set has the cardinality of the continuum.

26. Prove that

$$\tfrac{3}{4} = .202020202\ldots \qquad \text{and} \qquad \tfrac{1}{4} = .020202020\ldots.$$

27. Let S be the set of points y in the Cantor set K such that the ternary expansion of y has only finitely many occurrences of the digit 2. Prove that every point in K is the limit of a strictly increasing or a strictly decreasing sequence of points from S. Conclude that K, under the topology inherited from the real line, is a separable space in the sense that it has a countable dense subset.

28. A variant of the Cantor set may be constructed as follows. Divide the interval $[0, 1]$ into four intervals of equal length, and remove the open third fourth $(\tfrac{2}{4}, \tfrac{3}{4})$; divide each of the three remaining intervals into four intervals of equal length, and remove the open third fourths from them; and so on. Show that the intersection of the resulting sequence of sets, as a subspace of $[0, 1]$, is homeomorphic to the space 3^I under the product topology, where $3 = \{0, 1, 2\}$ is given the discrete topology, and I is the set of positive integers.

29. Suppose I is the set of positive integers. Use the homeomorphism between the Cantor space 2^I and the Cantor set to show that 2^I is a metric space. (Compare this exercise with Exercises 32.11 and 32.12.)

30. Suppose I is the set of positive integers. Use the homeomorphism between the Cantor space 2^I and the Cantor set to show that 2^I is separable. (Compare this exercise with Exercise 32.13.)

31. (Harder.) The *Stone–Čech compactification* of a Hausdorff space X is a compact Hausdorff space Y such that X is a dense subspace of Y and every continuous mapping from X to a compact Hausdorff space Z can be extended to a continuous mapping from Y to Z. (This notion is due independently to Stone [67] and Čech [14].) Prove that if Y is the Stone–Čech compactification of an infinite discrete space X, then the field of all clopen sets in Y is isomorphic to the field $\mathcal{P}(X)$. (It will be seen in Exercise 34.19 that Y is a Boolean space.)

32. (Harder.) Prove that the dual algebra of every Cantor space satisfies the countable chain condition.

33. (Harder.) Imitating the definition of a free Boolean algebra, define the concept of a free Boolean space, and prove existence and uniqueness theorems for free Boolean spaces.

34. (Harder.) If P and Q are disjoint closed subsets of a compact Hausdorff space X, prove that there is a continuous mapping from X into the interval $[0, 1]$ of real numbers that assumes the value 0 at every point in P and the value 1 at every point in Q. (This is a special case of a result due Urysohn [81], known as *Urysohn's lemma*.)

35. (Harder.) If P is a closed G_δ-set in a compact Hausdorff space X, prove that there is a continuous mapping ϕ from X into the interval $[0, 1]$ of real numbers such that

$$\phi(x) = 0 \qquad \text{if and only if} \qquad x \in P.$$

(This is a special case of a theorem due to Vedenisov [85].)

Chapter 34

Boolean Algebras and Boolean Spaces

The Stone representation theorem (Theorem 17, p. 189) describes a representation of an arbitrary Boolean algebra A as a field of subsets of the set X of all 2-valued homomorphisms on A. There is a natural topology on X, namely the one inherited from the Cantor space 2^A, and under this topology X becomes a Boolean space. It turns out that the field representing A in Theorem 17 is just the dual algebra of the space X.

Lemma 1. *The set of all 2-valued homomorphisms on a Boolean algebra A is a closed subset of the Cantor space 2^A of all 2-valued functions on A.*

Proof. For each element p in A, the 2-valued functions ϕ and ψ on 2^A defined by

$$\phi(x) = x(p') \qquad \text{and} \qquad \psi(x) = x(p)',$$

for x in 2^A, are continuous. For instance, the inverse images

$$\phi^{-1}(\{0\}) = \{x \in 2^A : x(p') = 0\}$$

and

$$\phi^{-1}(\{1\}) = \{x \in 2^A : x(p') = 1\}$$

are clopen in 2^A, and so are the inverse images

$$\psi^{-1}(\{0\}) = \{x \in 2^A : x(p)' = 0\} = \{x \in 2^A : x(p) = 1\}$$

and

S. Givant, P. Halmos, *Introduction to Boolean Algebras,*
Undergraduate Texts in Mathematics, DOI: 10.1007/978-0-387-68436-9_34,
© Springer Science+Business Media, LLC 2009

$$\psi^{-1}(\{1\}) = \{x \in 2^A : x(p)' = 1\} = \{x \in 2^A : x(p) = 0\},$$

by the definition of the topology on 2^A. Since the set of points where the two continuous functions ϕ and ψ are equal is always closed (Lemma 33.4), it follows that the set

$$\{x \in 2^A : x(p') = x(p)'\}$$

is closed in 2^A for each p in A. Consequently, the intersection of these sets, over all p in A, is a closed subset of 2^A, and it consists of just those 2-valued functions on A that preserve complementation. A similar argument involving 2-valued functions ϕ and ψ on 2^A defined by

$$\phi(x) = x(p \vee q) \qquad \text{and} \qquad \psi(x) = x(p) \vee x(q)$$

(for p and q in A), and sets such as

$$\{x \in 2^A : x(p \vee q) = x(p) \vee x(q)\},$$

justifies the same conclusion for the set of join-preserving functions. The set of 2-valued homomorphisms on A is the intersection of these two closed sets, so it is closed as well.

The preceding observation and Lemma 32.2 imply that the set X of all 2-valued homomorphisms on a Boolean algebra A has the structure of a Boolean space in a natural way; we shall call it the *dual space* of A. The description of the topology that X inherits from 2^A admits some simplifications. The basic clopen subsets of X are defined to be the intersections with X of the basic clopen subsets of 2^A; they have the form

$$U_\delta = \{x \in X : x(p) = \delta(p) \text{ for } p \in S\},$$

where S ranges over the finite subsets of A, and δ ranges over the 2-valued functions on S. Suppose an element p and its complement p' are both in S. If the set U_δ is not empty — say it contains a homomorphism x — then the definition of U_δ and the homomorphism properties of x imply that

$$\delta(p)' = x(p)' = x(p') = \delta(p').$$

It is therefore unnecessary to use both of the conditions

$$x(p) = \delta(p) \qquad \text{and} \qquad x(p') = \delta(p')$$

in defining U_δ, for each condition implies the other. Said differently, U_δ can be defined as the set of x in X that agree with δ on $S - \{p'\}$. It may be assumed, then, that the subsets S of A used to define the basic clopen sets in X do not contain both an element and its complement.

A condition of the form $x(p) = 0$ can always be replaced by the equivalent condition $x(p') = 1$, so it may also be assumed that δ always has the constant value 1 on S. Furthermore, the homomorphism properties of x imply that the conjunction of the conditions $x(p) = 1$, for p in S, is equivalent to the single condition $x(q) = 1$ when q is the meet of the elements in S. It may therefore be assumed that S contains just one element. The preceding observations may be summarized by saying that the basic clopen subsets of X have the form

$$(1) \qquad\qquad U_p = \{x \in X : x(p) = 1\},$$

where p ranges over the elements in A. (To obtain the empty set take $p = 0$.)

An arbitrary clopen subset of X is a finite union of basic clopen sets, and therefore has the form

$$V = \{x \in X : x(p) = 1 \text{ for some } p \in S\}$$

for some finite subset S of A. The disjunction of the conditions $x(p) = 1$ is equivalent to the single condition $x(q) = 1$ when q is the join of the elements in S. Conclusion: the clopen subsets of X are precisely the sets of the form (1), and the open subsets of X are the arbitrary unions of sets of the form (1).

Given a Boolean algebra A, we may form the dual space X of 2-valued homomorphisms on A. The field of clopen subsets of X is the dual algebra of X; it shall be called the *second dual* of A. The Stone representation theorem just says that every Boolean algebra is isomorphic to its second dual. Indeed, the canonical embedding f of A into $\mathcal{P}(X)$ defined in Theorem 17 is the correspondence that takes each element p in A to the clopen set U_p. The remarks of the preceding paragraphs show that f maps A onto the algebra of clopen subsets of X. The Stone representation theorem may therefore be reformulated as a result — the most fundamental result — about the relation between Boolean algebras and Boolean spaces.

Theorem 31. *The second dual of every Boolean algebra A is isomorphic to A. More explicitly, if B is the dual algebra of the dual space X of A, and if*

$$f(p) = U_p = \{x \in X : x(p) = 1\}$$

for each p in A, then f is an isomorphism from A onto B.

As was mentioned in Chapter 22, the set X of 2-valued homomorphisms on A may be replaced in Theorem 17 by the set Z of ultrafilters in A. The mapping ψ that takes each homomorphism x in X to its cokernel is a bijection from X to Z. It is natural to endow Z with a topology that turns ψ into a homeomorphism: the clopen subsets of Z are defined to be the sets

$$V_p = \{N \in Z : p \in N\},$$

for p in A; in other words, they are the images under ψ of the clopen subsets of X, given in (1). The open subsets of Z are defined to be the arbitrary unions of clopen sets. It is common in expositions of the theory of Boolean algebras to define the dual space of a Boolean algebra to be the space of ultrafilters instead of the space of 2-valued homomorphisms.

Theorem 31 also has a topological analogue. Given a Boolean space X, form its dual algebra A of clopen subsets of X. The set of 2-valued homomorphisms on A, endowed with the topology inherited from the Cantor space 2^A, is the dual space of A; it is called the *second dual* of X.

Theorem 32. *The second dual of every Boolean space X is homeomorphic to X. More explicitly, if Y is the dual space of the dual algebra A of X, and if $\phi(x)$ is the 2-valued homomorphism that sends each element P of A to 1 or 0 according as x is, or is not, in P, then ϕ is a homeomorphism from X onto Y.*

Proof. The function $\phi(x)$ that sends each element P of A to 1 or 0 according as x is, or is not, in P is a 2-valued homomorphism; in fact, it is the homomorphism on A induced by x (see Chapter 12). Consequently, ϕ is a well-defined mapping of X into Y. To prove that ϕ is one-to-one, assume that x_1 and x_2 are distinct points in X. There must then be a clopen set P that contains x_1, but not x_2, since the clopen sets separate points in X. The definition of ϕ implies that

$$\phi(x_1)(P) = 1 \neq 0 = \phi(x_2)(P),$$

so that $\phi(x_1) \neq \phi(x_2)$.

For the proof that ϕ maps X onto Y, consider an arbitrary point y in Y. By definition, y is a 2-valued homomorphism on the algebra A of clopen subsets of X. The cokernel of y, say N — the class of elements mapped to 1 by y — is an ultrafilter in A, and is therefore a maximal class of clopen sets in X with the finite intersection property (that is to say, with the property that the intersection of any finite family of sets in N is non-empty). The space X is compact, so the intersection of the (clopen) sets in N must be non-empty, say x is a point in that intersection. The image $\phi(x)$ is, by definition, the 2-valued homomorphism on A determined by

$$\phi(x)(P) = 1 \qquad \text{if and only if} \qquad x \in P$$

for each P in A. Since x is an element of every set in N, the cokernel of $\phi(x)$ contains every set in N, and must therefore coincide with N, by the maximality of N. In other words, the homomorphisms $\phi(x)$ and y have the same cokernel. Every 2-valued homomorphism on A is uniquely determined by its cokernel, so we must have $\phi(x) = y$.

To prove that the bijection ϕ is a homeomorphism, it suffices to establish its continuity (Lemma 33.5), and in fact it suffices to show that the inverse image of every clopen subset of Y is clopen in X (Lemma 33.1). A clopen subset of Y has the form

$$(2) \qquad\qquad \{y \in Y : y(P) = 1\}$$

for some element P in A (see the discussion following Lemma 1). The equality $\phi(x) = y$ is equivalent to the assertion that x is in P if and only if P is in the cokernel of y, by the observations of the preceding paragraph; in other words, it is equivalent to the assertion that x is in P if and only if $y(P) = 1$. The inverse image of (2) is therefore the set

$$\{x \in X : x \in P\};$$

that is to say, it is exactly the clopen set P. The proof of the theorem is complete.

It is sometimes convenient to indicate the relation between Boolean algebras and Boolean spaces by some special terminology and notation. By a *pairing* of a Boolean algebra A and a Boolean space X we shall mean a function that associates with every pair (p, x), where p is in A and x is in X, an element of 2 in a certain particular way. If the value of the function is denoted by $\langle p, x \rangle$, then the requirements on the function can be expressed by the following two *pairing conditions*: (1) for every element p in A, the correspondence that takes each element x to $\langle p, x \rangle$ is a 2-valued continuous function on X, and, by suitable choice of p, every 2-valued continuous function on X has this form; (2) for every element x in X, the correspondence that takes each element p to $\langle p, x \rangle$ is a 2-valued homomorphism on A, and, by suitable choice of x, every 2-valued homomorphism on A has this form.

Here is a typical example of a pairing. Let X be the dual space of a Boolean algebra A — the space of 2-valued homomorphisms on A — and write

$$(3) \qquad\qquad \langle p, x \rangle = x(p).$$

The requirements in pairing condition (2) are automatically satisfied: for each point x in X, the correspondence that takes every p to $\langle p, x \rangle$ is just x, which is a 2-valued homomorphism on A; and if y is any 2-valued homomorphism on A, then y is in X, by definition. The verification of pairing condition (1) takes a bit more work. For a given element p in A, write ψ_p for the correspondence that takes every x in X to $\langle p, x \rangle$. The set

$$U_p = \{x \in X : x(p) = 1\}$$

is a clopen subset of X (see (1)), and ψ_p is its characteristic function:

$$\psi_p(x) = 1 \qquad \text{if and only if} \qquad \langle p, x \rangle = 1,$$
$$\text{if and only if} \qquad x(p) = 1,$$
$$\text{if and only if} \qquad x \in U_p.$$

Therefore, ψ_p is a 2-valued continuous function on X, by Lemma 33.2. Conversely, every 2-valued continuous function on X is the characteristic function of some clopen subset of X, by the same lemma, and every clopen subset of X has the form U_p for some p in A (see, for instance, Theorem 31). Therefore, every 2-valued continuous function on X coincides with ψ_p for some p. The requirements of pairing condition (1) are thus satisfied.

Here is another typical example of a pairing. Let A be the dual algebra of a Boolean space X; write

$$(4) \qquad \langle P, x \rangle = \begin{cases} 1 & \text{if } x \in P, \\ 0 & \text{if } x \notin P. \end{cases}$$

Each element P in A is a clopen subset of X. The correspondence that takes every x to $\langle P, x \rangle$ is the characteristic function of P, by the definition of $\langle P, x \rangle$, so this correspondence is a 2-valued continuous function on X, by Lemma 33.2. An arbitrary 2-valued continuous function on X is necessarily the characteristic function of some clopen subset P of X, by the same lemma, so it must be the correspondence that takes every x to $\langle P, x \rangle$. The requirements of pairing condition (1) are thus seen to hold. For each x in X, the correspondence that takes every clopen set P to $\langle P, x \rangle$ is precisely the 2-valued homomorphism on A induced by x (see Chapter 12); in fact it is the homomorphism $\phi(x)$ from Theorem 32. If y is any 2-valued homomorphism on A, then y is an element of the dual space Y of A, by the definition of that space. By Theorem 32, there must be an element x in X such that $\phi(x) = y$. It follows from the preceding remarks that y is the homomorphism on A induced by x; consequently, y maps every clopen set P in A to $\langle P, x \rangle$, as desired. This completes the verification of pairing condition (2).

The next lemma says that in some sense the two examples of pairings just described are the only possible ones.

Lemma 2. *If a Boolean algebra A and a Boolean space X are paired, then A is isomorphic to the dual algebra of X, and X is homeomorphic to the dual space of A.*

Proof. Suppose A and X are paired via a function that takes each pair (p, x) to $\langle p, x \rangle$. Let Y be the dual space of A — the space of all 2-valued homomorphisms on A. The goal is to construct a homeomorphism ψ from X to Y. For each x in X, write $\psi(x)$ for the function on A that maps every element p to $\langle p, x \rangle$:

$$(5) \qquad\qquad \psi(x)(p) = \langle p, x \rangle.$$

The first part of pairing condition (2) ensures that $\psi(x)$ is a 2-valued homomorphism on A, and therefore an element of Y. The second part of condition (2) ensures that for every element y in Y — that is, for every 2-valued homomorphism on A — there is an element x in X such that $\psi(x) = y$. Consequently ψ maps X onto Y.

To show that ψ is one-to-one, consider distinct points x_1 and x_2 in X. The space X is Boolean, so there is a clopen subset P of X that separates the two points, say x_1 is in P and x_2 is in P'. The characteristic function of P — call it θ — is a 2-valued continuous function on X, by Lemma 33.2; the second part of pairing condition (1) ensures the existence of an element p in A such that

$$(6) \qquad\qquad \theta(x) = \langle p, x \rangle$$

for every x in X. Since θ maps x_1 to 1 and x_2 to 0, we must have

$$\langle p, x_1 \rangle = 1 \qquad \text{and} \qquad \langle p, x_2 \rangle = 0.$$

But then

$$\psi(x_1)(p) = \langle p, x_1 \rangle = 1 \qquad \text{and} \qquad \psi(x_2)(p) = \langle p, x_2 \rangle = 0,$$

by (5). It follows that the homomorphisms $\psi(x_1)$ and $\psi(x_2)$ disagree on the element p, and are therefore distinct. Consequently, ψ is one-to-one.

It remains to prove that ψ is continuous. Consider an arbitrary clopen subset of Y; it has the form

$$U_p = \{ y \in Y : y(p) = 1 \}$$

for some element p in A, by the definition of the topology of a dual space and the remarks following Lemma 1. The inverse image under ψ of this set is

$$\psi^{-1}(U_p) = \{x \in X : \psi(x)(p) = 1\} = \{x \in X : \langle p, x \rangle = 1\}.$$

The first part of pairing condition (1) ensures that the correspondence θ defined by (6) (for the specific p in question) is a 2-valued continuous function on X. The inverse image under θ of the clopen set $\{1\}$ (in the discrete space 2) is therefore a clopen subset of X. Since this inverse image is just

$$\{x \in X : \langle p, x \rangle = 1\},$$

it follows that $\psi^{-1}(U_p)$ is a clopen subset of X. It has been shown that the inverse image under ψ of every clopen subset of Y is a clopen subset of X, so ψ is a homeomorphism, by Lemmas 33.1 and 33.5. The second assertion of the lemma has been proved.

To establish the first assertion of the lemma, let B be the dual algebra of X. Since X and Y are homeomorphic, by the first part of the proof, their dual algebras are isomorphic (see Exercise 8). Consequently, B is isomorphic to the dual algebra of Y. The dual algebra of Y is, in turn, isomorphic to A, by Theorem 31. It follows that A and B are isomorphic, as was to be shown.

Notice that the pairing $\langle p, x \rangle$ of the lemma satisfies (5), and is therefore essentially of the form (3), modulo the homeomorphism ψ. A direct proof of the first assertion of the lemma proceeds by constructing an isomorphism f from A to the dual algebra B of X with the property

$$\langle p, x \rangle = \begin{cases} 1 & \text{if } x \in f(p), \\ 0 & \text{if } x \notin f(p). \end{cases}$$

The pairing of the lemma is therefore essentially also of the form (4), modulo the isomorphism f (see Exercise 10).

It follows from the proof of the lemma that the element x required in the second part of pairing condition (2) is unique: if

$$\langle p, x_1 \rangle = \langle p, x_2 \rangle$$

holds identically for all p in A, then $\psi(x_1) = \psi(x_2)$ and therefore $x_1 = x_2$. Similarly, the element p required in the second part of pairing condition (1) is unique: if

$$\langle p_1, x \rangle = \langle p_2, x \rangle$$

holds identically for all x in X, then $f(p_1) = f(p_2)$ and therefore $p_1 = p_2$.

The fundamental duality between Boolean algebras and Boolean spaces was discovered by Stone. Theorems 31 and 32 are from [67]. The notion of a pairing of a Boolean algebra and a Boolean space is from Halmos [23].

Exercises

1. Complete the proof of Lemma 1 by showing that the set of 2-valued functions on A that preserve join is a closed subset of the space 2^A.

2. Prove Lemma 1 directly, without using Lemma 33.4.

3. Describe explicitly the dual and the second dual of the four-element Boolean algebra. Prove directly (without using the Stone representation theorem) that this algebra and its second dual are isomorphic.

4. Let Z be the set of ultrafilters in a Boolean algebra A. The basic open subsets of Z are, by definition, the sets of the form

$$V_p = \{N \in Z : p \in N\},$$

and the open sets are, by definition, the unions of basic open sets. Prove directly that every basic open set is also closed, and therefore clopen. Conclude that a subset of Z is clopen if and only if it is a basic open set.

5. Describe explicitly the dual and the second dual of the field of finite and cofinite sets of the natural numbers. Prove directly (without using the Stone representation theorem) that this field and its second dual are isomorphic.

6. The set 3^I, endowed with the product topology, is a Boolean space (Exercise 32.9). Is it (up to homeomorphism) a subspace of a Cantor space?

7. Is every Boolean space a subspace of a Cantor space? (The answer to this question is due to Stone [67].)

8. Prove that the dual algebras of homeomorphic Boolean spaces are isomorphic. (The observations in this and the next exercise are due to Stone [67].)

9. Prove that the dual spaces of isomorphic Boolean algebras are homeomorphic.

10. Prove directly (without using Theorem 31) that if a Boolean algebra A and a Boolean space X are paired, then A is isomorphic to the dual algebra of X.

11. A theory of duality between Boolean rings without units and locally compact Boolean spaces (analogous to the theory of duality between Boolean algebras and Boolean spaces) was developed by Stone in [67]. The purpose of the next few exercises is to give some indication of this theory. A topological space is *locally compact* if for every point x, there is a compact set K whose interior contains x. A locally compact Hausdorff space in which the clopen sets form a base is called a *locally compact Boolean space*. Prove that in such a space the compact clopen sets form a base for the topology, and every compact open set is clopen.

12. Prove that the class of compact clopen subsets of a topological space is a Boolean ring (possibly without a unit) under the set-theoretic operations of symmetric difference and intersection.

13. The ring of all compact clopen subsets of a locally compact Boolean space X is called the *dual ring* of X. Prove that this dual ring has a unit if and only if X is compact.

14. (Harder.) Let A be a Boolean ring without a unit, and let X be the set of non-trivial 2-valued homomorphisms on A. (A ring homomorphism is *trivial* if it maps all elements to zero.)

    (a) Show that the class of sets of the form

    $$U_p = \{x \in X : x(p) = 1\},$$

    where p ranges over the elements of A, is closed under the (set-theoretic) operations of symmetric difference, intersection, and union. In fact, if p and q are in A, then

    $$U_p + U_q = U_{p+q}, \qquad U_p \cap U_q = U_{p \cdot q}, \qquad U_p \cup U_q = U_{p \vee q}.$$

    (b) Show that the class of all possible unions of sets of the form U_p, with p in A, is a topology for X. Prove that a set is open in this topology if and only if it can be written in the form

    $$U_M = \{x \in X : x(p) = 1 \text{ for some } p \in M\}$$

    for some ideal M in A.

(c) Show that if M is an ideal in A, and p an element of A, then

$$U_p \subseteq U_M \qquad \text{if and only if} \qquad p \in M.$$

Conclude that for ideals M and N in A,

$$U_M \subseteq U_N \qquad \text{if and only if} \qquad M \subseteq N.$$

(d) Prove that the correspondence taking each ideal M to the open set U_M is an isomorphism from the lattice of ideals of A to the lattice of open subsets of X.

(e) Prove that X is a Hausdorff space under the defined topology.

(f) Prove that each set U_p is compact. Conclude that each such set is clopen.

(g) Show, conversely, that every compact open subset of X is equal to U_p for some p in A.

(h) Prove that X is a locally compact Boolean space under the defined topology.

(i) Conclude that the topology defined on X in (b) is just the topology X inherits as a subset of the Cantor space 2^A. We shall call X the *dual space* of the ring A.

The reader may wonder why we did not immediately define the topology on X to be the one inherited from the Cantor space 2^A. The reason is that the sets U_p, though obviously open in the inherited topology, are not obviously closed. This only follows once we know that these sets are compact.

It is worth making one more observation. Lemma 1 continues to hold for Boolean rings without unit, but it seems to be of little use. The dual space of a Boolean ring A without unit does not consist of all 2-valued homomorphisms on A, since it does not contain the trivial homomorphism. In fact, the dual space is not a closed subset of 2^A; if it were closed, it would be compact (because 2^A is compact), and therefore A would have a unit, by Exercise 13.

15. (Harder.) Formulate and prove the analogue of Theorem 31 for Boolean rings without a unit.

16. (Harder.) Formulate and prove the analogue of Theorem 32 for locally compact Boolean spaces.

17. Prove that the dual space of a countable Boolean algebra is separable in the sense that it has a countable dense subset.

18. Prove that every countable Boolean algebra is isomorphic to a field of subsets of a countable set.

19. (Harder.) Prove that the Stone–Čech compactification (Exercise 33.31) of an infinite discrete space X is a Boolean space that is homeomorphic to the dual space of the field $\mathcal{P}(X)$.

Chapter 35

Duality for Ideals

The topological duality theory of Boolean algebras, introduced in the preceding chapter, pervades and enriches the entire subject. Each of the two halves of the theory (algebras and spaces) suggests interesting questions about the other half. By means of the theory it is in principle possible to dualize every fact and every concept, converting algebraic facts and concepts into topological ones, and vice versa. In almost every case the dualization is worthwhile; it is often useful and illuminating, and, at the very least, it is amusing.

The following example serves to illustrate the meaning of topological duality. Question: what can be said about the dual of a finite Boolean algebra? Answer: a Boolean algebra is finite if and only if its dual space is finite, or what amounts to the same thing, if and only if its dual space is discrete. Reason: a finite Boolean algebra has only finitely many 2-valued homomorphisms, so its dual space is finite; conversely, a finite Boolean space has only finitely many clopen sets, so its dual algebra is finite. For compact Hausdorff spaces, discreteness is the same as finiteness (Exercise 32.3).

Finite Boolean algebras are atomic. A natural generalization of the problem of dualizing finiteness, and one that is somewhat less trivial, is the problem of dualizing atomicity. If, as before, X is a Boolean space and A is its dual algebra, then, by definition, an atom of A is a non-empty clopen subset of X that does not include any properly smaller non-empty clopen set. No clopen set U containing at least two points can be an atom: if x and y are distinct points in U, take V to be a clopen set that contains x but not y; the intersection $U \cap V$ is a non-empty clopen set that is properly included in U. This implies that the atoms of A are precisely the clopen subsets of X that are singletons. A point x in a topological space is said to be *isolated* if $\{x\}$

S. Givant, P. Halmos, *Introduction to Boolean Algebras,*
Undergraduate Texts in Mathematics, DOI: 10.1007/978-0-387-68436-9_35,
© Springer Science+Business Media, LLC 2009

is an open set. Using this terminology, we may describe the atoms of A as the singletons of isolated points of X. To say that A is atomic is to say that every clopen subset of X contains an isolated point. Since the clopen sets form a base, it follows that A is atomic if and only if the isolated points are dense in X (in the sense that every open set contains an isolated point). The other extreme has an equally satisfactory dual: A is atomless if and only if X is *perfect* in the sense that it has no isolated points.

The concept of countability has an interesting dual: the dual algebra A of a Boolean space X is countable if and only if X is *metrizable* in the sense that its topology comes from some metric. (More precisely, a topological space X is metrizable if it is possible to define a metric on X such that the topology of the metric space coincides with the topology of X.) For Boolean spaces, and more generally for compact Hausdorff spaces, metrizability is the same as the possession of a countable base (Exercises 13 and 14), so it suffices to prove that A is countable if and only if X has a countable base.

If A is countable, then the sets in A — the clopen subsets of X — constitute a countable base for X. Suppose, conversely, that X has a countable base, say $V_1, V_2, V_3, \ldots$. It is to be shown that X has only countably many clopen subsets. We begin by showing that each set V_n is the union of a countable family of clopen sets. The clopen sets form a base for the topology of X, by the definition of a Boolean space, so V_n is certainly the union of a family of clopen sets, say

$$V_n = \bigcup_{i \in I} U_i.$$

Similarly, the family $\{V_j\}$ is a base for the topology of X, so each of the clopen sets U_i is the union of some of these base sets, say

$$U_i = \bigcup_{j \in J_i} V_j.$$

The union of the index sets J_i, over all i in I, is a set J of positive integers, and

$$\bigcup_{j \in J} V_j = \bigcup_{i \in I} \bigcup_{j \in J_i} V_j = \bigcup_{i \in I} U_i = V_n.$$

For each integer j in J, there is at least one index i in I such that j belongs to J_i; select one such index i, and observe that $V_j \subseteq U_i$. If I_0 is the set of selected indices, then I_0 is countable (because J is countable), and

$$V_n = \bigcup_{j \in J} V_j \subseteq \bigcup_{i \in I_0} U_i \subseteq V_n.$$

Thus, V_n is the union of the countable family $\{U_i\}_{i \in I_0}$ of clopen sets.

Apply this argument to every base set V_n to arrive at a countable family of clopen sets with the property that every set V_n can be written as a union of some of the clopen sets in the family. It follows that the clopen sets in the family form a countable base for the topology of X. The field generated by this base is still countable (Exercise 11.19), and it coincides with A, by Corollary 32.1.

The duality theory for subsets of a Boolean algebra (for example, ideals and filters) is both more interesting and more useful than the duality theory for elements. The following definitions are the basic ones. Suppose X is a Boolean space with dual algebra A. If M is an ideal in A, then the union of the clopen sets belonging to M (equally correctly and more simply, the union of M) is an open subset of X; it is called the *dual* of M.

Conversely, if U is an open subset of X, then the class M of clopen subsets that are included in U is an ideal in A; it is called the *dual* of U. The proof that M is an ideal is easy. The empty set is obviously a clopen subset of X that is included in U, so it is an element of M. The union of two clopen sets that are included in U is again a clopen set that is included in U, so M is closed under union. The intersection of an arbitrary clopen set with a clopen set that is included in U is a clopen set that is included in U, so the intersection of an arbitrary element in A with an element in M is again an element in M.

Suppose M is the dual of an open set U. The dual of M is its union, and this union is just U, since every open subset of X is the union of the clopen sets that are included in it. Conclusion: the second dual of every open set is itself. Conversely, if U is the dual of an ideal M in A, then the dual of U coincides with M. Indeed, every clopen set in M is obviously included in U, because U is the union of M; consequently, M is included in the dual of U. Consider now an element P in the dual of U. As a clopen, and hence closed, subset of a compact space, P is compact. Also, P is included in U, and the union of M is equal to U, so M is an open cover of P. Compactness implies that P is covered by finitely many of the sets in M. It follows that P belongs to the ideal M, by Theorem 11 (p. 155).

If M and N are ideals in A, and if U and V are their respective duals, then

$$M \subseteq N \qquad \text{if and only if} \qquad U \subseteq V.$$

Indeed, if M is included in N, then obviously the union of M is included in the union of N; in other words, U is included in V. On the other hand, if U

is included in V, then every clopen set that is included in U is also included in V; in other words, M is included in N.

These principal facts about the duality between ideals and open sets, which essentially go back to Stone [67], are summarized in the following duality theorem for ideals.

Theorem 33. *The dual of every open subset of a Boolean space X is an ideal in the dual algebra A, and the dual of every ideal in A is an open subset of X. The second dual of every ideal and of every open set is itself. Duality between ideals and open sets is an isomorphism between the lattice of ideals in A and the lattice of open sets in X.*

It is easy to examine the duals of various special concepts in ideal theory. The dual of the trivial ideal is $\varnothing$, and the dual of the improper ideal is X. Thus, all the ideals of A are either trivial or improper if and only if all the open subsets of X are either $\varnothing$ or X. In other words, the unique simple algebra 2 is the dual of a singleton set endowed with the discrete topology. The dual of a principal ideal is a clopen set, namely the generator (the largest clopen set in the ideal). The dual of a maximal ideal is a maximal open set, that is, the complement of a singleton. If M and N are ideals with duals U and V, respectively, then the dual of $M \cap N$ is $U \cap V$. If, for each i in a certain index set, M_i is an ideal with dual U_i, then the union $\bigcup_i U_i$ is the dual of the ideal generated by $\bigcup_i M_i$. Indeed, the ideal generated by the union of the ideals is the smallest ideal that includes each M_i, and it is mapped by the lattice isomorphism to the smallest open set that includes each U_i. That open set is clearly just the union of the sets U_i.

If M is an ideal in A, then $\{p : p' \in M\}$ is a filter in A; if U is an open set in X, then U' is a closed set in X. It follows that the duality between ideals and open sets induces a similar duality between filters and closed sets. The open duality is order-preserving; the closed duality is order-reversing. Thus, for example, the closed set corresponding to a maximal filter is a minimal closed set, that is, a singleton.

It is illuminating to look at the duality between ideals and open sets from the perspective of an arbitrary Boolean algebra A and its dual space X (instead of from the perspective of an arbitrary Boolean space X and its dual algebra A). Every element p in A is identified, via the canonical isomorphism from Theorem 31 (p. 328) with an element in the second dual of A, namely the clopen set

$$U_p = \{x \in X : x(p) = 1\}.$$

Every ideal M in A is consequently identified with the ideal of clopen sets

$$M_0 = \{U_p : p \in M\}$$

in the second dual. The ideal M in A determines an open subset of the dual space X, namely the union

$$U_M = \bigcup \{U_p : p \in M\} = \{x \in X : x(p) = 1 \text{ for some } p \in M\}$$

of the clopen sets belonging to M_0. Furthermore, every open subset V in X has the form $V = U_M$ for some ideal M in A. Indeed, the class M_0 of clopen subsets of V is an ideal in the dual algebra of X, and the union of the sets in M_0 is just the open set V, by the preceding theorem. If M is the ideal in A that corresponds to M_0 (under the canonical isomorphism), then

$$V = \bigcup M_0 = \bigcup \{U_p : p \in M\} = U_M.$$

Conclusion: the open sets in X are precisely the sets U_M, where M ranges over the ideals of A.

The canonical isomorphism from A to its second dual obviously induces an isomorphism between the corresponding lattices of ideals. The lattice of ideals of the second dual is isomorphic to the lattice of open subsets of X, by Theorem 33. Conclusion: the correspondence that takes each ideal M in A to the open set U_M in X is an isomorphism from the lattice of ideals in A to the lattice of open sets in X. In this formulation of the duality between ideals and open sets, the assertion that the second dual of every ideal is itself is not literally true; one must first identify the algebra A with its second dual before the assertion becomes true.

Exercises

1. Suppose A is a Boolean algebra and X its dual space. Prove directly (without using Theorem 33) that the opens sets in X are precisely the sets

   $$U_M = \bigcup \{U_p : p \in M\} = \{x \in X : x(p) = 1 \text{ for some } p \in M\},$$

   where M ranges over the ideals in A.

2. Suppose A is a Boolean algebra and X is its dual space. Prove directly (without using Theorem 33) that the correspondence taking each ideal M in A to the open set

$$U_M = \bigcup \{U_p : p \in M\}$$

is an isomorphism from the lattice of ideals in A to the lattice of open subsets of X.

3. Suppose A is a Boolean algebra and X its dual space. Prove directly (without using Theorem 33 or Exercises 1 and 2) that every filter N in A determines a closed subset of X, namely the set

$$V_N = \bigcap \{U_p : p \in N\} = \{x \in X : x(p) = 1 \text{ for all } p \in N\},$$

and that every closed subset of X is determined by a filter in just this manner. Show also that for filters M and N in A,

$$M \subseteq N \qquad \text{if and only if} \qquad V_N \subseteq V_M.$$

Conclude that the correspondence taking each filter N to the closed set V_N is a dual isomorphism (Exercise 12.14) from the lattice of filters in A to lattice of closed subsets of X.

4. Let M be an ideal and $N = \{p' : p \in M\}$ its dual filter in a Boolean algebra A, and let X be the dual space of A. Prove that the closed set

$$V_N = \{x \in X : x(p) = 1 \text{ for all } p \in N\} = \bigcap \{U_p : p \in N\}$$

is the complement of the open set

$$U_M = \{x \in X : x(p) = 1 \text{ for some } p \in M\} = \bigcup \{U_p : p \in M\}.$$

Use this result to derive the assertions in Exercise 3 from those of Exercises 1 and 2.

5. Prove directly that if X is a Boolean space with dual algebra A, then the dual of every ultrafilter is a closed set with just one point (namely, the unique point that belongs to every set in the ultrafilter), and the dual of a closed set with just one point is an ultrafilter (namely, the class of all clopen sets that contain the point). Show that the second dual of every ultrafilter is itself, and the second dual of every closed set with just one point is itself.

6. Let X be a Boolean space and A its dual algebra. Prove that the dual of every clopen set in X is the principal ideal in A generated by this clopen set. Prove further that every principal ideal in A is the dual of the clopen set that generates the ideal.

7. Formulate and prove the version of Exercise 6 that applies to a Boolean algebra A and its dual space X.

8. (Harder.) If "compact" is replaced by "locally compact" in the definition of Boolean spaces, most of the theory remains true. The dual of a locally compact but not compact Boolean space is a Boolean ring without unit, and conversely (Exercises 34.15 and 34.16). A typical example of a locally compact Boolean space is obtained by omitting one point from a compact one (Exercise 43.17). Prove that the act of restoring such an omitted point, that is, the one-point compactification, is the dual of the process of adjoining a unit (see Exercise 1.10 and Exercises 20.18–20.20). (This theorem is due to Stone [67].) One consequence of this observation is that the dual of the empty Boolean space is the one-element (zero) Boolean ring (without a unit). (If Y is a topological space, and if X is obtained from Y by adjoining a single point x_0, then the *one-point compactification* of Y is the set X with the following topology: the open subsets of X are defined to be the open subsets of Y and the complements in X of the closed compact subsets of Y. Note that in Hausdorff spaces, compact sets are automatically closed, but this is not true in arbitrary topological spaces.)

9. The duality theory of ideals rests ultimately on the two relatively deep theorems of Chapter 34. This explains the fact that dualization can sometimes convert a non-trivial assertion into a complete triviality. For an example, dualize the maximal ideal theorem (Theorem 12, p. 172).

10. Give a topological proof of the theorem (Exercise 20.3) that every ideal in a Boolean algebra is the intersection of the maximal ideals that include it.

11. The nowhere dense closed sets are of interest in a Boolean space, and so therefore are their complements, the dense open sets. (Recall that an open set U is dense if its closure is the entire space, or, equivalently, if every non-empty open set has a non-empty intersection with U.) Prove that the dual of a dense open set is a dense ideal, defined as in Exercise 18.8.

12. Prove that if a topological space has a countable base, then every family of open sets that is a base for the topology has a countable subfamily that is also a base.

13. (Harder.) Prove that every Boolean space X with a countable base is metrizable. In other words, there is a metric on X such that the resulting metric topology coincides with the original topology. (This result is a special case of a more general theorem, due to Urysohn [82], according to which every compact Hausdorff space with a countable base is metrizable.)

14. (Harder.) Prove that a compact metric space is separable (that is, it has a countable, dense subset) and has a countable base.

15. (Harder.) The dual algebra of a Boolean space with a countable base and no isolated points is a countable, atomless Boolean algebra, by the remarks in this chapter. Since two such algebras are always isomorphic (Theorem 10, p. 134), it follows (Exercise 34.9 and Theorem 32, p. 329) that two such spaces are always homeomorphic. Prove this topological theorem directly, and use it to give a topological proof that two countable, atomless Boolean algebras are always isomorphic. (The topological theorem is a special case of a more general result due to Kuratowski [37]. Urysohn [83] showed earlier that every zero-dimensional space is homeomorphic to a subspace of the Cantor middle-third set.)

16. (Harder.) What is the algebraic dual of separability?

17. (Harder.) The first countability axiom requires of a topological space that there be a countable *local base* at each point. This means that for each point x in the space, there is a countable family of open sets containing x with the property that every open set containing x includes at least one of the sets from the family. What is the algebraic dual of this axiom in the case of Boolean spaces?

18. (Harder.) What is the topological dual of a complete ideal? (The answer is due to Stone [67]. Complete ideals are discussed in Chapter 24.)

19. Let X be a Boolean space and P a clopen subset of X. Prove that the dual algebra of the subspace P (under the inherited topology) is just the relativization of the dual algebra of X to P.

20. Let A be a Boolean algebra and X its dual space. The algebraic dual of Exercise 19 says that if p is any element in A, then the dual space of the relativization $A(p)$ is homeomorphic to the subspace
$$U_p = \{x \in X : x_p = 1\}$$

of X (under the inherited topology). Derive this assertion as a corollary to Exercise 19.

21. Give a direct proof of the assertion in Exercise 20, without using Exercise 19.

Chapter 36

Duality for Homomorphisms

There is a dual correspondence between structure-preserving mappings of Boolean algebras and Boolean spaces, that is, between homomorphisms and continuous functions. Consider two Boolean spaces X and Y, and their respective dual algebras A and B. The correspondence takes each continuous mapping ϕ from X to Y to a homomorphism f from B to A, and is determined by the equivalence

$$(1) \qquad \phi(x) \in Q \qquad \text{if and only if} \qquad x \in f(Q)$$

for all clopen sets Q in B and all elements x in X. Since $\phi(x)$ belongs to Q just in case x belongs to $\phi^{-1}(Q)$, the equivalence in (1) may be rewritten in the form

$$x \in \phi^{-1}(Q) \qquad \text{if and only if} \qquad x \in f(Q),$$

and may therefore be equivalently expressed by the identity

$$(2) \qquad f(Q) = \phi^{-1}(Q)$$

for all Q in B.

Here are the details. If ϕ is a continuous function from X into Y, then for each clopen subset Q of Y the inverse image $\phi^{-1}(Q)$ is a clopen subset of X. The function f defined by (2) therefore maps B into A. It is easy to check that f is a homomorphism: if Q_1 and Q_2 are clopen subsets of Y, then

$$f(Q_1 \cap Q_2) = \phi^{-1}(Q_1 \cap Q_2) = \phi^{-1}(Q_1) \cap \phi^{-1}(Q_2) = f(Q_1) \cap f(Q_2)$$

and

$$f(Q_1') = \phi^{-1}(Q_1') = \phi^{-1}(Q_1)' = f(Q_1)'$$

S. Givant, P. Halmos, *Introduction to Boolean Algebras*,
Undergraduate Texts in Mathematics, DOI: 10.1007/978-0-387-68436-9_36,
© Springer Science+Business Media, LLC 2009

(see Exercise 33.1). Each continuous function from X to Y thus induces a *dual homomorphism* from B to A via definition (2).

The converse is also true: if f is a homomorphism from B into A, then the equivalence in (1) uniquely determines a continuous function ϕ from X into Y. Indeed, for each element x in X, the right side of (1) determines an ultrafilter in B, and there is a unique point y in Y that belongs to every set in this ultrafilter; consequently, there is just one way to define $\phi(x)$ so that (1) holds, namely

$$\phi(x) = y.$$

In more detail, the class of sets in A that contain x is an ultrafilter M in A (by the dual of Exercise 20.2), and the inverse image of M under f, the set

$$N = \{Q \in B : f(Q) \in M\} = \{Q \in B : x \in f(Q)\},$$

is an ultrafilter in B (by the dual of Exercise 20.5). The sets in any ultrafilter in A or in B have exactly one point in common (see Exercise 35.5 or the remarks at the end of Chapter 35). For M that point is clearly x, and for N it is some point y in Y; define $\phi(x)$ to be this point y. It is easy to verify that condition (1) is satisfied: if Q is any clopen set in Y, then $\phi(x)$ is in Q just in case Q belongs to N, by the definition of $\phi(x)$, so that

$$\{Q \in B : \phi(x) \in Q\} = N = \{Q \in B : x \in f(Q)\}.$$

To establish the continuity of ϕ, it suffices to prove that the inverse image under ϕ of any clopen set in Y is clopen in X. A clopen set Q in Y belongs to B, so its image $f(Q)$ is an element of A and hence a clopen set in X; consequently, $\phi^{-1}(Q)$ is a clopen set in X, by (2). Every homomorphism from B to A thus determines via condition (1) a unique *dual continuous function* from X to Y.

If f is a homomorphism from B into A with dual ϕ, then ϕ and f satisfy condition (1), and hence also condition (2), by the remark at the end of the preceding paragraph. The dual of ϕ is the function g from B to A defined by condition (2), with g in place of f. Consequently,

$$g(Q) = \phi^{-1}(Q) = f(Q)$$

for every Q in B, so that f and g are equal. Thus, f is its own second dual.

Similarly, if ϕ is a continuous function from X to Y with dual f, then ϕ and f satisfy condition (1). The dual of f is the unique continuous function ψ from X to Y that satisfies condition (1), with ψ in place of ϕ. Consequently,

$$\phi(x) \in Q \qquad \text{if and only if} \qquad \psi(x) \in Q$$

for every Q in B and every x in X. The space Y is Boolean, so distinct points can be separated by clopen sets. It follows that $\phi(x)$ and $\psi(x)$ must be equal for every x in X, so that ϕ and ψ are equal. Thus, ϕ is its own second dual.

A straightforward argument based on the preceding paragraphs shows that the correspondence that takes each continuous function to its dual maps the class of continuous functions from X to Y bijectively to the class of homomorphisms from B to A. The correspondence carries one-to-one functions to onto functions, and conversely. To prove this last assertion, consider a continuous mapping ϕ from X to Y, and let f be its dual. We formulate five assertions, each of which will be shown to be equivalent to its neighbors: (1) ϕ maps X onto Y; (2) $Y - \phi(X) = \varnothing$; (3) every clopen subset of $Y - \phi(X)$ is empty; (4) if a clopen subset Q of Y is such that $\phi^{-1}(Q) = \varnothing$, then $Q = \varnothing$; (5) f is one-to-one. It is obvious that (1) is equivalent to (2), and it is equally obvious that (2) implies (3). To see that (3) implies (2), notice that $\phi(X)$ is a compact, and hence a closed, subset of Y, by Lemmas 33.3 and 29.1. The set $Y - \phi(X)$ is therefore open, so it is the union of its clopen subsets. If each of these subsets is empty, then so is their union. Assertion (4) is really just a rephrasing of (3): to say of a subset Q (of Y) that $\phi^{-1}(Q)$ is empty is to say that Q does not contain any elements in the range of ϕ, or, what amounts to the same thing, that Q is a subset of $Y - \phi(X)$. As regards the equivalence of (4) and (5), recall that a Boolean homomorphism is one-to-one just in case its kernel contains only zero. The kernel of f is the class of clopen subsets Q such that $\phi^{-1}(Q)$ is empty, by the definition of f as the dual of ϕ. Therefore, f is one-to-one just in case assertion (4) holds. The equivalence of the five assertions, and in particular of (1) and (5), has now been established.

Consider, finally, the following four assertions: (1) ϕ is one-to-one; (2) the inverse images under ϕ of clopen subsets of Y separate points in X; (3) every clopen subset of X is the inverse image under ϕ of some clopen subset of Y; (4) f maps B onto A. Each assertion is equivalent to its neighbors. To establish the equivalence of (1) and (2), let x and y be distinct points in X. If (1) holds, then $\phi(x)$ and $\phi(y)$ are distinct points in Y. The clopen subsets of Y separate points (since Y is a Boolean space), so there is a clopen subset Q containing $\phi(x)$, but not $\phi(y)$. The inverse image $\phi^{-1}(Q)$ contains x, but not y. If (2) holds, then there is a clopen subset Q of Y such that $\phi^{-1}(Q)$ contains x, but not y. It follows that Q contains $\phi(x)$, but not $\phi(y)$, so these two points must be distinct. To go from (2) to (3) recall that the inverse images under ϕ of the clopen subsets of Y constitute a field of subsets of X (Exercise 33.1); it must be the field of all clopen sets in X, by (2) and Lemma 32.1. The reverse implication is an obvious consequence of the fact that the clopen subsets of X separate points (since X is a Boolean space). Finally, the equivalence of (3) and (4) follows directly from the definition of f as the dual of ϕ, and from the fact that the elements of A and B are just the clopen subsets of X and Y respectively.

The following fundamental duality theorem for homomorphisms has been proved.

Theorem 34. *Let X and Y be Boolean spaces, and A and B their respective dual algebras. There is a bijective correspondence between all continuous functions ϕ from X into Y and all homomorphisms f from B into A such that the equivalence*

$$\phi(x) \in Q \qquad \text{if and only if} \qquad x \in f(Q)$$

holds for all Q in B and all x in X. Each of ϕ and f is its own second dual. The homomorphism f is one-to-one if and only if ϕ maps X onto Y; the continuous function ϕ is one-to-one if and only if f maps B onto A.

As was pointed out in the preceding argument, the complement of $\phi(X)$ (in Y) is an open set, and as such it is the union of its clopen subsets. A clopen set Q (in Y) is included in this complement just in case $\phi^{-1}(Q)$ is empty, that is, just in case Q belongs to the kernel of f. The kernel of f is an ideal, and its dual is, by definition, the union of the clopen sets that belong to the ideal. Conclusion: the dual of the kernel of f is the complement of $\phi(X)$.

Corollary 1. *If ϕ is a continuous mapping from a Boolean space X into a Boolean space Y, and if f is the dual homomorphism of ϕ, then the dual of the kernel of f is the complement of the range of ϕ.*

In loose language the corollary can be expressed as follows: to divide an algebra by an ideal is the same as to discard an open set from a space.

The duality between homomorphisms and continuous functions preserves the operation of composition between functions: the composition of duals is the dual of compositions.

Corollary 2. *Let X, Y, and Z be Boolean spaces, and suppose ϕ is a continuous mapping from X to Y, and ψ a continuous mapping from Y to Z. The dual homomorphism of the composition $\psi \circ \phi$ is the composition of the dual of ϕ with the dual of ψ.*

Proof. Suppose A, B, and C are the dual algebras of the spaces X, Y, and Z respectively. The dual of ϕ is the homomorphism f from B to A defined by

$$f(Q) = \phi^{-1}(Q)$$

for all Q in B, and the dual of ψ is the homomorphism g from C to B defined by

$$g(P) = \psi^{-1}(P)$$

for all P in C. The composition $f \circ g$ is a homomorphism from C to A, and

$$(f \circ g)(P) = f(g(P)) = \phi^{-1}(\psi^{-1}(P)) = (\psi \circ \phi)^{-1}(P)$$

for all P in C. Consequently, $f \circ g$ is the dual of $\psi \circ \phi$, by the definition of that dual.

Thus far, we have looked at the duality between homomorphisms and continuous functions from the perspective of Boolean spaces and their dual algebras. It is also illuminating to look at this duality from the perspective of Boolean algebras and their dual spaces. Suppose A and B are Boolean algebras with dual spaces X and Y. If f is a homomorphism from B to A, and if x is a 2-valued homomorphism on A — that is, an element of X — then the composition $x \circ f$ is a 2-valued homomorphism on B — that is, an element of Y. The dual of f is the function ϕ from X to Y defined by

(3) $$\phi(x) = x \circ f,$$

and it is easily seen to be continuous. It suffices to check that the inverse image under ϕ of a clopen set is clopen. If Q is a clopen subset of Y, then

$$Q = \{y \in Y : y(q) = 1\}$$

for some element q in B. Write $p = f(q)$; a straightforward calculation gives

$$\phi^{-1}(Q) = \{x \in X : \phi(x) \in Q\} = \{x \in X : x \circ f \in Q\}$$
$$= \{x \in X : (x \circ f)(q) = 1\} = \{x \in X : x(p) = 1\},$$

so $\phi^{-1}(Q)$ is a clopen subset of X.

The close relationship between definitions (2) and (3) can be made more apparent if for the dual spaces X and Y, we take the spaces of ultrafilters instead of the spaces of 2-valued homomorphisms. In this case, the 2-valued homomorphisms x on A and $x \circ f$ on B are replaced by their cokernels. If M is the cokernel of x, then $f^{-1}(M)$ is the cokernel of $x \circ f$. Proof: an element q is in $f^{-1}(M)$ just in case $f(q)$ is in M, and the latter happens exactly when $x(f(q)) = 1$, that is to say, it happens exactly when q is in the cokernel of $x \circ f$. In the notation of ultrafilters, condition (3) therefore assumes the form

$$\phi(M) = f^{-1}(M)$$

for every M in X.

In discussing the duality between continuous mappings on Boolean spaces and homomorphisms on Boolean algebras, a good way to stay neutral and avoid giving preferential treatment to either is to use the concept of a pairing introduced in Chapter 34. Suppose, accordingly, that A is a Boolean algebra and X is a Boolean space, and suppose that $\langle p, x \rangle$ represents all continuous 2-valued functions on X and all 2-valued homomorphisms on A. Suppose, moreover, that B and Y are a similarly paired pair.

Fix a continuous function ϕ from X to Y, and consider $\langle q, \phi(x) \rangle$. As a function of q, for fixed x, it is a 2-valued homomorphism on B, by pairing condition (2) in Chapter 34. (Thus, it corresponds to an element of Y, namely $\phi(x)$.) This yields nothing new. The novelty comes from considering $\langle q, \phi(x) \rangle$ as a function of x for fixed q. Since it is the composite of the continuous function $x \to \phi(x)$ from X to Y, and the continuous function $y \to \langle q, y \rangle$ from Y to 2, it is a continuous 2-valued function on X. (The function $y \to \langle q, y \rangle$ is continuous by pairing condition (1).) As such, it is given by a unique element p of A, by pairing condition (1) and the remarks at the end of Chapter 34, so that

$$\langle q, \phi(x) \rangle = \langle p, x \rangle$$

identically in x. Denote the passage from q to p by f, that is, write $p = f(q)$. The proof that f is a Boolean homomorphism is a mechanical computation. Here, for instance, is the proof that f preserves complementation:

$$\langle f(q'), x \rangle = \langle q', \phi(x) \rangle = \langle q, \phi(x) \rangle' = \langle f(q), x \rangle' = \langle f(q)', x \rangle.$$

The first and third steps use the definition of f; the second and fourth steps use the fact that $\langle p, x \rangle$ and $\langle q, \phi(x) \rangle$, as functions of p and q (for fixed x), are 2-valued homomorphisms on A and B respectively, by pairing condition (2), and therefore preserve complementation. Since these equations hold for all x in X, we conclude that $f(q') = f(q)'$, by the remarks at the end of Chapter 34. The homomorphism f is the *dual* of ϕ.

Now fix a homomorphism f from B to A, and consider $\langle f(q), x \rangle$. As a function of x, for fixed q, it is a continuous 2-valued function on X, by pairing condition (1). (Thus, it corresponds to an element of A, namely $f(q)$.) In particular, it determines the clopen set

$$\{ x \in X : \langle f(q), x \rangle = 1 \}$$

in X. This yields nothing new. The novelty comes from considering $\langle f(q), x \rangle$ as a function of q for fixed x. Since it is the composite of the homomorphism $q \to f(q)$ from B to A, and the homomorphism $p \to \langle p, x \rangle$ from A to 2, it is a 2-valued homomorphism on B. (The function $p \to \langle p, x \rangle$ is a homomorphism by pairing condition (2).) As such, it is given by a unique element y of Y, by pairing condition (2) and the remarks at the end of Chapter 34, so that

$$\langle f(q), x \rangle = \langle q, y \rangle$$

identically in q. Denote the passage from x to y by ϕ, that is, write $y = \phi(x)$. The definition of ϕ implies that

$$\phi^{-1}(\{y : \langle q, y \rangle = 1\}) = \{x : \langle q, \phi(x) \rangle = 1\} = \{x : \langle f(q), x \rangle = 1\}.$$

Every clopen subset Q of Y is given by some element q in B in the sense that

$$Q = \{y \in Y : \langle q, y \rangle = 1\};$$

indeed, every clopen subset of Y is given by a unique 2-valued continuous function on Y (Lemma 33.2) and every continuous 2-valued function on Y is given by some q in B (pairing condition (1)). It follows from the preceding equations that the inverse image under ϕ of every clopen set in Y is a clopen set in X, and hence that ϕ is continuous. The mapping ϕ is the *dual* of f.

If f is the dual of ϕ, and ψ is the dual of f, then

$$\langle q, \phi(x) \rangle = \langle f(q), x \rangle = \langle q, \psi(x) \rangle$$

holds identically for x in X and q in B, by the definitions of these duals. Since the equations hold identically in q, it follows that $\phi(x) = \psi(x)$ for every x in X. In other words, ϕ is the dual of f, and is therefore its own second dual. If, finally, ϕ is the dual of f, and g is the dual of ϕ, then

$$\langle f(q), x \rangle = \langle q, \phi(x) \rangle = \langle g(q), x \rangle$$

holds identically. Since these equations hold identically in x, it follows that $f(q) = g(q)$ for every q in B. In other words, f is the dual of ϕ, and is therefore its own second dual.

It follows from the preceding remarks that the duality theorem for homomorphisms may be formulated in terms of pairings in the following way.

Theorem 35. *There is a bijective correspondence between all continuous functions ϕ from X into Y and all homomorphisms f from B into A such that*

$$\langle q, \phi(x) \rangle = \langle f(q), x \rangle$$

identically for all q in B and all x in X. Each of ϕ and f is its own second dual. The homomorphism f is one-to-one if and only if ϕ maps X onto Y; the function ϕ is one-to-one if and only if f maps B onto A.

To see a non-trivial application of the duality theorem for homomorphisms, we consider an alternative construction (due to Stone [67]) of free Boolean algebras (see Chapter 28).

Theorem 36. *For every set I, the dual algebra of the Cantor space 2^I is freely generated by the family of clopen sets of the form*

$$P_i = \{y \in 2^I : y_i = 1\}$$

for i in I (and this family has the same power as I).

Proof. Write $Y = 2^I$, and let B be the dual algebra of Y. The clopen sets P_i, for i in I, are distinct, so the family of these clopen sets,

$$E = \{P_i : i \in I\},$$

has the same power as I. The field generated by E — the class of finite unions of finite intersections of sets in E and their complements (Theorem 3, p. 82) — is a base for the topology of Y (by the definition of that topology) and therefore E generates B (Corollary 32.1). We shall prove that B is free on E.

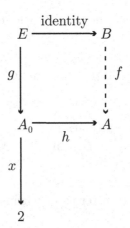

Consider an arbitrary Boolean algebra A_0, and an arbitrary mapping g from E into A_0 (see the diagram). Write $p_i = g(P_i)$. If X is the dual space of A_0 (the set of 2-valued homomorphisms on A_0), and if A is the dual algebra of X (the algebra of clopen subsets of X), then A_0 and A are isomorphic via the mapping h given by

$$h(p) = \{x \in X : x(p) = 1\}$$

(Theorem 31, p. 328). For each x in X, define a function $\phi(x)$ from I to 2 by

$$\phi(x)_i = x(g(P_i)) = x(p_i).$$

Clearly, ϕ maps X into Y, because $\phi(x)$ is an element of Y for each x. The inverse image under ϕ of the clopen set P_i is a clopen subset of X:

$$\phi^{-1}(P_i) = \{x \in X : \phi(x) \in P_i\}$$
$$= \{x \in X : \phi(x)_i = 1\} = \{x \in X : x(p_i) = 1\}.$$

The same is true for the inverse image of the complement of P_i, by Exercise 33.1(c). Since the elements of E and their complements form a subbase for the topology of Y, it follows that ϕ is continuous, by Lemma 33.1.

Let f be the dual homomorphism from B to A. This means that

$$f(Q) = \phi^{-1}(Q)$$

for every Q in B. In particular,

$$f(P_i) = \phi^{-1}(P_i) = \{x \in X : x(p_i) = 1\} = h(p_i) = h(g(P_i))$$

for each i in I. In other words, f is a homomorphism from B into A that agrees with $h \circ g$ on E. Consequently, $h^{-1} \circ f$ is the desired homomorphism from B into A_0 that agrees with g on E. The proof of the theorem is complete.

An explicit statement of the duality between Boolean homomorphisms and continuous mappings seems to have appeared for the first time in Sikorski's book [64]. It is derived there as a corollary of a theorem (due to Sikorski [60]) to the effect that every homomorphism between certain fields of subsets of sets X and Y is induced (in the sense of Chapter 12) by a mapping from Y to X. The formulation of the homomorphism–continuous-function duality in terms of a pairing of a Boolean algebra and a Boolean space (Theorem 35) goes back to Halmos [23].

Exercises

1. Suppose that X and Y are Boolean spaces with dual algebras A and B. Show that the correspondence given in Theorem 34 between continuous functions ϕ from X to Y and homomorphisms f from B to A is a bijection.

2. Suppose A and B are Boolean algebras with dual spaces X and Y. Show that for each continuous function ϕ from X into Y, there is a homomorphism f from B to A such that

$$\phi(x) = x \circ f$$

   for every x in X (cf. condition (3) in this chapter).

3. If X is a Boolean space and A its dual algebra, then the 2-valued function $\langle P, x \rangle$ defined for all P in A and all x in X by

$$\langle P, x \rangle = \begin{cases} 1 & \text{if} \quad x \in P, \\ 0 & \text{if} \quad x \notin P, \end{cases}$$

   is a pairing of A and X (see Chapter 34). Assume that a pairing of a Boolean space Y and its dual algebra B is also defined in an analogous fashion. According to Theorem 35, the dual of a continuous function ϕ from X to Y is the unique function f from B to A satisfying the pairing equation

$$\langle Q, \phi(x) \rangle = \langle f(Q), x \rangle$$

   for all Q in B and all x in X. Similarly, the dual of a homomorphism f from B to A is the unique function ϕ from X to Y satisfying the pairing equation. Show that these definitions are equivalent to the ones determined by the equivalence

$$\phi(x) \in Q \qquad \text{if and only if} \qquad x \in f(Q)$$

   in Theorem 34.

4. If A is a Boolean algebra and X its dual space, then the 2-valued function $\langle p, x \rangle$ defined for all p in A and all x in X by

$$\langle p, x \rangle = x(p)$$

   is a pairing of A and X (see Chapter 34). Assume that a pairing of a Boolean algebra B and its dual space Y is also defined in an analogous fashion. According to Theorem 35, the dual of a homomorphism f from B to A is the unique function ϕ from X to Y satisfying the pairing equation

$$\langle q, \phi(x) \rangle = \langle f(q), x \rangle$$

for all q in B and all x in X. Show that this definition is equivalent to the one given by the equation

$$\phi(x) = x \circ f$$

(cf. condition (3) in this chapter).

5. Formulate and prove the version of Corollary 2 that applies to the notion of pairings of Boolean algebras with Boolean spaces.

6. Let A be a Boolean algebra and X its dual space. Use the results in this chapter to give another proof of the theorem (Exercise 35.20) that the dual space of the relativization of A to an element p is homeomorphic to the subspace

$$U_p = \{x \in X : x_p = 1\}$$

of X.

7. Give a topological proof of the fact (Corollary 28.1) that a finite Boolean algebra is free if and only if the number of its atoms is a (finite) power of 2, or, equivalently, if and only if it has 2^{2^m} elements for some natural number m.

8. Prove that every free algebra satisfies the countable chain condition.

9. Use Exercises 8 to give another solution to Exercise 30.12.

10. Use Theorem 36 to give another proof of the fact (Corollary 28.2) that every infinite free algebra is atomless.

11. Derive Theorem 36 from Lemma 28.1.

12. Prove that Theorem 36 implies Lemma 28.1.

13. Prove that the group of automorphisms of a Boolean algebra is isomorphic to the group of homeomorphisms mapping the dual space to itself. (This theorem is due to Stone [67].)

14. (Harder.) Use Theorem 34 to prove that a Boolean space is separable if and only if its dual algebra is embeddable in the field of all subsets of the set of positive integers.

15. (Harder.) Give a topological solution of Exercise 22.4.

Chapter 37

Duality for Subalgebras

The epi–mono duality for structure-preserving maps implies a useful sub–quotient duality for the structures themselves. One half of this duality is a bijective correspondence between the closed subspaces of a Boolean space Y and the quotients of its dual algebra B. Consider first a closed subset X of the space Y. Under the inherited topology, X is a Boolean space, and its dual algebra A consists of the intersections with X of the clopen subsets of Y (Lemma 32.2). There is a natural mapping ϕ (namely the identity) from X into Y. Since ϕ is one-to-one, the dual homomorphism f (defined by

$$f(Q) = X \cap Q$$

for every clopen subset Q of Y) maps B onto A, by Theorem 34. As a homomorphic image, A is isomorphic to a quotient of B, and in fact it is isomorphic to the quotient B/M, where M is the kernel of f, by the first isomorphism theorem (Theorem 14, p. 179). It follows (Exercise 34.9) that X is homeomorphic to the dual space of B/M. The dual of the ideal M is the complement of the range of ϕ, by Corollary 36.1, that is to say, it is the open set

$$X' = Y - X = Y - \phi(X).$$

Thus, the closed subspace X corresponds to — and in fact is homeomorphic to the dual space of — the quotient B/M, where M is the ideal that is the dual of X'.

Consider now an arbitrary quotient of the algebra B, determined say by an ideal M. The dual of M is an open subset of Y, so the complement of this dual, say X, is a closed subset of Y; as such, it is a Boolean space under the inherited topology. The dual algebra of X — call it A — is isomorphic

S. Givant, P. Halmos, *Introduction to Boolean Algebras*,
Undergraduate Texts in Mathematics, DOI: 10.1007/978-0-387-68436-9_37,
© Springer Science+Business Media, LLC 2009

to the quotient B/M. Indeed, the projection mapping of B to B/M is an epimorphism, so its dual is a one-to-one continuous function ϕ from the dual space of B/M into Y, by Theorem 34. The range of ϕ is the complement of the dual of the kernel of the epimorphism (Corollary 36.1). Since the kernel of the projection is M, the range of ϕ is X. In other words, ϕ maps the dual space of B/M homeomorphically onto X. It follows (Exercise 34.8) that B/M is isomorphic to the dual algebra of X, which is just A. Thus, the quotient algebra B/M corresponds to — and in fact is isomorphic to the dual algebra of — the subspace X that is the complement of the dual of the ideal M. The preceding discussion is summarized in the following duality theorem for quotient algebras (due to Stone [67]).

Theorem 37. *There is a bijective correspondence between the closed subspaces of a Boolean space Y and the quotients of its dual algebra B. If M is an ideal in B, then the dual space of the quotient B/M is homeomorphic to the closed subspace X, where X' is the open set that is the dual of M. Inversely, if X is a closed subspace of Y, then the dual algebra of X is isomorphic to the quotient B/M, where M is the ideal that is the dual of X'.*

The second half of the sub–quotient duality is more complicated to formulate precisely, but here is an intuitive description. Suppose A and B are Boolean algebras with dual spaces X and Y respectively. If B is a subalgebra of A, then there is a natural homomorphism f (namely the identity) from B into A. Since f is one-to-one, the dual mapping ϕ maps X onto Y, so that Y is isomorphic to a certain quotient space of X. Inversely, every quotient of X of a certain type determines a subalgebra of A.

To make the preceding intuitive description precise, it is first necessary to clarify what is meant by a quotient of a topological space X. If ϕ is a function from X onto a set Y, then a topology can be defined on Y that makes ϕ continuous: declare a subset of Y to be open just in case its inverse image under ϕ is an open set in X. The resulting class of open sets is called the *quotient topology* on Y (with respect to the space X and the mapping ϕ), and it is clear that under this topology, ϕ becomes a continuous function from X to Y.

For a concrete example, consider any equivalence relation Θ on X, and let Y be the set of equivalence classes of Θ. The projection ϕ maps each element in X to its equivalence class modulo Θ. A subset V of Y is, by definition, open in the quotient topology (with respect to ϕ) if $\phi^{-1}(V)$ is an open subset of X. Since $\phi^{-1}(V)$ is the union of the equivalence classes in V

(equally correctly, it is the union of V), the set V is open in the quotient topology just in case $\bigcup V$ is an open set in X. The space Y is called the *quotient* of X modulo Θ, and we shall denote it by

$$Y = X/\Theta.$$

The example in the preceding paragraph typifies the general situation. Given an arbitrary mapping ϕ from X onto a set Y, define an equivalence relation Θ on X by requiring two elements x and y in X to be equivalent modulo Θ if they have the same image under ϕ:

$$x \equiv y \mod \Theta \qquad \text{if and only if} \qquad \phi(x) = \phi(y).$$

The topological analogue of the first isomorphism theorem says that Y, endowed with the quotient topology, is homeomorphic to the quotient space X/Θ, and in fact the correspondence that takes each equivalence class x/Θ to the element $\phi(x)$ in Y is a well-defined homeomorphism from X/Θ to Y.

A quotient of a compact space is always compact (Lemma 33.3). A quotient of a Hausdorff space or a Boolean space, however, need not be Hausdorff or Boolean (Exercises 5 and 6). In order to establish a duality between the subalgebras of a Boolean algebra and the quotients of its dual space, it is important to characterize the equivalence relations on the dual space that lead to quotients that are Boolean spaces.

Consider an equivalence relation Θ on a Boolean space X. A subset P of X is said to be *compatible with* Θ if it is a union of equivalence classes of Θ; in other words, if an element x is in P, then the entire equivalence class of x (modulo Θ) is included in P. The relation Θ is called *Boolean* if for any two equivalence classes of Θ, there is a clopen subset of X that is compatible with Θ and that includes one of the equivalence classes, but not the other.

To understand the significance of this definition, let Θ be an equivalence relation on X, and consider the class B of all those clopen subsets of X that are compatible with Θ. It is easy to check that B is a subalgebra of the dual algebra of X (Exercise 8). For each set P in B, the quotient set

$$P/\Theta = \{x/\Theta : x \in P\}$$

is a subset of the quotient space $Y = X/\Theta$. The inverse image of this subset under the projection (that is, the union of the equivalence classes in P/Θ) is just P, because P is assumed to be compatible with Θ. Since P is clopen in X (it belongs to B), the set P/Θ must be clopen in Y, by the definition of the quotient topology (Exercise 7). Thus, the quotient of every set in B is clopen in the quotient space.

If P and Q are sets in B, then

(1) $(P/\Theta) \cup (Q/\Theta) = \{x/\Theta : x \in P \cup Q\} = (P \cup Q)/\Theta$

and

(2) $(P/\Theta)' = \{x/\Theta : x \notin P\} = \{x/\Theta : x \in P'\} = (P')/\Theta.$

It follows from these equations, and from the closure of B under union and complement, that the class

$$B_\Theta = \{P/\Theta : P \in B\}$$

is closed under union and complement. The class obviously contains the empty set, so it is a field of clopen subsets of Y. The mapping that takes each set P in B to its quotient P/Θ is a homomorphism from B onto B_Θ, by equations (1) and (2). The kernel of this homomorphism is the class of sets P such that P/Θ is empty, so the kernel contains only the empty set. The homomorphism is therefore an isomorphism from B to B_Θ.

The condition that the relation Θ be Boolean simply means that B_Θ is a separating field of clopen subsets of the quotient space Y. It follows from Lemma 32.1 that if Θ is a Boolean relation, then Y is a Boolean space and B_Θ is the field of all clopen subsets of Y. Conversely, if Y is a Boolean space, then the clopen sets in Y separate points. Since the union of a clopen set in Y is a clopen set in X that is compatible with Θ, the fact that the clopen sets in Y separate points just means that the relation Θ is Boolean. The following lemma has been proved.

Lemma 1. *Let X be a Boolean space and Θ an equivalence relation on X. The quotient space X/Θ is Boolean if and only if the relation Θ is Boolean. If Θ is Boolean, and if B is the field of all clopen subsets of X that are compatible with Θ, then B_Θ is the dual algebra of the quotient space X/Θ, and B is isomorphic to B_Θ.*

Each Boolean relation Θ on a Boolean space X determines a subalgebra of the dual algebra of X, namely the algebra B of clopen subsets of X that are compatible with Θ. The algebra B is called the *dual* of Θ because it is isomorphic to the dual algebra of the quotient space X/Θ (Lemma 1).

Inversely, each subalgebra B of the dual algebra of X determines a *dual* Boolean relation Θ on X such that the quotient space X/Θ is homeomorphic to the dual space of B. The relation Θ is defined by requiring

$$x \equiv y \mod \Theta$$

just in case x and y belong to the same clopen sets in B, that is, just in case, for every P in B,

$$x \in P \qquad \text{if and only if} \qquad y \in P.$$

The proof that Θ is a Boolean relation is not difficult. First, each set P in B is compatible with Θ: if x is a point in X, then the definition of Θ implies that the equivalence class of x is either entirely included in, or entirely disjoint from, P. Second, if x and y are points in X that are not equivalent modulo Θ, then the definition of Θ implies that there must be a set P in B that separates the points, say P contains x but not y. The equivalence class of x is then entirely included in P, while that of y is disjoint from P, because P is compatible with Θ.

The subalgebra B is its own second dual, that is to say, if Θ is the dual relation of B, and if C is the dual algebra of Θ — the field of all clopen subsets of X that are compatible with Θ — then $B = C$. For the proof, recall from the preceding paragraph that Θ is a Boolean relation on X, so that the quotient $Y = X/\Theta$ is a Boolean space with dual algebra C_Θ (Lemma 1). On the other hand, the final remarks of the preceding paragraph imply that B_Θ is a separating field of clopen sets in Y, so it must be the field of all clopen sets in Y (Lemma 32.1). Thus,

(3) $$B_\Theta = C_\Theta.$$

Each set P in B or in C is the union of the equivalence classes in P/Θ, so equation (3) implies that $B = C$, as required. One consequence of these observations is that B, as the dual of Θ, is isomorphic to the dual algebra of Y, by Lemma 1. The dual space of B is therefore homeomorphic to Y (Exercise 34.9).

Each Boolean relation on the space X is also equal to its second dual. For the proof, suppose B is the dual algebra of a Boolean relation Θ on X, and assume Ψ is the dual relation of B. It is to be shown that $\Theta = \Psi$. If two points x and y in X are equivalent modulo Θ, then they must belong to the same clopen sets that are compatible with Θ, by the definition of compatibility. Since the dual algebra B consists of all clopen sets that are compatible with Θ, the points x and y belong to the same sets in B, and are therefore equivalent modulo Ψ, by the definition of Ψ. If the two points are not equivalent modulo Θ, then there is a clopen subset P of X that is compatible with Θ and that contains x, but not y, by the assumption that Θ is a Boolean relation. The set P belongs to B, by the definition of B, so the points x and y do not belong to precisely the same sets in B, and are therefore not equivalent modulo Ψ, by the definition of Ψ. Thus, two points in X are equivalent modulo Θ if and only if they are equivalent modulo Ψ.

If X is a Boolean space with dual algebra A, then the correspondence f that takes each subalgebra of A to the dual relation on X is a bijection from the lattice of subalgebras of A to the lattice of Boolean relations on X. The mapping f is one-to-one because each subalgebra is its own second dual, and f is onto because each Boolean relation is its own second dual. In more detail, if

$$f(B) = \Theta = f(C),$$

then the dual of Θ is the second dual of both B and C, so that $B = C$. Consequently, f is one-to-one. If Θ is an arbitrary Boolean relation on X, and if B is its dual algebra, then $f(B)$ is the dual relation of B and therefore the second dual of Θ, so that $f(B) = \Theta$. Hence, f is onto.

The correspondence f is similar in nature to a lattice isomorphism, except that it reverses inclusions (and therefore it maps meets to joins, and joins to meets — cf. Exercise 12.12). To see this, let B and C be subalgebras of A with dual relations Θ and Ψ respectively. Suppose first that C is a subalgebra of B. If two points x and y from X belong to the same sets in B, then they certainly belong to the same sets in C. In other words,

$$x \equiv y \mod \Theta \qquad \text{implies} \qquad x \equiv y \mod \Psi,$$

so that Θ is included in Ψ. Conversely, if Θ is included in Ψ, then every equivalence class of Ψ is a union of equivalence classes of Θ. Consequently, every clopen set in X that is compatible with Ψ is also compatible with Θ. It follows that C must be included in B. An order-reversing bijection between lattices is called a *dual (lattice) isomorphism*.

Here, finally, is the precise statement of the second half of the sub–quotient duality, the duality theorem for subalgebras.

Theorem 38. *Let X be a Boolean space and A its dual algebra. The correspondence that takes each subalgebra of A to its dual relation is a dual isomorphism from the lattice of subalgebras of A to the lattice of Boolean relations on X. The second dual of every subalgebra and of every Boolean relation is itself. If a relation Θ is the dual of a subalgebra B, then B is isomorphic to the dual algebra of the quotient space X/Θ, and X/Θ is homeomorphic to the dual space of B.*

The duality between subalgebras of a Boolean algebra and quotients of the dual space that are Boolean spaces was first observed by Stone [67]. The characterization of the equivalence relations on the dual space that lead to quotients that are Boolean spaces is from [45].

Exercises

1. Suppose X is a topological space, and ϕ a function from X onto a set Y. Prove that the quotient topology on Y is in fact a topology, and the mapping ϕ is continuous under this topology.

2. Suppose X is a topological space, and ϕ a function from X onto a set Y. Prove that the quotient topology on Y is the largest topology under which ϕ becomes continuous. In other words, prove that if T is any topology on Y under which ϕ is continuous, then every open set in T is also open in the quotient topology.

3. Prove the topological analogue of the first isomorphism theorem (see p. 361).

4. Show directly (without using Lemma 33.3) that the quotient of a compact space is compact.

5. Give an example to show that the quotient of a Hausdorff space need not be Hausdorff.

6. Give an example to show that the quotient of a Boolean space need not be Boolean.

7. Suppose X is a topological space, and ϕ a mapping from X onto a set Y. Prove that a subset of Y is clopen in the quotient topology if and only if its inverse image under ϕ is clopen in X.

8. Prove that the clopen sets in a Boolean space X that are compatible with a given equivalence relation on X form a subfield of the dual algebra of X.

9. Prove that the class of Boolean relations on a Boolean space is a complete lattice under the partial ordering of set-theoretic inclusion.

10. (Harder.) Suppose X is a Boolean space and A its dual algebra. If B is a subalgebra of A, and if Θ is the corresponding Boolean relation on X, then the dual space of B is homeomorphic to the quotient space X/Θ, by Theorem 38. Show that if the dual space of B is taken to be the

space of ultrafilters in B (see the remarks following Theorem 31, p. 328), then the function that maps each ultrafilter N to the intersection of the sets in N is such a homeomorphism.

11. (Harder.) Let X be a Boolean space and A its dual algebra. If M is an ideal in A, and if B is a subalgebra of A with dual Boolean relation Θ, what is the dual of the ideal $M \cap B$ (of B) in the space X/Θ? (Recall that X/Θ is a homeomorphic copy of the dual space of B.)

12. Suppose X is Boolean space, Y a non-empty clopen subset of X, and y_0 a point in Y. Show that the mapping ϕ from X onto Y defined by

$$\phi(x) = \begin{cases} x & \text{if} \quad x \in Y, \\ y_0 & \text{if} \quad x \notin Y, \end{cases}$$

is continuous, and prove directly that the equivalence relation Θ on X defined by

$$x \equiv z \mod \Theta \qquad \text{if and only if} \qquad \phi(x) = \phi(z)$$

is Boolean.

13. The formulation of the sub–epi duality in Theorem 37 is not exactly parallel to that of Theorem 38. It makes no mention of a dual lattice isomorphism or of second duals. Formulate a version of Theorem 37 that is parallel to Theorem 38.

14. A *pseudo-metric* on a set X is a non-negative real-valued function d of two arguments that satisfies all the conditions for being a metric (see Chapter 9, p. 55) except for strict positivity; the condition of strict positivity,

$$d(x,y) \geq 0, \qquad \text{and} \qquad d(x,y) = 0 \quad \text{if and only if} \quad x = y,$$

is replaced by the weaker condition

$$d(x,y) \geq 0 \qquad \text{and} \qquad d(x,x) = 0.$$

A *pseudo-metric space* is a set together with a pseudo-metric. In terms of the pseudo-metric, a topology may be defined in exactly the same way as for metric spaces (see Chapter 9).

Suppose d is a pseudo-metric on X. Prove that the binary relation Θ defined by

$$x \equiv y \mod \Theta \qquad \text{if and only if} \qquad d(x,y) = 0$$

is an equivalence relation on X. On the set X/Θ of equivalence classes, define a non-negative real-valued function e of two arguments by writing

$$e(u, v) = d(x, y)$$

whenever x is in u, and y in v. Show that e is a well-defined metric on X/Θ, and that under this metric the projection from X to X/Θ is a continuous function.

Chapter 38

Duality for Completeness

By now we have seen the dual of almost every significant finite algebraic concept that was introduced before; it is time to turn to the infinite ones. What topological property, for instance, characterizes a Boolean space whose dual algebra is known to be complete? The answer is a weird but interesting part of pathological topology.

A Boolean space is called *complete* if the closure of every open set is open. (It is not difficult to show that every compact Hausdorff space with this property is automatically a Boolean space; see Exercise 1.) Complete Boolean spaces are sometimes called *extremally disconnected* spaces. Completeness is a self-dual property: a space is complete if and only if the interior of every closed set is closed (see Exercise 2). At first glance it is not at all obvious that non-trivial (that is, non-discrete) complete spaces exist. It turns out, however, that they exist in profusion; there are as many of them as there are complete Boolean algebras.

The brunt of the major theorem in this direction is carried by an auxiliary result that has other applications also. It is in effect a topological characterization of the suprema that happen to be formable in a not necessarily complete Boolean algebra.

Lemma 1. *If* $\{P_i\}$ *is a family of elements (clopen sets) in the dual algebra* A *of a Boolean space* X, *and if* $U = \bigcup_i P_i$, *then a necessary and sufficient condition that* $\{P_i\}$ *have a supremum in* A *is that* U^- *be open. If the condition is satisfied, then*

$$\bigvee_i P_i = U^-;$$

that is, the algebraic supremum is the closure of the set-theoretic union.

S. Givant, P. Halmos, *Introduction to Boolean Algebras*,
Undergraduate Texts in Mathematics, DOI: 10.1007/978-0-387-68436-9_38,
© Springer Science+Business Media, LLC 2009

Proof. Assume first that $P = \bigvee_i P_i$. Since P is closed (it belongs to A, by assumption) and includes each P_i, it must include the union U, and therefore also the closure U^-. Since P is open, the set $P - U^-$ is open. If $P - U^-$ is not empty, then it includes a non-empty clopen set Q. The set $P - Q$ is clopen, it is properly included in P (because Q is not empty), and it includes all the sets P_i, because

$$U^- = P \cap U^- = P - (P - U^-) \subseteq P - Q$$

(see Exercise 6.2(d)). This contradicts the assumption that P is the supremum (least upper bound) of the family $\{P_i\}$. It follows that $U^- = P$, and hence that U^- is open.

If, conversely, U^- is open, then it is clopen, and of course it includes all the sets P_i. If P is any clopen set that includes each P_i, then P includes U and therefore also U^-, since P is closed. This implies that the family $\{P_i\}$ does have a supremum in A, namely U^-.

Corollary 1. *If a family of elements in the dual algebra of a Boolean space has a supremum, then that supremum differs from the set-theoretic union by a nowhere dense set.*

Proof. Suppose U is the union of a given family of elements in the dual algebra. If the family has a supremum, then that supremum is U^-, by Lemma 1. The difference $U^- - U$ is exactly the boundary of U, since U is open (see Chapter 9). Apply Lemma 10.5 to conclude that this boundary is nowhere dense.

Theorem 39. *The dual algebra A of a Boolean space X is complete if and only if X is complete.*

Proof. Assume A is complete. If U is an arbitrary open set in X, and if $\{P_i\}$ is the family of its clopen subsets, then $U = \bigcup_i P_i$ (because the clopen sets form a base for the topology of X) and $\{P_i\}$ has a supremum in A, by the assumed completeness of A. Apply Lemma 1 to conclude that U^- is open. It follows that the space X is complete.

Suppose now that X is complete, and consider an arbitrary family $\{P_i\}$ of elements of A. The union U of this family is an open set, so its closure U^- is also open, by the assumed completeness of X. Apply Lemma 1 to conclude that U^- is the supremum of $\{P_i\}$ in A. It follows that A is complete.

Recall (Chapter 25) that the completion of a Boolean algebra A is a Boolean algebra B with the following properties: A is a subalgebra of B that is dense in B (every non-zero element in B is above a non-zero element in A), and every family of elements in A has a supremum in B. The existence of the completion was established in Chapter 25 using algebraic methods. It is also possible to established this existence using topological methods, by combining two steps each of which separately is familiar by now. Use duality to associate A with a topological space, and then use some general topology to associate with that space the algebra of regular open sets; the result is, in a natural way, the completion of A.

Theorem 40. *If A is the dual algebra of a Boolean space X, and if B is the algebra of regular open sets in X, then B is the completion of A.*

Proof. Observe, first of all, that a clopen set P is always regular:

$$P^{\perp\perp} = P^{-\prime-\prime} = P^{\prime-\prime} = P'' = P.$$

(The second and third equalities use the fact that P and its complement are closed.) Consequently, every element in A belongs to B. To verify that A is a subalgebra of B, we need merely to observe that, for clopen sets, the Boolean operations of B reduce to the ordinary set-theoretic operations. For example, if P and Q are elements in A, then their union is clopen and therefore

$$(P \cup Q)^{\perp\perp} = P \cup Q.$$

The left side of this equation is just the join of P in Q in B, by Theorem 1 (p. 66), while the right side is their join in A.

The clopen sets form a base for the topology of X, so every non-empty regular open set includes a non-zero element of A. Thus, A is dense in B. Finally, a family $\{P_i\}$ of elements in A always has a supremum in B, because B is complete, by Theorem 1.

It was shown in Chapter 25 that a dense subalgebra of a Boolean algebra is always a regular subalgebra. It is interesting to see a direct proof of this fact for the algebras A and B in the previous theorem. Suppose $\{P_i\}$ is a family of elements in A. The supremum of this family in B is the set $U^{\perp\perp}$, where $U = \bigcup_i P_i$, by Theorem 1. If the family also has a supremum in A, then that supremum is U^-, and U^- is an open set, by Lemma 1. In this case

$$U^{\perp\perp} = U^{-\prime-\prime} = U^{-\prime\prime} = U^-,$$

since the complement of U^- is closed; the two suprema are therefore equal. It follows that the suprema formable in A stay the same in B.

We conclude this discussion with some historical and motivational remarks. The completion of a Boolean algebra A is (to within isomorphism) the Boolean algebra of complete ideals in A, by Theorem 22 (p. 216). The topological dual of a complete ideal in A is a regular open set in the dual space of A, as was noted by Stone [67] (see Exercise 35.18). The Boolean algebra of complete ideals in A should therefore be mapped isomorphically to the Boolean algebra of regular open sets in the dual space by the function that takes each ideal to its dual open set; see Theorem 33 (p. 341). This isomorphism does not, however, follow immediately from that theorem, because the operations of the two Boolean algebras do not, in general, coincide with the operations of the two lattices in Theorem 33 (the lattice of ideals and the lattice of open sets).

It should be noted that Theorem 39 is due to Stone [68].

Exercises

1. Show that a complete and compact Hausdorff space must be Boolean.

2. Prove that a topological space is complete if and only if the interior of every closed set is closed.

3. Formulate the dual of Lemma 1 for infima, and prove it directly, without using the lemma.

4. Prove that the dual of a complete Boolean space X coincides with the regular open algebra of X. Conclude that every complete Boolean algebra is isomorphic to the regular open algebra of some compact Hausdorff space. (This conclusion is due to Tarski [75]; a related result can be found in Stone [68]. The conclusion can be viewed as a kind of representation theorem for complete Boolean algebras.)

5. If the regular open sets of a Boolean space constitute a field of sets, does it follow that the space is complete?

6. Show that if I is infinite and if X is the Cantor space 2^I, then $\mathcal{P}(X)$ is not a completion of the dual algebra of X.

7. Let A be the Boolean algebra generated by the left half-closed intervals in $[0,1]$, let B be the quotient of the algebra of Lebesgue measurable sets in $[0,1]$ modulo the ideal of sets of measure zero, and let C be the quotient of the algebra of Borel sets in $[0,1]$ modulo the ideal of meager

sets. Prove that there is a complete embedding f of A into B, and a complete embedding g of A into C, such that the images $f(A)$ and $g(A)$ completely generate B and C respectively. (The algebras B and C are not isomorphic; see Chapter 31. Consequently, at least one of these algebras is not the completion of A. It follows that condition (3) in the definition of a completion (p. 214) cannot be weakened to say that A completely generates B.)

8. Does the completion of an algebra satisfying the countable chain condition satisfy that condition also?

Chapter 39

Boolean σ-spaces

A *Baire set* in a Boolean space is a set belonging to the σ-field generated by the class of all clopen sets. Clearly, every Baire set in a Boolean space is a Borel set; the converse is not true in general. A trivial way to manufacture open Baire sets is to form the union of a countable class of clopen sets. The converse is true but not trivial. The converse implies that every open Baire set is an F_σ (that is, the union of a countable class of closed sets), and, consequently, every closed Baire set is a G_δ (that is, the intersection of a countable class of open sets). We shall prove the main result about the structure of open Baire sets by proving first that every closed Baire set is a G_δ. Observe that in a metric space every closed set is a G_δ (Exercise 29.21); in a general topological space this not so. The proof of the following auxiliary result uses the fact about metric spaces just mentioned; the trick is to construct a suitable metric space associated with each given closed Baire set.

Lemma 1. *Every closed Baire set is a G_δ.*

Proof. Let F be a closed Baire set in the Boolean space X, and let $\{P_n\}$ be a sequence of clopen sets such that F belongs to the σ-field generated by $\{P_n\}$ (see Exercise 29.6). Let p_n be the characteristic function of P_n and write

$$d(x, y) = \sum_{n=1}^{\infty} \frac{1}{2^n} |p_n(x) - p_n(y)|$$

for all x and y in X. The function d is a metric except perhaps for strict positivity (see Exercise 2). If, in other words, two points x and y are defined to be equivalent, $x \equiv y$, in case $d(x, y) = 0$, then the equivalence classes may be more than singletons. (It is trivial that the relation so defined is an equivalence.)

S. Givant, P. Halmos, *Introduction to Boolean Algebras*, Undergraduate Texts in Mathematics, DOI: 10.1007/978-0-387-68436-9_39, © Springer Science+Business Media, LLC 2009

For any given point x in X, every set of the form

(1) $W = \{y \in X : d(x, y) < \epsilon\}$

is open in the topology of X. To see this, consider a point y in W, and write $\rho = d(x, y)$. Since $\rho < \epsilon$, there is a positive integer m such that $1/2^m < \epsilon - \rho$. For each integer n with $1 \leq n \leq m$, put

$$Q_n = \begin{cases} P_n & \text{if } y \in P_n, \\ P_n' & \text{if } y \notin P_n. \end{cases}$$

Each set Q_n is clopen and contains y, so the intersection

$$Q = \bigcap_{n=1}^{m} Q_n$$

is clopen and contains y. If z is in Q, then

$$p_n(y) - p_n(z) = 0$$

for $n = 1, 2, \ldots, m$, and therefore

$$d(y, z) \leq \sum_{n=m+1}^{\infty} 1/2^n = 1/2^m.$$

It follows that

$$d(x, z) \leq d(x, y) + d(y, z) \leq \rho + 1/2^m < \epsilon,$$

so that z is in W. Conclusion: the clopen set Q contains y and is included in W, so every point in W belongs to a clopen set that is included in W. Consequently, W is open.

If $x_1 \equiv x_2$ and $y_1 \equiv y_2$, then

$$d(x_1, y_1) \leq d(x_1, x_2) + d(x_2, y_2) + d(y_2, y_1) = 0 + d(x_2, y_2) + 0 = d(x_2, y_2),$$

and, by symmetry, the reverse inequality is also true, so that

(2) $d(x_1, y_1) = d(x_2, y_2)$.

This implies that writing

(3) $e(u, v) = d(x, y)$,

whenever u and v are the equivalence classes of x and y respectively, un-ambiguously defines a metric e on the set U of equivalence classes (Exercise 37.14).

Let ϕ be the projection from X onto U: for each x in X, the value of $\phi(x)$ is the equivalence class of x. If u is the equivalence class of x, and ϵ any positive real number, then the inverse image of the open ball

$$\{v \in U : e(u, v) < \epsilon\}$$

in U is the open set (1) in X, by (2) and (3). The open balls form a base for the metric topology on U, so the function ϕ is continuous, by Lemma 33.1.

The inverse image (under ϕ) of an arbitrary subset V of U is the union of a set of equivalence classes, namely the set of equivalence classes that belong to V; conversely, every subset Y of X that is a union of equivalence classes is the inverse image of a subset of U, namely the subset consisting of those equivalence classes that are included in Y. The class of subsets of X that are unions of equivalence classes is a σ-field. (In fact, it is a complete field.) It is not difficult to check that each set P_n belongs to this σ-field. Indeed, if $x \equiv y$, that is, if $d(x, y) = 0$, then $p_n(x) = p_n(y)$ for all n, so that x and y belong to the same sets P_n. Thus, an equivalence class is either entirely included in P_n or disjoint from it, so that P_n is a union of equivalence classes.

The set F belongs to the σ-field generated by the family $\{P_n\}$, so it must belong to the σ-field of unions of equivalence classes, by the observations of the preceding paragraph and the definition of the generated σ-field. It follows that $F = \phi^{-1}(V)$ for some subset V of U. The set F is closed, and hence compact. Since

$$\phi(F) = \phi(\phi^{-1}(V)) = V,$$

and since the continuous image of a compact set is compact (Lemma 33.3), we infer that V is compact and therefore closed (Lemma 29.1). Conclusion: V is a closed subset of the metric space U. It follows from the remarks preceding the lemma that V is a G_δ. The inverse images, under ϕ, of a countable class of open sets whose intersection is V form a countable class of open sets whose intersection is F (Exercise 33.12); consequently, F is also a G_δ.

Corollary 1. *Every open Baire set in a Boolean space is the union of a countable class of clopen sets.*

Proof. Let G be an open Baire set. Its complement is a closed Baire set and is therefore the intersection of a countable class of open sets, by Lemma 1. Form complements to conclude that G is the union of a countable class of closed sets, say $G = \bigcup_n F_n$. Since the clopen sets form a base, each F_n is covered by the clopen sets that are included in G, and hence, by compactness, each F_n is covered by a finite number of such clopen sets. The class that consists, for each n, of a finite collection of clopen subsets of G that cover F_n is a countable collection of clopen sets, and its union is G.

We shall say that a Boolean space is a *σ-space* in case the closure of every open Baire set is open. The role of Boolean σ-spaces in the theory of σ-algebras is the same as the role of complete spaces in the theory of complete algebras.

Theorem 41. *The dual algebra A of a Boolean space X is a σ-algebra if and only if X is a σ-space.*

Proof. (Compare Theorem 39, p. 369.) Assume first that A is a σ-algebra. If U is an open Baire set in X, then (by Corollary 1) U is the union of a countable class of clopen sets; since this class has a supremum in A, by assumption, it follows (Lemma 38.1) that U^- is open. Thus, X is a σ-space.

Assume now that X is a σ-space. If $\{P_n\}$ is any countable class of elements (clopen sets) in A, then $\bigcup_n P_n$ is an open Baire set in X; since the closure of that set is open, by assumption, it follows (Lemma 38.1) that $\{P_n\}$ has a supremum in A. Consequently, A is a σ-algebra.

It is important to distinguish between the dual algebra A of clopen sets and the σ-field B of Baire sets in a σ-space X. Both are σ-algebras and both contain all clopen sets (in X). In particular, A is a subfield of B, but it is not in general a σ-subfield. In B, the supremum of a sequence $\{P_n\}$ of clopen sets is the union of the sequence; in A it is the closure of that union (Lemma 38.1).

Exercises

1. Let I be an uncountable discrete space; give examples of open sets that are not Baire sets in the one-point compactification of I and in the Cantor space 2^I.

2. Let $\{P_n\}$ be any sequence of subsets of a set X, and let p_n be the characteristic function of P_n. Prove that the function defined by

$$d(x, y) = \sum_{n=1}^{\infty} \frac{1}{2^n} |p_n(x) - p_n(y)|$$

   for all x and y in X is a pseudo-metric; see Exercise 37.14. (The pseudo-metric just defined is very similar to the metric defined in Exercise 32.11. To see this, observe that the value of $|p_n(x) - p_n(y)|$

is exactly the same as the value of the sum of $p_n(x) + p_n(y)$ in the two-element Boolean ring.)

3. Prove that a subset of a Boolean space is a Baire set if and only if it belongs to the σ-field generated by the class of all open F_σ-sets. (This last condition is used to define the notion of a Baire set in topological spaces that are not Boolean spaces.)

4. Prove that in a Boolean space with a countable base, every Borel set is a Baire set.

5. Prove that in a metric space with a countable base, every Borel set is a Baire set in the sense of Exercise 3.

6. Is it true that in every topological space with a countable base, every Borel set is a Baire set (in the sense of Exercise 3)?

7. Is Lemma 1 true in arbitrary compact Hausdorff spaces? (See Exercise 3 for the appropriate definition of a Baire set.)

8. (Harder.) Prove that the closure of a Baire set in a Boolean space need not be a Baire set. What if the space is a σ-space?

Chapter 40

The Representation of σ-algebras

We know that every Boolean algebra is isomorphic to a field, whereas a complete Boolean algebra need not be isomorphic to a complete field (since, for instance, it need not be atomic). It is natural to ask the intermediate question: is every σ-algebra isomorphic to a σ-field? The answer (due to Tarski [75]) is no. We shall see, in fact, that if A is an atomless σ-algebra satisfying the countable chain condition, then A cannot be isomorphic to a σ-field. For an example of such an algebra consider the regular open algebra of any Hausdorff space (such as $\mathbb{R}^n$) with no isolated points and with a countable base (see Exercise 29.31 and the remarks preceding Lemma 30.1). Alternatively, consider either the reduced Borel algebra or the reduced measure algebra of the unit interval (see the remarks preceding Lemma 31.4).

To prove the negative result promised above, suppose that A is an atomless σ-algebra satisfying the countable chain condition. We shall make use of the fact (Corollary 30.1) that A is complete. Assume, for contradiction, that A is isomorphic to a σ-field; we may as well assume that A is a σ-field of subsets of a set X. Select a point x of X and consider the class E of all those sets in A that contain x. Since A is complete, E has an infimum in A, say P; since A satisfies the countable chain condition, E has a countable subclass $\{P_n\}$ such that $P = \bigwedge_n P_n$, by the dual of Lemma 30.1. The fact that A is a σ-field implies that $P = \bigcap_n P_n$; since each P_n contains x, it follows that $P \neq 0$. As a non-zero element of the atomless algebra A, the set P has a non-empty proper subset Q in A. Either Q or $P - Q$ contains x; we may assume that Q does. This means that Q is in E, and implies therefore that P

S. Givant, P. Halmos, *Introduction to Boolean Algebras*,
Undergraduate Texts in Mathematics, DOI: 10.1007/978-0-387-68436-9_40,
© Springer Science+Business Media, LLC 2009

is included in Q (recall that $P = \bigwedge E$). This in turn implies that $P = Q$, and, since Q was supposed to be a proper subset of P, the contradiction has arrived.

If a class of Boolean algebras is not large enough to represent every algebra of a certain kind, the next best thing to hope is that the homomorphic images of the algebras of the class will suffice for the purpose. We have just seen that the class of σ-fields is not large enough to represent every σ-algebra; next we shall see that the class of homomorphic images of σ-fields (and, in fact, σ-homomorphic images) is quite large enough. The theorem we shall prove resembles Theorem 29 (p. 277) in many details, in both statement and proof. It is almost certain that the two results are special cases of a common generalization; it is far from certain whether the formulation and proof of such a generalization would yield any new information or save any time.

The class of Baire sets of a σ-space is a σ-algebra, by definition. The subclass of meager Baire sets is naturally a σ-ideal in that algebra. It turns out that Baire sets are almost clopen sets in the sense that each Baire set differs symmetrically from a uniquely determined clopen set by a meager Baire set. It is helpful to formulate this result another way, using the notion of congruence modulo an ideal that was discussed in Chapter 18, so as to underscore the connection with Lemma 29.2.

Lemma 1. *Every Baire set in a Boolean σ-space is congruent to a unique clopen set modulo the σ-ideal of meager Baire sets.*

Proof. Let X be a Boolean σ-space and M the σ-ideal of meager Baire sets. Write $S \equiv U$ to mean that the sets S and U are congruent modulo M, that is, the symmetric difference $S + U$ is an element of M. Consider the class of all those subsets of X that are congruent (modulo M) to a clopen set. The first step is to show that this class is a σ-field that includes the clopen sets. Every clopen set U is in the class because $U + U = \varnothing$, and the empty set is a meager Baire set. Suppose $\{S_n\}$ is a sequence of sets in the class, say $\{U_n\}$ is a sequence of clopen sets such that $S_n \equiv U_n$ for all n. Each sum $S_n + U_n$ is then in M, and M is a σ-ideal, so the union of the sums is in M. Write

$$S = \bigcup_n S_n \qquad \text{and} \qquad U = \bigcup_n U_n.$$

The sum $S + U$ is included in the union of the sums $S_n + U_n$, by Exercise 8.26, so it, too, belongs to M, by (18.18). It follows that $S \equiv U$. As a union of a countable sequence of clopen sets, U is an open Baire set. Since the space X

is assumed to be Boolean, the closure of each open Baire set is open, and therefore clopen; in particular, U^- is clopen. The difference

$$U^- + U = U^- - U$$

is a nowhere dense set (Lemma 10.5), so it is meager. It follows that $U \equiv U^-$, and therefore that $S \equiv U^-$. Conclusion: S is congruent to a clopen set, so it is in the class.

The proof that the class is closed under complements is easy. If S is in the class, say $S \equiv U$ and U is clopen, then $S' \equiv U'$ (because $S' + U' = S + U$, by Exercise 6.2(g)) and U' is clopen. Consequently, S' is in the class.

It has been shown that the class of subsets of X congruent to clopen sets is a σ-field that contains all clopen sets. The class must therefore contain all Baire sets, since the clopen sets generate the σ-field of Baire sets. To establish the uniqueness assertion of the lemma, suppose a Baire set is congruent to clopen sets U and V. The sets U and V are then congruent to each other, so their sum

$$U + V = (U \cap V') \cup (U' \cap V)$$

is meager. This sum is also clopen (because U and V are clopen), so the Baire category theorem (Theorem 28, p. 273) implies that the sum is empty, and consequently that $U = V$. The proof of the lemma is complete.

In view of the preceding lemma, the function f that takes each Baire set S to the unique clopen set $f(S)$ such that $S \equiv f(S)$ is a well-defined mapping from the σ-field B of Baire sets into dual algebra A of clopen sets of the σ-space X. The function maps B onto A because every clopen set is mapped to itself. It has been shown (see the proof of the lemma) that if $S \equiv U$ and U is clopen, then $S' \equiv U'$; in other words,

$$f(S') = f(S)'.$$

It has also been shown that if $S_n \equiv U_n$ for every n, if

$$S = \bigcup_n S_n \quad \text{and} \quad U = \bigcup_n U_n,$$

and if each set U_n is clopen, then

$$S \equiv U^-.$$

The set U^- is the join of the family $\{U_n\}$ in A, by Lemma 38.1 and Theorem 41 (p. 376). Therefore,

$$f(S) = U^- = \bigvee_n U_n = \bigvee_n f(S_n).$$

These assertions together imply that f is a σ-homomorphism from B onto A. Its kernel consists of the Baire sets that are congruent to the empty set, and this is just the ideal M of meager Baire sets. The quotient B/M is therefore isomorphic to A, by the first isomorphism theorem (Theorem 14, p. 179). We summarize what has been accomplished in the following theorem.

Theorem 42. *Suppose that B is the σ-field of Baire sets and M is the σ-ideal of meager Baire sets in a Boolean σ-space X. The correspondence f that takes each Baire set S to the clopen set $f(S)$ such that*

$$S \equiv f(S) \mod M$$

is a σ-homomorphism from B onto the dual algebra A of X, with kernel M, so that A is isomorphic to B/M.

The following corollary is known as the *Loomis–Sikorski theorem*, after its discoverers, Loomis [40] and Sikorski [59].

Corollary 1. *Every σ-algebra is isomorphic to some σ-field modulo a σ-ideal.*

Proof. Every σ-algebra is isomorphic to the dual algebra of some σ-space, by Theorem 41 (p. 376) and the Stone representation theorem (p. 328). The dual of a σ-space is, in turn, isomorphic to the quotient of a σ-field modulo a σ-ideal, by the preceding theorem.

For σ-algebras, just as for plain Boolean algebras, the representation and duality theory yields an elegant proof of the existence and representation of free algebras. (The existence of free σ-algebras was proved by Rieger [51].)

Theorem 43. *For every set I, there exists a free σ-algebra generated by I, and, in fact, that algebra is isomorphic to the σ-field of all Baire sets in the Cantor space 2^I.*

Proof. Write $Y = 2^I$, let B be the dual algebra of clopen sets in Y, and let B^* be the σ-field of Baire sets in Y. We have seen (Theorem 36, p. 354) that B is the free Boolean algebra generated by the set E of clopen sets

$$P_i = \{y \in Y : y_i = 1\},$$

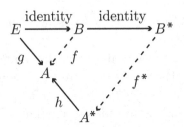

for i in I. We are to prove that B^* is the free σ-algebra generated by E. For this purpose we need to prove that every mapping g from E to an arbitrary σ-algebra A can be extended to a σ-homomorphism from B^* to A. We may and do assume (Theorem 41) that A is the dual algebra of a σ-space X.

Let A^* be the σ-field of Baire sets in X. Notice that A is a subalgebra (but not in general a σ-subalgebra) of A^*. By Theorem 42, there is a σ-epimorphism, say h, from A^* to A, and it maps every element of A (every clopen set) to itself. Since B is free on E, there exists a Boolean homomorphism f from B to A that agrees with g on E. The homomorphism f is the dual of a continuous mapping ϕ from X into Y; this implies that

$$f(Q) = \phi^{-1}(Q)$$

for every Q in B (see Theorem 34, p. 350, and (36.2)). Let f^* be the mapping from B^* into A^* defined by

$$f^*(S) = \phi^{-1}(S)$$

for every S in B^*. (See the diagram.) Observe that f^* agrees with f on B, and consequently agrees with g on E. It is easy to check (with the help of Exercise 33.5) that f^* is a σ-homomorphism. The promised extension is the composition $h \circ f^*$. Indeed, this composition is a σ-homomorphism, by Exercise 29.14, and

$$h(f^*(P_i)) = h(g(P_i)) = g(P_i).$$

The first equality holds because f^* agrees with g on E, while the second holds because $g(P_i)$ is clopen (it belongs to A), and h maps clopen sets to themselves.

Exercises

1. Prove that the class of meager Baire sets is a σ-ideal in the σ-algebra of all Baire sets of a Boolean space.

2. Prove that every σ-algebra is the homomorphic image of a free σ-algebra.

3. Let X and Y be Boolean spaces, and A^* and B^* the σ-fields of Baire sets in X and Y respectively. If ϕ is a continuous mapping of X into Y, prove that the function f^* from B^* to A^* defined by

$$f^*(S) = \phi^{-1}(S)$$

for every S in B^* is a σ-homomorphism.

4. Derive the Loomis–Sikorski theorem from Theorem 43.

5. (Harder.) Is the natural generalization of the homomorphism extension criterion (Theorem 4, p. 107) to a σ-homomorphism extension criterion (for σ-algebras) true? (The answer is due to Sikorski [59].)

6. (Harder.) Is the generalization of Lemma 28.1 to free σ-algebras true?

7. (Harder.) Find an example of an m-algebra (for some infinite cardinal m) that is not isomorphic to any m-field modulo an m-ideal. (The existence of such algebras was first observed by Tarski [75].)

Chapter 41

Boolean Measure Spaces

A *Boolean measure space* is a Boolean σ-space X, together with a normalized measure on the σ-field of Borel sets in X, such that non-empty open sets have positive measure and nowhere dense Borel sets have measure zero. The last condition is a very strange one. At first glance it might seem that since a nowhere dense set is topologically small and a set of measure zero is measure-theoretically small, it is fitting and proper that the one should imply the other. A little measure-theoretic experience (with Lebesgue measure in Euclidean spaces, for instance) shows, however, that the implication is not at all likely to hold. The results of this chapter will show that Boolean measure spaces, in which the implication is assumed to hold, have rather pathological and almost paradoxical properties. The reason for considering them anyway is that measure algebras are important, and, as it turns out, Boolean measure spaces are exactly the duals of measure algebras.

We proceed to establish the notation that will be used in this chapter. Let X be a Boolean σ-space with dual algebra A (which is, therefore, a σ-algebra; see Theorem 41, p. 376). Let B be the σ-field of Borel sets in X, let M be the σ-ideal of meager Borel sets, and let f be the natural σ-epimorphism (Theorem 29, p. 277) from B onto the regular open algebra of X with kernel M.

Lemma 1. *If ν is a normalized measure on B such that non-empty open sets have positive measure and nowhere dense Borel sets have measure zero, and if μ is the restriction of ν to A, then μ is a positive, normalized measure on A (so that A together with μ is a measure algebra).*

Proof. The only thing that needs proof is that μ is countably additive on A. (The countable additivity of μ does not follow automatically from that of ν

S. Givant, P. Halmos, *Introduction to Boolean Algebras*,
Undergraduate Texts in Mathematics, DOI: 10.1007/978-0-387-68436-9_41,
© Springer Science+Business Media, LLC 2009

because the supremum of an infinite sequence of elements in A is not, in general, the same as the supremum (union) of the sequence in B.) Suppose that $\{P_n\}$ is a disjoint sequence of elements of A (clopen sets in X); write

$$U = \bigcup_{n=1}^{\infty} P_n \quad \text{and} \quad P = \bigvee_{n=1}^{\infty} P_n.$$

Then $P = U^-$, by Lemma 38.1. The difference $U^- - U$ is nowhere dense, by Lemma 10.5, and therefore

$$\nu(U^- - U) = 0,$$

by assumption. Consequently,

$$\mu(P) = \nu(P) = \nu(U^-) = \nu(U) + \nu(U^- - U)$$

$$= \nu(U) = \sum_{n=1}^{\infty} \nu(P_n) = \sum_{n=1}^{\infty} \mu(P_n).$$

The first and last equalities follow from the definition of μ, while the third and fifth equalities use the additivity and the countable additivity of ν respectively. It follows that μ is countably additive.

Lemma 2. *If μ is a positive, normalized measure on A, then f maps B onto A. If*

$$\nu(S) = \mu(f(S))$$

for every S in B, then ν is a normalized measure on B such that non-empty open sets have positive measure and such that the sets of measure zero are exactly the meager sets. The restriction of ν to A coincides with μ.

Proof. The algebra A together with the measure μ is a measure algebra, by assumption, and therefore A is a complete Boolean algebra, by Corollary 31.1. It follows that the space X is a complete Boolean space, by Theorem 39 (p. 369), and hence that A coincides with the algebra of regular open sets in X, by Exercise 38.4. The function f maps B onto the regular open algebra of X, by definition, so it maps B onto A. This proves the first sentence of the lemma. The second sentence is an immediate consequence of Lemma 31.2. The third sentence follows from the fact that f maps all regular open sets (and hence all clopen sets) to themselves.

Corollary 1. *The dual algebra A of a Boolean space X is a measure algebra if and only if X is a Boolean measure space.*

Proof. Assume first that X is a Boolean measure space, say ν is a normalized measure on the σ-field of Borel sets in X such that non-empty open sets have positive measure and nowhere dense Borel sets have measure zero. The dual algebra A, together with the restriction of ν to A, is a measure algebra, by Lemma 1.

Now suppose that A is a measure algebra, say μ is a positive normalized measure on A. Since A is, by definition, a σ-algebra, the space X must be a σ-space (Theorem 41, p. 376). The function ν defined on the σ-field of Borel sets by

$$\nu(S) = \mu(f(S))$$

for every Borel set S is a normalized measure such that non-empty open sets have positive measure and such that the sets of measure zero are exactly the meager sets, by Lemma 2. Thus, X together with the measure ν is a Boolean measure space.

In the rest of the chapter we shall assume that X and A have not only the topological and algebraic properties originally required, but also the measure-theoretic structure (the measures ν and μ) described in Lemmas 1 and 2. In particular, the algebra A is complete and satisfies the countable chain condition, and the space X is complete. The additional structure has profound and surprising effects on the topology of X. Thus, for instance, every open set is included in a clopen set of the same measure (namely its own closure). In other words, every open set is almost clopen; next we shall see that something like this is true for arbitrary Borel sets also.

Consider, indeed, those Borel sets whose measure can be approximated arbitrarily closely by clopen sets, from both inside and outside. More precisely, we shall say (temporarily) that a Borel set S is *regular* in case

$$\sup \mu(P) = \nu(S),$$

where the supremum is extended over all clopen sets P included in S, and

$$\inf \mu(Q) = \nu(S),$$

where the infimum is extended over all clopen sets Q including S.

Lemma 3. *Every Borel set is regular.*

Proof. We have already noted that an open set U has the same measure as its closure U^- (the boundary $U^- - U$ is nowhere dense and therefore has measure zero), and U^- is clopen (because the space X is complete). Consequently,

$$\inf \mu(Q) = \mu(U^-) = \nu(U^-) = \nu(U)$$

(where Q ranges over the clopen sets that include U). The first equality holds because U^- is the smallest clopen set that includes U: if Q is clopen, then

$$U \subseteq Q \qquad \text{implies} \qquad U^- \subseteq Q.$$

The second equality follows from the final assertion of Lemma 2.

To approximate U from below, consider the class of all clopen sets included in U. Since A is complete, the class has a supremum in A and that supremum is U^-, by Lemma 38.1. Since A satisfies the countable chain condition, there is a countable subclass with the same supremum, by Lemma 30.1; consequently, there is an increasing sequence $\{P_n\}$ of clopen subsets of U with the same supremum U^-, by Exercise 8.21. It follows from Exercise 31.5 that

$$\lim_{n \to \infty} \mu(P_n) = \mu(U^-),$$

and therefore

$$\sup \mu(P) = \mu(U^-) = \nu(U^-) = \nu(U).$$

The preceding argument shows that every open set is regular. The self-dual character of the definition of regularity implies that the complement of a regular Borel set is regular, so that, in particular, every closed set is regular.

Next, we show that a meager Borel set S is included in a clopen set of small measure, say less than a given real number $\epsilon > 0$. The set S is, by definition, the union of a sequence $\{S_n\}$ of nowhere dense sets (indexed by the set of positive integers). The closures S_n^- are also nowhere dense (by the definition of a nowhere dense set) and they are Borel (they are closed), so they have measure zero, by the assumptions on ν. For each n, let Q_n be a clopen set including S_n^- with the property that

$$\mu(Q_n) < \epsilon/2^n;$$

such a set exists by the observations of the previous paragraph. Take U to be the union of these clopen sets. Since

$$S = \bigcup_n S_n \subseteq \bigcup_n S_n^- \subseteq \bigcup_n Q_n = U,$$

t follows from the monotony and countable subadditivity of ν (see Chapter 31), and from the final assertion of Lemma 2, that

$$\nu(S) \leq \nu(U) \leq \sum_n \nu(Q_n) = \sum_n \mu(Q_n) < \sum_n \epsilon/2^n = \epsilon.$$

The closure U^- is a clopen set with the same measure as U, since U is open (see the remarks at the beginning of the proof). Consequently, U^- is a clopen set including S that has measure less than ϵ. The argument also shows that each meager Borel set S has measure zero, since its measure is less than ϵ for each $\epsilon > 0$.

Every Borel set S is congruent modulo a meager set to some clopen set P, by Lemma 29.2 and Exercise 38.4. The sets S and P have the same measure:

$$\nu(S) = \nu(S - P) + \nu(S \cap P) = 0 + \nu(S \cap P) = \nu(S \cap P)$$

and

$$\nu(P) = \nu(P - S) + \nu(P \cap S) = 0 + \nu(P \cap S) = \nu(P \cap S).$$

Suppose Q is a clopen set of small measure that includes the meager set $S + P$. The union $P \cup Q$ is a clopen set that approximates S from above:

$$S \subseteq P \cup (S + P) \subseteq P \cup Q,$$

so that

$$\mu(P \cup Q) = \nu(P \cup Q) \geq \nu(S),$$

and

$$\mu(P \cup Q) - \nu(S) = \nu(P \cup Q) - \nu(S)$$
$$\leq \nu(P) + \nu(Q) - \nu(S) = \nu(Q) = \mu(Q).$$

The same argument applied to S' yields a clopen set that approximates S' from above; the complement of that clopen set is a clopen set approximating S from below. The proof of the lemma is complete.

Lemma 3 says something very strong about the measure ν; the property it ascribes to ν is considerably stronger than the standard measure-theoretic properties of regularity and completion regularity.

Lemma 4. *Every Borel set has the same measure as its closure.*

Proof. If S is a Borel set, then, by Lemma 3, there exist clopen sets Q_n including S such that $\nu(Q_n - S) < 1/n$ for $n = 1, 2, 3, \ldots$. The intersection of these clopen sets is a closed set that includes S and has the same measure as S.

Lemma 5. *A Borel set of measure zero is nowhere dense.*

Proof. If S is a Borel set with the property that $\nu(S) = 0$, then $\nu(S^-) = 0$, by Lemma 4. This implies that S^- includes no non-empty open set, since such sets are assumed to have positive measure (and the measure ν is monotone). Thus, S is nowhere dense.

Lemma 6. *Every meager set is nowhere dense.*

Proof. Suppose $S = \bigcup_n S_n$, where each S_n is nowhere dense. The closed set S_n^- is nowhere dense (by the definition of a nowhere dense set), so it has measure zero (by the assumptions about the measure space). The union $\bigcup_n S_n^-$ is a Borel set of measure zero (by countable subadditivity), and therefore nowhere dense, by Lemma 5. Since S is included in this union, it, too, is nowhere dense.

The reason for forming S_n^- in the preceding proof is that S_n is not known to be measurable (and neither is S).

Exercises

1. Fill in the details of the proof that in a Boolean measure space, every open set is included in a clopen set with the same measure (namely its closure).

2. Justify the claim, made at the end of the proof of Lemma 3, that if the complement of a Borel set S is approximated by a clopen set Q from above, then S is approximated by the clopen set Q' from below.

3. Use the regularity of open sets and algebraic duality to prove that a closed set in a Boolean measure space is regular. Then give a (longer) direct proof of this same fact that does not make use of the regularity of open sets.

4. Prove that in a Boolean measure space the boundary of every Borel set has measure zero.

5. Prove that the dual space of the reduced measure algebra of $[0, 1]$ is not separable.

6. Use Exercise 5 to show that the reduced measure algebra and the reduced Borel algebra of the interval $[0, 1]$ are not isomorphic by showing that the dual space of the latter is separable.

Chapter 42

Incomplete Algebras

The quotient of a Boolean algebra modulo an ideal may turn out to have a higher degree of completeness than one has a right to expect. Thus, for instance, the reduced Borel algebra and the reduced measure algebra of the unit interval are not only σ-algebras, which is all that the general theory can predict, but even complete. A few observations of this kind are likely to tip the balance of expectations too far over to the optimistic side. The purpose of this chapter is to provide a counterbalance in the form of some counterexamples. In other words, we shall obtain a few negative results: we shall see that certain quotient algebras are not complete.

The natural questions in this direction are obtained from the ones already answered by changing either the algebra or the ideal. The Borel sets modulo meager Borel sets in $[0, 1]$ constitute a complete Boolean algebra; what about the Borel sets modulo countable sets, and what about all sets modulo meager sets? In deriving some of the answers we shall make use of the continuum hypothesis. This is sometimes avoidable; since, however, it simplifies and shortens the argument in any case, and especially since the purpose of the discussion is not to build the theory but merely to give warning of some danger spots, the effort of avoidance is not worth the trouble.

A typical result is that if X is an uncountable set, then the field B of all subsets of X modulo the ideal M of countable sets in X is not a complete Boolean algebra. To illustrate the argument, consider the special case in which X is the Cartesian plane. The proof exhibits a concrete subset E of B/M that has no supremum. Let f be the projection from B to B/M; let E be the set of all those elements of B/M that have the form $f(S)$ for some vertical line S. The best way to prove that E has no supremum is to

S. Givant, P. Halmos, *Introduction to Boolean Algebras*,
Undergraduate Texts in Mathematics, DOI: 10.1007/978-0-387-68436-9_42,
© Springer Science+Business Media, LLC 2009

show that to every upper bound of E there corresponds a strictly smaller upper bound. Suppose, accordingly, that $f(S) \leq p$ for all vertical lines S. Since f maps B onto B/M, there exists a subset P of X such that $p = f(P)$. To say $f(S) \leq f(P)$, or, equivalently, $f(S - P) = 0$, means that $S - P$ belongs to M (Exercise 1), and therefore P contains all but countably many of the points of S. Since each S under consideration is uncountable, it follows that P contains at least one point in each S. Let Q be a subset of P that contains exactly one point in each S; then $P - Q$ contains all but countably many of the points in each S (it contains every point of S that is in P, with one exception), so $S - (P - Q)$ is in the ideal M, and therefore

$$f(S) \leq f(P - Q),$$

for each S. In other words, $f(P - Q)$ is an upper bound of E. Since, however, Q is uncountable (there are uncountably many vertical lines, one for each real number, and these lines are disjoint), it follows that $f(Q) \neq 0$ and hence (Exercise 7.5) that

$$f(P) - f(Q) \neq f(P).$$

Thus, $f(P - Q)$ is an upper bound of E that is strictly smaller than p.

The proof proves more than the statement states. Clearly the plane has nothing to do with the matter; any set in a one-to-one correspondence with the plane would do as well. The exact cardinality of X is also immaterial; all that matters is that X be uncountable. Indeed, since $m^2 = m$ for every infinite cardinal number m, there is always a one-to-one correspondence between X and X^2 (provided only that X is infinite), and the proof works again. Still another glance at the proof shows that the countability of the sets in the ideal M did not play a very great role; what mattered was that singletons belong to M and sets in one-to-one correspondence with X do not. On the basis of this last observation even the assumption that X is uncountable can be dropped; here is what remains.

Lemma 1. *If B is the field of all subsets of an infinite set X, and if M is an ideal in B containing all singletons and not containing any set in one-to-one correspondence with X, then the algebra B/M is not complete.*

The lemma includes the statement we started with as a special case; as another special case it contains the statement that the algebra of all subsets of an infinite set modulo the ideal of finite sets is never complete, not even if the basic set is merely countable.

Lemma 1 is a relatively crude result, but its proof contains, in skeletal form, the two constructions that yield the more delicate results obtainable along these lines. The first step is to construct the set of vertical lines; in abstract terms, the problem is to construct a large disjoint class of sets none of which belongs to the prescribed ideal. The second step is to cut across the vertical lines; here the problem is to construct a large set whose intersection with each of the sets constructed before does belong to the ideal. The first of these constructions is the harder one; it is based on the following result of Ulam [80].

Lemma 2. *If X is the set of all ordinal numbers less than the first uncountable ordinal number ω_1, then, corresponding to each natural number n and to each ordinal number α less than ω_1, there exists a subset $S(n, \alpha)$ in X such that the sets in each row of the array*

$$
\begin{array}{llllllll}
S(0,0), & S(0,1), & S(0,2), & \dots, & S(0,\omega), & \dots, & S(0,\alpha), & \dots \\
S(1,0), & S(1,1), & S(1,2), & \dots, & S(1,\omega), & \dots, & S(1,\alpha), & \dots \\
S(2,0), & S(2,1), & S(2,2), & \dots, & S(2,\omega), & \dots, & S(2,\alpha), & \dots \\
\vdots & \vdots & \vdots & & \vdots & & \vdots & \\
S(n,0), & S(n,1), & S(n,2), & \dots, & S(n,\omega), & \dots, & S(n,\alpha), & \dots \\
\vdots & \vdots & \vdots & & \vdots & & \vdots &
\end{array}
$$

are pairwise disjoint, and the union of the sets in each column is cocountable.

Proof. For each β in X select a sequence $\{k(\alpha, \beta)\}$ of type β (that is, indexed by the ordinals $\alpha < \beta$) whose terms are distinct natural numbers; this is possible because β is a countable ordinal. These sequences can be laid out in a triangular array, as follows.

$$
\begin{array}{c|llllll}
 & 0 & 1 & 2 & \dots & \alpha & \dots \\
\hline
0 & & & & & & \\
1 & k(0,1), & & & & & \\
2 & k(0,2), & k(1,2), & & & & \\
3 & k(0,3), & k(1,3), & k(2,3), & & & \\
\vdots & \vdots & \vdots & \vdots & & & \\
\beta & k(0,\beta), & k(1,\beta), & k(2,\beta), & \dots, & k(a,\beta), & \dots \\
\vdots & \vdots & \vdots & \vdots & & \vdots &
\end{array}
$$

Let $S(n, \alpha)$ be the set of all those β for which $k(\alpha, \beta) = n$. For example, to get $S(5, 2)$, consider the column labeled 2, and collect all those elements β for which the β entry in that column has the value 5. If β is in $S(n, \alpha)$, then $k(\alpha, \beta) = n$; since $k(\alpha_1, \beta) \neq k(\alpha_2, \beta)$ unless $\alpha_1 = \alpha_2$, it follows that for each n, the sets $S(n, \alpha)$ are indeed pairwise disjoint. If $\alpha < \beta$, then

$$\beta \in S(k(\alpha, \beta), \alpha) \subseteq \bigcup_n S(n, \alpha),$$

so that the union $\bigcup_n S(n, \alpha)$ contains every β greater than α; it follows that each such union is indeed cocountable.

To apply Lemma 2 we assume the continuum hypothesis.

Corollary 1. *The unit interval is the union of a disjoint class of power $\aleph_1$ consisting of sets none of which is meager.*

Proof. The continuum hypothesis implies that there is a bijective correspondence between $[0, 1]$ and the set of all ordinal numbers less than ω_1. Thus, we may and do assume that the sets $S(n, \alpha)$ described in Lemma 2 are subsets of $[0, 1]$. Each column of the square array of S's consists of countably many sets whose union is cocountable in $[0, 1]$; that is, the union contains all but countably many of the numbers in $[0, 1]$, and is therefore not meager. Since a countable union of meager sets is meager, it follows that at least one of the sets in each column must not be meager. There are uncountably many columns but only countably many rows, so some row must contain uncountably many non-meager sets, and those sets, by Lemma 2, are pairwise disjoint. In case the union of the non-meager sets so obtained is not the entire interval, adjoin the complement of that union to one of the sets.

Corollary 2. *The unit interval is the union of a disjoint class of power $\aleph_1$ consisting of sets none of which has measure zero.*

Proof. Same as for Corollary 1; just interpret the word "meager" to mean "having measure zero".

We are now ready to imitate the argument (vertical lines) that led to Lemma 1. This time let X be the unit interval, let B be the field of all subsets of X, and let M be the ideal of meager sets. By Corollary 1 there exists a disjoint family $\{S_i\}$ of power $\aleph_1$ consisting of sets not in M. Let f be the projection from B to B/M; let E be the set of all those elements of B/M that have the form $f(S_i)$ for some i. We shall show that the set E has no supremum by showing that to every upper bound of E there corresponds a strictly smaller upper bound.

The preceding paragraph has the analogues of the vertical lines; the next problem is to cut across them. The technique here is based on the fact that every meager set is included in some meager F_σ. (Proof: a meager set S is the union of a sequence $\{S_n\}$ of nowhere dense sets; the closures S_n^- are nowhere dense, and their union is a meager F_σ that includes S.) Since an easy argument shows that the cardinal number of the class of F_σ-sets is the power of the continuum (see Exercise 3), and since we have already assumed the continuum hypothesis, we may assume that all meager F_σ-sets occur as the terms of a family $\{R_i\}$ with the same set of indices as the family $\{S_i\}$.

Suppose now that p is an upper bound of E. Since f maps B onto B/M, there is a subset P of X such that $f(P) = p$. To say $f(S_i) \le f(P)$, or, equivalently, $f(S_i - P) = 0$, means that $S_i - P$ belongs to M, and therefore P includes all but a meager subset of S_i. Since R_i is meager but S_i is not, it follows that P contains at least one point in each $S_i - R_i$. Let Q be a subset of P that contains exactly one point in each $S_i - R_i$; then $P - Q$ includes all but a meager subset of S_i (it contains every point of S_i that is in P, with one exception), so $S_i - (P - Q)$ is in the ideal M, and therefore

$$f(S_i) \le f(P - Q),$$

for each i. In other words, $f(P-Q)$ is an upper bound of E. Since, however, Q is not included in any R_i (because it contains a point in $S_i - R_i$), and since every meager set is included in at least one of the R_i, the set Q cannot be meager. Consequently, $f(Q) \ne 0$ and hence

$$f(P) - f(Q) \ne f(P).$$

Thus, $f(P - Q)$ is an upper bound of E that is strictly smaller than p.

The proof is over; the time has come to see what it proves. The following statement (due to Sikorski [61]) is a suitably general formulation of what the technique can be made to yield.

Lemma 3. *Suppose that B is the field of all subsets of a set X, that M is an ideal containing all singletons, and that $\{R_i\}$ is a family of sets in M with the property that every set in M is included in some R_i. If there exists a disjoint family $\{S_i\}$, with the same set of indices, consisting of sets not in M, then the algebra B/M is not complete.*

A special case of the lemma, different from the one proved above, is that the algebra of all subsets of $[0, 1]$ modulo the ideal of sets of measure zero is not complete. To deduce this conclusion from the lemma, let $\{R_i\}$ be the family of G_δ-sets of measure zero.

Exercises

1. Let f be the projection from a Boolean algebra B to a quotient B/M. Prove that for any elements p and q in B, the following statements are equivalent: (a) $f(p) \leq f(q)$; (b) $f(p - q) = 0$; (c) $p - q \in M$.

2. Write out a detailed proof of Lemma 1.

3. Prove that the number of F_σ-sets in the unit interval is $2^{\aleph_0}$. Conclude that the number of meager F_σ-sets is also $2^{\aleph_0}$.

4. Prove that the number of G_δ-sets in the unit interval is $2^{\aleph_0}$. Conclude that the number of G_δ-sets of measure zero is also $2^{\aleph_0}$.

5. (Harder.) Prove Corollary 2 without assuming the continuum hypothesis. (This result is due to Lusin and Sierpiński [42].)

6. Prove Lemma 3.

7. Give a detailed proof that the algebra of all subsets of $[0, 1]$ modulo the ideal of sets of measure zero is not complete.

8. (Harder.) Let B be the field of all subsets of $[0, 1]$ and let M be the ideal of countable sets. Is there a normalized measure on the algebra B/M? (The answer to this question, under the assumption of the continuum hypothesis, dates back to Banach and Kuratowski [3].)

Chapter 43

Duality for Products

There is a dual correspondence between certain subalgebras of products of
Boolean algebras and the compactifications of unions of Boolean spaces. For
instance, the dual space of the product of two Boolean algebras is homeomorphic to the disjoint union of the dual spaces of the algebras. Since this
special case is useful and easier to describe than the general case, we treat it
first.

Consider two disjoint topological spaces Y and Z, and write

$$X = Y \cup Z.$$

Every subset S of X can be written in a unique way as the union of a subset
of Y with a subset of Z (see Chapter 26); in fact,

$$S = P \cup Q,$$

where

$$P = S \cap Y \qquad \text{and} \qquad Q = S \cap Z.$$

Call P and Q the *components* of S, and define S to be open in X if its
components are open in Y and Z respectively. It is easy to see that the
resulting class of open sets is a topology for X. For instance, if $\{S_i\}$ is a
family of open sets in X, say S_i has the open components P_i and Q_i, then

$$\bigcup_i S_i = \left(\bigcup_i P_i \right) \cup \left(\bigcup_i Q_i \right);$$

since a union of open sets in Y is again open, and similarly for Z, it follows
that $\bigcup_i S_i$ is open in X. The space X is called the *sum* (or the *union*) of the
spaces Y and Z. Motivated by the additive terminology, we write

S. Givant, P. Halmos, *Introduction to Boolean Algebras*,
Undergraduate Texts in Mathematics, DOI: 10.1007/978-0-387-68436-9_43,
© Springer Science+Business Media, LLC 2009

$$X = Y + Z.$$

To form the sum of topological spaces that are not disjoint, pass first to disjoint copies of the spaces and then form the sum of these copies.

The complement of a set $S = P \cup Q$ in the sum space X is formed componentwise:

$$S' = P' \cup Q'$$

(because the underlying component spaces Y and Z are disjoint). It follows that S is closed in X just in case P and Q are closed in Y and Z respectively. In particular, S is clopen just in case P and Q are clopen. It is not difficult to check that X is a Boolean space just in case Y and Z are Boolean spaces. For example, if X is compact, then the clopen subsets Y and Z are certainly compact, because they are closed subsets of a compact space. Conversely, if Y and Z are compact, then any open cover $\{S_i\}$ of X decomposes into component open covers $\{P_i\}$ of Y and $\{Q_i\}$ of Z, each of which has a finite subcover. If $\{P_i\}_{i \in J_1}$ is a finite subcover of Y, and $\{Q_i\}_{i \in J_2}$ is a finite subcover of Z, and if $J = J_1 \cup J_2$, then $\{S_i\}_{i \in J}$ is a finite subcover of X.

Theorem 44. *The dual algebra of the sum of disjoint Boolean spaces Y and Z is the internal product of the dual algebras of Y and Z.*

Proof. Assume X is the sum of two disjoint Boolean spaces Y and Z, and let A, B, and C be the respective dual algebras. It is to be shown that A is the internal product of B and C. Recall from Chapter 26 that this internal product consists of those elements S that can be written in the form

$$S = P \cup Q$$

for some P in B and some Q in C. The operations of the internal product are performed componentwise; for example, if

$$S_1 = P_1 \cup Q_1 \qquad \text{and} \qquad S_2 = P_2 \cup Q_2,$$

then

(1) $\qquad S_1 \cap S_2 = (P_1 \cap P_2) \cup (Q_1 \cap Q_2) \qquad$ and $\qquad S_1' = P_1' \cup Q_1',$

where the component operations of intersection and complement on the right sides of the equations are performed in B and in C.

The elements of A are the clopen subsets of X, and these are just the unions of clopen subsets of Y with clopen subsets of Z, by the remarks preceding the theorem. The universe of A therefore coincides with the universe of the internal product. The operations of A are the set-theoretic operations of union, intersection, and complement inherited from the field of all subsets of X. Since X is the disjoint union of Y and Z, these set-theoretic operations are performed componentwise, just as in (1) (see Chapter 26). It follows that A is the internal product of B and C, as desired.

The algebraic version of the preceding theorem (which dates back to the 1959 notes of Halmos [23]) says that the dual space of the product of two Boolean algebras B and C is homeomorphic to the (disjoint) sum of the dual spaces of B and C (see Exercises 7 and 8).

Everything that has been said about the union of two Boolean spaces carries over easily to the union of finitely many Boolean spaces (see Exercises 5 and 6). The case of infinitely many spaces is somewhat more intricate. Consider an infinite disjoint family $\{X_i\}$ of topological spaces, and let X be the union of the sets X_i. As in the finite case, each subset P of X can be written in a unique way as the union of the components $P_i = P \cap X_i$. The set P is declared to be open in X just in case each component P_i is open in X_i. A straightforward argument shows that the class of open sets so defined constitutes a topology on X (see Exercise 1). The space X is called the *(disjoint) union* of the family $\{X_i\}$.

Suppose each of the component spaces X_i is Boolean. It is an easy matter to check that the union space X is Hausdorff, and that each subset of X_i is open, closed, clopen, or compact in the topology of X just in case it is open, closed, clopen, or compact in the topology of X_i (see Exercises 2 and 3). Consequently, X has a base consisting of clopen sets, namely the class of clopen subsets of the various component spaces. Furthermore, X is *locally compact* in the sense that each of its points belongs to the interior of some compact subset of X; in fact, if x is any point in X, then x belongs to X_i for some i, and X_i is a compact open subset of X. The property of compactness, however, is not inherited by the union: if each of the component spaces is non-empty, then the family $\{X_i\}$ is itself an open cover of X that has no finite subcover (and in fact no proper subcover at all). To arrive at a Boolean space, it is necessary to pass to a compactification of X.

A *compactification* of a locally compact Hausdorff space X is a compact Hausdorff space Y such that X is a dense subspace of Y. In particular, the open subsets of X are precisely the intersections with X of the open subsets of Y, and every non-empty open subset of Y has a non-empty intersection with X. A *Boolean compactification* of X is a compactification of X that is simultaneously a Boolean space. (The terminology *zero-dimensional compactification* is also frequently employed.)

A simple example is the *one-point compactification* of a locally compact, but not compact, Hausdorff space X (due to Alexandroff — see [1]). This is a topological space Y that is obtained by adjoining a single new point to X and declaring a subset of Y to be open just in case it is either an open subset of X or else the complement in Y of a closed compact subset of X. Notice that X is itself an open subset of Y.

In general, a locally compact Hausdorff space is open in any of its compactifications; this statement is an easy corollary of the following lemma.

Lemma 1. *If X is a locally compact subspace of a Hausdorff space Y, then there is an open set U and a closed set F in Y such that $X = U \cap F$.*

Proof. Each point x in X belongs to some open set V_x that has a compact closure in X, by the assumption that X is locally compact (see Exercise 13). Write W_x for that compact closure in X, and write V_x^- for the closure of V_x in Y. It follows (by Exercise 9.9 applied to the set V_x) that

$$W_x = X \cap V_x^-.$$

The set W_x is also compact in Y (by Exercise 29.28), and is therefore closed in Y, by Lemma 29.1.

The set V_x is open in X, so there is an open subset U_x of Y such that

$$V_x = X \cap U_x,$$

by the definition of the inherited topology. The union

$$U = \bigcup_{x \in X} U_x$$

is an open set in Y that includes X.

It is easy to see that V_x is closed in the space U_x (under the topology inherited by U_x from Y): the set W_x is closed in Y, and

$$V_x = V_x \cap W_x = (X \cap U_x) \cap (X \cap V_x^-) = U_x \cap X \cap V_x^- = U_x \cap W_x.$$

Consequently, the set $U_x - V_x$ is open in U_x; since U_x is open in Y, it follows that $U_x - V_x$ must also be open in Y (Exercise 9.7). The union

$$\bigcup_{x \in X} (U_x - V_x)$$

is therefore an open subset of Y. The difference $U_x - V_x$ can be written in the form

$$U_x - V_x = U_x - (X \cap U_x) = U_x - X,$$

so that

$$\bigcup_{x \in X} (U_x - V_x) = \bigcup_{x \in X} (U_x - X) = U - X.$$

Conclusion: $U - X$ is an open subset of Y. If F is the complement of $U - X$ in Y, then F is closed in Y, and

$$U \cap F = U \cap (U - X)' = U \cap (U' \cup X)$$
$$= (U \cap U') \cup (U \cap X) = U \cap X = X,$$

as desired.

Corollary 1. *A dense locally compact subspace of a Hausdorff space Y is open in Y.*

Proof. Suppose X is a dense, locally compact subspace of Y. There is then an open set U and a closed set F in Y such that $X = U \cap F$, by the preceding lemma. Since X is included in F, and F is closed, the closure of X (in Y) must also be included in F. The assumption that X is dense implies that its closure coincides with Y, so $F = Y$ and therefore $X = U$.

The next task is to describe the relationship between compactifications of infinite unions of Boolean spaces and subalgebras of infinite products of Boolean algebras. We proceed to establish the notation that will be used in this description. Let $\{X_i\}$ be a disjoint family of Boolean spaces with union X, and let A_i be the dual algebra of X_i. The internal product A of the family $\{A_i\}$ is the field of subsets of X of the form $P = \bigcup_i P_i$, where each P_i is an element in A_i, that is, a clopen subset of X_i (see Chapter 26). The operations of A are performed coordinatewise: if

$$P = \bigcup_i P_i \qquad \text{and} \qquad Q = \bigcup_i Q_i$$

are elements of A, then

$$P \cup Q = \bigcup_i (P_i \cup Q_i), \qquad P \cap Q = \bigcup_i (P_i \cap Q_i), \qquad P' = \bigcup_i (X_i - P_i).$$

There is a subalgebra of A that plays a special role in the discussion, namely the one generated by the union $\bigcup_i A_i$. (Recall from Chapter 26 that each factor A_i is the relativization of A to X_i. In particular, each A_i, as a set, is included in A, so that it makes sense to speak of the subalgebra generated by the union of these factors.) The elements of this subalgebra are precisely those sets $P = \bigcup_i P_i$ in A such that either $P_i = \varnothing$ for all but finitely many i, or else $P_i = X_i$ for all but finitely many i (see Exercise 27). We denote this subalgebra by D and call it the *weak internal product* of the family $\{A_i\}$.

The goal is to establish a correspondence between the Boolean compactifications of the space X and the Boolean algebras that lie between the internal product A and the weak internal product D (in the sense of being subalgebras of A that include D). The following observation will be useful: if B is a subalgebra of A, then B includes D (so that $D \subseteq B \subseteq A$) if and only if for each i, the set X_i belongs to B and the relativization of B to X_i is A_i. The implication from right to left is immediate: if B contains each set X_i, and if the relativization of B to X_i is A_i, then B includes the set of generators of D, and consequently it includes D as a subalgebra. On the other hand, if B includes D, then it certainly includes the set of generators of D, namely the union of the factors A_i. In particular, B contains each set X_i (since X_i belongs to A_i), and the relativization of B to X_i equals A_i, since

$$A_i \subseteq B(X_i) \subseteq A(X_i) = A_i.$$

The next lemma says that the dual algebra of each compactification of X is isomorphic to an algebra between the internal product A and the weak internal product D.

Lemma 2. *If B is the dual algebra of a Boolean compactification of X, then the relativization of B to X is an algebra between A and D, and B is isomorphic to this relativization via the relativizing mapping induced by X.*

Proof. Suppose Y is a Boolean compactification of X, and B is the dual algebra of Y. The mapping g defined by

$$g(P) = P \cap X$$

for each subset P of Y is a homomorphism from the field $\mathcal{P}(Y)$ of all subsets of Y onto the field $\mathcal{P}(X)$ of all subsets of X. In fact, g is the relativizing homomorphism induced by X, and $\mathcal{P}(X)$ is the relativization of $\mathcal{P}(Y)$ to X (see Chapter 12, p. 92). The dual algebra B is a subfield of $\mathcal{P}(Y)$, so g maps B homomorphically onto some subfield of $\mathcal{P}(X)$, namely the relativization of B to X. (Note that X need not be an element of B in order for this relativization to be formed; see Exercise 12.19.) Write B_0 for this relativization.

The first step is to show that B_0 is a subalgebra of A. Both algebras are subfields of $\mathcal{P}(X)$, so it suffices to prove that every element in B_0 — every image under g of an element in B — belongs to A. Consider a set P in B, and write

$$P_i = P \cap X_i.$$

The set P_i is clopen in X_i (it is the intersection of the clopen subset P of Y with X_i — see Exercise 9.6), so it is an element of A_i. Since

$$g(P) = P \cap X = P \cap \left(\bigcup_i X_i \right) = \bigcup_i (P \cap X_i) = \bigcup_i P_i,$$

it follows that $g(P)$ is an element of A, as desired.

The next step is to show that D is a subalgebra of B_0. Observe that a subset P of X_i is clopen in Y just in case it is clopen in X_i. One direction of this assertion is obvious: if P is clopen in Y, then it must be clopen in X_i (by the definition of the inherited topology), since

$$P = P \cap X_i.$$

To prove the converse, assume that P is clopen in X_i. Since P is open in X_i, and X_i is open in X (Exercise 2), and X is open in Y (Corollary 1), it follows (Exercise 9.7) that P must be open in Y. Since P is closed in X_i, it is compact in X_i, and therefore compact in Y (Exercise 29.28); hence (Lemma 29.1) P is closed in Y.

Another way of phrasing the preceding observation is that a subset of X_i belongs to B (the dual algebra of Y, and hence the field of clopen subsets of Y) if and only if it belongs to A_i (the dual algebra of X_i, and hence the field of clopen subsets of X_i). In particular, X_i belongs to B (since X_i is clopen in itself) and the relativization of B to X_i is just A_i. If P is in A_i, then

$$g(P) = P \cap X = P,$$

so that g is the identity function on A_i. Thus, A_i is also included in the image algebra B_0 for each i. The subalgebra D is generated by the union of the sets A_i, so it too must be included in B_0.

It remains to prove that the mapping g is one-to-one on B. Suppose P and Q are elements in B. Observe that

$$(P \cap X)^- = P \qquad \text{and} \qquad (Q \cap X)^- = Q,$$

where these closures are formed in Y. For instance,

$$(P \cap X)^- = P \cap X^- = P \cap Y = P;$$

the first equality follows from Exercise 9.13(e), and the second from the density of X in Y. If

$$g(P) = g(Q)$$

then

$$P \cap X = Q \cap X$$

by the definition of g, and therefore

$$P = (P \cap X)^- = (Q \cap X)^- = Q,$$

as required. The proof of the lemma is complete.

The next task is to determine how subalgebras that correspond to various compactifications of the union space X relate to one another.

Lemma 3. *Suppose Y and Z are Boolean compactifications of X, with dual algebras B and C respectively. The relativization of C to X is a subalgebra of the relativization of B to X if and only if there is a continuous mapping from Y onto Z that is the identity on X.*

Proof. Let B_0 be the relativization of B to X, and g the relativizing mapping from B to B_0 defined by

$$g(P) = P \cap X$$

for each P in B. Similarly, let C_0 be the relativization of C to X, and h the relativizing mapping from C to C_0 defined by

$$h(P) = P \cap X$$

for each P in C. The mappings g and h are isomorphisms, by the preceding lemma.

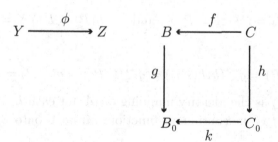

Consider first a continuous function ϕ from Y onto Z that maps each element in X to itself (see the diagram). The dual of ϕ is the monomorphism f from C into B that is defined by

$$f(P) = \phi^{-1}(P)$$

for every P in C, by Theorem 34 (p. 350). The composition

$$k = g \circ f \circ h^{-1}$$

is therefore a monomorphism from C_0 into B_0. We proceed to show that k is the identity mapping on C_0, from which it follows that C_0 is a subalgebra of B_0, as desired. For each Q in C_0, there is a P in C such that

$$h(P) = P \cap X = Q.$$

A simple computation yields

$$k(Q) = g(f(h^{-1}(Q))) = g(f(P)) = g(\phi^{-1}(P))$$
$$= \phi^{-1}(P) \cap X = P \cap X = Q.$$

The first equality follows from the definition of k, the second and fourth from the definitions of h and g as relativizing mappings, and the third from the definition of f as the dual of ϕ. For the fifth equality, observe that an element x is in the intersection $\phi^{-1}(P) \cap X$ just in case $\phi(x)$ is in P and x is in X, and this happens exactly when x is in $P \cap X$, because ϕ is the identity on X.

To prove the converse direction of the lemma, suppose C_0 is a subalgebra of B_0, and let k be the identity mapping from C_0 into B_0. The composition

$$f = g^{-1} \circ k \circ h$$

is a monomorphism from C into B. If P is an element in any one of the factor algebras A_i, then P is a subset of X, so that

$$g(P) = P \cap X = P \qquad \text{and} \qquad h(P) = P \cap X = P,$$

and therefore

$$f(P) = g^{-1}(k(h(P))) = g^{-1}(k(P)) = g^{-1}(P) = P.$$

In other words, f is the identity mapping on A_i for each i.

The dual of f is the continuous function ϕ from Y onto Z determined by the condition

$$\phi(x) \in P \qquad \text{if and only if} \qquad x \in f(P),$$

where P ranges over the elements of C — the clopen subsets of Z (Theorem 34). It must be shown that ϕ maps every element of X to itself. A point x in X necessarily belongs to one of the spaces X_i, and the function f

is the identity on the dual algebra of X_i, by the observations of the preceding paragraph. The equivalence determining ϕ therefore assumes the form

$$\phi(x) \in P \qquad \text{if and only if} \qquad x \in P$$

for sets P that are in A_i. The class N of all sets in A_i that contain x is an ultrafilter in A_i, and x is the only point that belongs to each of those sets (see Exercise 35.5 or the remarks following Theorem 33, p. 341). Since the sets in N also contain $\phi(x)$, by the preceding equivalence, it follows that $\phi(x) = x$, as desired.

Notice that the argument in the final paragraph proves a somewhat stronger assertion: if the dual of a continuous function ϕ from Y to Z is the identity mapping on A_i for each i, then ϕ is the identity mapping on X.

Corollary 2. *Suppose Y and Z are Boolean compactifications of X, with dual algebras B and C respectively. The spaces Y and Z are homeomorphic via a mapping that is the identity on X if and only if the relativizations of B and C to X are equal, or, equivalently, if and only if B and C are isomorphic via a mapping that is the identity on each field A_i.*

Proof. Let B_0 and C_0 be the relativizations of B and C to the set X. If there is a homeomorphism from Y to Z that is the identity mapping on X, then B_0 and C_0 are subalgebras of one another, and hence equal, by Lemma 3. On the other hand, if B_0 and C_0 are equal, then there are continuous mappings ϕ from Y onto Z, and ψ from Z onto Y, that are the identity on X, again by Lemma 3. The composition $\psi \circ \phi$ is a continuous function from Y onto itself that is the identity on the dense subset X. It follows from Corollary 33.2 and the subsequent remark that $\psi \circ \phi$ is the identity mapping on Y. A similar argument shows that $\phi \circ \psi$ is the identity mapping on Z. Consequently, ϕ is a bijection from Y to Z, and ψ is its inverse (see Exercise 12.32 or the section on bijections in Appendix A). Since both functions are continuous, ϕ must be a homeomorphism from Y to Z.

To establish the second equivalence, suppose first that C is isomorphic to B via a mapping that is the identity on each field A_i. The dual of the isomorphism is a homeomorphism from Y to Z, by Theorem 34, and it is the identity on X by the remark preceding the corollary.

Assume now that Y and Z are homeomorphic via a mapping that is the identity on X. The relativizations B_0 and C_0 are then equal, by the observations of the first paragraph. The relativizing mapping g from B to B_0 is an isomorphism, as is the relativizing mapping h from C to C_0, by Lemma 2. The composition

$$f = g^{-1} \circ h$$

is therefore an isomorphism from B to C, and it is easy to check that f is the identity on each field A_i. Indeed, an element P in A_i is a subset of X_i, and thus also a subset of X. Consequently,

$$g(P) = P \cap X = P \qquad \text{and} \qquad h(P) = P \cap X = P,$$

so that

$$f(P) = g^{-1}(h(P)) = g^{-1}(P) = P.$$

The final lemma says that every algebra between A and D comes from the dual of some Boolean compactification of X.

Lemma 4. *Every algebra between A and D is the relativization to X of the dual algebra of some Boolean compactification of X.*

Proof. We begin with a sketch of the main ideas of the proof. Consider an algebra C between D and A. Its dual space Y will be shown to be a Boolean compactification of a topological space V that is a homeomorphic image of X under a certain mapping ϕ. The mapping ϕ induces an isomorphism from C (which is a field of subsets of X) to a field C_0 of subsets of V. The relativization to V of the dual algebra of Y coincides with C_0. If each point in X is identified with its image point in V (using a topological version of the exchange principle), the dual space Y becomes a Boolean compactification of X, and the intermediate algebra C coincides with C_0. Consequently, the relativization to X of the dual algebra of Y coincides with C.

Here are the details of the proof. The space X is the topological union of the disjoint component spaces X_i. We proceed to construct a subspace V of Y by forming a homeomorphic image in Y of each X_i, and then taking the (disjoint) union of these images. Each space X_i is homeomorphic to its second dual — the dual space of the dual algebra A_i — via the mapping that takes each point x in X_i to the 2-valued homomorphism z_x on A_i defined by

$$z_x(P) = \begin{cases} 1 & \text{if } x \in P, \\ 0 & \text{if } x \notin P, \end{cases}$$

for every P in A_i, by Theorem 32 (p. 329). The dual space of A_i is, in turn, homeomorphic to the clopen subspace of Y determined by X_i, namely the subspace

$$V_i = \{y \in Y : y(X_i) = 1\}.$$

This last assertion is a direct consequence of Exercise 35.20, since the assumption that D is a subalgebra of C implies that the factor A_i is equal to the relativization of C to X_i (see the remarks preceding Lemma 2). In fact, if y_x is the 2-valued homomorphism on C defined by

$$y_x(P) = \begin{cases} 1 & \text{if } x \in P, \\ 0 & \text{if } x \notin P, \end{cases}$$

for every P in C, then the correspondence that takes z_x to y_x for each x in X_i is a homeomorphism from the dual space of A_i to the space V_i (Exercise 35.21). The composite function ϕ_i that maps each x in X_i to y_x is therefore a homeomorphism from X_i to V_i.

Take V to be the union of the family of subspaces $\{V_i\}$, endowed with the topology inherited from Y. We proceed to show that X is homeomorphic to V, and that V is a dense subspace of Y. The argument uses the properties of the canonical isomorphism from C to its second dual — the dual algebra of Y. Recall from Theorem 31 (p. 328) that this isomorphism maps each set P in C to the clopen subset

$$U_P = \{y \in Y : y(P) = 1\}$$

of Y.

The family of homeomorphisms $\{\phi_i\}$ has a common extension to a function ϕ from X to V that is defined by

$$\phi(x) = y_x$$

for x in X. Notice that ϕ is well defined because the sets X_i are assumed to be mutually disjoint; it is onto because each ϕ_i maps X_i onto V_i; and it is one-to-one because the mappings ϕ_i are one-to-one and the sets V_i are mutually disjoint. (The image of each X_i under the canonical isomorphism is just the set

$$U_{X_i} = V_i,$$

and images of disjoint sets under an isomorphism are always disjoint.) To prove that ϕ is continuous, it must be shown that the inverse image under ϕ

of each open set in V is open in X. An open set in V has the form $W \cap V$ for some open set W in Y. Of course,

$$W \cap V = W \cap \left(\bigcup_i V_i \right) = \bigcup_i (W \cap V_i),$$

and therefore

$$\phi^{-1}(W \cap V) = \bigcup_i \phi^{-1}(W \cap V_i) = \bigcup_i \phi_i^{-1}(W \cap V_i),$$

by Exercise 33.5 and the definition of ϕ. The set $W \cap V_i$ is open in the subspace topology of V_i (because W is open in Y), and therefore its inverse image $\phi_i^{-1}(W \cap V_i)$ is open in X_i, by the continuity of ϕ_i. This inverse image is the component of $\phi^{-1}(W \cap V)$ in the component space X_i (because the domain of ϕ_i is X_i), so each of the components of $\phi^{-1}(W \cap V)$ is open in its component space. Consequently, $\phi^{-1}(W \cap V)$ is open in X, by the definition of the union topology. Conclusion (Lemma 33.5): ϕ is a homeomorphism from X to V.

To prove that V is dense in Y, it must be shown that every non-empty clopen set in Y has a non-empty intersection with V. For any such clopen set Q, there is a non-empty set P in C such that $Q = U_P$ (since the canonical isomorphism maps C onto the dual algebra of Y). An easy computation yields

$$Q \cap V = Q \cap \left(\bigcup_i V_i \right) = \bigcup_i (Q \cap V_i) = \bigcup_i (U_P \cap U_{X_i}) = \bigcup_i U_{P \cap X_i}.$$

(The final equality holds because the canonical isomorphism preserves intersection.) Since P is a non-empty subset of X, it must have a non-empty intersection with one of the sets X_i. The corresponding set $U_{P \cap X_i}$ is then a non-empty clopen subset of $Q \cap V$, as desired. (Here, use is being made of the fact that the image of a non-empty set under the canonical isomorphism is non-empty.) Conclusion: V is a dense subspace of the Boolean space Y, and therefore Y is a Boolean compactification of V.

The homeomorphism ϕ naturally induces a bijection between the class of subsets of X and the class of subsets of V: each subset P of X is mapped to the subset

$$\phi(P) = \{\phi(x) : x \in P\}$$

of V. Consequently, every field of subsets of X is mapped isomorphically to a corresponding field of subsets of V. In particular, C is mapped to the field

$$C_0 = \{\phi(P) : P \in C\}.$$

Consider now the dual algebra of Y — call it B. It is the second dual of C and the image of C under the canonical isomorphism (Theorem 31). We proceed to demonstrate that the relativization of B to V is just the field C_0. First, a preliminary observation:

$$(1) \qquad U_P \cap \phi(X) = \phi(P)$$

for each set P in C. To prove (1), let y be any point in $\phi(X)$, say

$$y = \phi(x) = y_x,$$

where x is in X. If y is in U_P, then

$$1 = y(P) = y_x(P),$$

by the definition of U_P; hence, x is in P, by the definition of y_x, and therefore y is in $\phi(P)$, by the definition of ϕ. On the other hand, if y is in $\phi(P)$, so that x is in P, then

$$1 = y_x(P) = y(P),$$

by the definition of y_x; hence, y is in U_P, by the definition of U_P. This argument shows that the two sets U_P and $\phi(P)$ have the same intersection with $\phi(X)$. But $\phi(P)$ is included in $\phi(X)$, since P is included in X, so we arrive at (1).

As B is the second dual of C, each element Q in B may be written one and only one way in the form $Q = U_P$ for some P in C. If g is the relativizing homomorphism that maps every set in B to its intersection with V, then

$$g(Q) = Q \cap V = U_P \cap \phi(X) = \phi(P);$$

the first equality uses the definition of g, and the last uses (1). The correspondence between the sets P in C and the sets $Q = U_P$ in B is bijective, as is the correspondence between the sets P in C and the sets $\phi(P)$ in C_0. The preceding string of equalities therefore implies that the relativizing homomorphism g is a bijection, and hence an isomorphism, from B to C_0. In particular, the relativization of B to V is C_0.

Here is a summary of what has been accomplished so far. First, X has been shown to be homeomorphic, under a mapping ϕ, to a dense subspace V of Y (where Y is the dual space of the given intermediate algebra C). In particular, Y is a Boolean compactification of V. Second, ϕ induces an isomorphism between C and a field C_0 of subsets of V, and B (the dual algebra of Y), when relativized to V, coincides with C_0. (In fact, the relativizing mapping that takes each Q in B to the set $Q \cap V$ is an isomorphism from B

onto C_0.) All that is needed to complete the proof is to identify X with V, and specifically to identify each point x in X with its image $\phi(x)$. Under this identification, the space Y becomes a Boolean compactification of X; the algebra C coincides with C_0; the dual algebra B, when relativized to X, coincides with C; and the relativizing homomorphism that maps each element Q in B to the set $Q \cap X$ is an isomorphism from B to C.

The technical tool for carrying out this identification is a topological version of the exchange principle described in Chapter 12. In brief, a homeomorphic copy of the space Y is created in the following manner: the elements in V are replaced by the corresponding elements in X, and the elements in $Y - V$ are replaced by new elements that do not occur in X. This leads to a set Y^* that includes X, and to a bijection ψ from Y^* to Y that coincides with ϕ on the elements in X. A subset of Y^* is declared to be open just in case it is the inverse image under ψ of an open set in Y. The resulting space Y^* is homeomorphic to Y via the mapping ψ. The subspace V is dense in Y, and ψ^{-1} extends the homeomorphism ϕ^{-1} that maps V to X; consequently, X, under its own topology, is a dense subspace of Y^*. In other words, Y^* is a Boolean compactification of X. (See Exercise 33.21.)

If B^* is the dual algebra of Y^*, and B the dual algebra of Y, then B^* is obtained from B by replacing each set in B — each clopen subset of Y — with its inverse image under ψ. The restriction of ψ^{-1} to V is just ϕ^{-1}, so each set in the relativization of B to V — that is to say, each set in C_0 — is replaced by its inverse image under ϕ. It is obvious from the definition of C_0 that the resulting algebra is just C. Consequently, C is the relativization of B^* to X. The proof of the lemma is complete.

Lemmas 2 and 4 describe a correspondence between the class of Boolean compactifications of the space X and the class of algebras intermediate between D and A. In general, the correspondence is not one-to-one, as Corollary 2 makes clear; distinct compactifications may correspond to the same intermediate algebra. Such compactifications do not differ from one another in any material way, and it is natural to identify them by grouping them all together into one class. Motivated by these considerations, we define two Boolean compactifications of X to be *equivalent* if there is a homeomorphism between them that maps each element of X identically to itself. It is easy to check that this defines an equivalence relation on the class of all Boolean compactifications of X. Equivalent compactifications have dual algebras that are isomorphic via a mapping that is the identity on the elements of each set A_i, by Corollary 2; one may therefore speak, with some justification, of the dual algebra of the equivalence class.

Two Boolean compactifications of X are equivalent if and only if the relativizations (to X) of their dual algebras are equal, by Corollary 2. (Keep in mind that these relativizations are natural isomorphic copies of the dual algebras, copies that lie between D and A; see Lemma 2.) The correspondence that maps the equivalence class of such a compactification to the relativization of its dual algebra is therefore a well-defined bijection from the set of equivalence classes of Boolean compactifications of X to the set of algebras between D and A. As it turns out, this bijection is actually a lattice isomorphism.

The set of algebras between D and A is partially ordered by the relation of being a subalgebra, and under this ordering the set becomes a complete lattice with zero D and unit A. The meet of a family of such algebras is their intersection, and the join is the subalgebra generated by their union.

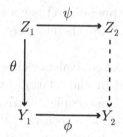

The partial ordering on the set of equivalence classes of Boolean compactifications is more complicated to describe, but it is equally natural. We begin by defining a binary relation $\leq$ between Boolean compactifications of X. For two such compactifications Y and Z, write

$$Y \leq Z$$

just in case there is a continuous mapping from Z onto Y that is the identity on X. This relation is preserved by equivalence in the following sense: if Y_1 and Y_2 are equivalent, and also Z_1 and Z_2, then

$$Y_1 \leq Z_1 \qquad \text{implies} \qquad Y_2 \leq Z_2.$$

For the proof, suppose ϕ is a homeomorphism from Y_1 to Y_2, and ψ a homeomorphism from Z_1 to Z_2, and θ a continuous function from Z_1 onto Y_1; and assume that all three functions are the identity on X. The composition

$$\phi \circ \theta \circ \psi^{-1}$$

is then a continuous function from Z_2 onto Y_2 that is the identity on X (see the diagram).

We shall say that the equivalence class of Y is less than or equal to the equivalence class of Z if $Y \leq Z$. The remarks of the preceding paragraph imply that the relation so defined does not depend on any particular choice of the representatives of the equivalence classes concerned. In other words, the relation is well defined. It is not difficult to prove that the relation is actually a partial ordering. The only non-trivial part is showing that it is antisymmetric. Suppose, accordingly, that $Y \leq Z$ and $Z \leq Y$, say ϕ maps Z continuously onto Y, and ψ maps Y continuously onto Z, and both functions are the identity on X. The composition $\phi \circ \psi$ maps Y continuously onto itself and is the identity on the dense subset X. The composition is therefore the identity function on Y, by Corollary 33.2 and the subsequent remark. Similarly, the composition $\psi \circ \phi$ is the identity function on Z. It follows that ϕ is a bijection from Z to Y with ψ as its inverse (see Exercise 12.32 or the section on bijections in Appendix A), so ϕ must be a homeomorphism. The compactifications Y and Z are therefore equivalent, so their equivalence classes are equal.

We have seen that the set of equivalence classes of Boolean compactifications of X is partially ordered by the relation just defined, and that the set of algebras between D and A is a complete lattice under the partial ordering of being a subalgebra. We have also seen that the correspondence mapping each equivalence class to the relativization of its dual algebra is a bijection between the two classes, and it is a direct consequence of Lemma 3 that this bijection preserves the partial orderings in both directions (in the sense that it satisfies the order-preserving equivalence of Exercise 12.13)(c)). It follows that the set of equivalence classes must be a lattice under its partial ordering, and that the correspondence between the two lattices is a lattice isomorphism. The following theorem summarizes what has been proved.

Theorem 45. *Let X be the union of a disjoint family $\{X_i\}$ of Boolean spaces, and for each i, let A_i be the dual algebra of X_i. Equivalent Boolean compactifications of X have the same dual algebra (up to isomorphisms that are the identity on the sets A_i), and that dual algebra is isomorphic (via relativization) to an algebra between the internal product A and the weak internal product D of the family $\{A_i\}$. The function that maps each equivalence class of Boolean compactifications of X to the corresponding algebra between A and D is a lattice isomorphism.*

The preceding theorem implies that the dual space of the product $\Pi_i A_i$

is, up to homeomorphic copies, the maximum element in the class of Boolean compactifications of X. We proceed to show that it is the Stone–Čech compactification of X. It is not important to know the precise details of the construction of the Stone–Čech compactification in order to prove this assertion. All that is needed is the definition: a compactification Y of a locally compact Hausdorff space X is a Stone–Čech compactification of X if every continuous mapping from X into a compact Hausdorff space Z can be extended to a continuous mapping from Y into Z. Two such compactifications are necessarily equivalent (Exercise 25), so one may justifiably speak of *the* Stone–Čech compactification of X. In keeping with the preceding development, we prove the dual version of the asserted theorem.

Theorem 46. *Let X be the union of a disjoint family $\{X_i\}$ of Boolean spaces, and for each i, let A_i be the dual algebra of X_i. The Stone–Čech compactification of X exists and is a Boolean space. Its dual algebra is isomorphic to the internal product of the family $\{A_i\}$ (via the relativizing mapping).*

Proof. There is, by Lemma 4, a Boolean compactification of X — call it Y — whose dual algebra is isomorphic to the internal product of the family $\{A_i\}$ via the relativizing mapping. We shall show that Y is the Stone–Čech compactification of X.

The proof makes use of the following preliminary observation: two disjoint closed subsets of X, say F_1 and F_2, are always separated by a clopen subset of Y. Indeed, for each index i, the intersections

$$F_1 \cap X_i \qquad \text{and} \qquad F_2 \cap X_i$$

are disjoint closed subsets of X_i, by the definition of the topology of X. A rather straightforward compactness argument, using the fact that X_i is a Boolean space, produces a clopen subset P_i of X_i such that

$$F_1 \cap X_i \subseteq P_i \qquad \text{and} \qquad F_2 \cap P_i = \varnothing$$

(Exercise 32.5). For each i, the set P_i belongs to the dual algebra A_i, so the family $\{P_i\}$ has a supremum in the internal product of the family $\{A_i\}$, by the definition of an internal product (see p. 233). The dual algebra of Y — call it B — is isomorphic to the internal product via a mapping that is the identity on the sets A_i, by Theorem 45, so the family $\{P_i\}$ also has a supremum in B. In fact, if P is the union of the family $\{P_i\}$, and if P^- is the closure of P in Y, then P^- is a clopen subset of Y and

$$P^- = \bigvee_i P_i$$

in B, by Lemma 38.1. An easy computation shows that P^- separates F_1 and F_2:

$$F_1 = F_1 \cap X = F_1 \cap \left(\bigcup_i X_i\right) = \bigcup_i (F_1 \cap X_i) \subseteq \bigcup_i P_i = P \subseteq P^-$$

and

$$F_2 \cap P^- = F_2 \cap \left(\bigvee_i P_i\right) = \bigvee_i (F_2 \cap P_i) = \bigvee_i \varnothing = \varnothing.$$

(Notice that the existence of this separating clopen set depends on the fact that B is isomorphic to the internal product of the family $\{A_i\}$ via a mapping that is the identity on the sets A_i.)

In order to prove that Y is the Stone–Čech compactification of X, consider an arbitrary continuous function ϕ from X into a compact Hausdorff space Z. It is to be shown that ϕ can be extended to a continuous function ψ from Y into Z. Every point in Y is completely determined by the clopen subsets of Y that contain it. In other words, if y is in Y, and if N_y is the class of clopen subsets of Y that contain y, then N_y is an ultrafilter (in B) and the intersection of the sets in N_y contains exactly one point, namely y. (Recall that clopen sets separate points in Y; see also Exercise 35.5.) It is natural to define $\psi(y)$ to be a point in the intersection of the class of image sets $\phi(P)$, for P in N_y. Two difficulties arise. First, ϕ is defined only on points in X, so it is necessary to use $\phi(P \cap X)$ instead of $\phi(P)$. Second, the sets $\phi(P \cap X)$ may not be closed, so there is no assurance that the intersection of all these sets is non-empty. The solution is to pass to the closures of these sets. Given any point y in Y, let K_y be the class of non-empty closed subsets of Z defined by

$$K_y = \{\phi(P \cap X)^- : P \in N_y\}.$$

We shall prove that the intersection of the sets in K_y contains exactly one point.

To show that this intersection is not empty, it suffices to prove that K_y has the finite intersection property; the desired conclusion then follows by compactness. The finite intersection property is a direct consequence of two observations. First, the sets in K_y are non-empty: if P is in N_y, then $P \cap X$ is non-empty, because X is dense, and P is clopen, in Y; hence, the closure of the image $\phi(P \cap X)$ is not empty. Second, the intersection of any finite sequence of sets from K_y includes a set from K_y and is therefore not empty. For the proof, consider a finite sequence $\{P_n\}$ of sets in N_y, and write P

for the intersection of the sequence. The set P is clopen and contains y, so it belongs to N_y. Consequently, the set $\phi(P \cap X)^-$ belongs to K_y; it is obviously included in the intersection of the sets $\phi(P_n \cap X)^-$, because P is included in each P_n.

To show that the intersection of the sets in K_y cannot contain two points, argue by contradiction: assume z_1 and z_2 are distinct points that belong to every set in K_y. Since Z is a compact Hausdorff space, there exist open sets U_1 and U_2 in Z, containing z_1 and z_2 respectively, such that the closures U_1^- and U_2^- are disjoint (Corollary 29.2). The inverse images $\phi^{-1}(U_1^-)$ and $\phi^{-1}(U_2^-)$ are closed subsets of X, by the continuity of ϕ, and they are obviously disjoint, since U_1^- and U_2^- are disjoint. The preliminary observation made at the beginning of the proof now implies that these inverse images are separated by a clopen set P in Y, say P includes $\phi^{-1}(U_1^-)$ and is disjoint from $\phi^{-1}(U_2^-)$. The point y is in exactly one of the sets P and P'; assume it is in P. In this case, P belongs to N_y; consequently, the set $\phi(P \cap X)^-$ belongs to K_y, so it contains both z_1 and z_2. Because z_2 belongs to the closure of $\phi(P \cap X)$, and also to the open set U_2, the intersection of U_2 with $\phi(P \cap X)$ is not empty, by a basic fact about closures (see p. 57). It follows that the intersection of $\phi^{-1}(U_2)$ with P is not empty (Exercise 33.2), which contradicts the disjointness of these two sets. The argument when y is in P' is completely analogous.

Define $\psi(y)$ to be the unique point that belongs to every set in K_y. To prove that ψ is an extension of ϕ, assume y is in X. Then y is in the intersection $P \cap X$ for every P in N_y, so $\phi(y)$ belongs to every set in K_y, by the definition of K_y. It follows from the definition of ψ that $\psi(y) = \phi(y)$.

It remains to demonstrate that ψ is continuous. Let U be an open set in Z. To show that the inverse image $\psi^{-1}(U)$ is open in Y, it suffices to show that for each point y in the inverse image, there is a clopen subset of Y that contains y and is included in $\psi^{-1}(U)$. By definition, $\psi(y)$ is the only point that belongs to the intersection of the sets in the class K_y. Since $\psi(y)$ belongs to U, the intersection of the sets in K_y is included in U. A rather straightforward compactness argument (Exercise 29.24) shows that the intersection of some finite family of sets in K_y is already included in U. Every such intersection includes a set from K_y (see the earlier remarks concerning the finite intersection property), so there is a clopen set P in N_y such that $\phi(P \cap X)^-$ is included in U. The point y belongs to P, by the definition of N_y. To prove that P is included in $\psi^{-1}(U)$, consider any point w in P. The set P belongs to N_w, by definition, and therefore $\phi(P \cap X)^-$ is in K_w. It follows that the intersection of the sets in K_w is included in $\phi(P \cap X)^-$, and hence also in U. The point $\psi(w)$ is, by definition, the unique element in this intersection, so $\psi(w)$ is in U; consequently, w is in $\psi^{-1}(U)$. The proof of the theorem is complete.

In the preceding theorem, the existence of the Stone–Čech compactification of the space X is not assumed; it is proved. However, even if its existence were assumed, the statement that this compactification is homeomorphic to the dual space of the product $\prod_i A_i$ would still not be an immediate consequence of Theorem 45. One would have to demonstrate first that the compactification has a base consisting of clopen sets and is therefore a Boolean space. This difficulty is circumvented in the preceding proof by starting, instead, with the dual space of the product — which is automatically a Boolean space — and showing that it possesses the defining properties of the Stone–Čech compactification of (a copy of) X.

For an application of the theorem, consider an infinite disjoint family $\{X_i\}$ of one-point spaces (singletons). The union of this family is an infinite space X endowed with the discrete topology. The dual algebra of each space X_i is a two-element Boolean algebra A_i, and the internal product of the family $\{A_i\}$ is isomorphic to the power 2^X and also to the field $\mathcal{P}(X)$ of all subsets of X. The preceding theorem says, in this case, that the dual spaces of the algebras 2^X and $\mathcal{P}(X)$ are (homeomorphic to) the Stone–Čech compactification of X (see Exercises 33.31 and 34.19). This shows, in particular, that an infinite product of even the simplest non-degenerate Boolean algebras has something as unruly as the Stone–Čech compactification of an infinite discrete space as its dual.

As we have seen, if $\{X_i\}$ is a family of Boolean spaces, and if Y is the Stone–Čech compactification of the disjoint union of the spaces, then Y is (to within a homeomorphism) the dual space of the product of the family of dual algebras. In the case of a finite family of spaces, Y is just the disjoint union of the family and is therefore called the sum of the spaces. It seems reasonable to extend the additive terminology and notation to the infinite case as well, and in all cases to call Y the *sum* of the family $\{X_i\}$, and to write

$$Y = \sum_i X_i.$$

We close with some historical remarks. Dwinger [16] proved that the dual space of a (direct) product A of Boolean algebras is homeomorphic to the Stone–Čech compactification of the (disjoint) union X of the dual spaces of the factor algebras. Bernardi [4] proved that the dual algebra of every Boolean compactification of X is isomorphic to a subalgebra of A that includes the weak (direct) product, and, conversely, the dual space of every subalgebra of A that includes the weak product is homeomorphic to a Boolean compactification of X. (The weak product is the subalgebra of A consisting of those elements p such that $p_i = 0$ for all but finitely many i, or else $p_i = 1$ for all but finitely many i.) Our presentation differs from Dwinger's and Bernardi's, and has been inspired by the presentation in [45].

Exercises

1. Prove that the "union topology", defined on the union of an arbitrary family of disjoint topological spaces, really is a topology.

2. Suppose X is the union of a family $\{X_i\}$ of disjoint spaces. Prove that a subset P of X is closed or clopen in X just in case $P \cap X_i$ is closed or clopen in X_i for each i. Conclude that a subset of X_i is open, closed, or clopen in the topology of X just in case it is open, closed, or clopen in the topology of X_i.

3. Prove that the union of a family of disjoint spaces is Hausdorff if and only if each component space is Hausdorff.

4. Prove that the union of a family of disjoint spaces has a base consisting of clopen sets if and only if each component space has a base consisting of clopen sets.

5. Prove that the union of a finite family of disjoint spaces is compact if and only if each component space is compact. Conclude that the union of the family is a Boolean space if and only if each component space is Boolean.

6. Formulate and prove the generalization of Theorem 44 to any finite disjoint family of Boolean spaces.

7. Use Theorem 44 to derive its dual (algebraic) version: if B and C are Boolean algebras, and if $A = B \times C$, then the dual space of A is homeomorphic to the (disjoint) sum of the dual spaces of B and C.

8. Prove the dual (algebraic) version of Theorem 44 directly, without using Theorem 44.

9. Let X be the space of all ordinals up to and including the second limit ordinal $\omega 2$, under the order topology (see Exercise 32.15). Suppose

$$X_0 = \omega \cup \{\omega\} = \{0, 1, 2, \dots, \} \cup \{\omega\}$$

is the space of all ordinals up to and including the first infinite ordinal, under the order topology, and

$$X_1 = \{\omega + 1, \omega + 2, \dots \} \cup \{\omega 2\}$$

is the space of the remaining ordinals in X, again under the order topology. (Thus, X_0 is the one-point compactification of the space of finite ordinals with the discrete topology, while X_1 is the one-point compactification of the space $Y = \{\omega + 1, \omega + 2, \dots \}$ with the discrete topology (see Exercise 32.16). The two spaces X_0 and X_1 are obviously homeomorphic via the function that maps n to $\omega + n + 1$ for each natural number n, and that maps ω to $\omega 2$.) Prove that $X = X_0 + X_1$. Conclude that if A is the field of finite and cofinite subsets of the natural numbers, then the dual algebra of X is isomorphic to $A \times A$.

10. Generalize the results of the preceding problem.

11. Let A be the field of finite and cofinite sets of natural numbers. Give a topological proof that $A \times A$ is not isomorphic to A.

12. (Harder.) Prove directly (without using duality and Exercise 26.34) that if X is a Boolean space with a countable base and infinitely many isolated points, and if $1 = \{0\}$ is the one-point space, then

$$X = X + 1$$

(where the equality sign is to be interpreted as asserting that the two spaces are homeomorphic). Use this result to give a topological proof of the theorem (Exercise 26.34) that if A is a countable Boolean algebra with infinitely many atoms, then A is isomorphic to $A \times 2$.

13. Prove that a Hausdorff space is locally compact if and only if each point belongs to some open set whose closure is compact.

14. Prove the following converse to Lemma 1: if Y is a compact Hausdorff space, then for every open set U and every closed set F, the intersection $X = U \cap F$ is a locally compact subspace of Y.

15. Suppose X is a topological space and Y a one-point compactification of X. Prove that the class of open sets in Y really is a topology, that Y is compact under this topology, and that X is a subspace of Y. (The results of this and the next exercise are due to Alexandroff — see [1].)

16. Prove that two one-point compactifications of a topological space X are homeomorphic via a mapping that is the identity on X. (This justifies speaking of *the* one-point compactification of X.)

17. Prove that the subspace of a compact Hausdorff space obtained by removing any single point is locally compact and Hausdorff.

18. Prove that the one-point compactification of a locally compact Hausdorff space is Hausdorff. Conclude that the one-point compactification of a space X is Hausdorff if and only if X is locally compact and Hausdorff.

19. Prove that the one-point compactification of a topological space X is a Boolean space if and only if X is a locally compact Boolean space.

20. Suppose X is a locally compact Hausdorff space, and Y is its one-point compactification. Prove that the following statements are equivalent: (1) X is not compact; (2) X is dense in Y; (3) the point in $Y - X$ is not isolated.

21. Prove the following stronger version of Exercise 16 for Hausdorff spaces. If Y and Z are any two compact Hausdorff spaces that include X as a subspace, and if $Y - X$ and $Z - X$ each have just one point, then Y and Z are homeomorphic via a mapping that is the identity on X.

22. Prove that the relation of equivalence between Boolean compactifications of a space X is an equivalence relation.

23. Suppose Y and Z are compactifications of a locally compact Hausdorff space X. Prove that any continuous mapping of Y into Z that is the identity function on X must map Y onto Z.

24. If Y is a Stone–Čech compactification of a locally compact Hausdorff space X, prove that Y is the largest compactification of X in the sense that for any compactification Z of X, there is a continuous mapping from Y onto Z that is the identity on X.

25. Show that two Stone–Čech compactifications of a locally compact Hausdorff space X are equivalent in the sense that they are homeomorphic via a mapping that is the identity on X. (This justifies speaking of *the* Stone–Čech compactification of X.)

26. Show that $[0, 1]$ is the one-point compactification of $(0, 1]$, but it is not the Stone–Čech compactification of $(0, 1]$.

27. Let X be the union of a disjoint family of sets $\{X_i\}$. Suppose A_i is a field of subsets of X_i for each i, and A is the internal product of the family $\{A_i\}$. Prove that the class of all sets P in A such that $P \cap X_i = \varnothing$ for all but finitely many i, or else $P \cap X_i = X_i$ for all but finitely many i, is a subfield of A. Prove further that this subfield is generated by the union $\bigcup_i A_i$, that is, it is generated by the class of all elements of the various factor algebras.

28. Let X be the union of an infinite family $\{X_i\}$ of non-empty disjoint Boolean spaces. Prove that the one-point compactification of X is the smallest Boolean compactification of X in the sense of the partial ordering on Boolean compactifications of X introduced in the chapter. Conclude that the dual algebra of the one-point compactification of X is (isomorphic to) the weak internal product of the dual algebras of the spaces X_i.

29. If X is the union of an infinite family $\{X_i\}$ of non-empty disjoint Boolean spaces, prove directly (without using Theorem 45) that the dual algebra of the one-point compactification of X is (isomorphic to) the weak internal product of the dual algebras of the spaces X_i.

30. Let X be an infinite set endowed with the discrete topology. Use Exercise 28 to give another proof that the field of finite and cofinite subsets of X is (up to isomorphic copies) the dual algebra of the one-point compactification of X. (Compare Exercise 32.16.)

31. If Y is the space of ordinals up to and including ω^2, under the order topology, what is the dual algebra of Y?

32. As was pointed out in Exercise 26.37, the product $\prod_i A_i$ of an infinite family of non-degenerate Boolean algebras includes two subalgebras, each of which might deserve some consideration as a kind of weak product of the family $\{A_i\}$. One subalgebra consists of those elements p for

which p_i is in 2 for all but finitely many i; the other, smaller, subalgebra consists of those elements p for which either $p_i = 0$ for all but finitely many i or else $p_i = 1$ for all but finitely many i. What can be said about the duals of these algebras?

Chapter 44

Sums of Algebras

The arrow diagram characterizing products in Chapter 26 says that whenever $\{A_i\}$ is a family of Boolean algebras, there exists a Boolean algebra A (the product) and, for each i, there exists an epimorphism f_i from A to A_i (the projection) such that if B is an arbitrary Boolean algebra, and if, for each i, there is a homomorphism g_i from B to A_i, then there is a unique homomorphism g from B to A satisfying $f_i \circ g = g_i$ for all i (see the diagram).

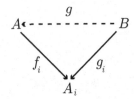

There are two ways to dualize such arrow diagrams. What, for instance, does the diagram for products of algebras imply about the corresponding dual spaces? That is the first question; the answer — a consequence of the duality between homomorphisms and continuous functions (Theorem 34, p. 350, and Corollary 36.2) — is given by a corresponding diagram for sums of spaces that says: whenever $\{X_i\}$ is a family of Boolean spaces, there exists a Boolean space X (the Stone–Čech compactification of the disjoint union), and, for each i, there exists a continuous one-to-one mapping ϕ_i from X_i into X (the identity), such that if Y is any Boolean space, and if, for each i, there exists a continuous mapping ψ_i from X_i to Y, then there exists a unique continuous mapping ψ from X to Y satisfying $\psi \circ \phi_i = \psi_i$ for all i (see the diagram).

S. Givant, P. Halmos, *Introduction to Boolean Algebras*,
Undergraduate Texts in Mathematics, DOI: 10.1007/978-0-387-68436-9_44,
© Springer Science+Business Media, LLC 2009

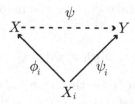

An equally natural question is this: What does the diagram for products of algebras become if the algebras and homomorphisms involved in it are replaced by spaces and continuous mappings? Two similar questions can be asked about the dualization of the diagram for sums of spaces. One of them leads back to products of algebras, and the other is the algebraic dual of the space question just asked. The purpose of the present chapter is to answer the two as yet unanswered questions.

We proceed to the precise formulations. Suppose that $\{X_i\}$ is a family of Boolean spaces. Does there exist a Boolean space X, and does there exist, for each i, a continuous mapping ϕ_i from X into X_i such that the requisite lifting condition is satisfied? The lifting condition says that if Y is a Boolean

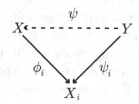

space and if, for each i, there exists a continuous mapping ψ_i from Y to X_i, then there exists a unique continuous mapping ψ from Y to X such that

(1) $$\phi_i \circ \psi = \psi_i$$

for all i. The answer is yes; if X is the Cartesian product of the family $\{X_i\}$, with the product topology (Exercise 32.20), and if the ϕ_i are the usual projections from a product space to its factors, then all the requirements are fulfilled. The proof that X is a Boolean space parallels the proof that Cantor spaces are Boolean (see Chapter 32 and Exercise 32.21). The continuity of the projections is an almost immediate consequence of the definition of the product topology (see Exercise 33.7). To verify the lifting condition, suppose Y is a Boolean space such that for each index i there is a continuous mapping ψ_i from Y to X_i. Take ψ to be the mapping from Y to X defined by

$$\psi(y)_i = \psi_i(y)$$

for every i. In other words, $\psi(y)$ is that element of X whose ith coordinate is $\psi_i(y)$, for each i. This definition just says that equation (1) holds for every i. Consequently, there can be only one function ψ that satisfies this equation for every i.

An argument similar to the one that proved the uniqueness of the product of Boolean algebras (to within an isomorphism) proves that there is a unique Boolean space (to within a homeomorphism) that, together with a suitable family of mappings, satisfies the lifting condition. Here are the details. Suppose that Y is a Boolean space and that, for each i, there is a continuous mapping ψ_i from Y into X_i such that the lifting condition is satisfied. It is to be shown that there is a homeomorphism ψ from Y to X satisfying the condition

(2) $$\phi_i \circ \psi = \psi_i$$

for all i. Since the space X and the projections ϕ_i satisfy the lifting condition (in particular, with respect to the space Y and the mappings ψ_i), by the observations of the preceding paragraph, there is a unique continuous mapping ψ from Y into X such that (2) holds. The space Y and the mappings ψ_i also satisfy the lifting condition (in particular, with respect to the space X and the projections ϕ_i), by assumption, so there is a unique continuous mapping ϕ from X into Y satisfying

(3) $$\psi_i \circ \phi = \phi_i$$

for all i.

Equations (2) and (3) combine to give

(4) $$\psi_i \circ \phi \circ \psi = \psi_i \qquad \text{and} \qquad \phi_i \circ \psi \circ \phi = \phi_i$$

for each i. Since the space Y and mappings ψ_i satisfy the lifting condition (in particular, with respect to themselves), there must be a unique continuous mapping θ from Y into itself such that

$$\psi_i \circ \theta = \psi_i$$

for each i. This system of equations is obviously satisfied if θ is taken to be the identity homeomorphism on Y, and it is also satisfied if θ is taken to be the composition $\phi \circ \psi$, by the first system of equations in (4). The assumed uniqueness of θ implies that $\phi \circ \psi$ must be the identity homeomorphism on Y. A similar argument, using the second system of equations in (4), shows that $\psi \circ \phi$ is the identity homeomorphism on X. It follows that ϕ and ψ are bijections and inverses of one another (see Exercise 12.32 or the section on bijections in Appendix A); thus, ψ is a homeomorphism from Y to X that satisfies condition (2), as desired.

It is natural to call the space we constructed the *product* of the given family of spaces and to use the multiplicative notation ($\prod_i X_i$, $X_1 \times X_2$, etc.) that this terminology suggests.

One consequence of the uniqueness of product spaces is that when the factor spaces X_i are all non-empty, the continuous functions ϕ_i in the definition of the product X must map X onto, and not just into, X_i. This is certainly the case when X is the Cartesian product of the family $\{X_i\}$ and the mappings ϕ_i are the projections. It follows from uniqueness, and in particular from equation (2), that it must be the case for every Boolean space and family of continuous mappings that satisfies the lifting condition for the family $\{X_i\}$.

Suppose next that $\{A_i\}$ is a family of Boolean algebras. Does there exist a Boolean algebra A, and does there exist, for each i, a homomorphism f_i from A_i to A such that the transfer condition is satisfied? By the transfer condition we mean that if B is a Boolean algebra and if, for each i, there exists a homomorphism g_i from A_i to B, then there exists a unique homomorphism g from A to B such that $g \circ f_i = g_i$ for all i. The answer by now is obviously yes; just dualize the theory of products of Boolean spaces. To be more precise,

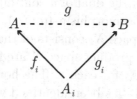

let X_i be the dual space of A_i, and identify A_i with its second dual (via the canonical isomorphism from A_i to the dual algebra of X_i). In this way, A_i may be thought of as the dual algebra of X_i. The product space

$$X = \prod_i X_i$$

and the projections ϕ_i from X to X_i satisfy the lifting condition, by the argument presented in the preceding paragraphs. Take A to be the dual algebra of X, and for each i, take f_i to be the dual of ϕ_i. Then f_i is a homomorphism from A_i into A, by Theorem 34 (p. 350). To verify that the transfer condition is satisfied, consider a Boolean algebra B such that, for each i, there is a homomorphism g_i from A_i into B. Take Y to be the dual space of B, and for each i take ψ_i to be the dual of g_i. Thus, ψ_i is a continuous mapping of Y into X_i, by the dual version of Theorem 34 (see Theorem 35, p. 350). The space X and the mappings ϕ_i together satisfy the lifting condition, so there is a unique continuous mapping ψ from Y into X such that $\phi_i \circ \psi = \psi_i$ for all i. Identify B with its second dual via the canonical isomorphism, and take g to be the dual of ψ. Then g is a homomorphism from B into A, by Theorem 34, and $g \circ f_i = g_i$ for all i, by Corollary 36.2. No other homomorphism from B to A can satisfy this system of equations, for the dual of such a homomorphism would present a counterexample to uniqueness of the continuous mapping ψ.

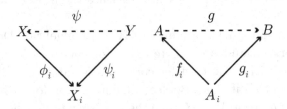

An explicit definition of the homomorphism f_i from A_i to A may be obtained from the definition of the dual of a continuous mapping and from the definition of the canonical isomorphism from A_i to its second dual (Theorem 31, p. 328). The dual space X_i consists of the 2-valued homomorphisms on A_i, and the points of the product space X are families $\{x_i\}$ of such homomorphisms, where x_i is in X_i for each i. The homomorphism f_i is the dual of the projection ϕ_i, after A_i has been identified with its second dual via the canonical isomorphism. The canonical isomorphism maps each point p in A_i to the set

$$U_p = \{y \in X_i : y(p) = 1\},$$

so

$$f_i(p) = \phi_i^{-1}(U_p) = \{x \in X : \phi_i(x) \in U_p\}$$

$$= \{x \in X : x_i \in U_p\} = \{x \in X : x_i(p) = 1\}.$$

The first equality uses (36.2), the second uses the definition of an inverse image, the third uses the definition of the projections, and the fourth uses the definition of U_p.

In analogy with the use of the term "sum" to denote the dual space of a product of Boolean algebras, we may call A (which is the dual algebra of a product of Boolean spaces) the (*direct*) *sum* of the given family of algebras, and we may use the additive notation ($\sum_i A_i$, $A_1 + A_2$, etc.) that this terminology suggests. An argument similar to the one given in the preceding paragraphs, based on duality and on the uniqueness of the product of Boolean spaces (to within a homeomorphism), proves that the transfer condition uniquely determines the sum of a family of algebras, to within an isomorphism. (Alternatively, one can prove this uniqueness directly, using the standard argument.)

It follows from uniqueness that if each algebra A_i in a family of Boolean algebras is non-degenerate, then the homomorphisms f_i in the definition of the sum of the family are actually monomorphisms. Indeed, the dual spaces X_i are non-empty (since the algebras A_i are non-degenerate), so the projections ϕ_i map the product space X onto (and not just into) the factor spaces X_i. Consequently, the dual homomorphisms f_i must in fact be one-to-one (Theorem 34). An application of uniqueness now shows that this remains true for every Boolean algebra and family of homomorphisms that satisfy the transfer condition for the family $\{A_i\}$.

An important insight into the intuition behind the sum construction can be gained by identifying each algebra A_i with its image under f_i (so that f_i becomes the identity homomorphism on A_i). The sum of the family $\{A_i\}$ is then an algebra A with the following properties: first, each A_i is a subalgebra of A; second, if B is any Boolean algebra such that for each i, there is a homomorphism g_i from A_i into B, then there is a unique homomorphism g from A into B that is a common extension of the g_i. We shall refer to A as the *internal sum* of the family $\{A_i\}$.

The requirement in the definition of an internal sum that the extension homomorphism be unique is equivalent to the condition that the union $\bigcup_i A_i$ generates A. One direction of this assertion is clear. If the union of the sets A_i generates A, then two homomorphisms from A to B that agree with g_i on A_i for each i, agree with each other on a generating set of A, and consequently agree with each other on all of A, by Lemma 13.2. To prove the converse direction of the assertion, argue by contraposition. Suppose the union of the sets A_i generates a proper subalgebra C of A. There is then a 2-valued homomorphism g on C that can be extended in two different ways to a

2-valued homomorphism on A, by Exercise 21.10. For each i, take g_i to be the restriction of g to A_i. The extension of the family $\{g_i\}$ to a homomorphism from A into 2 is not unique, so the uniqueness condition is not satisfied. An analogous remark applies to the general definition of the sum of Boolean algebras: the condition that the required homomorphism g from A to B satisfying the equations $g \circ f_i = g_i$ be unique is equivalent to the condition that the union $\bigcup_i f_i(A_i)$ generate A.

The preceding observations imply that a Boolean algebra A is the internal sum of a family $\{A_i\}$ of subalgebras just in case the union of the subalgebras generates A, and the following simplified version of the transfer condition is satisfied: whenever B is a Boolean algebra such that for each i, there is a homomorphism g_i from A_i into B, then there is a homomorphism g from A into B that extends each of the mappings g_i. An informal way of expressing this transfer condition is to say that every interaction in A between elements of distinct subalgebras A_i (that is, any equation that holds between such elements) can be transferred (or copied) via a homomorphism g to any Boolean algebra B in which there are homomorphic copies of the algebras A_i (the images of the A_i under the homomorphisms g_i). In other words, the interactions in A between the elements of distinct subalgebras A_i are as free as possible.

This remark suggests a connection between sums and free algebras (Chapter 28). Recall that an algebra A is freely generated by a subset E if any interaction in A between distinct elements of E (that is, any equation that holds between such elements) can be transferred (via an extension homomorphism) to any Boolean algebra B in which a (not necessarily one-to-one) copy of E has been made. In an internal sum, the elements of the free generating set E are replaced by subalgebras A_i, and the function mapping the elements of E into a Boolean algebra B is replaced by a family of structure-preserving functions (homomorphisms) that map the subalgebras A_i to subalgebras of B. The analogy between sums of Boolean algebras and free Boolean algebras is the reason why the sums are often called *free products*.

The characterization given in Lemma 28.1 of sets that freely generate a Boolean algebra can be modified to give a characterization of families of subalgebras that generate an internal sum.

Lemma 1. *A Boolean algebra A is the internal sum of a family $\{A_i\}$ of its subalgebras just in case the union of these subalgebras generates A, and whenever J is a finite, non-empty subset of the indices, and p_i is a non-zero element in A_i for each i in J, then*

$$\bigwedge_{i \in J} p_i \neq 0.$$

Proof. Suppose first that A is the internal sum of the family $\{A_i\}$ of subalgebras. The union of the sets A_i generates A, by the remarks preceding the lemma. Consider a finite, non-empty family $\{p_i\}_{i \in J}$ of non-zero elements in A such that p_i is in A_i for each i in J, and let p be the infimum of the family. It is to be shown that p is not zero.

The given family of elements is non-empty, so it contains at least one non-zero element. Consequently, A and each of its subalgebras is non-degenerate. For each index i in J, the element p_i is not zero, so Corollary 20.3 guarantees the existence of a maximal ideal M_i in A_i that does not contain p_i. For each index i not in J, take M_i to be any maximal ideal in A_i; such an ideal exists, by the maximal ideal theorem (Theorem 12, p. 172), because A_i is not degenerate. Let g_i be the 2-valued homomorphism on A_i with kernel M_i, and observe that if i is in J, then $g_i(p_i) = 1$. The transfer condition, applied to the Boolean algebra 2 and the homomorphisms g_i, implies the existence of a homomorphism g from A into 2 that extends g_i for each i. In particular, $g(p_i) = 1$ for every i in J, and therefore

$$g(p) = \bigwedge_{i \in J} g(p_i) = 1.$$

Since g is a homomorphism, the element p must be different from zero, as was to be shown.

Assume now that $\{A_i\}$ is a family of subalgebras in A whose union generates A and for which the condition formulated in the lemma holds. It must be shown that the transfer condition is satisfied. Notice first that if $i \neq j$, then the subalgebras A_i and A_j have only the elements 0 and 1 in common. Indeed, if they contained a common element q different from 0 and 1, then by putting $p_i = q$ and $p_j = q'$, we would arrive at a finite family $\{p_i, p_j\}$ of non-zero elements whose infimum is zero, in contradiction to the condition of the lemma.

Let B be any Boolean algebra such that for each i, there is a homomorphism g_i from A_i into B. Write

$$E = \bigcup_i A_i,$$

and define a function g from E into B by

$$g(p) = g_i(p) \quad \text{if} \quad p \in A_i.$$

Since an element p in E cannot occur in different subalgebras of the family unless it is 0 or 1, and since the homomorphisms g_i all agree on 0 and 1, the mapping g is well defined.

The next step is to show that g satisfies the homomorphism extension criterion formulated in Theorem 4 (p. 107). Suppose a is a 2-valued function on some finite subset F of E such that

$$(1) \qquad\qquad \bigwedge_{q \in F} p(q, a(q)) = 0$$

(where, as usual, $p(q, j)$ is q or q', according as j is 1 or 0). Let F_i be that part of F that is included in A_i,

$$F_i = F \cap A_i,$$

and let J be a finite set of indices i such that

$$F = \bigcup_{i \in J} F_i.$$

(The set J exists because F is finite.) The element

$$(2) \qquad\qquad p_i = \bigwedge_{q \in F_i} p(q, a(q))$$

belongs to A_i, for each i in J, because A_i is a subalgebra of A. The preceding definitions and (1) imply that

$$\bigwedge_{i \in J} p_i = \bigwedge_{q \in F} p(q, a(q)) = 0.$$

Consequently, $p_i = 0$ for some i in J, by the assumed condition of the lemma. For such an i, we have $g_i(p_i) = 0$, and therefore

$$\bigwedge_{q \in F_i} p(g_i(q), a(q)) = 0,$$

by (2) and the homomorphism properties of g_i. It follows from this equation and from the definition of g that

$$\bigwedge_{q \in F} p(g(q), a(q)) = \bigwedge_{i \in J} \bigwedge_{q \in F_i} p(g_i(q), a(q)) = 0,$$

so the homomorphism criterion is satisfied.

Invoke Theorem 4 to conclude that g can be extended to a homomorphism from A into B. Since that homomorphism extends g, it extends g_i for each i. The proof of the lemma is complete.

If A is the internal sum of a family of subalgebras, then these subalgebras are mutually disjoint, except for 0 and 1, by the remarks in the proof of the preceding lemma. In the general sum construction, the algebras A_i may have many elements in common. The homomorphisms f_i (which are always monomorphisms when each of the algebras A_i is non-degenerate) are needed in order to pass to copies of these algebras that are mutually disjoint, except for 0 and 1. The homomorphisms may be dispensed with if one assumes that the intersection of two members of the family $\{A_i\}$ (with different indices) is always just the subalgebra $\{0, 1\}$. Under this additional assumption, the internal sum of the family always exists (by the general existence proof given above, and a suitable version of the exchange principal), and it is unique to within an isomorphism that is the identity on each A_i (by the corresponding uniqueness argument — see Exercise 2).

The relationship between sums and internal sums is akin to the relationship between products and internal products. The product of an arbitrary family of Boolean algebras always exists, while the internal product exists only when the algebras are mutually disjoint except for a common zero. Similarly, the sum of an arbitrary family of Boolean algebras always exists, while the internal sum exists only when the algebras are mutually disjoint except for a common zero and unit. In restricting oneself to internal sums, no real generality is lost: every sum is isomorphic to an internal one. The advantage of internal sums is that they involve less notation, so that it is easier to understand the formulations and proofs of theorems. Once the essential ideas are understood in the context of internal sums, the formulations and proofs of the analogous results for arbitrary sums is rather straightforward. As an example, here is the formulation of Lemma 1 for arbitrary sums.

Corollary 1. *A Boolean algebra A is the sum of a family of Boolean algebras $\{A_i\}$ just in case there are homomorphisms f_i from A_i into A such that the union $\bigcup_i f_i(A_i)$ generates A, and whenever J is a finite, non-empty subset of the indices, and p_i is an element in A_i such that $f_i(p_i) \neq 0$, for each i in J, then*

$$\bigwedge_{i \in J} f_i(p_i) \neq 0.$$

If the algebra A in the preceding corollary is non-degenerate, then the word "homomorphisms" may be replaced by the word "monomorphisms", and the condition $f_i(p_i) \neq 0$ may be replaced by the condition $p_i \neq 0$, in the formulation of the corollary.

So far, our view of sums has been rather abstract. If A is the (internal) sum of a family of Boolean algebras $\{A_i\}$, what do the elements in A look like? According to Theorem 3 (p. 82), they are the finite joins of finite meets of elements and complements of elements from the generating set

$$E = \bigcup_i A_i.$$

The set E is closed under complementation: if p is in A_i, then so is p'. Also, the meet of a finite family of elements from one of the subalgebras A_i is again an element in that subalgebra. Consequently, the elements in A are just the finite joins of meets of the form

$$\bigwedge_{i \in J} p_i,$$

where J is some finite subset of the index set and, for each i in J, the element p_i is in A_i. Two such meets are always distinct as long as the elements p_i are different from 0 and 1. More precisely, if p_j is an element in A_j and $p_j \neq 0, 1$ for each j in a finite subset J of indices, and if q_k is an element in A_k and $q_k \neq 0, 1$ for each k in a finite subset K of indices, then

$$\bigwedge_{j \in J} p_j = \bigwedge_{k \in K} q_k$$

implies that $J = K$ and $p_j = q_j$ for each j in J (see Exercise 17).

Two examples may serve to illustrate the sum construction. In the first, let A be a Boolean algebra freely generated by a set E, and let $\{E_i\}$ be a partition of E, that is, a family of mutually disjoint subsets whose union is E. For each i, take A_i to be the subalgebra of A generated by E_i. It is not difficult to see that the algebra A is the internal sum of the family $\{A_i\}$. For the proof, consider an arbitrary Boolean algebra B such that for each i, there is a homomorphism g_i from A_i into B. Define a function h from E into B by putting

$$h(p) = g_i(p)$$

if p is in E_i. Since E freely generates A, the mapping h can be extended to a homomorphism g from A into B. The homomorphisms g and g_i agree with h on the set E_i, so they must agree with each other on the generated subalgebra A_i, by Lemma 13.2.

The preceding example underscores the close connection between sums of families of Boolean algebras and free Boolean algebras. The next example is more generic and therefore more illuminating; it gives an explicit construction of the sum of two arbitrary fields of sets. Suppose A_1 and A_2 are fields of subsets of sets X_1 and X_2 respectively. Form the Cartesian product $X_1 \times X_2$ of the two sets, and write h_1 and h_2 for the left and right projections on this product:

$$h_1(x_1, x_2) = x_1 \quad \text{and} \quad h_2(x_1, x_2) = x_2$$

for all x_1 in X_1 and all x_2 in X_2. Define mappings f_1 and f_2 from A_1 and A_2 into the field $\mathcal{P}(X_1 \times X_2)$ of all subsets of $X_1 \times X_2$ by writing

$$f_1(P) = h_1^{-1}(P) = P \times X_2 \quad \text{and} \quad f_2(Q) = h_2^{-1}(Q) = X_1 \times Q$$

for all P in A_1 and all Q in A_2. It is an easy matter to check that f_1 and f_2 are homomorphisms (see Exercise 33.1), and actually monomorphisms when the underlying sets X_1 and X_2 are non-empty. The sum of the algebras A_1 and A_2 is just the subfield of $\mathcal{P}(X_1 \times X_2)$ generated by the union $f_1(A_1) \cup f_2(A_2)$. Indeed, if P is in A_1, and Q in A_2, then

$$f_1(P) \cap f_2(Q) = (P \times X_2) \cap (X_1 \times Q) = P \times Q,$$

and this set is non-empty whenever P and Q are nonempty; thus, the conditions of Corollary 1 are satisfied. The elements in the sum are precisely the finite unions of *rectangles* $P \times Q$ with *sides* P in A_1 and Q in A_2, by the remarks preceding the examples. Notice, in particular, that the elements in the image algebra $f_1(A_1)$ are the rectangles of the form $P \times X_2$, for P in A_1, and, similarly, the elements in $f_2(A_2)$ are the rectangles of the form $X_1 \times Q$, for Q in A_2.

The example just given is more general than might appear at first glance: it can be modified to give a concrete description of the sum of two arbitrary Boolean algebras A_1 and A_2. In fact, by the Stone representation theorem (Theorem 17, p. 189), A_1 is isomorphic to a subfield of $\mathcal{P}(X_1)$ for some set X_1, and A_2 is isomorphic to a subfield of $\mathcal{P}(X_2)$ for some set X_2. Suppose the isomorphisms involved are k_1 and k_2 respectively. The mappings f_1 and f_2 from A_1 and A_2 into $\mathcal{P}(X_1 \times X_2)$ defined by

$$f_1(p) = h_1^{-1}(k_1(p)) = k_1(p) \times X_2 \quad \text{and} \quad f_2(q) = h_2^{-1}(k_2(q)) = X_1 \times k_2(q)$$

are homomorphisms, and the sum of A_1 and A_2 is, to within an isomorphism, just the subfield of $\mathcal{P}(X_1 \times X_2)$ generated by the union $f_1(A_1) \cup f_2(A_2)$.

The example may be generalized still further. Consider an arbitrary family $\{A_i\}$ of Boolean algebras. Each algebra A_i can be represented as a field of subsets of some set X_i via an isomorphism k_i. Take X to be the Cartesian product of the family of sets $\{X_i\}$, and let h_i be the projection from X to X_i. The mapping f_i defined by

$$f_i(p) = h_i^{-1}(k_i(p))$$

for each p in A_i is a homomorphism from A_i into $\mathcal{P}(X)$. The sum of the family $\{A_i\}$ is just the subalgebra of $\mathcal{P}(X)$ generated by the union $\bigcup_i f_i(A_i)$. When the underlying sets X_i are non-empty, this observation follows readily from Corollary 1. This construction gives another proof of the existence of the sum of an arbitrary family of Boolean algebras.

Sum and product constructions similar to the Boolean ones introduced above are useful for every known mathematical category, and, almost as a consequence of their universality, they are called by many different names. The terminology adopted above clashes head-on with some terms in common usage, but even so it is as nearly consistent with all existing terminologies as any systematic usage could possibly be. No one will argue about products of topological spaces; that terminology is universally accepted. Products of algebras are almost as good (but not quite); the terminology is in harmony with accepted usage for groups, modules, and rings. Instead of "product" a group-theorist would perhaps say "direct product", or, in the infinite case, "strong direct product", but that is close enough. Disagreements begin when group-theorists speak of "direct sum" or "strong direct sum". Even our "product" of Boolean algebras is sometimes called "direct sum", or, worse yet, "direct union". Our "sum" of Boolean spaces is not in common usage, but it does not seriously conflict with anything either; its sole competitor is "disjoint union", and that in the finite case only. The most radical departure is our "sum" of algebras. The word is in harmony with "weak direct sum" for modules, which, however, has also been called "weak direct product". The word is completely out of harmony with the usage in non-abelian group theory; the corresponding concept there is called "free product", and this terminology is also frequently used in Boolean algebra. Whether the word has the right intuitive connotations is perhaps arguable; at the very least a good case can be made for it.

The notion of the sum of a family of Boolean algebras goes back to Sikorski [62] (though the terminology and definition he used are different from what has been used in this chapter). He proved the existence and uniqueness of the sum of any family of Boolean algebras. (His existence proof uses the method described two paragraphs above.) Corollary 1 is also due to him, and shows the equivalence of his definition with the one given in this chapter.

Exercises

1. Suppose X is the sum of a family $\{X_i\}$ of Boolean spaces. Prove that X, together with a family of continuous mappings ϕ_i from X_i into X that satisfies the transfer condition, is unique to within a homeomorphism. The transfer condition says that if Y is any Boolean space, and if, for each i, there is a continuous mapping ψ_i from X_i into Y, then there is a unique continuous mapping ψ from X into Y such that $\psi \circ \phi_i = \psi_i$ for all i.

2. Suppose A is the sum of a family $\{A_i\}$ of Boolean algebras. Prove that A, together with a family of homomorphisms f_i from A_i into A that satisfies the transfer condition, is unique to within an isomorphism. The transfer condition says that if B is any Boolean algebra, and if, for each i, there is a homomorphism g_i from A_i into B, then there is a unique homomorphism g from A into B such that $g \circ f_i = g_i$ for all i.

3. Prove that the requirement, in the definition of the sum of a family of Boolean algebras, that the homomorphism g in the transfer condition be unique is equivalent to the requirement that the union $\bigcup_i f_i(A_i)$ generate A.

4. Prove that the sum of a family of Boolean algebras is degenerate if and only if one of the algebras in the family is degenerate.

5. Prove that the sum of the empty family of Boolean algebras is the two-element Boolean algebra.

6. Prove that a Boolean algebra A is the sum of a family $\{A_i\}$ of Boolean algebras with respect to homomorphisms f_i from A_i into A if and only if A is the internal sum of the family $\{f_i(A_i)\}$ of image algebras.

7. Prove Corollary 1.

8. Prove the observation made after Corollary 1 that if the algebra A (in the corollary) is non-degenerate, then the word "homomorphisms" may be replaced by "monomorphisms", and the condition $f_i(p_i) \neq 0$ may be replaced by the condition $p_i \neq 0$. Why is it necessary to require that A be non-degenerate?

9. Formulate and prove simplified versions of Lemma 1 and Corollary 1 that apply to finite, non-empty families of Boolean algebras.

10. Prove that a Boolean algebra A is the internal sum of a family $\{A_i\}_{i \in I}$ of subalgebras if and only if A is generated by the union of the family and, for each finite, non-empty subset J of the indices, the subalgebra generated by the union of the subfamily $\{A_i\}_{i \in J}$ is the internal sum of this subfamily. (This exercise says that, in essence, the sum of an infinite family of Boolean algebras is the union of the sums of the finite subfamilies.)

11. Formulate and prove an appropriate version of Exercise 10 for infinite sequences of subalgebras.

12. Suppose A is the internal sum of subalgebras A_1 and A_2. Prove directly, without using Theorem 3, that the elements of A are precisely the finite joins of elements of the form $p \wedge q$, with p in A_1 and q in A_2.

13. Suppose A is the internal sum of a finite sequence $A_1, A_2, \ldots, A_n$ of subalgebras. Describe the elements of A.

14. Use Exercise 12 to give a description of the elements in the sum of two arbitrary Boolean algebras.

15. Suppose A is the sum of an arbitrary family $\{A_i\}$ of Boolean algebras. Describe the elements of A.

16. Suppose A is the internal sum of subalgebras A_1 and A_2. Show that the following equivalences hold for all non-zero elements p_1, q_1 in A_1, and p_2, q_2 in A_2.

    (a) $p_1 \wedge p_2 \leq q_1 \wedge q_2$ if and only if $p_1 \leq q_1$ and $p_2 \leq q_2$.

    (b) $p_1 \wedge p_2 = q_1 \wedge q_2$ if and only if $p_1 = q_1$ and $p_2 = q_2$.

17. Suppose A is the internal sum of a family of subalgebras $\{A_i\}$, and let J and K be finite subsets of the set of indices. Show that the following equivalences hold whenever $\{p_i\}_{i \in J}$ and $\{q_i\}_{i \in K}$ are families of non-zero elements in A such that p_i, respectively q_i, is in A_i for every i in J, respectively K.

    (a) $\bigwedge_{i \in J} p_i \leq \bigwedge_{i \in K} q_i$ if and only if $p_i \leq q_i$ for i in $J \cap K$, and $q_i = 1$ for i in $K - J$.

    (b) $\bigwedge_{i \in J} p_i = \bigwedge_{i \in K} q_i$ if and only if $p_i = q_i$ for i in $J \cap K$, while $q_i = 1$ for i in $K - J$, and $p_i = 1$ for i in $J - K$.

18. If $A_i = 2$ for every element i in a set I, what is $\sum_i A_i$?

19. Prove that if $A_i = 2 \times 2$ for every element i in a set I, then $\sum_i A_i$ is isomorphic to the free algebra on m generators, where m is the cardinality of I.

20. Prove that
$$A + (B \times C) = (A + B) \times (A + C).$$
(The equality sign here should be interpreted to mean isomorphism.)

21. For an arbitrary Boolean algebra B, what is $B + 2$?

22. For an arbitrary Boolean algebra B, and any natural number n, what is $B + 2^n$?

23. If B is a finite Boolean algebra, and I an arbitrary set, what is $B + 2^I$?

24. If B is an arbitrary Boolean algebra, and I an arbitrary set, what is $B + 2^I$?

25. Suppose B is an arbitrary Boolean algebra, I an arbitrary set, and C the subalgebra of 2^I consisting of those elements x such that $x(i) = 1$ for all but finitely many i, or else $x(i) = 0$ for all but finitely many i. (See Exercise 26.37.) What is $B + C$?

26. Suppose Boolean algebras A_1 and A_2 are isomorphic to fields of subsets of X_1 and X_2 via mappings k_1 and k_2 respectively. Let h_1 and h_2 be the left and right projections of $X_1 \times X_2$ to X_1 and X_2, and define mappings f_1 and f_2 from A_1 and A_2 into $\mathcal{P}(X_1 \times X_2)$ by
$$f_1(p) = h_1^{-1}(k_1(p)) = k_1(p) \times X_2,$$
$$f_2(q) = h_2^{-1}(k_2(q)) = X_1 \times k_2(q),$$
for p in A_1 and q in A_2. Show that these mappings are homomorphisms, and prove that the sum $A_1 + A_2$ is the subalgebra of $\mathcal{P}(X_1 \times X_2)$ generated by the union $f_1(A_1) \cup f_2(A_2)$.

27. Characterize the atoms of a sum $A_1 + A_2$ in terms of the atoms of A_1 and A_2.

28. When is the sum of two non-degenerate Boolean algebras atomic? When is it atomless?

29. If, in an infinite family of non-degenerate Boolean algebras, there are infinitely many algebras of cardinality greater than two, prove that the sum of the family is always atomless.

30. Show that a sum of complete Boolean algebras need not be complete. What about a finite sum? What about σ-algebras?

31. (Harder.) A Boolean σ-algebra A is called the σ-*sum* of a family $\{A_i\}$ of σ-algebras if for each i, there is a σ-homomorphism f_i from A_i to A such that the transfer condition for σ-algebras is satisfied. This transfer condition says that if B is a σ-algebra and if, for each i, there exists a σ-homomorphism g_i from A_i to B, then there exists a unique σ-homomorphism g from A to B such that $g \circ f_i = g_i$ for all i. Is the natural generalization of Corollary 1 to σ-sums of σ-algebras true? (The answer is due to Sikorski [59].)

Chapter 45

Isomorphisms of Countable Factors

The purpose of this chapter is to show (following Hanf [24], as simplified, orally, by Dana Scott) that there exist countable Boolean algebras A and B such that

$$A = A \times B \times B, \qquad \text{but} \qquad A \neq A \times B.$$

The method of attack is topological; in fact, we shall construct Boolean spaces X and Y, each with a countable base, such that

$$X = X + Y + Y, \qquad \text{but} \qquad X \neq X + Y.$$

(The equal sign denotes homeomorphism here.) The countability of the bases implies that the corresponding fields of clopen sets — the dual algebras of X and Y — are countable (see p. 339). Take A and B to be these dual algebras; the topological result then implies the algebraic one, by Theorem 44 (p. 397) and Exercises 34.8 and 34.9.

We begin by constructing for each natural number n $(= 0, 1, 2, \ldots)$ a Boolean space U_n with a countable base, and a distinguished point u_n of U_n, such that no neighborhood of u_n is homeomorphic to any neighborhood of any other point in any U_m (not even in U_n itself). Here is one way to do this: let U_n consist of a sequence of type ω^n in $[-1, 0]$ converging to 0, together with the Cantor set in $[0, 1]$, under the topology inherited from the space of real numbers.

In more detail, let U_0 be the Cantor set (see Chapter 33). To construct the set U_1, define inductively a sequence $\{x_n\}$ of points in $[-1, 0]$ by repeated

S. Givant, P. Halmos, *Introduction to Boolean Algebras*,
Undergraduate Texts in Mathematics, DOI: 10.1007/978-0-387-68436-9_45,
© Springer Science+Business Media, LLC 2009

(right-hand) bisection of the interval $[-1, 0]$: put $x_0 = -1$, take x_1 to be the midpoint of $[-1, 0]$, and in general take x_{n+1} to be the midpoint of $[x_n, 0]$ (see the diagram). Obviously, $x_n = -1/2^n$. The sequence so defined is said to be of *type* ω; it is strictly increasing and converges to 0. The set U_1 consists of the points x_n and the points in the Cantor set. The isolated points in U_1 are just the points x_n.

The construction of the set U_2 is analogous, but uses a sequence of type ω^2 in the interval $[-1, 0]$ instead of a sequence of type ω. In other words, it uses a strictly increasing sequence $\{y_\alpha\}$ of real numbers between -1 and 0, indexed by the ordinals $\alpha < \omega^2$, with the property that whenever $\{\alpha_n\}_n$ is a strictly increasing sequence of ordinals (less than ω^2) with supremum α, the subsequence $\{y_{\alpha_n}\}_n$ converges to y_α, or to 0, according as α is less than, or equal to, ω^2. In particular, the subsequence $\{y_{\omega n + m}\}_m$ converges to $y_{\omega(n+1)}$, and the subsequence $\{y_{\omega n}\}_n$ converges to 0. To construct such a sequence, apply the method of bisection to each of the intervals $[x_n, x_{n+1}]$ (where $\{x_n\}$ is the sequence of type ω defined in the preceding paragraph): put $y_{\omega n} = x_n$, take $y_{\omega n + 1}$ to be the midpoint of $[x_n, x_{n+1}]$, and in general take $y_{\omega n + m + 1}$ to be the midpoint of $[y_{\omega n + m}, x_{n+1}]$ (see the diagram). The set U_2 consists of the points in this sequence and the points in the Cantor set. The isolated points in U_2 are the points y_α for non-limit ordinals α.

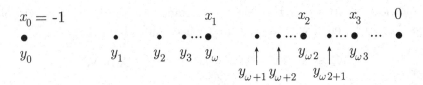

The construction of U_3 uses a sequence of type ω^3 in $[-1, 0]$ that has 0 as the supremum of its limit points. One way to define such a sequence is to use the method of bisection to create, for each $\alpha < \omega^2$, a strictly increasing sequence of type ω in the interval $[y_\alpha, y_{\alpha+1}]$ that starts at y_α and has $y_{\alpha+1}$ as its limit point. (Alternatively, in each of the intervals $[x_n, x_{n+1}]$ construct a sequence of type ω^2 that starts at x_n and has x_{n+1} as the supremum of its limit points.) The set U_3 consists of the points in this sequence and the points in the Cantor set. Continue in the fashion to define each of the sets U_n.

Take u_n to be the point 0 in U_n. The proof that no neighborhood of u_n is homeomorphic to a neighborhood of any other point in any U_m uses the notion of the derivative of a set S of real numbers. The *derivative* of S is, by definition, the set of real numbers that are limit points of one-to-one sequences of points (numbers) in S. Equivalently, it is the set S with the isolated points of S removed. The second derivative of S is defined to be the derivative of the derivative of S, the third derivative of S is the derivative of the second derivative of S, and so on. The 0th derivative of S is defined to be S itself.

The derivative of the set U_0 is itself, since every point in the Cantor set is the limit of a strictly increasing or strictly decreasing sequence (Exercise 33.27); consequently, for every natural number n, the set U_0 is its own nth derivative. The derivative of U_1 is the set U_0, since the points x_n are all isolated in U_1. The nth derivative of U_1 is therefore equal to U_0 for all $n > 0$. The derivative of U_2 is the set $U_1 - \{x_0\}$, since, for each positive integer n, a strictly increasing sequence of isolated points converging to x_n was adjoined to U_1 to obtain the set U_2. The nth derivative of U_2 is equal to U_0 for every $n > 1$. In other words, the zeroth, first, and second derivatives of U_2 are distinct from each other, and all higher-order derivatives coincide with the second derivative. In general, the zeroth through the nth derivatives of U_n form a strictly decreasing sequence of sets (in the sense of inclusion), and all derivatives of order greater than n coincide with the nth derivative, which is U_0.

The argument just given also applies to arbitrary neighborhoods of the point $u_n = 0$ in the space U_n: the derivatives of order less than n of every neighborhood of u_n contain isolated points, whereas the nth derivative is perfect in the sense that it contains no isolated points. No other point in any of the spaces under consideration can make that claim.

Next, we form the union of a disjoint class consisting of exactly one copy of each of the spaces U_k with $k \geq n$; let Y_n be the one-point compactification, by y, of that union. The neighborhoods of y are, by definition, the complements of closed compact subsets of the union. Since a closed subset of the union is compact just in case it has a non-empty intersection with only finitely many of the component spaces U_k, each neighborhood of y must include all but finitely many of the component spaces of the union. Schematically, Y_n may be represented in the form

$$ n, \quad n+1, \quad n+2, \quad \cdots \longrightarrow \quad y, $$

where, for the sake of brevity, we have used the symbol for the integer n to denote the space U_n. As the one-point compactification of a countable union of Boolean spaces with countable bases, each space Y_n is Boolean and has a countable basis (Exercise 43.19).

We form also the union of a disjoint class consisting of exactly two copies of each of the spaces U_k with $k \geq n$; let Z_n be the one-point compactification, by z, of that union. Each neighborhood of z includes all but finitely many of the component spaces from the union. Again, Z_n is a Boolean space with a countable base; it may be represented schematically in the form

$$n, \quad n+1, \quad n+2, \quad \cdots \longrightarrow \quad z \quad \longleftarrow \cdots, \quad n+2, \quad n+1, \quad n.$$

We go on to form the union of a countably infinite disjoint class consisting of copies of Z_0, and compactify it by one point z^*. The result is a Boolean

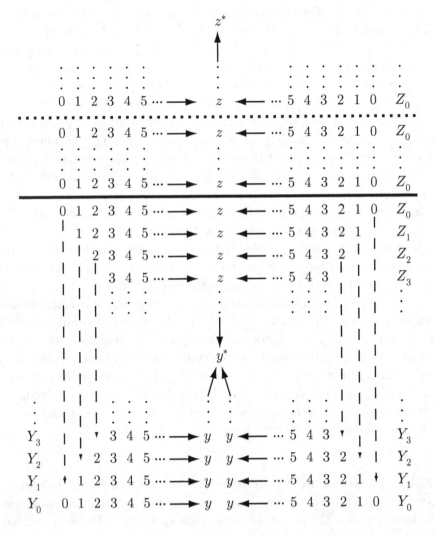

space with a countable base; it is represented schematically by the part of the subjoined diagram that lies above the unbroken dividing line.

The part of the diagram below that line is a schematic representation of the union of a disjoint class consisting of exactly one copy of each Z_n and of exactly two copies of each Y_n, all compactified by one point y^*; it, too, is a Boolean space with a countable base. Let X be the disjoint union of the two grand unions formed before, so that the whole diagram represents X. Clearly, X is a Boolean space with a countable base.

Each copy of each u_n in X has a neighborhood that contains no other copy of that u_n or of any other. (Take the copy of U_n to which the given copy of u_n belongs.) The u_n's are the only points of X with this property.

Every neighborhood of each copy of y in X contains a copy of almost all the u's (that is, all but a finite number), and some neighborhood of each y contains exactly one copy of each u. (Take the copy of Y_n to which the given copy of y belongs, and form its union with one copy of each of the sets U_k for $k = 0, \ldots, n - 1$, taken, say, from some copy of Y_0. Since the sets U_k are open, the union of these sets with the copy of Y_n is a neighborhood of y.) The y's are the only points in X with this property.

Every neighborhood of each copy of z in X contains at least two copies of almost all the u's, and some neighborhood of each z contains exactly two copies of each u. (Take the copy of Z_n to which the given copy of z belongs, and form its union with two distinct copies of each of the sets U_k for $k = 0, \ldots, n - 1$, taken, say, from some copy of Z_0.) The z's are the only points in X with this property.

Every neighborhood of y^* contains almost all y's and infinitely many z's. The point y^* is the only point in X with this property.

Every neighborhood of z^* contains infinitely many z's, and some neighborhood of z^* contains no y's. (Take the part of the diagram that lies above the unbroken dividing line.) The point z^* is the only point in X with this property.

The preceding paragraphs imply that for each n, the set U_n^* of all u_n's (in X) is definable in topological terms. Similarly, the set Y^* of all y's, the set Z^* of all z's, and the points y^* and z^* are all definable in purely topological terms. It follows that if ϕ is a homeomorphism of X onto X, then the points y^* and z^* are invariant under ϕ in the sense that they are mapped to themselves by ϕ, and the sets Y^*, Z^*, and U_n^* (for all n) are also invariant under ϕ in the sense that each of these sets is mapped to itself by ϕ.

Let the space Y be the Y_0 already defined above. We are to prove that two copies of Y can be adjoined to X with impunity; we shall prove the equivalent assertion that two copies of Y can be discarded from X with impunity. Suppose, indeed, that the bottom row of the diagram is erased, and write $\widetilde{X}$ for the resulting space. To reconstruct the space X, take the 0's from the lowest Z_0 and give them to Y_1 to make a copy of Y_0, take the 1's from Z_1 and give them to Y_2 to make a copy of Y_1, and so on, as indicated by the long dashed vertical arrows in the diagram; leave all other parts of X alone. (More precisely, map each Y_1 in the deleted space $\widetilde{X}$ to the corresponding part of Y_0 in X, and map the 0's from the lowest Z_0 in $\widetilde{X}$ to the 0's of the corresponding Y_0 in X. Map each Y_2 in $\widetilde{X}$ to the corresponding part of Y_1 in X, and map the 1's from Z_1 in $\widetilde{X}$ to the 1's of the corresponding Y_1 in X. Continue in this fashion. Then map the remaining part of the lowest Z_0 in $\widetilde{X}$ to Z_1 in X, map the remaining part of Z_1 in $\widetilde{X}$ to Z_2 in X, and so on. Finally, map each Z_0 above the unbroken dividing line in $\widetilde{X}$ to the Z_0 directly below it in X. The points y^* and z^* are mapped to themselves.) The transformation so defined is a homeomorphism from the deleted space $\widetilde{X}$ to the original X. Indeed, since the transformation is obviously a bijection, the proof that it is a homeomorphism amounts to establishing its continuity (Lemma 33.5). A Boolean space with a countable base is naturally a metric space (Exercise 35.13); a proof of continuity therefore reduces to a proof of sequential continuity (Exercises 33.10 and 33.11). The verification of the latter is routine. The only excitement can come from a sequence chosen from the moving parts and converging to y^*; the construction guarantees that the transform of such a sequence still converges to the (fixed) point y^*.

The next and last thing to prove is that the adjunction of one copy of Y to X makes a difference; we shall prove the equivalent assertion that if X is diminished by discarding one copy of Y, say the right half of the bottom line, then the resulting space $\widetilde{X}$ is topologically distinguishable from X. Indeed, X has an involution that leaves fixed each point of the definable subset

$$(1) \qquad\qquad\qquad Z^* \cup \{y^*\} \cup \{z^*\}$$

of X, and nothing else; just reflect the diagram about the central vertical axis. (By an *involution* of X is meant a homeomorphism of X to X such that the composition of the homeomorphism with itself is the identity function on X.) We shall prove that $\widetilde{X}$ has no such homeomorphism. Suppose that, on the contrary, ϕ is an involution of $\widetilde{X}$ whose set of fixed points is exactly the definable subset (1) of $\widetilde{X}$. Our remaining task is to derive a contradiction from this supposition.

The involution ϕ fixes the point z^*, and therefore must map the subspace $\widetilde{X} - \{z^*\}$ homeomorphically to itself. This subspace is not compact: if V is the part of $\widetilde{X}$ represented by the part of the diagram below the unbroken horizontal line (with the bottom-right copy of Y removed), then V and the different copies of Z_0 represented in the part of the diagram above the unbroken horizontal line constitute together an infinite open cover of the subspace that has no proper subcover at all, much less a finite one. The subspace V itself is compact (it is the one-point compactification of a disjoint union of Boolean spaces) and so are the different copies of Z_0; the union of V with finitely many of the copies of Z_0 is therefore compact (Exercise 43.5).

Since V is a compact subspace of $\widetilde{X} - \{z^*\}$, the same is true of $\phi(V)$. This implies that there exists a dotted horizontal line above finitely many of the copies of Z_0 (as indicated) such that $\phi(V)$ is below it. (The linguistic identification of parts of the diagram with corresponding parts of the space $\widetilde{X}$ is obvious and harmless.) The involution ϕ maps U_0^* into itself, and by assumption it leaves no copy of u_0 fixed; it therefore induces a pairing of the various copies of u_0. Below the dotted line there are an odd number of 0's: two in each of the finitely many copies of Z_0 and one in the unique copy of Y_0. One of those 0's (or, to be a little more precise, the copy of u_0 belonging to one of those 0's) is therefore mapped above the dotted line. The 0 (or 0's) to which this happens cannot be in V (since $\phi(V)$ is below the dotted line). Conclusion: one of the 0's between the two horizontal lines gets mapped above the dotted line. There are also an odd number of 1's below the dotted line: two in each copy of Z_0, two in the unique copy of Z_1, one in each of the two copies of Y_1, and one in the single copy of Y_0. As in the case of 0, one of the 1's between the two horizontal lines must get mapped above the dotted line. What was just argued about the 0's and the 1's is just as true about the 2's, the 3's, etc. Since there are only a finite number of rows between the two horizontal lines, it follows that there is at least one such row with the property that infinitely many of its parts get mapped above the dotted line. Since from those parts a sequence of points converging to some z (between the lines) can be selected, the continuity of ϕ implies that ϕ moves some z from between the lines to above the dotted line. The contradiction has arrived: the z's are assumed to be fixed under ϕ.

Exercises

1. (Harder.) Prove that each of the spaces U_n is Boolean and has a countable basis.

2. It is proved in this chapter that if two copies of Y are removed from the space X, the resulting space is homeomorphic to X. Show that this

implies $X + Y + Y = X$.

3. Prove that there exist countable Boolean algebras A and B such that

$$A = A \times B \times B \times B,$$

but

$$A \neq A \times B \quad \text{and} \quad A \neq A \times B \times B.$$

(This result is due to Hanf [24].)

4. Generalize Exercise 3.

5. Show that a topological version of the Schröder–Bernstein theorem fails, even when the spaces in question are Boolean spaces with countable bases. More precisely, show that there exist two Boolean spaces with countable bases such that each space is homeomorphic to an (open) subspace of the other, but the two spaces are not homeomorphic to one another. (This theorem is due to Kinoshita [33] and predates Hanf's result. It improves an earlier theorem of Kuratowski [38].)

Epilogue

There is much more to Boolean algebras than is covered in this volume. The reader who wants to learn more should consult the comprehensive three-volume *Handbook of Boolean Algebras* [45], edited by Monk and Bonnet, or Sikorski's scholarly book [65]. Both works have excellent bibliographies.

Appendix A

Set Theory

In parts of this book, familiarity with some of the basic notions and theorems of set theory is needed. The purpose of this appendix is to present the requisite material.

Sets and Subsets

Intuitively speaking, a *set* is a collection of objects, and these objects are called the *elements* of the set. A line, for instance, is a set of points, and the points are the elements of the line; the set of lines in a plane is an example of a set of sets. The principal concept of set theory is that of *belonging*, or being an element of. If x belongs to X, we write

$$x \in X.$$

Other phrases used to express this notion are, for instance, "x is an element of X", and "x is a member of X", and "x is contained in X".

If X and Y are sets, and if every element of Y is an element of X, we say that Y is a *subset* of X, or that Y is *included* in X, or that X *includes* Y, and we write

$$Y \subseteq X \quad \text{or} \quad X \supseteq Y.$$

For example, the points inside the unit circle in the Cartesian plane are a subset of the set of all points in the plane. Two sets X and Y are *equal* if they have the same elements, that is, if

$$X \subseteq Y \quad \text{and} \quad Y \subseteq X.$$

A subset of X that is not equal to X is said to be a *proper* subset.

A subset of a given set X is often described by specifying a defining property of its elements. For instance, if X is the set of real numbers, then the subset

$$Y = \{x \in X : |x| \leq 1\}$$

is the closed interval of all real numbers between -1 and 1:

$$Y = \{x \in X : -1 \leq x \leq 1\} = [-1, 1].$$

The subset

$$Z = \{x \in X : |x| = 1\}$$

consists of just two numbers, -1 and 1.

Unordered Pairs and their Relatives

A set with just two elements, say x and y, is called an (*unordered*) *pair* and is denoted by

$$\{x, y\}.$$

Analogously, a set with exactly three elements, say x, y, and z, is called an (*unordered*) *triple* and is denoted by

$$\{x, y, z\},$$

and so on. A set with exactly one element, say x, is called a *singleton* and is denoted by

$$\{x\}.$$

The (unique) set with no elements is called the *empty set*, and is denoted by $\varnothing$. The sets $\varnothing$ and $\{\varnothing\}$ are not the same; the first has no elements, while the second has exactly one element, namely $\varnothing$. The sets

$$\{\varnothing\}, \qquad \{\{\varnothing\}\}, \qquad \{\{\{\varnothing\}\}\}$$

are all singletons, but they are different from one another; the first is the singleton of $\varnothing$, the second the singleton of $\{\varnothing\}$, and the third the singleton of $\{\{\varnothing\}\}$.

Operations on Sets

Given two sets X and Y, we can form their *union*, the set of those elements that belong either to X or to Y (see the diagram):

$$X \cup Y = \{x : x \in X \text{ or } x \in Y\};$$

we can also form their *intersection*, the set of those elements that belong both to X and to Y (see the diagram):

$$X \cap Y = \{x : x \in X \text{ and } x \in Y\}.$$

For example, the union of the intervals $[-1, 1]$ and $[0, 2]$ (of real numbers) is

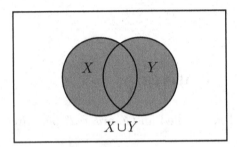

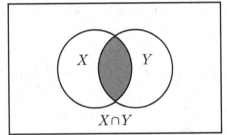

the interval $[-1, 2]$, while the intersection of the two intervals is $[0, 1]$. The *union* of a set X of sets is the set of those elements that belong to at least one of the sets in X, in symbols

$$\bigcup X = \{x : x \in P \text{ for some } P \in X\}.$$

When X is non-empty, the *intersection* of X is the set of those elements that belong to all of the sets in X, in symbols

$$\bigcap X = \{x : x \in P \text{ for all } P \in X\}.$$

For example, if X is the set of intervals $[n, n + 1]$, for $n = 0, 1, 2, 3, \ldots$, then the union of X is the interval $[0, \infty)$ of all non-negative real numbers. If X is the set of intervals $[-1/n, 1/n]$, for $n = 1, 2, 3, \ldots$, then the intersection of X is the singleton $\{0\}$.

The *difference* of two sets X and Y, often called the *complement of Y in X* (or the complement of Y *with respect to X*), is the set of elements that belong to X, but not to Y (see the diagram below):

$$X - Y = \{x \in X : x \notin Y\}.$$

If all sets under discussion are subsets of a given set X, then the complement

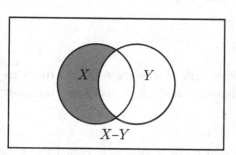

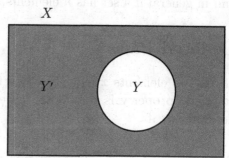

of Y in X is usually written as Y'.

The *symmetric difference* of two sets X and Y is the set of elements that are in one of the two sets, but not in the other (see the diagram below):

$$X + Y = (X - Y) \cup (Y - X) = (X \cup Y) - (X \cap Y).$$

For example, the difference of the intervals $[-1, 1]$ and $[0, 2]$ is the half-closed

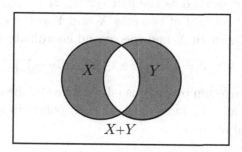

interval

$$[-1, 0) = \{x : -1 \le x < 0\},$$

while the symmetric difference of the two intervals is the union

$$[-1, 0) \cup (1, 2] = \{x : -1 \le x < 0 \text{ or } 1 < x \le 2\}.$$

The *power set* of a set X is the set of all subsets of X, and is denoted by $\mathcal{P}(X)$. For instance, the power set of the pair $\{x, y\}$ consists of the four subsets

$$\varnothing, \qquad \{x\}, \qquad \{y\}, \qquad \{x, y\}.$$

Similarly, if a set has three elements, then its power set has eight elements, and in general if a set has n elements, then its power set has 2^n elements.

Ordered Pairs

Given two elements x and y, we can form the *ordered pair* (x, y). Its characteristic property is that both the elements and their order are uniquely determined: if

$$(x, y) = (z, w),$$

then $x = z$ and $y = w$. The elements x and y are respectively called the *first* and *second coordinates* of the pair. There are several ways of defining this notion in set theory so as to achieve the desired property. The most common definition (due to Kuratowski [35]) is to put

$$(x, y) = \{\{x\}, \{x, y\}\}.$$

The notions of an *ordered triple*, an *ordered quadruple*, and so on can be defined in terms of the notion of an ordered pair. For instance, the ordered triple (x, y, z) can be defined as the pair $((x, y), z)$.

The (*Cartesian*) *product* of two sets X and Y is the set of ordered pairs with the first coordinate in X and the second coordinate in Y:

$$X \times Y = \{(x, y) : x \in X \text{ and } y \in Y\}.$$

For example, the Cartesian plane is usually defined as the set $\mathbb{R} \times \mathbb{R}$ (where $\mathbb{R}$ denotes the set of all real numbers), so that points in the plane are just ordered pairs of real numbers.

Relations

A (*binary*) *relation* is, by definition, a set of ordered pairs. For example, from a set-theoretic point of view the relation between people of "being a parent of" is the set of pairs (x, y) such that x and y are human beings, and x is a parent of y. The *domain* and *range* of a relation Θ are, respectively, the sets

$$X = \{x : (x, y) \in \Theta \text{ for some } y\} \quad \text{and} \quad Y = \{y : (x, y) \in \Theta \text{ for some } x\}.$$

Notice that Θ is a subset of the product $X \times Y$. In fact, this product is itself a relation with domain X and range Y.

A subset of $X \times X$ is said to be relation *on* the set X. Such a relation Θ is said to be *reflexive* if (x, x) is in Θ for every x in X, *symmetric* if the presence of (x, y) in Θ implies that (y, x) is in Θ, *antisymmetric* if the presence of (x, y) and (y, x) in Θ implies that $x = y$, and *transitive* if the presence of (x, y) and (y, z) in Θ implies that (x, z) is in Θ.

Equivalence Relations

A relation on X that is reflexive, symmetric, and transitive is called an *equivalence relation* on X. If Θ is such a relation, and if (x, y) is in Θ, then we say that x and y are *equivalent modulo* Θ and write

$$x \equiv y \quad \mathrm{mod}\ \Theta.$$

The set

$$x/\Theta = \{y \in X : x \equiv y \quad \mathrm{mod}\ \Theta\}$$

(the set of elements that are equivalent to x modulo Θ) is called the *equivalence class* of x (modulo Θ).

Equivalent elements have the same equivalence class and conversely:

$$x/\Theta = y/\Theta \quad \text{if and only if} \quad x \equiv y \quad \mathrm{mod}\ \Theta.$$

For the proof, assume first that the left side of the equivalence holds. The reflexivity of Θ implies that y is in y/Θ and therefore also in x/Θ; consequently, the right side of the equivalence holds, by the definition of the equivalence class of x. Now suppose that the right side of the equivalence holds. If z is in y/Θ, then $y \equiv z$ mod Θ, by the definition of the equivalence class of y, and therefore $x \equiv z$ mod Θ, by the transitivity of Θ; hence, z is in x/Θ, by the definition of the equivalence class of x. This argument shows that y/Θ is included in x/Θ. An analogous argument, using also $y \equiv x$ mod Θ (which holds by symmetry) shows that x/Θ is included in y/Θ, so that the two equivalence classes are equal, as desired.

An easy consequence of the preceding observation is that two equivalence classes of Θ are either equal or disjoint. Indeed, suppose that x/Θ and y/Θ have at least one element in common, say z. Then

$$x \equiv z \quad \mathrm{mod}\ \Theta \quad \text{and} \quad y \equiv z \quad \mathrm{mod}\ \Theta,$$

by the definition of an equivalence class. An application of symmetry and transitivity yields $x \equiv y$ mod Θ, so that the equivalence classes x/Θ and y/Θ are equal, by the observations of the preceding paragraph. Conclusion: an equivalence relation on a set X induces a natural *partition* of X into equivalence classes; this means that the equivalence classes are *disjoint* (distinct classes have no elements in common), and that the union of the classes is all of X.

Conversely, every partition of X induces an equivalence relation Θ on X the equivalence classes of which are precisely the sets of the given partition; just define Θ to be the set of pairs (x, y) such that x and y belong to the same set of the partition.

Functions

A *function* from a set X to (or into) a set Y is a relation f with domain X, and range included in Y, such that for each x in X, there is a unique y for which the pair (x, y) is in f. In other words, f is a subset of $X \times Y$ with the property that

$$(x, y) \in f \qquad \text{and} \qquad (x, z) \in f \qquad \text{implies} \qquad y = z.$$

It is customary to write

$$f(x) = y$$

instead of $(x, y) \in f$. The element y is called the *value* that f assumes (or takes on) at the *argument* x; we often say that f *maps*, or *sends*, or *takes* x to y. The word *mapping* is used as a synonym for the word function, and if f is a function from X to Y, we say that f *maps* X *to* Y.

The domain of f is all of X, but the range may not be all of Y (though it is certainly included in Y). If it is all of Y, that is to say, if for each y in Y there is a least one x in X such that $f(x) = y$, then f is said to map X *onto* Y. If each value of f corresponds to exactly one argument, that is to say, if for each y in Y there is at most one x in X such that $f(x) = y$, then f is said to be *one-to-one*. Another way of phrasing this definition is to say that f is one-to-one if $f(x_1) = f(x_2)$ always implies that $x_1 = x_2$.

If f is a function from X to Y, and if g is a function from Y to Z, then a function h from X to Z may be defined by

$$h(x) = g(f(x)).$$

The function h is called the *composite*, or the *composition*, of f and g, and is denoted by $g \circ f$. It is easy to verify that if f and g are both one-to-one, or both onto, then their composition $g \circ f$ is also one-to-one, or onto, accordingly. For instance, assume that both functions are one-to-one. If

$$(g \circ f)(x_1) = (g \circ f)(x_2),$$

then $f(x_1) = f(x_2)$, because g is one-to-one, and therefore $x_1 = x_2$, because f is one-to-one. Consequently, $g \circ f$ is one-to-one. To prove the second assertion, assume that both functions are onto. Given any z in Z, there is a y in Y such that $g(y) = z$, because g is onto. There is also an x in X such that $f(x) = y$, because f is onto. Obviously,

$$(g \circ f)(x) = g(f(x)) = g(y) = z.$$

Consequently, $g \circ f$ is onto.

If f is a function from X to Y, and if P is a subset of X, then there is a natural way of constructing a function g from P to Y: define

$$g(x) = f(x)$$

for every x in P. The function g is called the *restriction* of f to P, and f is called an *extension* of g to X.

Bijections

A one-to-one function from X onto Y is called a *bijection* from X to Y. The *identity function* on X is an example of a bijection from X to X; it is defined by

$$f(x) = x$$

for all x in X. If f is a bijection from X to Y, then a bijection g from Y to X is defined by requiring

$$g(y) = x \qquad \text{if and only if} \qquad f(x) = y.$$

The function g is called the *inverse* of f, and is denoted by f^{-1}.

If f is a bijection from X to Y, then the composition $f^{-1} \circ f$ is the identity function on X, and $f \circ f^{-1}$ is the identity function on Y. This observation has a useful converse. Suppose f is a function from X to Y, and g a function from Y to X. If $g \circ f$ is the identity function on X, and if $f \circ g$ is the identity function on Y, then f is a bijection and g is its inverse. The proof is not difficult. To show that f is onto, consider any element y in Y; the value $x = g(y)$ is an element of X, and

$$f(x) = f(g(y)) = (f \circ g)(y) = y,$$

by the assumption that $f \circ g$ is the identity function on Y. To show that f is one-to-one, assume $f(x_1) = f(x_2)$; then

$$x_1 = (g \circ f)(x_1) = g(f(x_1)) = g(f(x_2)) = (g \circ f)(x_2) = x_2,$$

by the assumption that $g \circ f$ is the identity function on X. Consequently, f is a bijection. If $g(y) = x$, then

$$f(x) = f(g(y)) = (f \circ g)(y) = y;$$

conversely, if $f(x) = y$, then

$$g(y) = g(f(x)) = (g \circ f)(x) = x.$$

Thus, g satisfies the condition for being the inverse of f.

Images and Inverse Images

Suppose f is a function from X to Y. For each subset P of X, the *image of P under f* is defined to be the set of values of f at arguments in P, in symbols

$$f(P) = \{f(x) : x \in P\}.$$

The use of the suggestive notation $f(P)$ to denote the image set should not cause any confusion, even though the same notation is also used to denote the value of f at an argument; it will always be clear whether P is subset of the domain of f (in which case the image interpretation of the notation is intended) or an element of the domain (in which case the value interpretation is intended).

Analogously, for each subset Q of Y, the *inverse image of Q under f* is defined to be the set of arguments that are mapped into Q by f, in symbols

$$f^{-1}(Q) = \{x \in X : f(x) \in Q\}.$$

Notice that this notation makes sense even when f is not a bijection and the inverse function of f does not exist. The double use of the notation f^{-1} should not cause the reader any confusion; it will always be clear whether Q is a subset of the set Y (in which case the inverse image interpretation of the notation is intended) or an element of Y (in which case the inverse function interpretation is intended).

Families

There are occasions when the range of a function is deemed to be more important than the function itself, and in this case a different terminology is employed. Suppose, for instance, that x is a function from a set I to a set X.

An element of I is called an *index*, I is called the *index set*, the function x itself is called a *family* indexed by I, and the value of x at an index i is called a *term* of the family and is denoted by x_i. It is common to speak of the family $\{x_i\}$ in X; when necessary, the index set I is indicated by some notational device such as

$$\{x_i\}_{i\in I} \quad \text{or} \quad \{x_i : i \in I\},$$

or by some parenthetical phrase such as "(i in I)".

If $\{P_i\}$ is a family of sets, then the union of the range of the family is called the union of the family $\{P_i\}$, or the union of the sets P_i; the standard notation for this union is

$$\bigcup_{i\in I} P_i \quad \text{or} \quad \bigcup_i P_i,$$

according as it is or is not important to emphasize the index set I. An element x belongs to this union if and only if x belongs to P_i for at least one i. An empty union makes sense (and is empty), but an empty intersection does not make sense. Except for this triviality, the terminology and notation for intersections parallels that for unions in every respect. Thus, if $\{P_i\}$ is a non-empty family of sets, then the intersection of the range of the family is called the intersection of the family $\{P_i\}$, or the intersection of the sets P_i; the standard notation for this intersection is

$$\bigcap_{i\in I} P_i \quad \text{or} \quad \bigcap_i P_i.$$

(By a "non-empty family" is meant a family whose domain I is not empty.) A necessary and sufficient condition that an element x belong to this intersection is that x belong to P_i for all i.

The *Cartesian product* of a family $\{P_i\}$ of sets is, by definition, the set of functions

$$x = \{x_i\}$$

with domain I such that x_i belongs to P_i for each index i. The notation

$$\prod_{i\in I} P_i \quad \text{or} \quad \prod_i P_i$$

is commonly employed. If x is an element of this product, then x_i is called the ith *coordinate* of x. When all the terms P_i of the family are equal to the same set, say Q, the product of the family is denoted by

$$Q^I$$

and is called the Ith *power* of Q. It makes sense to form the product of an empty family of sets, and this product is not empty; it consists of the unique function with an empty domain, namely the empty set.

Order

The theory of order plays an important role throughout mathematics. A *partial order* (or a *partial ordering*) on a set X is a binary relation on X that is reflexive, antisymmetric, and transitive. Usually, the symbol $\leq$, or some relative of it, is used to denote a partial order. In terms of this symbol, the definition of a partial order on X assumes the following form: for all x, y, and z in X, we have $x \leq x$ (reflexivity), $x \leq y$ and $y \leq x$ implies $x = y$ (antisymmetry), and $x \leq y$ and $y \leq z$ implies $x \leq z$ (transitivity).

A partial order on X is said to be *total*, or *linear*, if for every x and y in X, either $x \leq y$ or $y \leq x$. The standard ordering of the natural numbers is an example of a total order. The inclusion relation on the power set of a set X is an example of a partial ordering that is, in general, not total (unless X is empty or has just one element). A totally ordered set is frequently called a *chain*.

A totally ordered set is said to be *well ordered* (and its ordering is called a *well ordering*) if every non-empty subset has a smallest element. The set of natural numbers is well ordered by its standard ordering; the set of integers, the set of rational numbers, and the set of real numbers are not.

Natural and Ordinal Numbers

It is occasionally useful to have a concrete set-theoretic definition of the *natural numbers*. One definition (due to von Neumann [47]) identifies each natural number with the set of its predecessors, so that

$$0 = \varnothing,$$
$$1 = 0 \cup \{0\} = \{0\},$$
$$2 = 1 \cup \{1\} = \{0, 1\},$$
$$3 = 2 \cup \{2\} = \{0, 1, 2\},$$
$$4 = 3 \cup \{3\} = \{0, 1, 2, 3\},$$

and in general

$$n + 1 = n \cup \{n\} = \{0, 1, 2, \ldots, n\}.$$

The set of all natural numbers is denoted by ω, so that

$$\omega = \{0, 1, 2, \ldots, n, \ldots\}.$$

This set may itself be viewed as a number, an infinite number that coincides with the set of its predecessors. The process can then be continued indefinitely to define what are called the *ordinal numbers*. The successor of ω is defined to be the set

$$\omega + 1 = \omega \cup \{\omega\} = \{0, 1, 2, \ldots, \omega\},$$

the successor of $\omega + 1$ is defined to be the set

$$\omega + 2 = (\omega + 1) \cup \{\omega + 1\} = \{0, 1, 2, \ldots, \omega, \omega + 1\},$$

and in general

$$\omega + (n + 1) = (\omega + n) \cup \{\omega + n\} = \{0, 1, 2, \ldots, \omega, \omega + 1, \omega + 2, \ldots, \omega + n\}.$$

The process does not stop here. We can define $\omega + \omega$ (or $\omega 2$) to be the set consisting of the natural numbers and the number of the form $\omega + n$, where n ranges over the natural numbers:

$$\omega 2 = \{0, 1, 2, \ldots, \omega, \omega + 1, \omega + 2, \ldots\}.$$

Next come

$$\omega 2 + 1 = \omega 2 \cup \{\omega 2\} = \{0, 1, 2, \ldots, \omega, \omega + 1, \omega + 2, \ldots, \omega 2\},$$
$$\omega 2 + 2 = (\omega 2 + 1) \cup \{\omega 2 + 1\} = \{0, 1, 2, \ldots, \omega, \omega + 1, \omega + 2, \ldots, \omega 2, \omega 2 + 1\},$$

and so on. After that comes $\omega + \omega + \omega = \omega 3$; it is the set of all the ordinals defined so far. In other words, it consists of the ordinal numbers of the form

$$n \; (= \omega 0 + n), \qquad \omega + n \; (= \omega 1 + n), \qquad \omega 2 + n,$$

where n ranges over the natural numbers. Next come

$$\omega 3 + 1, \; \omega 3 + 2, \; \omega 3 + 3, \ldots,$$

and after them comes $\omega 4$. In this fashion the ordinals of the form $\omega m + n$ are defined for all natural numbers m and n. The next ordinal is $\omega \omega = \omega^2$; it is defined as the set of all preceding ordinals, that is to say, the set of ordinals of the form $\omega m + n$. After that, the whole thing starts all over again:

$$\omega^2 + 1, \; \omega^2 + 2, \ldots, \omega^2 + \omega, \; \omega^2 + \omega + 1, \; \omega^2 + \omega + 2, \ldots,$$
$$\omega^2 + \omega 2, \; \omega^2 + \omega 2 + 1, \ldots, \omega^2 + \omega 3, \ldots, \omega^2 + \omega 4, \ldots,$$
$$\omega^2 2, \ldots, \omega^2 3, \ldots, \omega^2 4, \ldots, \omega^3, \ldots, \omega^4, \ldots,$$

$$\omega^\omega, \ldots, \omega^{(\omega^\omega)}, \ldots, \omega^{(\omega^{(\omega^\omega)})}, \ldots .$$

In general, an *ordinal number* is defined as a well-ordered set α such that each element β in α is equal to the set of it predecessors in α:

$$\beta = \{\xi \in \alpha : \xi < \beta\}.$$

The ordinal numbers explicitly mentioned above are just the tip of the ordinal iceberg.

Some ordinals are *finite*; they are just the natural numbers. The other ordinals are called *transfinite*. The smallest of these is ω. Each ordinal α has an *immediate successor*, namely $\alpha + 1 = \alpha \cup \{\alpha\}$. Each non-zero finite ordinal n also has an *immediate predecessor*, namely the ordinal m such that $n = m + 1$. This is not always true of the transfinite ordinals. A transfinite ordinal that has an immediate predecessor is called a *successor ordinal*; the others are called *limit ordinals*. The numbers ω, $\omega 5$, ω^2, and $\omega^3 + \omega 4$ are all examples of limit ordinals, while $\omega + 5$, $\omega 5 + 9$, $\omega^2 + 1$, and $\omega^3 + \omega 4 + 12$ are examples of successor ordinals.

Sequences

A family indexed by a natural number n (conceived as the set of its predecessors), or by the set ω of all natural numbers, is called a *sequence* (finite or infinite, respectively). It is also common in mathematics to start sequences at 1 instead of 0, and in this case an n-termed sequence ends at n instead of $n - 1$. For example, if $\{P_i\}$ is an n-termed sequence of sets, then the union of this sequence is written as

$$\bigcup_{i=0}^{n-1} P_i = P_0 \cup P_1 \cup \cdots \cup P_{n-1}$$

if the sequence is indexed by the natural number n (the set of natural numbers $0, 1, \ldots, n - 1$), and it is written as

$$\bigcup_{i=1}^{n} P_i = P_1 \cup P_2 \cup \cdots \cup P_n,$$

if the sequence is indexed by the set of natural numbers $1, 2, \ldots, n$. If the sequence is infinite, the notation employed is

$$\bigcup_{i=0}^{\infty} P_i = P_0 \cup P_1 \cup P_2 \cup \cdots$$

or

$$\bigcup_{i=1}^{\infty} P_i = P_1 \cup P_2 \cup P_3 \cup \cdots,$$

according as the sequence begins at 0 or at 1.

Similar notation is used for the intersection and the Cartesian product of a (finite or infinite) sequence of sets. For instance, for the product of an infinite sequence of sets $\{P_i\}$ we write

$$\prod_{i=0}^{\infty} P_i = P_0 \times P_1 \times P_2 \times \cdots$$

or

$$\prod_{i=1}^{\infty} P_i = P_1 \times P_2 \times P_3 \times \cdots,$$

according as the sequence begins at 0 or 1.

It is occasionally useful to consider sequences (families) indexed by arbitrary ordinal numbers instead of just the natural numbers and ω. When the ordinal is greater than ω, the sequence is said to be *transfinite*.

Induction

The *principle of (mathematical) induction* says that if a set S of natural numbers contains 0, and if it contains $n + 1$ whenever it contains n, then S must coincide with the set ω of all natural numbers. Usually, S is the set of all natural numbers that satisfy some specific property. For that reason, the principle of induction can be paraphrased informally in terms of properties: if a property of natural numbers holds for 0, and if it holds for $n+1$ whenever it holds for n, then it holds for all natural numbers.

There are several variants of the principle of induction. First of all, induction can start at 1 (or any natural number k) instead of 0. In this case the conclusion is that the set S coincides with the set of all positive natural numbers (or the set of all natural numbers greater than or equal to k).

Second, there is a principle of induction not only for ω, but for each ordinal α. It says that if a subset S of α has the property that it contains an ordinal β (in α) whenever it contains all ordinals less than β, then S must coincide with α itself. (Notice that this condition forces S to contain 0 when $\alpha \geq 1$, since the ordinals less than 0 — there are none — are vacuously contained in S.) For ordinals α greater than ω, the principle is called the *principle of transfinite induction*. As in the case of induction over the natural numbers, transfinite induction is usually phrased in terms of properties. A proof by transfinite induction on ordinals less than a given ordinal α often takes the following concrete form. First, it is shown that 0 has the desired property. This is called the *base case* of the proof. The argument concerning ordinals β such that $0 < \beta < \alpha$ breaks into two cases: in the first case, β is a successor ordinal, say $\beta = \xi + 1$, and it is assumed that ξ has the desired property; in the second case, β is a limit ordinal and it is assumed that all ordinals less than β have the desired property. In each case one then proves that β has the desired property; this is called the *induction step* of the proof. The text and the exercises contain several examples of proofs by transfinite induction.

There are also definitions that proceed by induction. To define an infinite sequence $\{x_n\}$ by induction, one first defines x_0 or x_1, according as the induction starts at 0 or 1, and then one defines x_{n+1} in terms of x_n. For example, the sequence $\{x_n\}$ defined inductively by

$$x_0 = \varnothing, \quad \text{and} \quad x_{n+1} = \mathcal{P}(x_n)$$

for every natural number n, has, as its first four terms,

$$
\begin{aligned}
x_0 &= \varnothing, \\
x_1 &= \mathcal{P}(x_0) = \mathcal{P}(\varnothing) = \{\varnothing\}, \\
x_2 &= \mathcal{P}(x_1) = \mathcal{P}(\{\varnothing\}) = \{\varnothing, \{\varnothing\}\}, \\
x_3 &= \mathcal{P}(x_2) = \mathcal{P}(\{\varnothing, \{\varnothing\}\}) = \{\varnothing, \{\varnothing\}, \{\{\varnothing\}\}, \{\varnothing, \{\varnothing\}\}\}.
\end{aligned}
$$

To define by transfinite induction a sequence $\{x_\beta\}$ indexed by an ordinal α (that is to say, indexed by the set of ordinals less than α), one defines each x_β (for $\beta < \alpha$) in terms of the x_ξ for $\xi < \beta$. Often, such definitions take the following concrete form. First, x_0 is defined. This is called the *base case* of the definition. The definition of x_β for $0 < \beta < \alpha$ breaks into two cases: in the first case, β is a successor ordinal, say $\beta = \xi + 1$, and x_β is defined in terms of x_ξ; in the second case, β is a limit ordinal, and x_β is defined in terms of the x_ξ for $\xi < \beta$. This is called the *induction step* of the definition. The text and the exercises contain several examples of definitions by transfinite induction.

Cardinality

A set X is said to have the *same cardinality* (or the same *power*) as a set Y if there is a bijection from X to Y. Intuitively, this means that the two sets have the same number of elements, the same "size". Every set X has of course the same cardinality as itself, because the identity function on X is a bijection; if X has the same cardinality as Y, then Y has the same cardinality as X, because the inverse of a bijection from X to Y is a bijection from Y to X; if X has the same cardinality as Y, and if Y has the same cardinality as Z, then X has the same cardinality as Z, because the composition of a bijection from X to Y with a bijection from Y to Z is a bijection from X to Z. This situation can be summarized by saying that the notion of two sets having the same cardinality possesses the three defining properties of an equivalence relation: it is reflexive, symmetric, and transitive.

The cardinality of X is said to be *at most* that of Y if X has the same cardinality as some subset of Y, that is to say, if there is a one-to-one mapping from X into Y. The *Schröder–Bernstein theorem* says that if X has cardinality at most that of Y, and if Y has cardinality at most that of X, then X and Y have the same cardinality. The following lemma contains the heart of the argument.

Lemma 1. *If there is a bijection from a set X to one of its subsets P, then for any set Q such that $P \subseteq Q \subseteq X$, there is a bijection from X to Q.*

Proof. Let f be a bijection from X to P, and suppose Q is a subset of X that includes P. Define two sequences of sets, $\{P_n\}$ and $\{Q_n\}$, by induction on n as follows:

$$P_1 = X \quad \text{and} \quad P_{n+1} = f(P_n),$$
$$Q_1 = Q \quad \text{and} \quad Q_{n+1} = f(Q_n)$$

(where $f(P_n)$ and $f(Q_n)$ are the respective images of the sets P_n and Q_n under the mapping f). In other words, the sets P_{n+1} and Q_{n+1} are the result of applying n times the function f to the sets X and Q respectively. The definitions of these two sequences imply that

$$P_2 = f(P_1) = f(X) = P \subseteq Q = Q_1 \subseteq X = P_1.$$

In particular,

$$P_2 \subseteq Q_1 \subseteq P_1.$$

Apply f to each of these sets to obtain

$$f(P_2) \subseteq f(Q_1) \subseteq f(P_1),$$

or, in other words,

$$P_3 \subseteq Q_2 \subseteq P_2.$$

Iterate this process repeatedly to arrive at

$$P_{n+1} \subseteq Q_n \subseteq P_n.$$

The preceding inclusions combine to give

$$X = P_1 \supseteq Q_1 \supseteq P_2 \supseteq Q_2 \supseteq P_3 \supseteq Q_3 \supseteq \cdots.$$

If R is the intersection of this sequence of sets, then

$$R = \bigcap_n P_n = \bigcap_n Q_n.$$

The sets

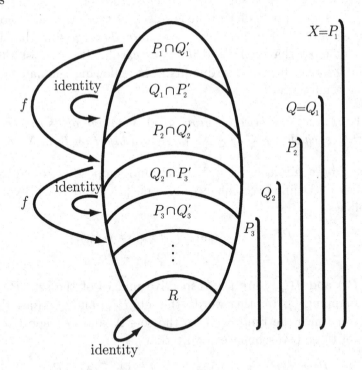

$$P_1 \cap Q_1', \qquad Q_1 \cap P_2', \qquad P_2 \cap Q_2', \qquad Q_2 \cap P_3', \qquad \ldots,$$

together with R, partition X (see the diagram). Similarly, the sets

$$Q_1 \cap P_2', \qquad P_2 \cap Q_2', \qquad Q_2 \cap P_3', \qquad \ldots,$$

together with R, partition Q (see the diagram). For each n, the function f maps the sets P_n and Q_n bijectively to the sets P_{n+1} and Q_{n+1} respectively, by the definition of these sets; since

$$Q_{n+1} \subseteq P_{n+1} \subseteq Q_n \subseteq P_n,$$

it follows that f maps $P_n \cap Q_n'$ bijectively to $P_{n+1} \cap Q_{n+1}'$. The identity function on X maps each set $Q_n \cap P_{n+1}'$, and also R, bijectively to itself. Combine these four observations to conclude that the function h defined, for each x in X, by

$$h(x) = \begin{cases} f(x) & \text{if } x \in P_n \cap Q_n' \text{ for some } n, \\ x & \text{if } x \in Q_n \cap P_{n+1}' \text{ for some } n, \text{ or else } x \in R, \end{cases}$$

is a bijection from X to Q.

To derive the Schröder–Bernstein theorem from the lemma, assume that f maps X one-to-one into Y, and g maps Y one-to-one into X. Write

$$P_0 = f(X), \qquad P = g(P_0), \qquad Q = g(Y).$$

The composition $g \circ f$ maps the set X bijectively to its subset P, and

$$P = g(P_0) \subseteq g(Y) = Q \subseteq X.$$

Apply the lemma (with "f" replaced by "$g \circ f$") to conclude that there is a bijection h from X to Q. The composition $g^{-1} \circ h$ is the desired bijection from X to Y,

$$X \xrightarrow{\ h\ } Q \xrightarrow{\ g^{-1}\ } Y.$$

If a set Y can be mapped onto X, then X always has cardinality at most that of Y. For the proof, consider such an onto mapping g. For each x in X, the inverse image under g of the singleton subset $\{x\}$ is not empty, because g is onto, so we can choose an element (any element) y_x in this inverse image. Define a mapping f from X into Y by

$$f(x) = y_x$$

for each x in X. Since the inverse images under g of distinct singletons are disjoint sets, the function f maps X one-to-one into Y, as desired.

The converse statement is also true in all but the trivial case: if there is a one-to-one function from X into Y, and if X is not empty, then Y can be mapped onto X. For the proof, consider such a one-to-one function f, and define a mapping g from Y to X as follows: fix any element z in X (this is where the non-emptiness of X is used), and write

$$g(y) = \begin{cases} x & \text{if } y \text{ is in the range of } f \text{ and } f(x) = y, \\ z & \text{otherwise.} \end{cases}$$

The function g is well defined because f is one-to-one, and it is onto because every element in X is mapped by f to some element in Y.

When the cardinality of a set X is at most that of a set Y, but the two sets do not have the same cardinality, then X is said to have *smaller cardinality* than Y (and Y has *larger cardinality* than X). Equivalently, X has smaller cardinality than Y if there exists a one-to-one mapping of X into Y, but there does not exist any mapping of X onto Y.

A set is called *finite* if it has the same cardinality as some natural number, and *infinite* otherwise. The ordinal number ω is an example of an infinite set. A family of sets is said to be finite or infinite according as the index set is finite or infinite. It is easy to check that the union of a finite family of finite sets is finite. Also, any subset of a finite set is finite.

Countable Sets

A set is said to be *countable* (or *denumerable*) if it has cardinality at most that of ω. In other words, a set is countable if has the same cardinality as some natural number or else the same cardinality as ω. If it has the same cardinality as ω, then the set is said to be *countably infinite* (or *denumerably infinite*). Notice that a non-empty set is countable just in case ω (or any countably infinite set) can be mapped onto it.

A proper subset of an infinite set can have the same cardinality as the original set. For example, the set of even natural numbers has the same cardinality as the set of all natural numbers (the requisite bijection maps every even number n to the quotient $n/2$), and the same is true of the set of odd natural numbers (in this case, the requisite bijection maps every odd number n to $(n-1)/2$). This observation implies the somewhat surprising fact that the union of two countable sets is always countable. If one of the two sets is empty, then the claim is trivially true. If both sets, say X_1 and X_2, are countable and non-empty, then there must be a mapping g_1 of the set of even natural numbers onto X_1, and a mapping g_2 of the set of odd natural numbers onto X_2. The function f defined by

$$f(n) = \begin{cases} g_1(n) & \text{if } n \text{ is even,} \\ g_2(n) & \text{if } n \text{ is odd,} \end{cases}$$

maps ω onto $X_1 \cup X_2$, so that this union must also be countable. A simple argument by mathematical induction now proves that the union of any finite family of countable sets is always countable. (A more general result is actually true: if each member of a finite family of sets has cardinality at most that of a given infinite set Y, then the union of the family has cardinality at most that of Y.)

A consequence of the preceding observations is that the set of *integers* (the natural numbers and their negatives) has the same cardinality as the set of natural numbers. In other words, the set of integers is countably infinite.

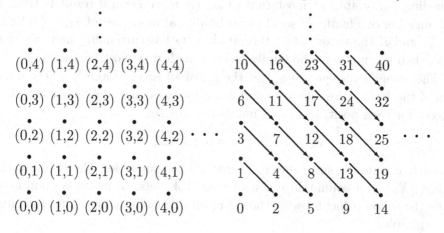

An even more surprising fact is that the Cartesian product $\omega \times \omega$ is countably infinite. One way to see this is to identify $\omega \times \omega$ with the set of those points in the Cartesian plane that have natural numbers as coordinates, and then to enumerate these points by counting down the diagonals (see the diagram). The bijection f from $\omega \times \omega$ to ω implied by this diagram can be defined by an explicit formula:

$$f(m, n) = [1 + 2 + \cdots + (m + n)] + m = \tfrac{1}{2}[(m + n)^2 + 3m + n].$$

An argument by mathematical induction now proves that the product of any finite family of countable sets is always countable. In particular, the set ω^n of all n-termed sequences of natural numbers is countable. (A more general result is actually true: for any infinite set X, the Cartesian product $X \times X$, and the set X^n of all n-termed sequences of elements in X, have the same cardinality as X.)

The observation of the preceding paragraph has a number of surprising and important consequences. The first is that the union of a countably infinite family of countable sets is always countable. Suppose, for the proof, that $\{X_i\}$ is a family of countable sets, indexed by a countably infinite set. Without loss of generality, it may be assumed that the index set is ω, and that none of the sets in the family is empty. For each index i in ω, there is a mapping g_i from ω onto X_i. Define a function f from $\omega \times \omega$ to the union $\bigcup_i X_i$ by

$$f(i, j) = g_i(j)$$

for all natural numbers i and j. The function f is onto, so the union has cardinality at most that of $\omega \times \omega$, and hence (by the observation of the preceding paragraph) at most that of ω. (A more general result is true: if each member of a family of sets has cardinality at most that of a given infinite set X, and if the index set of the family also has cardinality at most that of X, then the union of the family has cardinality at most that of X.)

The second consequence is that the set of all *rational numbers* (the numbers of the form m/n, where m and n are integers, and $n \neq 0$) is countable. Indeed, for each positive natural number n, the set

$$X_n = \{m/n : m = 0, \pm 1, \pm 2, \dots\}$$

is clearly countable, since the set of integers is countable. The union of the family $\{X_n\}$ is a countable union of countable sets, so it too is countable. Since the union coincides with the set of all rational numbers, that set must be countable.

The third consequence is that the set S of all finite sequences of natural numbers is countable. The proof is not difficult. It was already observed that for each natural number n, the set ω^n of n-termed sequences of natural numbers is countable. Since S is the union of the countable family $\{\omega^n\}_n$ of these sets, it, too, must be countable. (More generally, the set of all finite sequences of elements from an infinite set X has the same cardinality as X.)

Cantor's Theorem

Not all infinite sets are countable, however. In fact, *Cantor's theorem* asserts that for every set X, the power set $\mathcal{P}(X)$ has cardinality strictly larger than X. The cardinality of X is certainly at most that of $\mathcal{P}(X)$, because the function that maps each element x in X to the singleton $\{x\}$ is a one-to-one function from X into $\mathcal{P}(X)$. On the other hand, no function can map X

onto $\mathcal{P}(X)$, so the two sets do not have the same cardinality. Indeed, consider any function g from X into $\mathcal{P}(X)$; the value of g at each x in X is, by definition, a subset of X. The set P defined by

$$P = \{x \in X : x \notin g(x)\}$$

is certainly a subset of X. However, P cannot be in the range of g. Suppose, to the contrary, that $P = g(x)$ for some x in X. Either x is in P, or it is not. If x is in P, then x is in $g(x)$, and therefore x is not in P, by the definition of P. If, however, x is not in P, then x is not in $g(x)$, and therefore x is in P, again by the definition of P. Both possibilities lead to a contraction, so the assumption that $P = g(x)$ is impossible. Conclusion: g is not onto.

It follows from Cantor's theorem that there are infinite sets that are not countable. In fact, there is a whole hierarchy of ever bigger infinite sets: $\mathcal{P}(\omega)$ has larger cardinality than ω, and $\mathcal{P}(\mathcal{P}(\omega))$ has larger cardinality than $\mathcal{P}(\omega)$, and so on. Such sets are said to be *uncountable* (or *non-denumerable*).

Every subset P of a given set X uniquely determines a natural function from X to 2 called the *characteristic function* of P; it is the mapping ϕ defined by

$$\phi(x) = \begin{cases} 1 & \text{if } x \in P, \\ 0 & \text{if } x \in X - P. \end{cases}$$

Conversely, every mapping ϕ from X to 2 uniquely determines a subset of X called the *support* of ϕ; it is the set P defined by

$$P = \{x \in X : \phi(x) = 1\}.$$

The correspondence that maps each subset of X to its characteristic function is a bijection from $\mathcal{P}(X)$ to the set 2^X of all functions from X into 2; the inverse of this bijection maps each function in 2^X to its support. Conclusion: the sets $\mathcal{P}(X)$ and 2^X have the same cardinality. In particular, the cardinality of 2^X is larger than the cardinality of X.

It is natural to believe that the set $(2^X)^X$ must have cardinality larger than 2^X, but for infinite sets X this belief is incorrect. An element in $(2^X)^X$ is a function x from X to 2^X. In other words, for each i in X, the value $x(i)$ is a function with domain X such that for each j in X, the value $x(i)(j)$ is an element of 2. For every x in $(2^X)^X$ define a corresponding function y in $2^{X \times X}$ by

$$y(i, j) = x(i)(j)$$

for all i and j in X. The mapping that assigns to each function x in $(2^X)^X$ its corresponding function y is a bijection from $(2^X)^X$ to $2^{X \times X}$; consequently, these two sets have the same cardinality. The sets X and $X \times X$ have the same cardinality whenever X is infinite, as was previously observed; therefore, 2^X and $2^{X \times X}$ must also have the same cardinality. It follows that 2^X and $(2^X)^X$ have the same cardinality whenever X is infinite. (Notice, in passing, that the cardinality of $2^{(2^X)}$ is strictly larger than that of 2^X, by the remarks of the preceding paragraph.)

The Continuum

Every real number x in the interval $[0, 1]$ has a *binary expansion* (or a *binary representation*) of the form

$$x = \sum_{n=1}^{\infty} a_n / 2^n,$$

where each coefficient a_n is either 0 or 1. Usually, this expansion is written in the binary analogue of decimal notation:

$$x = 0.a_1 a_2 a_3 \dots.$$

For example, it follows from the rule for summing an infinite geometric series that

$$1/3 = .010101 \dots.$$

Indeed,

$$\sum_{n=0}^{\infty} (1/4)^n = 1/(1 - 1/4) = 4/3,$$

and therefore

$$.010101 \dots = \sum_{n=1}^{\infty} (1/2)^{2n} = \sum_{n=1}^{\infty} (1/4)^n$$

$$= (1/4) \cdot \sum_{n=0}^{\infty} (1/4)^n = (1/4) \cdot (4/3) = 1/3.$$

Most numbers in the interval $[0, 1]$ have a unique binary expansion. However, the rational numbers strictly between 0 and 1 with denominators that are a power of 2 have two such expansions. For example, the number $1/16$ obviously has the representation

$$1/16 = .0001000\ldots;$$

it also has the representation

$$1/16 = .0000111\ldots,$$

since

$$\sum_{n=5}^{\infty}(1/2)^n = (1/32)\cdot\sum_{n=0}^{\infty}(1/2)^n = (1/32)\cdot[1/(1-1/2)] = (1/32)\cdot 2 = 1/16.$$

For another example, the number 13/16 has the two representations

$$13/16 = .1101000\ldots \quad \text{and} \quad 13/16 = .1100111\ldots.$$

Let X be the set of positive natural numbers. Define a function f from the set 2^X to the interval $[0,1]$ as follows: for each x in X, the value $f(x)$ is the real number with the binary expansion $.x_1x_2x_3\ldots.$ The discussion of the preceding paragraph implies that the mapping f is onto and that it is almost one-to-one. The countably many exceptions to one-to-oneness are the rational numbers strictly between 0 and 1 with denominators that are some power of 2; in these cases, f maps two elements in 2^X to one real number. The cardinality of the interval $[0,1]$ is therefore equal to the cardinality of the set 2^X minus a countable subset. Since 2^X has uncountable cardinality, it follows from the earlier remarks about the cardinality of the union of two infinite sets that $[0,1]$ and 2^X have the same cardinality. The set ω of natural numbers has the same cardinality as X, so of course 2^ω and 2^X also have the same cardinality. Hence, 2^ω and $[0,1]$ have the same cardinality. In particular, the interval $[0,1]$ is uncountable.

The open interval

$$(0,1) = \{x : 0 < x < 1\}$$

is obtained from the closed interval $[0,1]$ by removing the two endpoints, so these two intervals also have the same cardinality. The interval $(0,1)$ is easily seen to have the same cardinality as the set $\mathbb{R}$ of all real numbers. In fact, the function

$$f(x) = \tan(\pi(2x-1)/2)$$

maps the interval $(0,1)$ bijectively to the set $\mathbb{R}$. Conclusion: the sets

$$(0,1), \quad [0,1,], \quad \mathbb{R}, \quad 2^\omega, \quad \text{and} \quad (2^\omega)^\omega$$

all have the same uncountable cardinality. It is called the *cardinality* (or the *power*) *of the continuum*, because the set $\mathbb{R}$ is sometimes referred to as the *continuum.* A set with the same cardinality as $\mathbb{R}$ is said to have *continuum many elements.*

Cardinal Numbers

We have explained when two sets have the same cardinality, and when one set has cardinality at most that of another, but we have not yet said what the cardinality of a set actually is. There are several possible answers; the one we shall give has become standard in most modern expositions of set theory.

It is a consequence of one of the basic axioms of set theory (the so-called *axiom of choice*) that every set can be put into bijective correspondence with some ordinal number. The intuitive idea of the proof goes as follows. Suppose a set X is given. Choose an element in X (if X is not empty) and denote it by x_0. Next, choose an element in $X - \{x_0\}$, if possible, and denote it by x_1. Then choose an element in $X - \{x_0, x_1\}$, if possible, and denote it by x_2. Continue on in this fashion until all of the elements of X have been used up. In this way, each element in X is associated with an ordinal number. The domain of the resulting function x is an ordinal number α, and x maps α bijectively to X.

A set X can be put into bijective correspondence with many ordinals. For example, a countably infinite set can be put into bijective correspondence with ω, with $\omega 2$, and with ω^2 (to name just a few). The only ordinal in bijective correspondence with X that seems to clamor for attention is the smallest one (in the sense of the natural ordering of the ordinals). We therefore define a *cardinal number* to be an ordinal α with the property that if β is any ordinal that can be put into bijective correspondence with α, then $\alpha \leq \beta$. The *cardinality* of a set X is defined to be the smallest ordinal that can be put into bijective correspondence with X.

The fact that cardinal numbers are special ordinal numbers introduces the possibility of some confusion with regard to the notation for the arithmetic operations on cardinal and ordinal numbers. If α and β are cardinal numbers, then they are also ordinal numbers, and the notation $\alpha + \beta$ has two possible meanings: the cardinal sum or the ordinal sum of α and β. In general, these two sums are not the same. Similarly, the cardinal and ordinal products of α and β are generally different from one another, and so are the cardinal and ordinal powers. For example, the ordinal number ω^ω is countable, while the cardinal number ω^ω is uncountable.

In practice it is easy to avoid confusion by introducing special symbols for the infinite cardinal numbers. One symbol that is frequently employed is $\aleph$ (*aleph*, the first letter of the Hebrew alphabet), followed by an appropriate subscript. For instance, the smallest transfinite ordinal is ω; it is also the first infinite cardinal number, and as such it is denoted by $\aleph_0$. The next infinite cardinal number is the least uncountable ordinal; it is denoted by $\aleph_1$. Then comes $\aleph_2$ (the least uncountable ordinal that cannot be put into bijective correspondence with $\aleph_1$), and so on.

The notation $2^{\aleph_0}$ is used to denote the cardinality of the continuum, the smallest ordinal that can be put into bijective correspondence with $\mathbb{R}$. Since $2^{\aleph_0}$ is uncountable (by Cantor's theorem), it is clear that

$$\aleph_1 \leq 2^{\aleph_0}.$$

Is the inequality strict? The celebrated *continuum hypothesis* asserts that the answer is no; in other words, it asserts that

$$\aleph_1 = 2^{\aleph_0}.$$

It is known that this assertion is independent of the standard axioms of set theory: neither the continuum hypothesis nor its negation can be proved on the basis of those axioms.

Appendix B

Hints to Selected Exercises

Chapter 1

9. The Boolean ring is assumed to have more than two elements, so there are distinct non-zero elements p and q. If $p \cdot q = 0$, then p and q are zero-divisors. If $p \cdot q \neq 0$, then $p \cdot q$ and $p + q$ have zero as their product, and are therefore zero-divisors.

10. To motivate the construction, let A be a Boolean ring (with or without a unit), and consider a Boolean ring B with a unit 1 that is not in A. Assume that B extends A in the sense that the operations of addition and multiplication in A are the restrictions of the corresponding operations of B, and both rings have the same zero. (The technical way of describing this situation is to say that A is a *subring* of B.)

    If p and q are distinct elements of A, then $1 + p$ and $1 + q$ are distinct elements of B, by the cancellation law, and neither one of them is in A. Indeed, if $r = 1 + p$ were in A, then $p + r$ would be in A, because of the closure of A under addition; it would then follow that 1 is in A, since

    $$p + r = p + 1 + p = p + p + 1 = 0 + 1 = 1,$$

    and this would contradict the assumption that 1 is not in A.

    Easy computations show how to add elements of the form $1 + p$ (for p in A) to other elements of this form and to elements of A:

    $$(1 + p) + (1 + q) = p + q,$$
    $$(1 + p) + q = p + (1 + q).$$

    A similar remark applies to multiplication:

    $$(1 + p) \cdot (1 + q) = 1 + p + q + p \cdot q,$$

$$(1+p) \cdot q = q + p \cdot q, \quad \text{and} \quad p \cdot (1+q) = p + p \cdot q.$$

The above motivation suggests the following procedure. Let A be an arbitrary Boolean ring with or without a unit. With each element p in A, associate a new element $\bar{p}$ that is not in A in such a manner that $p \neq q$ implies $\bar{p} \neq \bar{q}$. The new element $\bar{p}$ is the analogue of the element $1+p$ above. Take B to be the set of elements in A together with the set of new elements. Operations $\oplus$ and $\odot$ of addition and multiplication must be defined in B so that it becomes a Boolean ring with unit that extends A. The above calculations suggest how the operations should be defined. For instance, if p and q are in A, then define

$$p \oplus q = p + q,$$
$$p \oplus \bar{q} = \overline{p+q},$$
$$\bar{p} \oplus q = \overline{p+q},$$
$$\bar{p} \oplus \bar{q} = p + q.$$

Here, the operation on the right side is the operation of addition in A, whereas the operation on the left is the operation of addition that is being defined in B. The element $\overline{p+q}$ is the new element that is associated with the element $p+q$ in A. The operation of multiplication in B is defined in a similar fashion. It must be checked that B is a Boolean ring (with unit) under the operations so defined, and that the operations of B, when restricted to the elements of A, coincide with the operations of A.

11. The symmetric functions of the n arguments $v_1, v_2, \ldots, v_n$ are defined in any ring by

$$s_1 = v_1 + v_2 + \cdots + v_n = \sum_{i=1}^{n} v_i,$$

$$s_2 = \sum_{i<j} v_i \cdot v_j,$$

$$s_3 = \sum_{i<j<k} v_i \cdot v_j \cdot v_k,$$

$$\vdots$$

$$s_n = v_1 \cdot v_2 \cdot \cdots \cdot v_n.$$

Let $t_n = t_n(v_1, \ldots, v_n)$ be the sum of the symmetric functions of n arguments:

$$t_n = s_1 + s_2 + \cdots + s_n.$$

Derive the identity

$$t_n(v_1, v_2, \ldots, v_n) = v_k + t_{n-1}(v_1, \ldots, v_{k-1}, v_{k+1}, \ldots, v_n)$$
$$+ v_k \cdot t_{n-1}(v_1, \ldots, v_{k-1}, v_{k+1}, \ldots, v_n)$$

for each positive integer $k \le n$. Given a finite Boolean ring A with n elements, say $p_1, p_2, \ldots, p_n$, use the preceding identity to show that

$$p_k \cdot t_n(p_1, p_2, \ldots, p_n) = p_k$$

for $k = 1, 2, \ldots, n$. Conclude that $t_n(p_1, p_2, \ldots, p_n)$ is the unit of A.

12. Let X be an infinite set, and consider the class of functions in 2^X that assume the value 1 on at most finitely many arguments.

13. The answer is negative if one interprets the question as asking whether every non-degenerate Boolean ring with unit is an extension of some Boolean ring that has no unit whatsoever. The answer is positive if one interprets the question as asking whether every non-degenerate Boolean ring with unit is an extension of a Boolean ring that either does not have a unit, or else does not have the same unit.

14. View the group as a vector space over the two-element field 2. As such, it has a (possibly infinite) vector space basis, say I, and the elements of the group are the finite sums of the basis elements. Identify each element p in the group with the function in the Boolean ring 2^I that assumes the value 1 at an argument i (in I) just in case i appears in the basis representation of p. Use the ring multiplication in 2^I to define a ring multiplication in the Boolean group.

Chapter 2

2.   (a) Derive (11) from (13), (14), and (18).

      (b) Derive (16) from (13), (14), and (20).

      (c) Derive (12) from (13), (14), (18), and (20). Begin with the equations

$$0 = p \wedge p' = p \wedge (p' \vee 0).$$

      (d) Derive the absorption laws,

$$p \wedge (p \vee q) = p \quad \text{and} \quad p \vee (p \wedge q) = p,$$

from (12), (13), (18), and (20). Begin with the equation

$$p \wedge (p \vee q) = (p \vee 0) \wedge (p \vee q).$$

(e) Show, using (13), (14), (18), and (20), that complements are unique: if

$$p \wedge q = 0 \quad \text{and} \quad p \vee q = 1,$$

then $q = p'$. Use the fact that

$$q = q \wedge 1 = q \wedge (p \vee p') \quad \text{and} \quad p' = p' \wedge 1 = p' \wedge (p \vee q).$$

As a consequence, derive the double complement law (15).

(f) Derive the following cancellation law for meet from (13), (14), and (20):

$$q \wedge p = r \wedge p \quad \text{and} \quad q \wedge p' = r \wedge p'$$

imply $q = r$. Use the fact that

$$q = q \wedge (p \vee p').$$

Formulate and derive an analogous law for join.

(g) Derive (19) from (13), (14), (18), and (20). By the cancellation law, it suffices to derive the equations

$$(p \vee (q \vee r)) \wedge p = ((p \vee q) \vee r) \wedge p$$

and

$$(p \vee (q \vee r)) \wedge p' = ((p \vee q) \vee r) \wedge p'$$

in order to establish the associative law for join. The first equation can be derived by reducing both sides of the equation to p, using the absorption and distributive laws. The second equation can be derived by reducing both sides of the equation to $p' \wedge (q \vee r)$, using mainly the distributive laws.

(h) Derive the De Morgan laws (17) from (13), (14), (18), and (20). By the uniqueness of complements, the second De Morgan law is a consequence of the equations

$$(p \vee q) \wedge (p' \wedge q') = 0 \quad \text{and} \quad (p \vee q) \vee (p' \wedge q') = 1.$$

The form of these equations suggests using the distributive laws.

5. (a) As a model, take the two-element Boolean ring. Interpret join and meet as ring addition and multiplication respectively, interpret complement as the unary operation that interchanges 0 and 1, and interpret zero and one as 0 and 1 respectively. Show that all the identities in (13), (14), and (18), and the first identity in (20), are true in this model, but that the second identity in (20) fails. Why should this model be an "obvious" choice to show the independence of the distributive law for join over meet?

(b) As in part (a), take the model to be the two-element Boolean ring, but interpret join as ring multiplication, interpret meet as ring addition, interpret complement as before, interpret zero as 1, and interpret one as 0.

(c) Modify the two-element Boolean algebra by redefining the complement operation to be constantly zero or constantly one.

6. Show that each of the identity laws is derivable from the remaining seven axioms. Use part (c) of the hint for Exercise 2.

7. It is easy to check that the axioms and the definitions (of meet, zero, and one) formulated in the exercise are derivable from (13), (14), (18), and (20). It must be demonstrated that, conversely, the axioms and definitions formulated in the exercise entail (13), (14), (18), and (20). The derivations of these equations seem to require the derivation of nearly every equation in (11)–(20). A broad outline of the proof is sketched in a series of steps below.

It is helpful to gather together various instances of Huntington's axiom (H).

(a) $$(p' \vee p''')' \vee (p' \vee p'')' = p.$$
(b) $$(p'' \vee p''')' \vee (p'' \vee p'')' = p'.$$
(c) $$(p'' \vee p'')' \vee (p'' \vee p')' = p'.$$
(d) $$(p''' \vee p'')' \vee (p''' \vee p')' = p''.$$
(e) $$(p' \vee q'')' \vee (p' \vee q')' = p.$$
(f) $$(p'' \vee q'')' \vee (p'' \vee q')' = p'.$$
(g) $$(q' \vee p'')' \vee (q' \vee p')' = q.$$
(h) $$(q'' \vee p'')' \vee (q'' \vee p')' = q'.$$
(i) $$(p' \vee p')' \vee (p' \vee p)' = p.$$
(j) $$(p'' \vee p')' \vee (p'' \vee p)' = p'.$$

The first goal is the derivation of the double complement law in (15), because many of the equations in (11)–(20) follow easily from it. Begin with the preliminary step

(k) $$p \vee p' = p' \vee p''.$$

Use (a) and (b) to expand $p \vee p'$, and (c) and (d) to expand $p' \vee p''$. The identities (a), (d), and (k) entail the double complement law. One corollary of this law is that complementation is a one-to-one function:

(l) $$p' = q' \quad \text{implies} \quad p = q.$$

With the help of the double complement law and the definition of meet, it is easy to derive the commutative and associative laws for meet, and the De Morgan laws.

Before applying the definitions of zero and one, it must be shown that these constants are well defined, that is to say, it must be shown that the definitions do not depend on the particular choice of the element p:

$$p \vee p' = q \vee q'.$$

Use (e) and (f) to expand $p \vee p'$, and (g) and (h) to expand $q \vee q'$.

The definition of zero and one, and the double complement law, immediately imply the complement laws for zero and one in (11). It is almost as easy to derive the complement laws for meet and join in (14).

The next task is to derive the identity laws in (13). First, justify the following steps:

$$(1'' \vee 1'')' \vee (1'' \vee 1')' = 1',$$
$$1 = 1'' \vee 1',$$
$$1'' \vee 1'' = 1 \vee 1,$$
$$(1 \vee 1)' \vee 1' = 1',$$
$$1 = 1 \vee 1' = 1 \vee [(1 \vee 1)' \vee 1'] = (1 \vee 1') \vee (1 \vee 1)' = 1 \vee (1 \vee 1)',$$
$$1 \vee 1 = 1 \vee [1 \vee (1 \vee 1)'] = (1 \vee 1) \vee (1 \vee 1)' = 1,$$
$$1' = (1 \vee 1)' \vee 1' = 1' \vee 1',$$
$$(p' \vee p)' = (p' \vee p)' \vee (p' \vee p)'.$$

Then begin as follows:

$$p \vee 0 = p \vee 1' = p \vee (p' \vee p)' = \cdots,$$

and use (i).

At this point one can show that complements are unique. A preliminary observation sets up the strategy:

(m) $p' \vee q = 1$ and $q' \vee p = 1$ imply $p = q$.

Use (H) and the definition of 1 to expand p. Obtain a similar expansion of q, and prove that the two expansions are equal.

The uniqueness of complements is formulated as follows:

$p \vee q = 1$ and $p \wedge q = 0$ imply $q = p'$.

In the derivation, use the definition of meet and the double complement law.

Now derive the idempotent laws in (16), using the identity laws, the commutative laws, the definition of zero, the double complement law, and (j). The laws in (12) follow easily from the idempotent laws.

Before proceeding further, it is helpful to write (H) in a form that shows some relationship with the distributive laws:

(n) $$(p \wedge q) \vee (p \wedge q') = p.$$

The derivation of the distributive law for meet over join in (20) is more involved, and requires several intermediate steps. (The distributive law for join over meet follows from the distributive law for meet over join in the usual way.) The strategy is to show that

(o) $$[(p \wedge q) \vee (p \wedge r)] \vee [p \wedge (q \vee r)]' = 1$$

and

(p) $$[(p \wedge q) \vee (p \wedge r)] \wedge [p \wedge (q \vee r)]' = 0.$$

The uniqueness of complements then implies that

$$[p \wedge (q \vee r)]' = [(p \wedge q) \vee (p \wedge r)]'.$$

The distributive law is an immediate consequence of this equation and (1). For notational convenience, write

$$
\begin{aligned}
s_{111} &= p \wedge q \wedge r, & s_{110} &= p \wedge q \wedge r', \\
s_{101} &= p \wedge q' \wedge r, & s_{100} &= p \wedge q' \wedge r', \\
s_{011} &= p' \wedge q \wedge r, & s_{010} &= p' \wedge q \wedge r', \\
s_{001} &= p' \wedge q' \wedge r, & s_{000} &= p' \wedge q' \wedge r'.
\end{aligned}
$$

Show that the meet of any two of these terms is zero, and that the join of all of these terms is one,

(q) $$s_{111} \vee s_{110} \vee s_{101} \vee s_{100} \vee s_{011} \vee s_{010} \vee s_{001} \vee s_{000} = 1.$$

To prove (q), use (n) to expand p, and then use (n) again to expand both $p \wedge q$ and $p \wedge q'$.

The identity (n) is also used to prove that the two sides of the distributive law can be written as joins of terms of the form s_{ijk}. In connection with writing the complement of the left side of the distributive law as such a join, it is helpful to recall that

$$(q \vee r)' = (q'' \vee r'')' = q' \wedge r'.$$

What is the connection between the terms s_{ijk} and the distributive law? The right side of the distributive law is equal to the join of the first three of these terms,

$$(p \wedge q) \vee (p \wedge r) = s_{111} \vee s_{110} \vee s_{101},$$

and the complement of the left side of the distributive law is equal to the join of the last five terms s_{ijk},

$$[p \wedge (q \vee r)]' = s_{100} \vee s_{011} \vee s_{010} \vee s_{001} \vee s_{000}.$$

In view of these remarks, (q) shows that equation (o) is satisfied. One more observation — a special case of the distributive law — is needed before deriving (p):

(r) $p \wedge q = 0$ and $p \wedge r = 0$ imply $p \wedge (q \vee r) = 0.$

Show that the hypotheses of (r) imply $(p \wedge q) \vee (p \wedge r) = 0$, and use this to draw the desired conclusion of (r) from (o).

As was already observed,

$$s_{111} \wedge s_{100} = 0 \qquad \text{and} \qquad s_{111} \wedge s_{011} = 0.$$

An application of (r) gives

$$s_{111} \wedge (s_{100} \vee s_{011}) = 0.$$

Continue in this fashion to obtain

$$(s_{111} \vee s_{110} \vee s_{101}) \wedge (s_{100} \vee s_{011} \vee s_{010} \vee s_{001} \vee s_{000}) = 0.$$

8. The independence of the commutative law can be demonstrated in a model with the universe $\{0, 1, 2, 3, 4, 5\}$ and the arithmetic tables

$\vee$	0	1	2	3	4	5
0	0	1	2	3	4	5
1	1	1	1	1	1	1
2	2	1	2	1	1	2
3	3	1	1	3	3	1
4	4	1	1	4	4	1
5	5	1	5	1	1	5

and

$'$	
0	1
1	0
2	3
3	2
4	5
5	4

Consider the instance of the commutative law in which $p = 5$ and $q = 2$.

The independence of the associative law can be demonstrated in a model with the universe $\{0, 1, 2, 3\}$ and the arithmetic tables

$\vee$	0	1	2	3
0	0	1	2	3
1	1	1	2	0
2	2	2	2	1
3	3	0	1	3

and

$'$	
0	1
1	0
2	3
3	2

Consider the instance of the associative law in which $p = 2$, $q = 1$, and $r = 3$. The independence of Huntington's axiom (H) can be demonstrated in a model with the universe $\{0, 1, 2, 3, 4, 5\}$ and the arithmetic tables

∨	0	1	2	3	4	5
0	0	1	2	3	4	5
1	1	1	1	1	1	1
2	2	1	2	1	1	1
3	3	1	1	3	1	1
4	4	1	1	1	4	1
5	5	1	1	1	1	5

and

′	
0	1
1	0
2	3
3	2
4	5
5	4

.

Consider the instance of (H) in which $p = 3$ and $q = 5$.

9. It is not difficult to show that each of the identities

(a) $$p'' = p,$$
(b) $$p \vee (q \vee q')' = p,$$
(c) $$p \vee (q \vee r)' = ((q' \vee p)' \vee (r' \vee p)')',$$

is derivable from the set of Huntington axioms described in Exercise 7, since the Huntington axioms imply all of the laws in (11)–(20).

It requires more work to show that each of the Huntington axioms is derivable from (a)–(c). In terms of join and complement, define an operation of meet, and constants 0 and 1, in the usual way. In order to derive the Huntington axioms, it is helpful to derive first most of the laws in (11)–(20). Begin with the idempotent laws in (16). Justify the derivation

$$q' = q' \vee (q \vee q')' = ((q' \vee q')' \vee (q'' \vee q')')'$$
$$= ((q' \vee q')' \vee (q \vee q')')' = ((q' \vee q')')' = q' \vee q',$$

and use it to arrive at the idempotent law for join. The idempotent law for meet follows from the definition of meet, the idempotent law for join, and the double complement law.

Justify the following derivation of the commutative law for join:

$$p \vee q = p \vee (q')' = p \vee (q' \vee q')' = ((q'' \vee p)' \vee (q'' \vee p)')'$$
$$= ((q \vee p)' \vee (q \vee p)')' = ((q \vee p)')' = q \vee p.$$

The commutative law for meet is a consequence of the commutative law for join and the definition of meet.

Huntington's axiom

(d) $$(p' \vee q')' \vee (p' \vee q)' = p$$

can now be established. Begin with the string of equations

$$p' = p' \vee (q \vee q')' = ((q' \vee p')' \vee (q'' \vee p')')'$$
$$= ((q' \vee p')' \vee (q \vee p')')' = ((p' \vee q')' \vee (p' \vee q)')'.$$

Derive the identity

$$p \vee p' = q \vee q',$$

using (b), to show that the constants 0 and 1 are well defined. Here are the first few steps in the derivation:

$$(p \vee p')' = (p \vee p')' \vee (q \vee q')' = (q \vee q')' \vee (p \vee p')' = (q \vee q')'.$$

Derive the complement law for meet, using the definition of meet, the definition of 0, and the preceding identity.

The identity law for join is just a reformulation of (b), while the identity law for meet follows from the definitions of meet, zero, and one, and the identity law for join.

Derive the identity

$$p \vee q = (p' \wedge q')'$$

(an analogue for join of the definition of meet) with the help of the definition of meet and the double complement law. This identity and the definition of meet imply the De Morgan laws.

The derivation of the distributive law for join over meet uses the definition of meet and (c), and begins as follows:

$$p \vee (q \wedge r) = p \vee (q' \vee r')' = ((q'' \vee p)' \vee (r'' \vee p)')' = \cdots .$$

The distributive law for meet over join is a consequence of the distributive law for join over meet, the definition of meet, and the De Morgan laws.

The key steps in the derivation of the second law in (12) are

$$1 \vee p = ((1'' \vee p)')' = ((1'' \vee p)' \vee 0)'$$

$$= ((1'' \vee p)' \vee (p' \vee p)')' = p \vee (1' \vee p)'.$$

The corresponding law for meet is a consequence of the law for join.

The absorption laws (part (d) of the solution to Exercise 2) are consequences of the distributive and identity laws, and the second law in (12). The derivation of the absorption law for join begins as follows:

$$p \vee (p \wedge q) = (p \wedge 1) \vee (p \wedge q) = \cdots .$$

The absorption law for meet is a consequence of the distributive law for meet over join, and the absorption law for join.

The next observation is used to establish the uniqueness of complements:

$$p' \vee q = 1 \quad \text{and} \quad q' \vee p = 1 \quad \text{imply} \quad p = q.$$

The derivation proceeds by using (d) and the definition of meet to show that $p = p \wedge q$. An analogous argument shows that $q = q \wedge p$.

The uniqueness of complements may be formulated as follows:

(e) $\qquad p \wedge q = 0 \quad \text{and} \quad p \vee q = 1 \quad \text{imply} \quad q = p'.$

To establish (e), argue that $p'' \vee q = 1$ and $q' \vee p' = 1$, and conclude (with the help of the preceding observation) that $q = p'$.

Turn now to the derivation of the associative law for join. Write

$$s = (p \vee q) \vee r \quad \text{and} \quad t = p \vee (q \vee r).$$

It is to be shown that $s = t$. The proof involves three intermediate steps, the first of which is concerned with the derivation of the three identities

(f) $\qquad\qquad p \wedge s = p, \qquad q \wedge s = q, \qquad r \wedge s = r.$

The derivations use the definition of s, and the distributive and absorption laws.

The second intermediate step is concerned with the derivation of the three identities

(g) $\qquad\qquad p \wedge s' = 0, \qquad q \wedge s' = 0, \qquad r \wedge s' = 0.$

All three derivations are similar. The derivation of the first identity uses the first equation in (f) and begins as follows:

$$p' \vee s = 1 \wedge (p' \vee s) = (p' \vee p) \wedge (p' \vee s) = p' \vee (p \wedge s) = \cdots .$$

The third intermediate step is concerned with the derivation of the three identities

(h) $\qquad\qquad p' \vee t = 1, \qquad q' \vee t = 1, \qquad r' \vee t = 1.$

Again, all three derivations are similar; they use the definition of t and the third identity in (g) (with r replaced by p, and with the three variables p, q, and r that are involved in the definition of s replaced by q, r, and p respectively).

To derive the associative law, show that the hypotheses of (e) are satisfied when p and q are replaced by t and s'. The derivation of the equation $t \wedge s' = 0$ uses the definition of t, the identities in (g), and several applications of the distributive law for meet over join. The derivation of the equation $t \vee s' = 1$ uses the definition of s, the De Morgan laws, the distributive law for join over meet, and the three identities in (h).

10. The independence of the double complement law can be demonstrated in a model with the universe $\{0, 1, 2, 3, 4, 5\}$ and the arithmetic tables

∨	0	1	2	3	4	5
0	0	1	4	5	2	3
1	1	1	1	1	1	1
2	2	1	0	1	2	1
3	3	1	1	0	1	3
4	4	1	4	1	0	1
5	5	1	1	5	1	0

and

	′
0	1
1	0
2	3
3	4
4	5
5	2

.

Consider the instance of the double complement law in which $p = 2$.

The independence of the identity law for zero can be demonstrated in a model with the universe $\{0, 1, 2\}$ and the arithmetic tables

∨	0	1	2
0	0	1	2
1	1	1	1
2	2	1	2

and

	′
0	1
1	0
2	2

.

Consider the instance of the identity law in which $p = 0$ and $q = 2$.

The independence of the last law can be demonstrated in a model with the universe $\{0, 1, 2\}$ and the arithmetic tables

∨	0	1	2
0	0	2	1
1	1	1	0
2	2	0	2

and

	′
0	0
1	2
2	1

.

Consider the instance of the law in which $p = 1$, $q = 1$, and $r = 2$.

11. Show that the associative and commutative laws for join, together with the equivalence

(a) $$p \vee q' = r \vee r' \quad \text{if and only if} \quad p \vee q = p,$$

imply each of the Huntington axioms described in Exercise 7, and conversely. The most involved part of the proof is the derivation of the third axiom,

(b) $$(p' \vee q')' \vee (p' \vee q) = p,$$

from (a) and the associative and commutative laws.

It is helpful to establish first some auxiliary laws about join. The idempotent law (for join) in (13) is a consequence of (a) and the fact that $p \vee p' = p \vee p'$. The law

(c) $$p \vee p' = q \vee q'$$

is a consequence of (a) and the idempotent law.

The implication

(d) $\qquad p \vee q = p \qquad$ and $\qquad q \vee r = q \qquad$ imply $\qquad p \vee r = p$

follows from the associative law; begin the proof with the equation

$$p \vee r = (p \vee q) \vee r.$$

The most involved derivation concerns the double complement law. Use the equation

$$p'' \vee p' = r \vee r'$$

(which follows from (c)), and apply (a) to conclude that

(e) $$p'' \vee p = p''.$$

The equations

(f) $\qquad p''' \vee p' = p''' \qquad$ and $\qquad p'''' \vee p'' = p''''$

are both instances of (e). The second equation in (f) and the equation in (e) together yield

$$p'''' \vee p = p'''',$$

by (d). Invoke (a), with p and q replaced by p'''' and p, to conclude that

$$p'''' \vee p' = r \vee r'.$$

Apply (a) one more time (with p and q replaced by p' and p''') to arrive at

$$p' \vee p''' = p'.$$

Combine the preceding equation with the first equation in (f) to see that

$$p''' = p'.$$

In the equation

$$p \vee p' = r \vee r',$$

replace p' by p''' to get

$$p \vee p''' = r \vee r'.$$

Apply (a), with q replaced by p'', to conclude that

$$p \vee p'' = p.$$

Combine this last equation with (e) to arrive at the desired conclusion. The identity

(g) $$p \vee (q \vee q') = q \vee q'$$

is an easy consequence of (c) and the associative and idempotent laws. Next, derive the identity

(h) $$p \vee (p' \vee q)' = p;$$

first show, with the help of (g) and (c), that

$$p \vee (p' \vee q)'' = r \vee r',$$

and then invoke (a).

Now comes an identity,

(i) $$p \vee (p \vee q)' = p \vee q',$$

with a more involved derivation. First, prove that

$$(p \vee (p \vee q)') \vee q'' = r \vee r',$$

using the double complement law and (c) (with p and q replaced by $p \vee q$ and r). Apply (a) (with p replaced by $p \vee (p \vee q)'$, and q by q') to arrive at

$$(p \vee (p \vee q)') \vee q' = p \vee (p \vee q)',$$

which is equivalent to

(j) $$(p \vee q') \vee (p \vee q)' = p \vee (p \vee q)'.$$

Observe that

$$q' \vee (p \vee q)' = q' \vee (q'' \vee p)' = q',$$

by (h), and therefore

$$(p \vee q') \vee (p \vee q)' = p \vee q'.$$

Equation (i) follows from this equation and from (j).

Derive (b) as follows. By (h),

$$p \vee (p' \vee q)' = p \qquad \text{and} \qquad p \vee (p' \vee q')' = p,$$

so that

(k) $$p \vee (p' \vee q')' \vee (p' \vee q)' = p.$$

The identity (i) (with p and q replaced by p' and q') and the double complement law together yield

$$p' \vee (p' \vee q')' = p' \vee q.$$

Form the join of both sides with $(p' \vee q)'$, and use (c), to obtain

$$p' \vee (p' \vee q')' \vee (p' \vee q)' = r \vee r'.$$

Apply (a) (with p and q replaced by $(p' \vee q')' \vee (p' \vee q)'$ and p) to arrive at

$$p \vee (p' \vee q')' \vee (p' \vee q)' = (p' \vee q')' \vee (p' \vee q).$$

This equation and (k) together imply (b).

Chapter 3

2.   (a) This amounts to proving that the Boolean ring axioms (1.1)–(1.3), (1.5), (1.6), and (1.8)–(1.11) can be derived from the Boolean algebra axioms (2.11)–(2.20) using the definitions in (3). The derivations of the associative law (1.1) and the distributive laws (1.8) and (1.9) are more involved than the other derivations.

     (b) This amounts to proving that the Boolean algebra axioms can be derived from the Boolean ring axioms using the definitions in (4). It is sufficient to derive the pairs of axioms (2.13), (2.14), (2.18), and (2.20), since they imply the remaining ones. Only the derivation of the distributive law for joins over meets presents any difficulties.

     (c) This amounts to showing that the definitions in (4) can be derived from the Boolean algebra axioms and the definitions in (3).

     (d) This amounts to showing that the definitions in (3) can be derived from the Boolean algebra axioms and the definitions in (4).

4. The expressions for meet, join, and complement are

$$p \wedge q = pq, \qquad p \vee q = p + q - pq, \qquad \text{and} \qquad p' = 1 - p.$$

5. Use Exercise 1.10.

Chapter 5

3. There are four periodic sets of integers of period two, namely the empty set, the set of even integers, the set of odd integers, and the set of all integers. There are eight periodic sets of integers of period three; they are just the various possible unions of the sets

$$P_0 = \{3n : n \in X\}, \quad P_1 = \{3n + 1 : n \in X\}, \quad P_2 = \{3n + 2 : n \in X\}.$$

4. Fix a positive integer m, and for each integer $k = 0, 1, \ldots, m - 1$, let P_k be the set of those integers that when divided by m, leave remainder k:

$$P_k = \{mn + k : n \in X\}.$$

There are 2^m periodic sets of integers of period m, namely the sets

$$P_K = \bigcup_{k \in K} P_k,$$

where K ranges over the subsets of the set

$$M = \{0, 1, \ldots, m - 1\}.$$

Show that these sets satisfy the equations

$$P_K \cup P_L = P_{K \cup L}, \qquad P_K \cap P_L = P_{K \cap L}, \qquad P_K' = P_{M-K}.$$

5. Show that if P and Q are periodic sets of integers of periods m and n, then $P \cap Q$ is a periodic set of period k, where k is the least common multiple of m and n.

7. Prove the following lemma about the union of two left half-closed intervals $[c_1, d_1)$ and $[c_2, d_2)$, where c_1, c_2, d_1, and d_2 are extended real numbers satisfying $c_1 < d_1$ and $c_2 < d_2$. If there is a real number e such that $c_1, c_2 \leq e \leq d_1, d_2$, then

$$[c_1, d_1) \cup [c_2, d_2) = [a, b),$$

where

$$a = \min\{c_1, c_2\} \quad \text{and} \quad b = \max\{d_1, d_2\}.$$

If no such number e exists, then

$$c_1 < d_1 < c_2 < d_2 \quad \text{or} \quad c_2 < d_2 < c_1 < d_1.$$

Given an element of the interval algebra of real numbers, use the lemma to combine left half-closed intervals that satisfy the condition of the hypothesis of the lemma.

Chapter 6

10. On the basis of the definition of complement in (6), the three axioms for stroke and the definitions of meet and join in (7) and (8) may be rewritten in the following form:

(a) $$p' \mid p' = p,$$
(b) $$p \mid (q \mid q') = p',$$
(c) $$(p \mid (q \mid r))' = (q' \mid p) \mid (r' \mid p).$$
(d) $$p \wedge q = p' \mid q',$$
(e) $$p \vee q = (p \mid q)'.$$

Use (6) and (a) to derive the double complement law. Use the double complement law, (c), and the definition of complement in (6) to prove that stroke is commutative. The commutativity of stroke implies that of meet and join.

Derive the identity

$$p \mid p' = q \mid q',$$

and then define constants 0 and 1 by

$$0 = p \mid p' \quad \text{and} \quad 1 = (p \mid p')' = 0'.$$

The identity and complement laws follow readily.

To establish the distributive law for join over meet, justify the following steps:

$$p \vee (q \wedge r) = (p \mid (q \wedge r))' = (p \mid (q' \mid r'))' = (q'' \mid p) \mid (r'' \mid p)$$
$$= (q \mid p) \mid (r \mid p) = (p \mid q) \mid (p \mid r) = (p \mid q)'' \mid (p \mid r)''$$
$$= (p \vee q)' \mid (p \vee r)' = (p \vee q) \wedge (p \vee r).$$

The derivation of the distributive law for meet over join is similar. The identity in (5) is a consequence of (d).

In the second part of the proof, it must be shown that the axioms for stroke and the identities (6)–(8) are derivable from the axioms for Boolean algebra on the basis of the definition of stroke in (5). The identity in (6) follows from (5) and the idempotent law for meet. The identity in (7) is a consequence of (6), (5), and the double complement law. The identity in (8) is a consequence of (6), the De Morgan laws, and the double complement law.

The first axiom for stroke follows from (7) and the idempotent law for meet. The derivation of the second axiom for stroke uses (5) (with q replaced by $q \mid (q \mid q)$), (6), (3), the complement law for join, and the identity law for meet.

To derive the third axiom for stroke, justify the following steps:

$$((q \mid q) \mid p) \mid ((r \mid r) \mid p) = ((q \mid q) \mid p'') \mid ((r \mid r) \mid p'')$$
$$= ((q \mid q) \mid (p' \mid p')) \mid ((r \mid r) \mid (p' \mid p'))$$
$$= (q \wedge p') \mid (r \wedge p')$$
$$= (p' \wedge q) \mid (p' \wedge r)$$
$$= (p' \wedge q)' \wedge (p' \wedge r)'$$
$$= ((p' \wedge q) \vee (p' \wedge r))'$$
$$= (p' \wedge (q \vee r))'$$
$$= (p' \wedge (q \mid r)')'$$
$$= (p \mid (q \mid r))'$$
$$= (p \mid (q \mid r)) \mid (p \mid (q \mid r)).$$

13. Every operation on 2 that is definable in terms of $\Leftrightarrow$ alone assumes the value 1 when all arguments are 1. Not all operations on 2 have this property.

17. Experiment with the case of binary operations on 2 first.

19. If g is regarded as defining a binary operation for each q, by, say,

$$p(q)r = g(p, q, r),$$

then $p(0)r = p \wedge r$ and $p(1)r = p \vee r$. Boolean algebras can be axiomatized in terms of the operations g and $'$ by the equations

(b) $\qquad\qquad g(p, q, g(r, s, t)) = g(g(p, q, r), s, g(p, q, t)),$

(c_1) $\qquad\qquad\qquad g(p, q, q) = q,$

(c_2) $\qquad\qquad\qquad g(q, q, p) = q,$

(d_1) $\qquad\qquad\qquad g(p, q, q') = p,$

(d_2) $\qquad\qquad\qquad g(q', q, p) = p.$

The following variants of (c_1), (c_2), and (d_1), (d_2), follow readily from axiom (b) (with the help of the other axioms):

(c_3) $\qquad\qquad\qquad g(q, p, q) = q,$

(d_3) $\qquad\qquad\qquad g(p, q', q) = p,$

(d_4) $\qquad\qquad\qquad g(q, p, q') = p.$

The derivation of (c_3) begins as follows:

$$g(q, p, q) = g(g(q, p, p'), p, g(q, p, p')) = \cdots .$$

The two identities

$$g(p, q, r) = g(p, r, q),$$

$$g(p, q, r) = g(q, r, p),$$

also follow readily from (b), and they can be used to show that the operation g is *totally commutative* in the sense that

$$g(p, q, r) = g(\sigma(p), \sigma(r), \sigma(q))$$

for every permutation σ of $\{p, q, r\}$.

Given a model A of the preceding axioms, and an arbitrary element u in A, define

$$1 = u, \qquad 0 = u',$$
$$p \vee q = g(p, 1, q) = g(p, u, q), \qquad p \wedge q = g(p, 0, q) = g(p, u', q).$$

Under these operations, A becomes a Boolean algebra. To demonstrate this, it suffices to verify the identity laws (2.13), the complement laws (2.14), the commutative laws (2.18), and the distributive laws (2.20) (see Exercise 2.2). The identity laws follow from (d_3) and (d_1), the complement laws from (d_4), and the commutative laws from the total commutativity of the operation g. The distributive laws are particular instances of (b), and, in fact, (b) can be thought of as a general distributive law.

Chapter 7

1. Use part (1) of Lemma 3.

7. Both implications are false.

9. Use Exercise 8.

11. Use Exercise 10.

19. Assume that the first distributive law holds identically in a lattice. Use the first distributive law, the commutative laws, the absorption laws, and the associative laws to derive the second distributive law. Begin with the equation

$$(p \vee q) \wedge (p \vee r) = [(p \vee q) \wedge p] \vee [(p \vee q) \wedge r].$$

Chapter 8

14. Consider a finite field of subsets of an infinite set.

15. Chapter 10 presents an example.

16. Consider the Boolean algebra of finite and cofinite sets of integers.

23. Use Exercise 6.2(e) and induction.

24. Use Exercise 23.

28. Consider the family of half-closed intervals $[0, 1/n)$, for $n = 1, 2, 3, \ldots$.

29. Prove that the sequence of intervals $P_n = [2n, 2n+1)$, for $n = 1, 2, 3, \ldots$, has no supremum.

Chapter 9

10. The key is to establish the *Cauchy–Schwarz inequality* for finite sums of real numbers:

$$\left(\sum_{i=1}^{n} x_i y_i\right)^2 \leq \left(\sum_{i=1}^{n} x_i^2\right)\left(\sum_{i=1}^{n} y_i^2\right).$$

Write

$$r = \sum_{i=1}^{n} x_i y_i, \qquad s = \sum_{i=1}^{n} x_i^2, \qquad t = \sum_{i=1}^{n} y_i^2,$$

and prove that $r^2 \leq st$.

For each n-termed sequence $x = (x_1, \ldots, x_n)$ of real numbers, define the *norm* of x to be

$$\|x\| = \sqrt{x_1^2 + \cdots + x_n^2}.$$

In terms of the norm, the Cauchy–Schwarz inequality may be rewritten in the form

$$\left|\sum_i x_i y_i\right| \leq \|x\|\|y\|.$$

This inequality implies the *triangle inequality* for the norm:

$$\|x + y\| \leq \|x\| + \|y\|.$$

The distance function d in $\mathbb{R}^n$ may be defined in terms of the norm as follows:

$$d(x, y) = \|x - y\| = \sqrt{(x_1 - y_1)^2 + \cdots + (x_n - y_n)^2}.$$

13. Use the definitions of the closure and interior operations.

15. Consider the open intervals $P = (0, 1)$ and $Q = (1, 2)$ in the space of all real numbers.

16. Imitate the proof of Lemma 10.4

24. Observe that a set P in a topological space X is nowhere dense just in case

$$P^{-\prime-} = X.$$

Use this observation and Exercise 13(d) to prove that

$$Q^{-\prime} = P^{-\prime-} \cap Q^{-\prime} \subseteq (P \cup Q)^{-\prime-}.$$

Conclude that the union of two nowhere dense sets is nowhere dense.

25. Use (3), Exercise 13(b), and the definition of a nowhere dense set.

26. Use Exercises 19, 21, 24, and 25.

27. Use Exercise 25.

29. Use Exercises 19, 21, 27, and 28.

34. Define an equivalence relation between the points of an open set U as follows: points x and y in U are equivalent if every point on the segment connecting x and y is in U. Show that the equivalence classes of this relation are open intervals. Use the fact that the set of rational numbers is countable and dense to conclude that there are only countably many equivalence classes.

35. Use Exercise 34, together with the identity

$$(2^{\aleph_0})^{\aleph_0} = 2^{\aleph_0}$$

from cardinal arithmetic. The identity means that if a set X has the power of the continuum, then there are continuum many (countable) sequences whose terms belong to X.

36. Prove that the set $Q \cap P^{\perp}$ is open, and that

$$(Q \cap P^{\perp})^{\perp\perp} = Q \qquad \text{and} \qquad Q \neq Q \cap P^{\perp}.$$

To establish the equality, show that

$$X = \varnothing' = P^{\perp-},$$

and therefore

$$Q = Q \cap P^{\perp-} \subseteq (Q \cap P^{\perp})^{-} \subseteq Q^{-}.$$

Conclude that

$$Q^{-} = (Q \cap P^{\perp})^{-},$$

and consequently

$$Q = Q^{\perp\perp} = (Q \cap P^{\perp})^{\perp\perp}.$$

37. Open balls are regular open sets.

38. Consider an infinite space endowed with the cofinite topology.

40. The largest number of distinct sets obtainable from a subset of $\mathbb{R}^n$ by repeated applications of closure and complementation is 14. To prove that 14 is an upper bound, write

$$P^\top = P'^-,$$

and establish the analogues of Lemmas 1–3 in Chapter 10. To prove that the number 14 can actually be achieved, let Q be the set of rational numbers and consider the set

$$P = [0,1) \cup (1,2] \cup \{3\} \cup [4,5] \cup (Q \cap (6,7))$$

in the space of real numbers.

Chapter 10

4. For the first De Morgan law, use the assumption that P and Q are regular, and use equation (7). For the second De Morgan law, use the definition of join, together with (7).

5. Reduce the verification of the associative law for join to that of meet, using (7) and Lemma 3.

9. It is convenient to write $P^\top = P^{\circ-}$. Observe that P is a regular closed set just in case

$$P = P^{\top\top}.$$

If P and Q are regular closed sets, define

$$P \wedge Q = (P \cap Q)^{\top\top}, \quad P \vee Q = P \cup Q, \quad \text{and} \quad P' = P^\top.$$

10. To demonstrate that the given equation may fail when $\{P_i\}$ is a family of open sets that are not regular, take X to be the set of integers endowed with the cofinite topology, and put

$$P_i = X - \{i\}$$

for each integer i in X. Show that

$$\left(\bigcap_i P_i\right)^{-'-'} = \varnothing \quad \text{and} \quad \left(\bigcap_i P_i^-\right)^{'-'} = X.$$

Chapter 11

2. The field B is a subfield of A if and only if m divides n.

4. Consider the field $\mathcal{P}(X)$ of all subsets of an infinite set X and the class of all finite subsets of X.

5. Consider the Boolean algebra of all subsets of $\mathbb{R}^n$ and the class of regular open subsets of $\mathbb{R}^n$.

6. Consider the class of all finite subsets of the natural numbers, with the set of natural numbers adjoined as the unit.

8. Dualize the formulation and proof of Theorem 2. For each 2-valued function a on E, put

$$p_a = \bigvee_{i \in E} p(i, a(i)),$$

and let K be the set of functions a such that $p_a \neq 1$. For every subset X of K, write

$$p_X = \bigwedge_{a \in X} p_a.$$

Formulate and prove the duals of equations (4)–(9).

15. Consider the field of all subsets of an infinite set X and the subfield of the finite and cofinite subsets of X.

16. Consider the field A of finite and cofinite sets of integers, and the class B of those subsets of integers that are either finite sets of even integers, or else the complements of such sets.

17. Consider the field of all sets of non-negative integers, and the subfield of all finite subsets of positive integers and the complements of these subsets (with respect to the set of non-negative integers).

18. Take B to be any complete Boolean algebra — for instance the field of all subsets of an infinite set — and take A to be the canonical extension of B (see Chapter 23). Both algebras are complete, but every infinite join of B is "broken" in A.

19. Can m be finite? How many meets of finite subsets are there? How many finite joins of such meets are there? Use Theorem 3 and a counting argument.

22. Use Exercise 21 and Lemma 8.1 to prove that the stated condition is equivalent to the condition formulated in the final paragraph of the chapter.

24. The answer is affirmative. Show that if a family $\{P_i\}$ has a supremum in the field of finite and cofinite sets of integers, then that supremum must be the union $P = \bigcup_i P_i$. Consider three cases: P is finite; P is cofinite; and P is infinite but not cofinite.

25. Use the definitions involved and Exercise 20 to prove that a complete subalgebra is a regular subalgebra. A regular subalgebra need not be a complete subalgebra; see Exercise 24.

26. Use the definitions involved to prove that if B is a regular subalgebra of A that happens to be complete, then B must be a complete subalgebra of A.

28. Use Exercise 20 and the condition formulated in the final paragraph of the chapter to show that the answer is affirmative.

Chapter 12

1. Argue by cases to show that f preserves the fundamental operations. For instance, to prove that f preserves join, consider two cases: in the first case both arguments are finite, while in the second case at least one of the arguments is cofinite.

8. To answer the first question, consider the function that maps every element to zero. To answer the second question, use the equations of (3.4) in Chapter 3.

16. Neither $f \vee g$ nor $f + g$ is a homomorphism.

20. For a Boolean monomorphism from B to A, the image of an atom in B may not be an atom in A. Consider any set X with more than one element, and let f be the function from the Boolean algebra 2 into the field of sets $\mathcal{P}(X)$ that maps 0 to the empty set and 1 to the set X.

24. Prove that the answer is affirmative. Let O be the set of odd integers, and consider the function f from A to B that is defined by

$$f(P) = \begin{cases} \{2n : n \in P\} & \text{if } P \text{ is finite,} \\ \{2n : n \in P\} \cup O & \text{if } P \text{ is cofinite.} \end{cases}$$

25. The first algebra is atomic and the second algebra has no atoms whatsoever.

30. For the first part, use Theorem 3. Alternatively, let C be the set of elements p in B such that $f(p) = g(p)$. Prove that C is a subalgebra of B that includes E. For the second part, the assumption is that B is a complete Boolean algebra, and in fact the smallest complete subalgebra of itself that includes the set E. An analogue of Theorem 3 is not available for the proof. However, the alternative approach works: show that C is a complete subalgebra of B.

31. One possibility is to use Theorem 3. Alternatively, consider the set B_0 of those elements q in B such that the meet $q \wedge p_0$ is generated by F in the relativization $B(p_0)$; prove that $B_0 = B$.

35. Consider an arbitrary family $\{p_i\}$ of elements in B with a supremum p. Show that the family $\{f(p_i)\}$ has a supremum in A, and that supremum is $f(p)$.

40. For the first and third parts of the exercise, let X be an infinite set and consider a 2-valued homomorphism f on the field $\mathcal{P}(X)$ of all subsets of X that maps all finite sets to 0. For the second part of the exercise, let B be a complete Boolean algebra that is a subalgebra, but not a regular subalgebra, of a complete Boolean algebra A (see Exercises 11.18 and 11.26). Take f to be the identity mapping on B and use Lemma 1.

41. Use Lemma 1.

42. Use Lemma 1 (or Exercise 41) and Exercises 11.25 and 11.26.

44. Consider the 2-valued homomorphism on the finite–cofinite algebra of the integers that maps each finite set to zero.

Chapter 13

4. Use the following consequences of (11.9):

$$\bigwedge_{i \in F} p(i, a(i)) = \bigvee_{b \in L} \bigwedge_{i \in E} p(i, b(i))$$

and

$$\bigwedge_{i \in F} p(g(i), a(i)) = \bigvee_{b \in L} \bigwedge_{i \in E} p(g(i), b(i)),$$

where L is set of 2-valued functions b on E that extend a.

Chapter 14

6. Use Exercise 5 and Lemmas 1–3.

7. Choose an atom q below p, and consider the set E of all those elements in A that are above q. Define a 2-valued function f on A as follows: f maps the elements in E to 1 and the elements not in E to 0. Show that f is a homomorphism.

8. Call two points in a topological space *regularly separable* if there is a regular open set that contains one of the points, but not the other. The regular open algebra of a topological space is atomless if and only if every non-empty regular

open set contains two points that are regularly separable. The regular open algebra is atomic if and only if every non-empty regular open set includes a regular open subset in which no two points are regularly separable.

9. The answer is negative. Notice that a counterexample cannot be atomic (in an atomic Boolean algebra the unit is the supremum of all atoms), but it must have infinitely many atoms. Let Y be the set of negative real numbers, and B the interval algebra of Y. Let Z be the set of positive integers, and C the algebra of finite and cofinite subsets of Z. Write $X = Y \cup Z$, and consider the subalgebra A of $\mathcal{P}(X)$ generated by the class of singletons of positive integers together with the class of subintervals of Y of finite length. This subalgebra consists of the sets $S = P \cup Q$ such that P is in B, and Q is in C, and either P has finite length (as a set of real numbers) and Q is finite, or else P has infinite length and Q is cofinite.

10. To prove the implication from right to left, imitate the proof of Theorem 7. It must be shown, in addition, that for each function a from I to 2, the set $\{p(i, a(i)) : i \in I\}$ always has an infimum, and that infimum is either zero or an atom. To prove the implication from left to right, use Theorem 6, Lemma 12.1, and Exercises 8.3 and 11.29.

11. Use Lemma 12.1 and Exercises 8.3 and 11.29 for the first part of the exercise. Use the first part of the exercise, Theorem 6, and Exercise 10 for the second part of the exercise.

Chapter 15

1. For each 2-valued function x on m, define a 2-valued function z_x on n by

$$z_x(i) = \begin{cases} x(i) & \text{if } i < m, \\ x(m-1) & \text{if } m \le i < n. \end{cases}$$

The correspondence that maps the function x to the function z_x for each x in 2^m is a monomorphism from 2^m into 2^n.

4. If a function ϕ maps the set Y onto the set X, then the homomorphism induced by ϕ, that is to say, the mapping f from $\mathcal{P}(X)$ into $\mathcal{P}(Y)$ defined by

$$f(P) = \phi^{-1}(P)$$

for every subset P of X, is a monomorphism. Given an arbitrary monomorphism f of $\mathcal{P}(X)$ into $\mathcal{P}(Y)$, define a function ϕ from Y into X by

$$\phi(y) = x \quad \text{if and only if} \quad y \in f(\{x\}).$$

Show that ϕ is a well-defined mapping of Y onto X, and f is the monomorphism induced by ϕ.

5. Corollary 2 and the assumptions on f imply that algebras A and B are isomorphic. Corollary 1 implies that we may assume

$$A = \mathcal{P}(X) \quad \text{and} \quad B = \mathcal{P}(Y),$$

where X and Y are sets of cardinality n, the number of atoms in each algebra. In this context, the assumptions on f imply that, for two subsets P and Q of Y, if P is a proper subset of Q, then $f(P)$ is a proper subset of $f(Q)$. Use this to prove that P and $f(P)$ always have the same cardinality. Define a function ϕ from X into Y by

$$\phi(x) = y \quad \text{if and only if} \quad f(\{y\}) = \{x\},$$

for each x in X. Show that ϕ is a bijection and that f is the isomorphism induced by ϕ.

Chapter 16

5. For the first part, use Exercises 1 and 2. For the second part, use Exercises 1 and 4.

Chapter 17

3. Use Exercises 1 and 2.

5. The fundamental ring operations are definable in terms of the Boolean operations.

6. The fundamental Boolean operations are definable in terms of the ring operations.

11. Imitate the argument in the last example in the chapter, using Exercises 7.9(a), (b) and 6.2(g).

12. The argument is very similar to that for Exercise 11.

Chapter 18

5. Use Exercises 9.24 and 9.25.

8. To prove that an ideal M is dense if and only if its annihilator is trivial, use the first assertion of the exercise, together with Exercises 6 and 7.

9. Use the following properties of a measure μ (see Chapter 31): (1) μ is a non-negative real-valued function on a field of sets A; (2) $\mu(\varnothing) = 0$; (3) if P and Q are sets in A, then
$$\mu(P \cup Q) \le \mu(P) + \mu(Q);$$
(4) if P and Q are sets in A such that $Q \subseteq P$, then $\mu(Q) \le \mu(P)$.

10. Use the fact that the idempotent laws (2.16), the commutative laws (2.18), the associative laws (2.19), the distributive laws (2.20), and the absorption laws in Lemma 6.3 are true in a Boolean ring without unit, when meet and join are defined as in the exercise. These laws are easy to derive directly, but their validity also follows from the fact that every Boolean ring without unit can be extended to a Boolean ring with unit. See Exercises 1.10 and 3.5.

14. The ideal consists of those elements that can be written as finite joins of atoms.

17. The analogue of Theorem 11 for filters says that an element p of a Boolean algebra is in the filter generated by a set E if and only if there is a finite subset F of E such that $\bigwedge F \le p$. One way to prove this analogue is to dualize the proof of Theorem 11.

21. Use Theorem 11 and Exercise 17.

22. Use Exercise 21.

23. Use conditions (7) and (18).

25. For the second part, use Exercise 18.

27. The supremum of a family $\{M_i\}$ of ideals is the intersection of all ideals that include the set $\bigcup_i M_i$. See Chapter 19 for the proof that the set of all ideals in a Boolean algebra is a distributive lattice.

28. See Chapter 19.

30. Assume that M is generated by a countable set
$$E = \{q_1, q_2, q_3, \dots\}.$$
For each positive integer n, put
$$p_n = \bigvee_{m=1}^{n} q_m,$$
and use Theorem 11.

31. Use the identities established in Exercises 6.7, 6.8, and 7.11. For part (b), if N is the cokernel of a congruence Θ, then Θ can be defined in terms of N by means of the equivalence

$$p \equiv q \quad \text{mod } \Theta \qquad \text{if and only if} \qquad p \Leftrightarrow q \in N.$$

For part (c), if N is any Boolean filter, define a binary relation Θ by the preceding equivalence, and prove that Θ is a congruence relation with cokernel N. For part (d), the equivalence class p/Θ coincides with the "cokernel coset"

$$p \Leftrightarrow N = \{p \Leftrightarrow q : q \in N\}.$$

Chapter 19

3. Use Exercises 18.10 and 18.12.

5. Use the definition of f and Corollary 1.

8. If M and N are filters in a Boolean algebra, then

$$M \vee N = \{p \wedge q : p \in M \text{ and } q \in N\}$$

and

$$M \wedge N = \{p \vee q : p \in M \text{ and } q \in N\}.$$

The proof is the dual of the proof of Lemma 1.

9. The argument is the dual of the proof that the lattice of ideals is distributive. It uses Exercise 8 instead of Lemma 1.

Chapter 20

1. To prove the non-trivial direction of the assertion, assume that M satisfies the stated conditions and show that it is an ideal satisfying the criterion of Lemma 1. The key step is showing that M is downward closed: if p is in M, and if $q \leq p$, then q is in M. One of q and q' is in M, so it suffices to prove that q' cannot be in M. Argue by contradiction.

2. Show that the class of all sets in B that do not contain x_0 is an ideal that satisfies the criterion of Lemma 1.

3. Use Corollary 3.

4. Use Exercise 18.25 and Lemma 1.

5. Use Exercise 18.26 and Lemma 1.

8. The dual of Corollary 1 asserts that a filter is maximal if and only if it is prime in the sense that the presence of $p \vee q$ in the filter always implies the presence of at least one of p and q. One way to prove this assertion is to dualize the proof of Corollary 1. Another way is to use the isomorphism between the lattice of ideals and the lattice of filters that is discussed in Chapter 19.

9. The dual of Corollary 2 asserts that a principal filter generated by an element p is maximal if and only if p is an atom. One way to prove this assertion is to dualize the proof of Corollary 2, and another way is to use the isomorphism between the lattice of ideals and the lattice of filters that is discussed in Chapter 19.

10. Use Exercise 19.10 and Exercise 7.

11. Dualize the proof of Corollary 4, using Exercises 9 and 10. Alternatively, use Corollary 4, Exercise 19.11, and the isomorphism between the lattice of ideals and the lattice of filters.

12. Use Exercises 18.20 and Exercise 10.

13. A subset E of a Boolean algebra B has the *finite join property* if the join of every finite subset of E is different from the unit. Prove that if E has the finite join property, then it is included in a maximal ideal of B.

15. Show that if q is an element in A that is not in B, then the set

$$\{p \in B : p \le q\} \cup \{p \in B : p \le q'\}$$

has the finite join property, and consequently can be extended to a maximal ideal N in B (Exercise 13). Extend

$$N \cup \{q\} \quad \text{and} \quad N \cup \{q'\}$$

to (distinct) maximal ideals in A.

16. Let X be any infinite set and write $B = \mathcal{P}(X)$. Take M to be a non-principal, maximal ideal in B, and put $A = B/M$. The canonical epimorphism f from B onto A is not complete: it maps each atom in B to the zero element of A, but it maps the unit of B — which is the supremum of the set of atoms in B — to the unit of A.

17. For each ideal M in B write

$$N = \{q \in A : q \wedge p \in M\}.$$

Prove that N is an ideal in A, and in fact that it is the ideal generated by $M \cup \{p'\}$. Show that N is maximal (in A) just in case M is maximal (in B).

19. To prove the existence of the desired extension B use Exercise 1.10. The extension B is unique in the following sense: if C is a Boolean algebra that includes A as a maximal ideal, then there is an isomorphism from B to C that maps each element p in A to itself and maps the complement of p in B to the complement of p in C (see Exercise 22).

20. Use Exercise 1.10 and Exercises 18 and 19.

21. Use Exercise 1.10 and Exercise 19 to show that if the ring is finite, then it must have a unit.

22. Consider a ring homomorphism g from M into a Boolean algebra A, and define a mapping f from B into A by

$$f(p) = \begin{cases} g(p) & \text{if } p \in M, \\ g(p')' & \text{if } p \in B - M. \end{cases}$$

Show that f is a Boolean homomorphism. (The argument proceeds by cases.)

25. Follow the lines of the proof of Theorem 12, but with the following modifications. First, add p_i to M_{i+1} just in case the ideal generated by $M_i \cup \{p_i\}$ does not contain p. Second, show that the annihilator of p — the set of elements q in B such that $p \cdot q = 0$ — is included in M_α. Finally, prove that the element $q + p \cdot q$ is in the annihilator of p for every q in B. Use this to show that if N is an ideal that includes M_α, then $N = B$ when p is in N, and $N = M_\alpha$ when p is not in N.

26. Apply Exercise 25.

27. Follow the lines of the proof of Theorem 12, but with the following modifications. First, add p_i to M_{i+1} just in case the ideal generated by $M_i \cup \{p_i\}$ does not contain p. Second, use the distributive law to show that if $q \wedge r$ belongs to M_α, then one of q and r must belong to M_α. It is helpful to use a version of Lemma 18.1 that applies to distributive lattices.

28. Use the notion of a prime ideal (instead of a maximal ideal) in the formulation, and apply Exercise 27.

Chapter 21

1. Use Exercise 6.8.

3. Use the maximal ideal theorem and Corollary 3.

4. Use Corollary 20.3 to get a maximal ideal that does not contain $p + q$.

5. Use the maximal ideal theorem to get a maximal ideal that contains all finite subsets of X.

7. To formulate the correspondence theorem for filters, replace the word "ideal" by the word "filter" everywhere in the correspondence theorem for ideals. To prove the theorem, either imitate the proof of the correspondence theorem for ideals, or else derive it as a corollary of that theorem, using Exercise 18.31 and the isomorphism between the lattice of ideals and the lattice of filters.

9. The two cosets are not equal. The first is a set of elements in B, while the second is a set of subsets of B.

10. Use Exercise 20.15.

11. Use Exercises 20.23, 20.24, and 1.9.

Chapter 22

1. Use Exercise 18.23, Lemma 20.1, and Corollary 20.3.

2. Use Exercises 18.24 and 20.7, together with Corollary 20.3.

4. Suppose A is a Boolean algebra, and p_0 a non-zero element in A. Let x be any 2-valued homomorphism on the relativization $A(p_0)$, and in terms of x define a mapping f from $A(p_0)$ into A by

$$f(p) = \begin{cases} p \vee p_0' & \text{if } x(p) = 1, \\ p & \text{if } x(p) = 0, \end{cases}$$

for every $p \le p_0$. Show that f is a monomorphism. Notice that when p_0 is zero and A is not degenerate, no such monomorphism can exist.

5. A complete field of sets is necessarily atomic (Theorem 8, p. 124) and completely distributive (see, for example, Corollary 14.3, p. 123). However, not all complete Boolean algebras are atomic.

7. Imitate the proof of Theorem 17. To show that the homomorphism f is one-to-one, use Exercise 20.27 to obtain a prime ideal N that contains one of two given points, but not the other, and then show that N determines a 2-valued homomorphism of which it is the kernel.

Chapter 23

1. Show that a finite Boolean algebra satisfies the defining criteria for being a canonical extension of itself.

3. Consider any infinite complete and atomic Boolean algebra and its canonical extension.

4. Use the exchange principle and the uniqueness theorem for canonical extensions.

5. Use Exercise 3.

7. First, observe that

$$f(p) = \bigwedge \{g(s) : s \in B \text{ and } p \le s\}$$

for elements p in E, and

$$f(p) = \bigvee \{f(q) : q \in E \text{ and } q \leq p\}$$

for arbitrary elements p in B. Use these equations to check that f is order-preserving on elements in E and therefore also on elements in B. Next, prove the following lemma: for every element p in B_1 and every atom q in A_1, if $q \leq f(p)$, then there is an atom r in B_1 such that $r \leq p$ and $q \leq f(r)$. Argue that the set $\{s \in B : q \leq g(s)\}$ is an ultrafilter in B, and that the infimum of this ultrafilter is the desired atom r. Use this lemma, the order-preserving property of f, and the atomicity of A_1 to show that f preserves arbitrary joins. Finally, prove that if p and q are elements in E, then

$$f(p) \wedge f(q) \leq f(p \wedge q).$$

Argue that an atom that is not below the right side of this inequality cannot be below the left side. The proof involves the compactness property of canonical extensions. Use this inequality, the lemma, and the atomicity of A_1 to prove that f preserves meet. Conclude with the help of Exercise 12.9 that f is a complete homomorphism.

8. Apply Exercise 7 with the identity monomorphism on B as the homomorphism g, and use Lemma 12.1 and Exercise 11.26.

Chapter 24

4. One approach is to show that the function mapping each complete ideal to its dual filter is a bijection that preserves the constants zero and one, and the operations of meet, join, and complement. (See the remarks in the second paragraph of Chapter 12.) Use the appropriate restriction of the function f that maps each ideal to its dual filter (Chapter 19). Apply Exercise 3 and properties of f. Special attention must be given to the operations of join and complement. To prove that complement is preserved, show that if M is a complete ideal, and N is its dual filter, then the dual of the annihilator of M is just the annihilator of N.

5. Use Exercise 3 and Theorem 20.

6. Use Exercise 4.

7. One approach is to use Exercise 6 to obtain the desired result as a consequence of Lemma 1.

8. One approach is to use Exercise 6 to obtain the desired result as a consequence of Lemma 2.

12. Use Theorem 20.

13. For each integer i, let P_i be the singleton $\{i\}$. The family $\{P_i\}$ has a supremum, but that supremum does not belong to the ideal.

14. The join of the three ideals in the lattice of ideals of B is the ideal of the finite sets of integers. The join in the lattice of complete ideals of B is B itself.

17. One approach is to use Exercise 16 to prove that if p is in the annihilator of the set E, then p' is an upper bound of E, and hence also an upper bound of E^d and of the set of suprema of subsets of E^d. Apply Lemma 1.

19. Use Exercise 15 and Theorem 21.

20. Use Exercises 15 and 19, together with the fact that for subsets F and E of a Boolean algebra, if F is included in E, then the annihilator of E is included in the annihilator of F.

22. Use Exercise 16 and Lemma 2 to verify the double complement law in (2.15). Use Exercise 17 to verify the second De Morgan law in (2.17). The first De Morgan law in (2.17) follows from the second and the double complement law.

23. Use Corollary 1 and Lemma 12.1.

Chapter 25

1. Use Corollary 1.

2. Use the fact that each element in B is the supremum of the set of elements in A that it dominates.

3. Use Exercise 2 and Corollary 14.1.

5. Use Exercise 3.

6. Use Exercise 2.

7. Use Lemma 1.

9. The answer is negative; a Boolean algebra B is the completion of A if and only if B is complete and includes A as a dense subalgebra. (See the remarks in the first few paragraphs of the chapter.)

11. Use Exercise 23.5.

12. The key step is the following lemma: if p is the supremum in B_1 of a subset E of B, then
$$f(p) = \bigvee \{g(s) : s \in E\}.$$
To prove the lemma one shows for every s in B with $s \leq p$ that
$$s = \bigvee \{s \wedge q : q \in E\}$$
and therefore
$$g(s) = \bigvee \{g(s \wedge q) : q \in E\} \leq \bigvee \{g(q) : q \in E\}.$$
It then follows that $f(p) \leq \bigvee \{g(q) : q \in E\}$.

13. Apply Exercise 12 with the identity monomorphism as the homomorphism g, and use Lemma 12.1 and Exercise 11.26.

Chapter 26

4. Use Exercise 3.

5. For the second part of the problem, show that the subalgebra

$$B = \{(0,0), (1,1)\}$$

of A is not the product of subalgebras of A_1 and A_2.

6. For the second part of the problem, assume that L is an ideal in A, and show that the sets

$$M = \{p : (p,0) \in L\} \quad \text{and} \quad N = \{q : (0,q) \in L\}$$

are ideals in A with the property that $L = M \times N$.

7. An ideal L in A is maximal if and only if it can be written in the form $L = M \times N$, where M and N are ideals in B and C respectively, and one of the ideals M and N is maximal, while the other is improper.

8. Use Exercise 7 and Exercise 20.14.

10. Let C be a non-trivial Boolean algebra, and consider the endomorphism f of $C \times C$ defined by

$$f((p_1, p_2)) = (p_1, p_1)$$

for each pair (p_1, p_2). More generally, for each homomorphism h from a Boolean algebra B_1 into C, and for each non-trivial Boolean algebra B_2, consider the homomorphism f from $B_1 \times B_2$ into $C \times C$ defined by

$$f((p_1, p_2)) = (h(p_1), h(p_1)).$$

14. Use Exercise 12.19, an internal product version of Exercise 5, and Lemma 1 to show that the internal product of $B(r)$ and $B(r')$ is a subalgebra of A. The internal product includes the set $B \cup \{r\}$, and each of its elements is generated by this set.

16. Use Corollary 1 and Exercise 12.18.

17. Use Corollary 1 and Exercise 12.17.

18. Imitate part of the proof of Corollary 2.

19. If A and B are two Boolean algebras that satisfy the hypotheses of the exercise, and if p and q are the suprema of the sets of all atoms in A and B respectively, then

$$A = A(p) \times A(p') \qquad \text{and} \qquad B = B(q) \times B(q').$$

Use Corollary 15.1 and Theorem 10 to prove that $A(p)$ and $B(q)$ are isomorphic, as are $A(p')$ and $B(q')$.

20. Let B be a countably infinite atomic Boolean algebra, and C a countably infinite atomless Boolean algebra. Consider the algebra B and the product $A = B \times C$.

22. Take A to be the field of all subsets of an infinite set Y, or just the field of finite and cofinite subsets of Y. Use Exercise 3 (or 4) and the remarks in Chapter 12, p. 94 (or Exercise 12.22).

23. Take A to be any infinite, atomless Boolean algebra.

27. Use Exercise 26.

37. To show that the three algebras may be distinct, let I be the set of natural numbers, and take $A_i = 2 \times 2$ for each i in I. Consider the elements p and q in A defined by

$$p_i = \begin{cases} (1,0) & \text{if } i \text{ is even,} \\ (0,1) & \text{if } i \text{ is odd,} \end{cases} \qquad \text{and} \qquad q_i = \begin{cases} (0,0) & \text{if } i \text{ is even,} \\ (1,1) & \text{if } i \text{ is odd.} \end{cases}$$

38. Use Exercises 30 and 37.

40. Verify that the product of the canonical extensions is a complete, atomic extension of A that satisfies the atom separation and the compactness properties with respect to A.

43. Show that if B is the product of the completions of the factor algebras, then A is a dense subalgebra of B, and every subset of A has a supremum in B.

Chapter 27

1. Use Lemma 1 and some of the ideas from the proof of Theorem 25.

2. Use Corollary 26.1.

3. Use Corollary 26.1.

4. A one-to-one mapping f from a set X to a set Y induces an isomorphism from the field of all subsets of X to a relativization of the field of all subsets of Y, namely the relativization to the set

$$f(X) = \{f(x) : x \in X\}.$$

The isomorphism maps each subset P of X to the set

$$f(P) = \{f(x) : x \in P\}.$$

Use Lemma 26.1 to conclude that the field of all subsets of X is a factor of the field of all subsets of Y.

5. The mapping f from $A \times (B \times C)$ to $(A \times B) \times C$ defined by

$$f((p, (q, r))) = ((p, q), r)$$

for all p in A, q in B, and r in C is an isomorphism.

6. The mapping f from $A \times B$ to $B \times A$ defined by

$$f((p, q)) = (q, p)$$

for all p in A and q in B is an isomorphism.

8. The law is true. The required isomorphism maps each pair of functions (p, q) in $B^X \times B^Y$ to their union, that is, to the function r from $X \cup Y$ into B defined by

$$r(x) = \begin{cases} p(x) & \text{if } x \in X, \\ q(x) & \text{if } x \in Y. \end{cases}$$

9. The law is true. Each element p in $(B^X)^Y$ is a function with domain Y such that for each y in Y, the value $p(y)$ is a function from X into B. The required isomorphism maps p to the function r in $B^{X \times Y}$ that is defined by

$$r(x, y) = p(y)(x).$$

11. Let X be an infinite set, and consider the algebras

$$A = 2^X, \qquad B = 2^2, \qquad \text{and} \qquad C = 2.$$

12. Consider the field A of finite and cofinite subsets of an infinite set X, the field B of all subsets of a two-element set, and the field C of all subsets of a one-element set. Use Exercises 12.22 and 26.4.

13. Use a cardinality argument and Corollary 15.2 to show that the answer is positive.

20. For the first part, use a cardinality argument and Corollary 15.2 to show that the answer is positive. For the second part, use Theorem 25.

21. Let A be the Boolean algebra constructed in Hanf's counterexample, and put $B = A \times 2$.

22. Let A be the Boolean algebra constructed in Hanf's counterexample, and put $B = A \times 2$.

24. Let A and B be the Boolean algebras constructed in Hanf's counterexample and in Exercise 23 respectively, and put

$$A_1 = A \times 2 \qquad \text{and} \qquad A_2 = B \times 2 \times 2.$$

26. Show that if A has at least n atoms, and if $A \times 2 = A$, then A must have at least $n + 1$ atoms.

27. If $A \times B = A$ for some finite, non-degenerate Boolean algebra B, then A must have infinitely many atoms.

28. The proof involves a back-and-forth argument. Its general structure is similar to that of the proof of Theorem 10, p. 134 (which asserts that two countable, atomless Boolean algebras with more than one element are isomorphic). Write $B = A \times 2$. Define sequences

$$p_1, p_2, \ldots, p_n \qquad \text{and} \qquad q_1, q_2, \ldots, q_n$$

in A and B respectively such that for every 2-valued function b on $I_n = \{1, \ldots, n\}$,

$$B(q_b) = A(p_b) \qquad \text{or} \qquad B(q_b) = A(p_b) \times 2,$$

and in the latter case there are infinitely many atoms below p_b, where

$$p_b = \bigwedge_{i \in I_n} p(p_i, b(i)) \qquad \text{and} \qquad q_b = \bigwedge_{i \in I_n} p(q_i, b(i)).$$

29. Use Exercise 28 and the fact that every finite Boolean algebra is isomorphic to a finite power of 2.

30. Use Exercises 27 and 29.

Chapter 28

2. Use Exercise 12.5.

3. Use Corollary 1 and the isomorphism between the Boolean algebras $\mathcal{P}(X)$ and 2^X.

4. Use Corollary 2.

5. Use Theorem 26 and Corollary 1.

6. Use ideas from the proof of Corollary 1 and from the proof of Theorem 4 (p. 107).

7. The subalgebra generated by the set

$$E = \{p_i : i \in I\}$$

of projections is freely generated by E, by Theorem 26. The proof of the theorem shows that E does satisfy the criterion of Lemma 1.

9. The answer is negative. For example, consider the free Boolean algebra 2^4 and a subalgebra isomorphic to 2^3.

10. The answer is negative. Use Corollary 2 and the subsequent remark, together with Exercise 12.23.

12. Use Lemma 13.2 and Exercise 21.10.

13. Use an argument similar to the proof of Theorem 10, but without the back-and-forth component, to construct an infinite sequence of elements that generates the given atomless Boolean algebra and that satisfies the condition of Lemma 1.

Chapter 29

5. Imitate the proof of Lemma 11.1.

6. Imitate the proof of Corollary 11.1, using Exercise 5.

8. Use Lemma 13.1 and Exercise 11.20.

11. Imitate the proof of Lemma 13.2.

13. Imitate the proof of the analogous result for free Boolean algebras.

17. Imitate the argument in Chapter 11 that the subfield of A generated by the set of singleton subsets of X coincides with the field of finite and cofinite subsets of X.

18. Use Lemma 20.1.

19. Use an infinite distributive law for the intersection of two unions, and use an infinite associative law for unions.

21. For each positive integer n, let U_n be the union of the class of open balls of radius $1/n$ whose center is in the given closed set, and form the intersection of the family $\{U_n\}$.

22. Use Exercise 9.35 and the set-theoretic fact that $(2^{\aleph_0})^{\aleph_0} = 2^{\aleph_0}$; the latter implies that if a set X has the power of the continuum, then there are continuum many (countable) sequences whose terms are all in X.

31. Use Exercises 30 and Exercise 9.30 to show that the singleton of any isolated point is a regular open set. If U is a non-empty regular open set that is not a singleton, then it must contain distinct points x and y. Use open sets that separate these two points (and Exercise 9.14 and Lemma 10.2) to construct a regular open set that contains x but not y. Conclude that U cannot be an atom.

33. Use Lemma 1 and imitate part of its proof.

34. If a linear ordering on a set X is complete, then there is a smallest element 0 (the supremum of the empty set) and a largest element 1 (the supremum of X). To demonstrate the compactness of X, consider an open covering $\{U_i\}$, and let P be the set of points x such that the interval $[0, x]$ is covered by a finite subfamily of the open covering. Form the supremum y of the set P. There must be an open interval that contains y and is included in one of the sets U_i. Use this interval to show that there is a finite subcover of the interval $[0, y]$, and consequently there is a finite subcover of X.

    If the ordering of X is incomplete, then there is a subset P of X that has no supremum. For each element x in P let U_x be the open interval $(-\infty, x)$, and for each element x in the set Q of upper bounds of P, let U_x be the open interval (x, ∞). Show that the family $\{U_x\}_{x \in P \cup Q}$ is an open cover of X that has no finite subcover.

36. The existence of a set that has the Baire property and is not Borel can be shown by using the fact that there exists a nowhere dense set of real numbers with continuum many elements (for instance, the Cantor middle third set — see p. 318). Apply Exercise 9.25 to conclude that there are $2^{2^{\aleph_0}}$ nowhere dense sets of real numbers. On the other hand, there are $2^{\aleph_0}$ Borel sets of real numbers, by Exercise 22. The desired conclusion follows from the cardinal arithmetic inequality $2^{2^{\aleph_0}} > 2^{\aleph_0}$.

37. Consider the ideal of meager sets in the σ-algebra of Borel sets of an uncountable set endowed with the cocountable topology (see Exercise 9.32).

38. It suffices to extend Corollary 1 to locally compact Hausdorff spaces.

39. Consider the Euclidean space $X = \mathbb{R}^n$. The function f maps each singleton subset of X to the empty set, but it maps the improper subset X to itself.

43. Show that the union of the sets E_i is a σ-subalgebra of A that includes E.

44. One possibility is to use Exercise 43. Alternatively, consider the set A_0 of those elements q in A such that the meet $q \wedge p_0$ is σ-generated by F in the relativization of A to p_0. Prove that A_0 is a σ-subalgebra of A that includes E, so it must coincide with A.

Chapter 30

4. Any base for the discrete topology must include all singleton subsets of the space.

5. The intersection of a countable family of cofinite subsets of an uncountable set is not empty. Any non-empty, proper subset of that intersection cannot be written as a union of sets in the family.

7. Use Lemma 1.

10. Use Exercise 9.32(h) and Exercise 5 to show that the answer is negative.

11. Use Exercises 9 and 8.

12. Consider the algebra of all subsets of a countable set modulo the ideal of all finite sets. For ease in manipulation, let the countable set be the set of all rational numbers, and for each real number x, find an infinite set of rational numbers that has x as its unique limit point.

13. Every disjoint set of non-zero elements can be extended to a maximal set of that kind, and that maximal set necessarily has a supremum, namely 1.

Chapter 31

5. Use Exercise 8.24.

6. Apply Exercise 5 to the sequence $\{p_n'\}$, and then use Exercise 4.

7. Use Exercise 6.9.

8. Use Exercise 6.3 to prove the triangle inequality. If $\{p_n\}$ is a Cauchy sequence, then there is a subsequence $\{p_{n_k}\}$ such that

$$d(p_{n_k}, p_{n_{k+1}}) < 1/2^k$$

for each k. Put

$$r_i = \bigvee_{k=i}^{\infty} p_{n_k} \quad \text{and} \quad p = \bigwedge_{i=1}^{\infty} r_i,$$

and show that the sequence $\{p_{n_k}\}$ converges to p. Conclude that the original sequence $\{p_n\}$ converges to p. It may be helpful to establish the following auxiliary properties of the metric:

$$d(p, q) = \mu(p \vee q) - \mu(p \wedge q),$$
$$d(p, q) = d(r \vee p, r \vee q) + d(r \wedge p, r \wedge q).$$

They follow from the addition property of the measure and Exercise 7.

10. A positive normalized measure exists on the product A if and only if there are only countably many factor algebras. For the implication from right to left, assume that there are countably many factor algebras, indexed, say, by the set of positive integers. If μ_i is a positive normalized measure on A_i for each index i, show that the function μ defined on A by

$$\mu(p) = \sum_i \mu_i(p_i)/2^i$$

for each $p = \{p_i\}$ in A is a positive normalized measure on A.

For the implication from left to right, assume that there is a positive normalized measure on the product A, and observe that if a disjoint subset E of A consists entirely of elements with measure greater than $1/n$, then E must have cardinality less than n. This implies that the number of factors must be countable.

Chapter 32

3. Use Exercises 1 and 2.

5. Use a compactness argument with clopen sets (see, for instance, the proof of Lemma 1).

6. First, show that the empty set and the whole space are open. Next, show that the union of an arbitrary family of open sets is open by using Exercise 8.2. Finally, characterize the intersection of two basic clopen sets; there are two cases to consider. Use this characterization and Exercise 8.3 to show that the intersection of two open sets is open.

8. To prove that X is compact, observe that the constant function x in X whose value at each index i is the unit of Y belongs to every non-empty closed set. Consequently, the intersection of a class of closed sets with the finite intersection property cannot be empty. The space is not Hausdorff, since the constant function x cannot be separated from any other point in X by open sets.

11. The metric properties of d can be derived directly from the properties of the norm (see Exercise 10), without using the definition of that norm.

12. It suffices to prove that every basic open set in the product topology is an open set in the metric topology, and every open ball in the metric topology is an open set in the product topology.

13. Show that the class of characteristic functions of finite subsets of I is a countable, dense subset of 2^I.

15. Use Exercise 29.34.

16. The topology of the space X is the one-point compactification of the space of finite ordinals with the discrete topology.

18. Prove that in a compact Hausdorff space, the intersection of the clopen sets that contain a given point is a connected set.

19. The dual algebra of the Cantor space 2^I has cardinal number m. There are m sets of the form (1) and hence m basic clopen sets. Consequently, there are m clopen sets. The proofs of these assertions use some basic facts about the arithmetic of cardinal numbers (see Appendix A).

21. The answer is affirmative. Proceed as in the proof of Theorem 30. The main difficulty is the proof of Tychonoff's theorem that the product of the spaces is compact under the product topology.

Chapter 33

3. Use the function $\phi(x) = \sin x$ to show that the identities in (b)–(d) fail.

4. The identities (a)–(c) hold when ϕ is one-to-one, and all of the identities hold when ϕ is a bijection.

7. Use Lemma 1 to show that the projection ϕ_i is continuous. To show that it maps open set to open sets, it suffices to show that it maps basic clopen sets to clopen sets. If

$$U = \{x \in X : x_j \in P_j \text{ for all } j \in S\}$$

is a non-empty basic clopen set in X, show that

$$\phi_i(U) = P_i \quad \text{or} \quad \phi_i(U) = X_i,$$

according as i is, or is not, in S.

9. Use Exercise 8.

10. Let X and Y be metric spaces with metrics d and e respectively, and consider a function ϕ from X to Y. The standard ϵ, δ-definition of continuity, applied to ϕ, says that for every point x_0 in X and for $\epsilon > 0$, there is a $\delta > 0$ such that

$$d(x_0, x) < \delta \quad \text{implies} \quad e(\phi(x_0), \phi(x)) < \epsilon$$

for every x in X. The class of open balls in each space constitutes a base for the topology of the space, so it suffices to restrict one's attention to open balls.

11. Argue by contraposition: assume that ϕ is not continuous at a point x in X, and construct a sequence $\{x_n\}$ that converges to x, with the property that $\{\phi(x_n)\}$ does not converge to $\phi(x)$. Use the characterization of continuity given in Exercise 10.

14. The function $\phi(x) = e^x$ maps the (closed) set of all real numbers to an open interval. The identity function on the set $X = \{0, 1\}$ is a continuous mapping of the space X under the discrete topology to the space X under the Sierpiński topology. The former space is Hausdorff, but the latter is not.

15. Consider the sets $X = [0, 1] \cup \{2\}$ and $Y = [0, 1]$, endowed with the topology inherited from the real line, and the function $\phi(x) = \frac{1}{2}x$.

17. Argue by contraposition. Assume that ϕ is a continuous mapping between topological spaces. Show that if the image $\phi(P)$ of some subset P of the domain can be split by open sets V_1 and V_2, then P can be split by the open sets $\phi^{-1}(V_1)$ and $\phi^{-1}(V_2)$. Use Exercise 2.

18. Consider the subsets of the spaces $\mathbb{R}$ and $\mathbb{R}^2$ obtained by removing a single point from each space, and use Exercise 17.

20. Use Exercise 17.

23. The set K_n has measure $(2/3)^n$, and the Cantor set K is included in this set K_n for each n.

26. The geometric series $\sum\limits_{n=0}^{\infty} (\frac{1}{3})^n$ sums to $\frac{3}{2}$.

30. Use Exercise 27.

31. The mapping f that takes each clopen subset P of Y to the set $P \cap X$ is the desired isomorphism. Use Exercise 9.13(e) to show that f is one-to-one. Use the characteristic functions of subsets of X to show that f is onto. (The characteristic functions are continuous because X is a discrete space.)

32. Regard the Cantor space as a topological group and use Haar measure (see [20] or [22]) and Lemma 31.3.

33. To prove the existence of free Boolean spaces, use the Stone–Čech compactification of discrete spaces.

34. Let R be the set of *dyadic fractions* of the form

$$r = k/2^n,$$

where k and n are non-negative integers and $k \leq 2^n$. Construct, by induction on the exponent n of the dyadic fraction r, a family $\{U_r\}$ of open sets in X such that

$$P \subseteq U_0 \quad \text{and} \quad Q \cap U_1 = \varnothing,$$

and

$$r < s \qquad \text{implies} \qquad U_r^- \subseteq U_s$$

for all r and s in R. Define a mapping ϕ from X to $[0,1]$ by

$$\phi(x) = \inf\{r : x \in U_r\}$$

if $x \notin Q$, and $\phi(x) = 1$ if $x \in Q$, and prove, using Exercise 8, that ϕ is continuous.

35. If P is the intersection of a sequence $\{U_n\}$ of open sets, then P is disjoint from each of the closed sets U_n'. Invoke Exercise 34 to obtain a continuous function ϕ_n from X into the interval $[0,1]$ that assumes the value 0 on P, and the value 1 on U_n'. Define the mapping ϕ by

$$\phi(x) = \sum_{n=1}^{\infty} \frac{1}{2^n} \phi_n(x),$$

and show (with the help of Exercise 8) that ϕ is continuous.

Chapter 34

5. It may be easier to work with the dual space of ultrafilters in A instead of the dual space of 2-valued homomorphisms on A; see Exercise 4. Use Exercise 20.14. The dual space is essentially the one-point compactification of the space of natural numbers under the discrete topology.

7. The answer is affirmative; use Theorem 32.

8. Let ϕ be a homeomorphism from a Boolean space X to a Boolean space Y, and consider the function that maps each clopen set P in X to its image $\phi(P) = \{\phi(x) : x \in P\}$ in Y.

9. Let f be a Boolean isomorphism from A to B, and consider the function that maps each 2-valued homomorphism x on A to the 2-valued homomorphism $x \circ f^{-1}$ on B.

10. Construct an isomorphism f from A to the dual algebra B of X with the property that

$$\langle p, x \rangle = \begin{cases} 1 & \text{if } x \in f(p), \\ 0 & \text{if } x \notin f(p). \end{cases}$$

12. The union of two compact sets is compact, and so is a closed subset of a compact set.

14. For part (b) use part (a) and Exercise 18.12; for part (c) use Exercise 20.25; for part (d) use parts (b) and (c); for part (f) use parts (b)–(e), Exercise 18.12, and Lemma 29.1; and for part (h) use parts (e) and (f). In parts (b) and (e), the fact that X consists of non-trivial homomorphisms plays a role.

15. Imitate the proof of Theorem 31, and use Exercises 12 and 14, as well as Exercise 20.25.

16. Follow the lines of the proof of Theorem 32, and use Exercise 14 and Exercise 21.11.

17. The number of clopen subsets of the dual space is countable. Choose one point in each of these subsets, and consider the set of chosen points.

18. If X is the dual space of a countable Boolean algebra A, then X has a countable dense subset Y, by Exercise 17. Consider the function that maps each element p in A to the subset

$$U_p \cap Y = \{y \in Y : y(p) = 1\}$$

of Y.

19. It is helpful to think of the dual of the field $\mathcal{P}(X)$ as the space Y of all ultrafilters in $\mathcal{P}(X)$. Show that the subspace $\tilde{X}$ of principal ultrafilters is dense in Y and homeomorphic to X. Consider an arbitrary function ϕ from X into a compact Hausdorff space Z. For each point N in Y, show that there is exactly one point that is common to all of the sets in the class $\{\phi(S)^- : S \in N\}$, and define $\psi(N)$ to be that point. Show that ψ is a continuous function from Y into Z, and that the value of ψ at the ultrafilter generated by a point x in X is just $\phi(x)$.

Chapter 35

1. Use the definition of the topology of X and Theorem 11 (p. 155).

2. In view of Exercise 1, it suffices to prove for any ideals M and N in A that

$$M \subseteq N \quad \text{if and only if} \quad U_M \subseteq U_N$$

(see Exercise 12.13). To establish the implication from right to left, argue by contraposition and use Corollary 20.3.

8. Consider a Boolean ring B without unit. Extend B to a Boolean ring A with unit such that B is a maximal ideal in A (Exercise 20.19). The dual of A is the space X of all 2-valued homomorphisms on A, and the dual of B (as a Boolean ring) is the space Y of all non-trivial 2-valued homomorphisms on B (Exercise 34.14). Every 2-valued homomorphism y on B (including the trivial

homomorphism) can be extended in a unique way to a 2-valued homomorphism y^* on A, and every 2-valued homomorphism on A is the extension of
a unique such homomorphism on B (Exercise 20.22). The correspondence
that takes y to y^*, for each y in Y, is therefore a one-to-one function from Y
to X, and the only element in X that is not in the range of this function is
the 2-valued homomorphism on A with kernel B (the extension to A of the
trivial homomorphism on B). Endow Y^* with the topology induced by this
correspondence, so that the spaces Y and Y^* become homeomorphic. Show
that Y^* is a subspace of X, and X is the one-point compactification of Y^*.
(It may be helpful to use Exercise 43.21.)

11. Use the isomorphism between the lattice of ideals and the lattice of open sets,
    and apply Exercise 18.8.

12. Imitate the proof that a Boolean space with a countable base has a countable
    dual algebra.

13. Let $\{P_n\}$ be a base of clopen sets for the topology of X, indexed by the set of
    positive integers, and for each positive integer n, let p_n be the characteristic
    function of the set P_n. Define a real-valued function d on pairs of points in X
    by

    $$d(x, y) = \sum_{n=1}^{\infty} \frac{|p_n(x) - p_n(y)|}{2^n}$$

    for all x and y in X. Prove that d is a metric on X, and that the open
    subsets of X under this metric are open sets in the original topology of X and
    conversely. (The function d provides a way of encoding the base sets that two
    points of X have in common.)

14. For each positive integer n, the family of all open balls of radius $1/n$ cover
    the space. Compactness implies the existence of a finite subcover; take S_n to
    be the (finite) set of centers of the balls in this finite subcover. The union of
    the sets S_n, over all n, is a countable dense subset of the space. The class of
    balls of center x and radius $1/n$, for all points x in S and all positive integers
    n, is a countable base for the metric topology of the space.

15. Let X be a Boolean space without isolated points and with a countable base
    (which may be assumed to consist of clopen sets — see Exercise 32.14). Use
    Exercise 28.13 (or a parallel argument) to construct a sequence of clopen sets
    $Q_1, Q_2, Q_3, \ldots$ in X that together generate all of the clopen sets and that
    have the following additional property: for every positive integer m and every
    2-valued function a on the set $\{1, \ldots, m\}$,

    $$\bigcap_{n=1}^{m} p(Q_n, a(n)) \neq \varnothing,$$

where

$$p(i,j) = \begin{cases} i & \text{if } j = 1, \\ i' & \text{if } j = 0. \end{cases}$$

For each point x in X, let $\phi(x)$ be the function on the set I of positive integers such that the value of $\phi(x)$ at each n in I is either 1 or 0 according as x is, or is not, in Q_n. Show that ϕ is a homeomorphism from X to 2^I.

To draw the desired conclusion about countable, atomless Boolean algebras, use duality and Exercise 34.8.

16. A Boolean space is separable just in case its dual algebra has a countable sequence of maximal ideals with a trivial intersection (that is, with an intersection that contains just zero). Alternatively, the space is separable just in case the dual algebra can be embedded into 2^I, where I is the set of positive integers.

17. A Boolean space satisfies the first countability axiom just in case every maximal ideal in its dual algebra is countably generated. For the proof it is helpful to use Theorem 11 (p. 155) and Exercise 18.30.

18. The topological dual of a complete ideal is a regular open set. For the proof, use Exercise 9.39 and Exercises 24.18 and 24.19.

21. Use Exercise 20.17.

Chapter 36

1. Use the fact that the second dual of each mapping is itself.

2. For each element q in B, the set

$$Q = \{y \in Y : y(q) = 1\}$$

is clopen in Y, so the inverse image

$$P = \phi^{-1}(Q)$$

is clopen in X. Consequently, there is a unique element p in A such that

$$P = \{x \in X : x(p) = 1\}.$$

Define a function f from B to A by

$$f(q) = p.$$

Show that $\phi(x) = x \circ f$, and conclude, using the homomorphism properties of x and $\phi(x)$, that f is a homomorphism.

6. The relativizing homomorphism f on A defined by

$$f(q) = q \wedge p,$$

for each q in A, maps A onto the relativization $A(p)$. Its dual is a one-to-one continuous function ϕ from the dual space of $A(p)$ onto a subspace of X. Use Corollary 1 to show that this subspace is just U_p.

7. Use Theorem 36.

8. Use Theorem 36 and Exercise 33.32.

9. Start with any Boolean algebra A that does not satisfy the countable chain condition. The dual algebra of the Cantor space 2^A is free (Theorem 36) and therefore satisfies the countable chain condition (Exercise 8). Show that A is a homomorphic image of this dual algebra.

12. Let I be a set of generators of a Boolean algebra A. The dual algebra B of the Cantor space 2^I is freely generated by the family E of clopen sets

$$P_i = \{y \in 2^I : y_i = 1\},$$

for i in I (Theorem 36). The set E is easily seen to satisfy the criterion of Lemma 28.1. Define a bijection g from E into I by

$$g(P_i) = i,$$

for i in I, and extend g to an epimorphism f from B to A. Show that the following statements are equivalent: (1) I freely generates A; (2) f is one-to-one and therefore an isomorphism; (3) the kernel of f is trivial; (4) the condition formulated in Lemma 28.1 holds.

13. Use Theorem 34 and Corollary 2.

14. Let P be the set of positive integers, endowed with the discrete topology. The dual space X of the field of all subsets of P is the Stone–Čech compactification of the discrete space P (Exercise 34.19). Consequently, every continuous function from P to a compact Hausdorff space Y can be extended to a continuous function from X to Y. Use Theorem 34.

15. Let X be a Boolean space with dual algebra A. Given a non-degenerate relativization $A(P_0)$, fix a point y_0 in P_0 and show that the function ϕ from X onto P_0 defined by

$$\phi(x) = \begin{cases} x & \text{if } x \in P_0, \\ y_0 & \text{if } x \in P_0', \end{cases}$$

is continuous. Apply Exercise 35.19 and Theorem 34 to obtain a monomorphism from $A(P_0)$ into A.

Chapter 37

5. Find a Hausdorff space X such that some quotient of X is homeomorphic to the Sierpiński space (see Exercise 32.8).

6. Find a Boolean space X such that some quotient of X is homeomorphic to the Sierpiński space (see Exercise 32.8).

10. If the points of the dual space of B are the ultrafilters in B, prove that the intersection of the sets in such an ultrafilter is an equivalence class of Θ, and every equivalence class of Θ is the intersection of a uniquely determined ultrafilter in B. Conclude that the correspondence ϕ mapping each ultrafilter N in B to its intersection Q_N is a bijection from Y to X/Θ. Show that ϕ maps clopen sets in Y to clopen sets in X/Θ, and consequently is continuous.

11. Use Exercise 10 to prove that the dual of the ideal $M \cap B$ in the quotient space X/Θ is the quotient modulo Θ of the open set $\bigcup(M \cap B)$.

Chapter 38

1. Use Corollary 29.1.

2. Use the identity (9.1).

4. In every topological space, clopen sets are regular open sets; in a complete Boolean space, the converse is true.

5. If the regular open sets of a Boolean space X constitute a field of sets, then every regular open set is clopen, and consequently the field of regular open sets is just the dual algebra of A. Apply Theorem 1 (p. 66) and Theorem 39.

6. Use Theorem 36 (p. 354), Corollary 28.2, and Exercise 25.6.

7. It may be assumed that $B = D/M$, where D is the σ-field of Borel sets in the interval $[0,1)$, and M is the σ-ideal of Borel sets in A of measure zero. (See the remark preceding Lemma 31.4.) The quotient B is a measure algebra (Lemma 31.1), so the countable chain condition holds in it. The interval algebra A is a subfield of D that generates D as a σ-algebra. The projection f of D onto B is a σ-epimorphism, and its restriction to A is one-to-one. Consequently, A satisfies the countable chain condition (Exercise 30.6).

To prove that the restriction of f to A is a complete monomorphism, it suffices to show (Exercise 30.7) that f preserves all countable suprema in A that happen to exist. Consider a sequence $\{P_n\}$ of sets in A with a supremum, say P, in A. The union $Q = \bigcup_n P_n$ is the supremum of $\{P_n\}$ in D. Prove that the difference $P - Q$ is countable and therefore has measure zero. Conclude that $f(P) = f(Q) = \bigvee_n f(P_n)$.

Chapter 39

1. In the one-point compactification of the discrete space I, the Baire sets are just the countable and the cocountable sets. A subset of I that is neither countable nor cocountable is an example of an open set that is not a Baire set. In the Cantor space $X = 2^I$, every Baire set P has countable support in the sense that there is a countable subset J of I such that for all points x and y in X, if x is in P and if y agrees with x on J, then y is in P. For any point x in X, the set $X - \{x\}$ is an example of an open set that does not have countable support.

4. The dual algebra of a Boolean space with a countable base is countable, so the space has only countably may clopen sets. Consequently, every open set is a union of countably many clopen sets and is therefore a Baire set. (See Chapter 35.)

5. Use Exercise 35.12 to prove that the space has a countable base consisting of open balls, and show that every open ball is an F_σ.

6. What are the Baire sets and the Borel sets in the Sierpiński space? (See Exercise 32.8.)

7. Use the fact (Exercise 33.35) that for every closed subset P of a compact Hausdorff space X, there is a continuous function from X into the interval $[0,1]$ (of real numbers) that assumes the value 0 at x if and only if x is in P. Suitably chosen functions of this type can replace the characteristic functions p_n in the proof of Lemma 1.

8. Let X_1 be a copy of the set of all ordinals up to and including the first infinite ordinal, with the standard ordering. Let X_2 be a copy of the set of all ordinals up to and including the first uncountable ordinal, but with the ordering of its elements reversed. (Thus, the copy of zero is the largest element, and the copy of the first uncountable ordinal is the smallest element, in X_2.) Assume that the two orderings are disjoint, except that the largest element of X_1 (the copy of the first infinite ordinal) coincides with the smallest element of X_2 (the copy of first uncountable ordinal); call this common element w. The set $X = X_1 \cup X_2$ is linearly ordered, provided we assume that the elements in X_1 that are less than w precede the elements in X_2 that are greater than w. Under the order topology (Exercise 9.33), X becomes a Boolean space. Show that the interval

$$U = (-\infty, w)$$

is an open Baire set in X and that its closure,

$$U^- = (-\infty, w],$$

is a closed set that is not a G_δ, and consequently not a Baire set.

Chapter 40

5. The natural generalization of the homomorphism extension criterion says that a mapping g from a generating set E of a σ-algebra B into a σ-algebra A can be extended to a σ-homomorphism from B into A just in case

(1) $$\bigwedge_{i \in F} p(i, a(i)) = 0 \qquad \text{implies} \qquad \bigwedge_{i \in F} p(g(i), a(i)) = 0$$

for every countable subset F of E, and every 2-valued function a on F. This condition is obviously necessary, but it is not sufficient. Consider, for example, the Cantor space $X = 2^I$, where I is some countably infinite set. Take D to be the dual algebra of X and take C to be the σ-field of Baire sets in X, that is to say, the σ-field generated by D. Choose a Baire set P_0 in C that is not a G_δ-set, and let M be the σ-ideal in C generated by the closed sets in C that are disjoint from P_0. Show that the relativization $C(P_0)$ is a σ-algebra that is σ-generated by the relativization $D(P_0)$.

The mapping g from $D(P_0)$ to the quotient C/M defined by

$$g(P \cap P_0) = P/M$$

for all P in D is a well-defined monomorphism that satisfies condition (1) when $D(P_0)$ is taken for the set E, and $C(P_0)$ for the σ-algebra B, and C/M for the σ-algebra A. (To show that (1) holds, it suffices to verify that

$$\bigcap_i (P_i \cap P_0) = \varnothing \qquad \text{implies} \qquad \bigwedge_i g(P_i \cap P_0) = \varnothing$$

for every countable family $\{P_i\}$ of sets in D.) However, g cannot be extended to a σ-homomorphism from $C(P_0)$ into C/M. Assume, to the contrary, that f is such an extension. Let f_1 be the σ-homomorphism from C to C/M defined by

$$f_1(P) = f(P \cap P_0)$$

for each P in C, and let f_2 be the σ-homomorphism that projects C onto C/M. Show that f_1 and f_2 agree on the generating set D, but they do not agree on all of C; in fact, $f_1(P_0') \neq f_2(P_0')$.

7. Suppose that B is an m-field, M an m-ideal in B, and f the projection of B onto B/M, where m is greater than or equal to the power of the continuum. Prove that if $I = \{1, 2, 3, \dots\}$, then the equation

$$\bigvee_{a \in 2^I} \bigwedge_{i \in I} p(p_i, a(i)) = 1$$

holds for every sequence $\{p_i\}$ of elements of B/M, where

$$p(p_i, j) = \begin{cases} p_i & \text{if} \quad j = 1, \\ p_i' & \text{if} \quad j = 0. \end{cases}$$

The idea is that the above equation does hold in B, and it is preserved by f, so it must hold in every quotient of B modulo an m-ideal. The regular open algebra of the interval $(0, 1)$ of real numbers, on the other hand, is an example of an m-algebra that does not satisfy the above equation (see the remarks at the end of Chapter 10); consequently, it cannot be isomorphic to the quotient of an m-field modulo an m-ideal.

Chapter 41

1. Use Corollary 1, Corollary 31.1, and Theorem 39.

3. The direct proof uses Exercises 38.2 and 38.3, and the duals of Lemma 30.1 and Exercises 8.21 and 31.5.

5. Use Lemma 4 or Lemma 6. Use also the fact that the algebra is atomless, so its dual space has no isolated points; consequently, each singleton subset is nowhere dense.

6. Let A be the regular open algebra of the interval $[0, 1]$. The reduced Borel algebra of $[0, 1]$ is isomorphic to A (Theorem 29), so it suffices to show that the dual space of A — call it X — is separable. For each real number t in $[0, 1]$ define a proper filter N_t in A by

$$U \in N_t \qquad \text{if and only if} \qquad t \in U.$$

Since every proper filter is included in some ultrafilter, there exists a 2-valued homomorphism x_t on A such that if $U \in N_t$, then $x_t(U) = 1$. If f is the canonical isomorphism from A to its second dual (the dual algebra of X), then for every set U in A (that is, for every regular open set in X) and for every number t in U, the homomorphism x_t belongs to $f(U)$. The set of rational numbers in $[0, 1]$ is a countable dense subset of the interval, so the set of homomorphisms x_t, with t rational, is a countable dense subset of X.

Chapter 42

3. There are continuum many open subsets of $[0, 1]$, and hence continuum many closed subsets. It follows from the cardinal arithmetic identity

$$(2^{\aleph_0})^{\aleph_0} = 2^{\aleph_0}$$

that there are continuum many F_σ-sets. Every singleton subset is a meager F_σ, so there must be continuum many meager F_σ-sets.

4. Use Exercise 3 and the fact that every singleton subset of $[0, 1]$ is a G_δ of measure zero.

5. Enumerate the non-empty perfect subsets of $[0, 1]$ in a transfinite sequence

$$P_0, P_1, P_2, \ldots, P_\beta, \ldots$$

(indexed by the set X of ordinals of power less than $2^{\aleph_0}$) in such a way that each non-empty perfect set occurs $2^{\aleph_0}$ times in the sequence. For each β in X, define a transfinite sequence $\{k(\alpha, \beta)\}$ in $[0, 1]$ of type β (that is, indexed by the ordinals $\alpha < \beta$) with the properties that $k(\alpha, \beta)$ belongs to P_β whenever $\alpha < \beta$ (and β is in X), and k, as a function of the two arguments α and β, is one-to-one.

For each α in X, define S_α to be the set of those numbers $k(\alpha, \beta)$ with $\beta > \alpha$. Show that the sets so defined all have cardinality $2^{\aleph_0}$, are mutually disjoint, and have the unit interval as their union. Show further that each set S_α has a non-empty intersection with every non-empty perfect set. (If P is any non-empty perfect set, then there is a $\beta > \alpha$ such that $P = P_\beta$; the number $k(\alpha, \beta)$ belongs to P_β and to S_α.) Use this last property to conclude that the outer measure of each set S_α is one, and therefore S_α cannot have measure zero. (Assume, to the contrary, that S_α has outer measure less than one. There is then a countable sequence of open subintervals of $[0, 1]$ such that the sum of the lengths of these subintervals is less than one, and S_α is included in their union. The complement of this union is a closed set of positive measure, so it includes a non-empty perfect subset. This contradicts the fact that S_α has a non-empty intersection with every perfect set.)

8. Assume the continuum hypothesis, and observe that the proof of Corollary 2 goes through for any normalized measure that assigns measure zero to singletons. Use Lemma 31.2 to show that the existence of a normalized measure on B/M implies the existence of such a measure on B that assigns measure zero to singletons. This contradicts (the generalized version of) Corollary 2. (If E is an uncountable disjoint class of sets in B with non-zero measure, then each set in E must have positive measure, because the measure is defined on all subsets of $[0, 1]$. If E_n is the class of sets in E with measure at least $1/n$, then E_n cannot be finite for every n; otherwise, E would be countable. At least one of the classes E_n is therefore infinite, and this leads to a contradiction.)

Chapter 43

7. Let Y and Z be disjoint topological spaces that are homeomorphic to the dual spaces of B and C, respectively, and write $X = Y + Z$. Use Theorem 44 (together with Theorem 31, p. 328, and Exercises 34.8 and 26.9) to show that the dual algebra of X is isomorphic to A. The second dual of X is then homeomorphic to the dual of A.

8. Suppose $A = B \times C$. If $p = (1,0)$, then B and C are isomorphic to the relativizations $A(p)$ and $A(p')$ respectively, and A is the internal product of $A(p)$ and $A(p')$. The dual spaces of $A(p)$ and $A(p')$ are the disjoint subspaces

$$U_p = \{x \in X : x(p) = 1\} \quad \text{and} \quad U_{p'} = \{x \in X : x(p') = 1\}$$

respectively (Exercise 35.20). Prove that the dual space of A, say X, is the sum of U_p and $U_{p'}$, and then use Exercise 34.9.

9. The spaces X_0 and X_1 are homeomorphic. The dual algebra of X_0 is the field A of finite and cofinite subsets of natural numbers (Exercise 32.16). The dual algebra of the sum space

$$X = X_0 + X_1$$

must therefore be isomorphic to $A \times A$.

11. Dualize and use Exercise 9. The dual space of A has one isolated point, while the dual space of $A \times A$ has two.

12. The proof involves a back-and-forth argument, and is similar in structure to the proof of Theorem 10 (p. 134). Write $Y = X + 1$. Enumerate the clopen subsets of X and Y, using respectively the positive even and odd integers, say

$$P_2, P_4, P_6 \ldots \quad \text{and} \quad Q_1, Q_3, Q_5 \ldots.$$

Define clopen sets P_n in X for odd n, and Q_n in Y for even n, so that, for every 2-valued function a on $I_n = \{1, \ldots, n\}$, the sets

$$P_a = \bigcap_{i \in I_n} p(P_i, a(i)) \quad \text{and} \quad Q_a = \bigcap_{i \in I_n} p(Q_i, a(i))$$

are either both empty or both non-empty (where $p(S, j)$ is S or S' according as j is 1 or 0). The construction is carried out by induction on n, using the induction hypothesis that

$$Q_a = P_a \quad \text{or} \quad Q_a = P_a + 1,$$

and in the latter case there are infinitely many isolated points in P_a.

Once this construction is accomplished, a homeomorphism ϕ from X to Y can be defined as follows. Every point x in X uniquely determines a 2-valued function a on the set of positive integers that is defined by

$$a(i) = \begin{cases} 1 & \text{if } x \in P_i, \\ 0 & \text{if } x \notin P_i. \end{cases}$$

The intersections

$$\bigcap_i p(P_i, a(i)) \qquad \text{and} \qquad \bigcap_i p(Q_i, a(i))$$

each contain exactly one point; the first one contains x, and the second one contains some point y in Y. Write $\phi(x) = y$.

14. Given a point x, use Corollary 29.1 to find an open set V containing x such that the closure V^- is included in U. Show that $V^- \cap F$ is a compact set whose interior contains x.

15. Let Y be the one-point compactification of X obtained by adjoining a new point y. Notice that if a set U is open in Y, then the sets

$$U \cap X \qquad \text{and} \qquad X - U$$

are respectively open and closed in X. The proof that the union of a family of open sets is open splits into two cases, according to whether the union does, or does not, contain the point y. Similarly, the proof that the intersection of two open sets is open splits into two cases, according to whether the intersection does, or does not, contain the point y.

17. Let X be the subspace obtained from a compact Hausdorff space Y by removing a single point y. To prove that X is locally compact, consider any one of its points x and argue that x and y can be separated (in Y) by open sets with disjoint closures (Corollary 29.2).

18. Use Lemma 29.1 and Exercise 17.

23. Use Corollary 33.1.

24. Use Exercise 23.

25. Use Exercise 24.

26. To prove the second assertion, consider the function $\sin(1/x)$ from the interval $(0, 1]$ to the interval $[-1, 1]$.

28. Use Exercises 19 and 20 to prove that if Y is the one-point compactification of X, say by the point y, then Y is a Boolean compactification of X. To prove that Y is the smallest such compactification, consider any other Boolean compactification Z of X. The function ϕ defined by

$$\phi(x) = \begin{cases} x & \text{if } x \in X, \\ y & \text{if } x \in Z - X, \end{cases}$$

maps Z continuously onto Y. To derive the final conclusion of the exercise, apply Theorem 45.

29. Let Y be the dual space of the weak internal product of the dual algebras of the spaces X_i. It was shown in the first part of the proof of Lemma 4 that each space X_i is homeomorphic to the subspace

$$V_i = U_{X_i} = \{y \in Y : y(X_i) = 1\}$$

of Y via a mapping ϕ_i. It was also shown that if V is the union space of the family $\{V_i\}$, then V is a dense subspace of Y (so that Y is a Boolean compactification of V), and X is homeomorphic to V via a mapping that extends ϕ_i for each i. It therefore suffices to prove that $Y - V$ contains just one point (Exercise 21). The set N of those elements in the internal product whose ith coordinate is X_i for all but a finite number of indices i is an ultrafilter. The 2-valued homomorphism y with cokernel N is a point in Y that does not belong to V. If x is any other point in Y, then its cokernel must be an ultrafilter M that is different from N. Consequently, there must be an element P in M whose ith coordinate P_i is the empty set for all but finitely many indices i. This forces P_i to belong to M for some i, and consequently it forces x to belong to U_{X_i}, and hence also to V.

30. For each element i in X, let X_i be the discrete space with i as its only point. The union of the family of Boolean spaces $\{X_i\}$ is just X.

31. If A is the field of finite and cofinite sets of natural numbers, and if I is the set of natural numbers, then the dual algebra of Y is the weak direct power of A with exponent I. The proof is based on Exercise 32.16 and Exercise 28.

Chapter 44

3. The implication from right to left uses Lemma 13.2. To prove the reverse implication, argue by contraposition. Suppose the union of the images $f_i(A_i)$ generates a proper subalgebra C of A. Use Exercise 21.10 to get a 2-valued homomorphism f on C that can be extended in two different ways to a 2-valued homomorphism on A. The mappings g_i defined by

$$g_i = f \circ f_i$$

are homomorphisms from A_i into 2. The family of these mappings can be extended in two different ways to a homomorphism from A into 2 that satisfies the transfer condition (except for the assertion of uniqueness).

4. Use Exercise 12.5.

6. Suppose A is the internal sum of the family $\{f_i(A_i)\}$. Let B be a Boolean algebra such that for each i there is a homomorphism g_i from A_i into B. Define a homomorphism h_i from $f_i(A_i)$ into B by

$$h_i(f_i(p)) = g_i(p)$$

for each p in A_i. Use the simplified transfer condition to conclude that A is the sum of the family $\{A_i\}$ with respect to the homomorphisms f_i.

For the reverse implication, assume that A is the sum of the family $\{A_i\}$ with respect to homomorphisms f_i from A_i into A. Let B be a Boolean algebra such that for each i there is a homomorphism g_i from $f_i(A_i)$ into B. The composition $h_i = g_i \circ f_i$ is a homomorphism from A_i into B for each i. Use the transfer condition to conclude that A is the internal sum of the family $\{f_i(A_i)\}$.

7. Use Exercise 6.

9. A Boolean algebra A is the internal sum of a finite, non-empty family $\{A_i\}_{i \in I}$ of subalgebras just in case the union of these subalgebras generates A, and whenever p_i is a non-zero element in A_i for each i in the index set I, then

$$\bigwedge_{i \in I} p_i \neq 0.$$

10. Use Exercise 9 or Lemma 1.

11. A Boolean algebra A is the internal sum of an infinite sequence $\{A_i\}_{i=1}^{\infty}$ of subalgebras if and only if the union of the sequence generates A and, for each positive integer n, the subalgebra generated by the union of the subsequence $\{A_i\}_{i=1}^{n}$ is the internal sum of this subsequence.

13. The elements of A are the finite joins of meets of the form

$$p_1 \wedge p_2 \wedge \cdots \wedge p_n,$$

where p_i is in A_i for each i. One proof of this assertion proceeds by induction on n, and uses Exercises 9 and 12.

18. The sum is just 2.

20. Let X, Y, and Z be the dual spaces of A, B, and C respectively. Prove that

$$X \times (Y + Z) = (X \times Y) + (X \times Z),$$

and then use Exercise 34.8 and Theorem 44 (p. 397) to arrive at the desired conclusion.

22. Use Exercises 20 and 21, and proceed by induction on n.

24. The sum is the subalgebra of B^I consisting of those functions from I to B that have a finite range.

25. The sum is the subalgebra of B^I consisting of those functions from I to B that are constant on a cofinite subset of I.

27. If A is the sum of Boolean algebras A_1 and A_2 via homomorphisms f_1 and f_2, then the atoms of A are the elements of the form

$$r = f_1(p) \wedge f_2(q)$$

for some atoms p in A_1 and q in A_2. For the proof, use Exercise 12 or 14.

28. The sum is atomic if and only if both summands are atomic, and it is atomless if and only if at least one of the summands is atomless. For the proof, use Exercises 16 and 27.

30. Let X be the set of natural numbers, and take $A_1 = A_2 = \mathcal{P}(X)$ (the field of all sets of natural numbers). The sum of A_1 and A_2 is the subfield A of $\mathcal{P}(X \times X)$ consisting of all finite unions of rectangles $S \times T$, where S and T are subsets of X. The one-point rectangle

$$P_n = \{n\} \times \{n\} = \{(n, n)\}$$

belongs to A for each n in X; the family of these rectangles has no supremum in A.

31. The natural generalization of Corollary 1 to σ-sums of σ-algebras asserts that a σ-algebra A is the σ-sum of a family $\{A_i\}$ of σ-algebras with respect to σ-homomorphisms f_i from A_i into A just in case the union $\bigcup_i f_i(A_i)$ generates A as a σ-algebra, and whenever J is a countable, non-empty subset of the indices, and p_i is an element in A_i such that $f_i(p_i) \neq 0$, for each i in J, then

(1) $$\bigwedge_{i \in J} f_i(p_i) \neq 0.$$

This generalization is false, even when the family consists of just two σ-algebras.

Let X be an analytic set that is not Borel, and write

$$X_1 = \mathbb{R} - X \qquad \text{and} \qquad X_2 = \mathbb{R}.$$

Take A_2 to be the σ-field of all Borel sets in $\mathbb{R}$; take A_1 to be the relativization of A_2 to X_1; take B to be the σ-field of Borel sets in $\mathbb{R}^2$; and take A to be the relativization of B to $X_1 \times X_2$. The σ-field B is generated by the (countable) class E of open sets $P \times Q$, where P and Q are open intervals in $\mathbb{R}$ with rational endpoints (Exercise 30.3), and the relativization A is generated (as a σ-algebra) by the class

$$\{(P \times Q) \cap (X_1 \times X_2) : P \times Q \in E\}$$

(Exercise 29.44). It follows that A is generated by the class of sets of the form

$$(P \cap X_1) \times Q,$$

where P and Q are Borel sets in $\mathbb{R}$.

Define mappings f_1 and f_2 from A_1 and A_2 respectively into A, by

$$f_1(P \cap X_1) = h_1^{-1}(P \cap X_1) = (P \cap X_1) \times X_2$$

and

$$f_2(Q) = h_2^{-1}(Q) = X_1 \times Q$$

for all Borel sets P and Q in $\mathbb{R}$ (where h_1 and h_2 are the left and right projections of the Cartesian product $X_1 \times X_2$). Both functions are easily seen to be σ-monomorphisms. Observe that if P and Q are Borel sets in $\mathbb{R}$, then

$$f_1(P \cap X_1) \cap f_2(Q) = (P \cap X_1) \times Q.$$

Combine this remark with the observations of the preceding paragraph to conclude that A is generated (as a σ-algebra) by the union of the image algebras $f_1(A_1) \cup f_2(A_2)$ and A satisfies condition (1).

The strategy for showing that A is not the σ-sum of the σ-algebras A_1 and A_2 with respect to the σ-homomorphisms f_1 and f_2 is to construct a σ-ideal M in B, and σ-monomorphisms g_1 and g_2 from A_1 and A_2 respectively into the quotient B/M such that no σ-homomorphism g from A into B/M satisfies the transfer condition

$$g \circ f_1 = g_1 \quad \text{and} \quad g \circ f_2 = g_2.$$

Let N be the σ-ideal of Borel subsets of $\mathbb{R}$ that are disjoint from X_1, and define M to be the σ-ideal in B generated by

$$\{P \times \mathbb{R} : P \in N\}.$$

Check that

$$M = \{S \in B : S \subseteq P \times \mathbb{R} \text{ for some } P \in N\}.$$

Define the mappings g_1 and g_2 from A_1 and A_2 respectively into B/M, by

$$g_1(P \cap X_1) = (P \times \mathbb{R})/M \quad \text{and} \quad g_2(Q) = (\mathbb{R} \times Q)/M$$

for all Borel sets P and Q in $\mathbb{R}$, and show that g_1 and g_2 are well-defined σ-monomorphisms.

The proof that there is no σ-homomorphism g from A to B/M satisfying the transfer condition proceeds by contradiction. Assume, to the contrary, that such a σ-homomorphism g exists. Prove that

$$g(S \cap (X_1 \times X_2)) = S/M$$

for every set S in B. The idea is that the class of sets S in B for which this equation holds is a σ-subfield of B that includes a set of generators of B, and hence must coincide with B.

As an analytic set, X must be the left projection of some G_δ-subset T of $\mathbb{R}^2$. As a Borel set, T belongs to B, so its intersection with $X_1 \times X_2$ belongs to A. Consequently,

$$g(T \cap (X_1 \times X_2)) = T/M.$$

Argue that T is included in $X \times \mathbb{R}$, so

$$T \cap (X_1 \times X_2) = \varnothing,$$

and therefore

$$T/M = g(T \cap (X_1 \times X_2)) = \varnothing.$$

Thus, T belongs to the ideal M, and therefore

$$T \subseteq P \times \mathbb{R}$$

for some Borel subset P of $\mathbb{R}$ that is disjoint from X_1 (the complement of X). The left-projection of T is included in the left-projection of $P \times \mathbb{R}$, and this forces $X = P$, which contradicts the assumption that X is not a Borel set.

Chapter 45

1. The key point is showing that there is a countable basis of clopen sets for the topology of U_n. Call a real number b in U_n *right-clopen* if every right-closed interval in U_n with (right) endpoint b is also right-open in the sense that there is a $d > b$ such that

$$(a,b] \cap U_n = (a,d) \cap U_n \qquad \text{and} \qquad [a,b] \cap U_n = [a,d) \cap U_n$$

for every $a < b$. (The numbers a and d need not be in U_n.) Similarly, call a number a in U_n *left-clopen* if every left-closed interval in U_n with (left) endpoint a is also left-open; in other words, there is a $c < a$ such that

$$[a,b) \cap U_n = (c,b) \cap U_n \qquad \text{and} \qquad [a,b] \cap U_n = (c,b] \cap U_n$$

for every $b > a$. (The numbers b and c need not be in U_n.) If a is left-clopen and b is right-clopen, then the interval $[a,b] \cap U_n$ is a clopen subset of U_n, by the definition of the inherited topology, since

$$[a,b] \cap U_n = [a,d) \cap U_n = (c,d) \cap U_n$$

for some $c < a$ and some $d > b$.

Examples of right-clopen numbers include every number in the distinguished ω^n-sequence in U_n, and every number in the Cantor set whose ternary representation (in the sense of Chapter 33) has only finitely many occurrences of the digit 0 (and of course no occurrences of the digit 1). Let S_r be the set of right-clopen numbers of these two types, and observe that S_r is countable. Examples of left-closed numbers include every number in the distinguished ω^n-sequence in U_n whose index is not a limit ordinal, and every non-zero number in the Cantor set whose ternary representation has only finitely many occurrences of the digit 2. Let S_ℓ be the set of left-clopen numbers of these two types, and observe that S_ℓ is countable.

Every open set in U_n is the union of a class of (relativized) intervals of the form $(a, b) \cap U_n$ for some real numbers a and b. It suffices therefore to show that each such interval is the union of a family of clopen sets,

$$(a, b) \cap U_n = \bigcup_k \Big([a_k, b_k] \cap U_n \Big),$$

with endpoints a_k in S_ℓ and b_k in S_r. The construction of the sequence $\{b_k\}$ depends only on b, while that of $\{a_k\}$ depends only on a.

3. Show that there are Boolean spaces X and Y such that

$$X = X + Y + Y + Y,$$

but

$$X \neq X + Y \qquad \text{and} \qquad X \neq X + Y + Y.$$

The definitions of X and Y are analogous to those in the chapter, but "two" is replaced at all appropriate places by "three". More precisely, conjoin three copies of each of the spaces U_k (for $k \geq n$) when creating the space Z_n, and conjoin three copies of each of the spaces Y_n when creating the space X.

5. Show that each of the spaces X and $X + Y$ constructed in the chapter is homeomorphic to an open subspace of the other, but the two spaces are not homeomorphic to one another.

References

[1] Alexandroff, P. S. and Hopf, H., *Topologie I*, Springer-Verlag, Berlin, 1935.

[2] Baire, R.-L., Sur les fonctions de variables réelles, *Annali di Matematica*, ser. 3, vol. 3 (1899), pp. 1–123.

[3] Banach, S. and Kuratowski, K., Sur une généralization du problème de la mesure, *Fundamenta Mathematicae*, vol. 14 (1929), pp. 127–131.

[4] Bernardi, C., Lo spazio duale di un prodotto di algebre di Boole e le compattificazioni di Stone, *Annali di Matematica Pura ed Applicata*, vol. 126 (1980), pp. 253–266.

[5] Birkhoff, G., On the combination of subalgebras, *Proceedings of the Cambridge Philosophical Society*, vol. 29 (1933), pp. 441–464.

[6] Birkhoff, G., Order and the inclusion relation, *Comptes Rendus du Congrès des Mathématciens Oslo 1936*, vol. 2, *Conférences de Sections*, A. W. Brøggers Boktrykkeri A/S, Oslo, 1937, p. 37.

[7] Birkhoff, G., *Lattice Theory*, Colloquium Publications, vol. 25, American Mathematical Society, Providence, R. I., 1940.

[8] Birkhoff, G., *Lattice Theory*, third edition, Colloquium Publications, vol. 25, American Mathematical Society, Providence, R. I., 1967.

[9] Boole, G., *The Mathematical Analysis of Logic. Being an Essay Towards a Calculus of Deductive Reasoning*, Macmillan, Barclay, and Macmillan, Cambridge, 1847.

[10] Boole, G., *An Investigation of the Laws of Thought, on Which Are Founded the Mathematical Theories of Logic and Probabilities*, Walton and Maberly, London, 1854.

[11] Byrne, L., Two brief formulations of Boolean algebra, *Bulletin of the American Mathematical Society*, vol. 52 (1946), pp. 269–272.

[12] Cantor, G., Über unendliche, lineare Punktmannichfaltigkeiten, *Mathematische Annalen*, vol. 21 (1883), pp. 545–591.

[13] Cantor, G., Beiträge zur Begrundung der transfiniten Mengenlehre, *Mathematische Annalen*, vol. 46 (1895), pp. 481–512.

[14] Čech, E., On bicompact spaces, *Annals of Mathematics*, 2nd series, vol. 38 (1937), pp. 823–844.

[15] Dedekind, R., *Stetigkeit und irrationale Zahlen*, Vieweg Verlag, Braunschweig, 1872.

[16] Dwinger, P., Remarks on the field representations of Boolean algebras, *Koninklijke Nederlandse Akademie van Wetenschappen, Proceedings, Series A*, vol. 63 (=*Indagationes Mathematicae*, vol. 22) (1960), pp. 213–217.

[17] Gaifman, H., Infinite Boolean polynomials, *Fundamenta Mathematicae*, vol. 54 (1964), pp. 229–250.

[18] Gottschalk, W., The theory of quaternality, *Journal of Symbolic Logic*, vol. 18 (1953), pp. 193–196.

[19] Grau, A. A., Ternary Boolean algebra, *Bulletin of the American Mathematical Society*, vol. 53 (1947), pp. 567–572.

[20] Haar, A., Der Massbegriff in der Theorie der kontinuierlichen Gruppen, *Annals of Mathematics*, second series, vol. 34 (1933), pp. 147–169.

[21] Hales, A. W., On the existence of free complete Boolean algebras, *Fundamenta Mathematicae*, vol. 54 (1964), pp. 45–66.

[22] Halmos, P. R., *Measure Theory*, Graduate Texts in Mathematics, vol. 18, Springer-Verlag, New York, 1974. (Reprint of the edition published in 1950 by Van Nostrand, New York, in: The University Series in Higher Mathematics.)

[23] Halmos, P. R., *Lectures on Boolean Algebras*, Graduate Texts in Mathematics, vol. 18, Springer-Verlag, New York, 1974. (Reprint of the edition published in 1963 by Van Nostrand, New York, as: van Nostrand Mathematical Studies #1.)

[24] Hanf, W., On some fundamental problems concerning isomorphism of Boolean algebras, *Mathematica Scandinavica*, vol. 5 (1957), pp. 205–217.

[25] Hirsch, R. and Hodkinson, I., Complete representations in algebraic logic, *Journal of Symbolic Logic*, vol. 62 (1997), pp. 816–847.

[26] Hirsch, R. and Hodkinson, I., *Relation Algebras by Games*, Studies in Logic and the Foundations of Mathematics, vol. 147, Elsevier Science B.V., Amsterdam, 2002.

[27] Horn, A. and Tarski, A., Measures in Boolean algebras, *Transactions of the American Mathematical Society*, vol. 64 (1948), pp. 467–497.

[28] Huntington, E. V., Sets of independent postulates for the algebra of logic, *Transactions of the American Mathematical Society*, vol. 5 (1904), pp. 288–309.

[29] Huntington, E. V., The continuum as a type of order: an exposition of the modern theory, *Annals of Mathematics*, second series, vol. 6, no. 4 (1905), pp. 151–184.

[30] Huntington, E. V., New sets of independent postulates for the algebra of logic, with special reference to Whitehead and Russell's Principia Mathematica, *Transactions of the American Mathematical Society*, vol. 35 (1933), pp. 274–304. Correction. vol. 35 (1933), pp. 557–558.

[31] Jevons, W. S., *Pure Logic or the Logic of Quality Apart from Quantity with Remarks on Boole's System and the Relation of Logic and Mathematics*, E. Stanford, London, 1864.

[32] Jónsson, B. and Tarski, A., Boolean algebras with operators. Part I, *American Journal of Mathematics*, vol. 73 (1951), pp. 891–939.

[33] Kinoshita, S., A solution of a problem of R. Sikorski, *Fundamenta Mathematicae*, vol. 40 (1953), pp. 39–41.

[34] Ketonen, J., The structure of countable Boolean algebras, *Annals of Mathematics*, second series, vol. 108 (1978), pp. 41–89.

[35] Kuratowski, K., Sur la notion de l'ordre dans la théorie des ensembles, *Fundamenta Mathematicae*, vol. 2 (1921), pp. 161–171.

[36] Kuratowski, K., Sur l'opération $\bar{A}$ de l'analysis situs, *Fundamenta Mathematicae*, vol. 3 (1922), pp. 182–199.

[37] Kuratowski, K., Sur l'appliation des espaces fonctionnels à la théorie de la dimension, *Fundamenta Mathematicae*, vol. 18 (1932), p. 285.

[38] Kuratowski, K., On a topological problem connected with the Cantor-Bernstein theorem, *Fundamenta Mathematicae*, vol. 37 (1950), p. 213-216.

[39] Kuratowski, K. and Posament, T., Sur l'isomorphie algébro-logique et les ensembles relativement boréliens, *Fundamenta Mathematicae*, vol. 22 (1934), pp. 281–286.

[40] Loomis, L. H., On the representation of σ-complete Boolean algebras, *Bulletin of the American Mathematical Society*, vol. 53 (1947), pp. 757–760.

[41] Lusin, N., *Leçons sur les ensembles analytiques et leurs applications*, Gauthier-Villars, Paris, 1930.

[42] Lusin, N. and Sierpiński, W., Sur une décomposition d'un intervalle en une infinité non dénombrable d'ensembles non mesurables, *Comptes Rendus Hebdomadaire des Séances de l'Academie des Sciences, Paris*, vol. 165 (1917), pp. 422–424.

[43] MacNeille, H., Partially ordered sets, *Transactions of the American Mathematical Society*, vol. 42 (1937), pp. 416–460.

[44] Monk, J. D., Completions of Boolean algebras with operators, *Mathematische Nachrichten*, vol. 46 (1970), pp. 47–55.

[45] Monk, J. D. and Bonnet, R. (editors), *Handbook of Boolean Algebras*, North-Holland, Amsterdam and New York, 1989, three volumes.

[46] Mostowski, A. and Tarski, A., Boolesche Ringe mit geordneter Basis, *Fundamenta Mathematicae*, vol. 32 (1939), pp. 69–86.

[47] von Neumann, J., Zur Einführung der transfiniten Zahlen, *Acta Litterarum ac Scientiarum Regiae Universitatis Hungaricae Francisco-Josephinae. Sectio Scientiarum Mathematicarum*, vol. 1 (1923), pp. 199–208.

[48] Nikodym, O., Sur une généralisation des intégrales de M. J. Radon, *Fundamenta Mathematicae*, vol. 15 (1930), pp. 131–179.

[49] Peirce, C. S., A Boolean algebra with one constant, *Collected papers of Charles Sanders Peirce*, eds. C. Hartshorne and P. Weiss, vol. 4, Harvard University Press, Cambridge, 1933, pp. 12-20.

[50] Post, E. L., Introduction to a general theory of elementary propositions, *American Journal of Mathematics*, vol. 43 (1921), pp. 163–185.

[51] Rieger, L. S., On free $\aleph_\xi$-complete Boolean algebras, *Fundamenta Mathematicae*, vol. 38 (1951), pp. 35–52.

[52] Schröder, E., *Vorlesungen über die Algebra der Logik (Exakte Logik)*, vol. 2, part 1, B. G. Teubner, Leipzig, 1891, xxviii + 606 pp. [Reprinted by Chelsea Publishing Company, New York, 1966.]

[53] Sheffer, H. M., A set of five independent postulates for Boolean algebra, with application to logical constants, *Transactions of the American Mathematical Society*, vol. 14 (1913), pp. 481–488.

[54] Sheffer, H. M., Review of A survey of symbolic logic, by C. I. Lewis, *American Mathematical Monthly*, vol. 27 (1920), p. 310.

[55] Sierpiński, W., Sur une décomposition d'ensembles, *Monatshefte für Mathematik und Physik*, vol. 35 (1928), pp. 239–242.

[56] Sikorski, R., A generalization of theorems of Banach and Cantor-Bernstein, *Colloquium Mathematicum*, vol. 1 (1948), pp. 140–144.

[57] Sikorski, R., Problem 44, *Colloquium Mathematicum*, vol. 1 (1948), p. 242.

[58] Sikorski, R., A theorem on extension of homomorphisms, *Rocznik Polskiego Towarzystwa Matematycznego (Annales de la Société Polonaise de Mathématique)*, vol. 21 (1948), pp. 332–335.

[59] Sikorski, R., On the representation of Boolean algebras as fields of sets, *Fundamenta Mathematicae*, vol. 35 (1948), pp. 247–256.

[60] Sikorski, R., On the inducing of homomorphisms by mappings, *Fundamenta Mathematicae*, vol. 36 (1949), pp. 7–22.

[61] Sikorski, R., On an unsolved problem from the theory of Boolean algebras, *Colloquium Mathematicum*, vol. 2 (1949), pp. 27–29.

[62] Sikorski, R., Cartesian products of Boolean algebras, *Fundamenta Mathematicae*, vol. 37 (1950), pp. 25–54.

[63] Sikorski, R., On an analogy between measures and homomorphisms, *Rocznik Polskiego Towarzystwa Matematycznego (Annales de la Société Polonaise de Mathématique)*, vol. 23 (1950), pp. 1–20.

[64] Sikorski, R., *Boolean Algebras*, Ergebnisse der Mathematik und Ihrer Grenzgebiete, new series, vol. 25, Springer-Verlag, Berlin, Göttingen, and Heidelberg, 1960.

[65] Sikorski, R., *Boolean Algebras*, third edition, Ergebnisse der Mathematik und Ihrer Grenzgebiete, new series, vol. 25, Springer-Verlag, Berlin and New York, 1969.

[66] Stone, M., The theory of representations for Boolean algebras, *Transactions of the American Mathematical Society*, vol. 40 (1936), pp. 37–111.

[67] Stone, M., Applications of the theory of Boolean rings to general topology, *Transactions of the American Mathematical Society*, vol. 41 (1937), pp. 321–364.

[68] Stone, M., Algebraic characterizations of spcial Boolean rings, *Fundamenta Mathematicae*, vol. 29 (1937), pp. 223–303.

[69] Stone, M., The representation of Boolean algebras, *Bulletin of the American Mathematical Society*, vol. 44 (1938), pp. 807–816.

[70] Tarski, A., Une contribution à la théorie de la mesure, *Fundamenta Mathematicae*, vol. 15 (1930), pp. 42–50.

[71] Tarski, A., Zur Grundlegung der Bool'schen Algebra. I, *Fundamenta Mathematicae*, vol. 24 (1935), pp. 177–198.

[72] Tarski, A., Grundzüge des Systemskalküls. Erster Teil, *Fundamenta Mathematicae*, vol. 25 (1935), pp. 503–526.

[73] Tarski, A., Grundzüge des Systemskalküls. Zweiter Teil, *Fundamenta Mathematicae*, vol. 26 (1936), pp. 283–301.

[74] Tarski, A., Ideale in den Mengenkörpern, *Rocznik Polskiego Towarzystwa Matematycznego (Annales de la Société Polonaise de Mathématique)*, vol. 15 (1937), pp. 186–189.

[75] Tarski, A., Über additive und multiplikative Mengenkörper und Mengen-funktionen, *Sprawozdania z Posiedzeń Towarzystwa Naukowego Warsza-wskiego, Wydział III Nauk Matematyczno-fizycznych* (=*Comptes Rendus des Séances de la Société des Sciences et des Lettres de Varsovie, Classe III*), vol. 30 (1937), pp. 151–181.

[76] Tarski, A., Axiomatic and algebraic aspects of two theorems on sums of cardinals, *Fundamenta Mathematicae*, vol. 35 (1948), pp. 79-104.

[77] Tarski, A., *Cardinal Algebras*, Oxford University Press, New York, 1949.

[78] Tychonoff, A., Über die topologische Erweiterung von Räumen, *Mathema-tische Annalen*, vol. 102 (1929), pp. 544–561.

[79] Tychonoff, A., Über einen Funktionenraum, *Mathematische Annalen*, vol. 111 (1935), pp. 762–766.

[80] Ulam, S., Zur Maßtheorie in der allgemeinen Mengenlehre, *Fundamenta Mathematicae*, vol. 16 (1930), pp. 140–150.

[81] Urysohn, P., Über die Mächtigkeit der zusammenhängenden Mengen, *Math-ematishe Annalen*, vol. 94 (1925), p. 290.

[82] Urysohn, P., Zum Metrisationsproblem, *Mathematishe Annalen*, vol. 94 (1925), p. 310.

[83] Urysohn, P., Mémoire sur les multiplicités Cantoriennes, *Fundamenta Math-ematicae*, vol. 7 (1925), pp. 30–137, and vol. 8 (1926), pp. 225–351.

[84] Vaught, R. L., *Topics in the Theory of Arithmetical Classes and Boolean Algebras*, doctoral dissertation, University of California at Berkeley, Berkeley CA, 1954.

[85] Vedenisov, N. B., Sur les fonctions continues dans des espaces topologiques, *Fundamenta Mathematicae*, vol. 27 (1936), pp. 234–238.

Index